Rates and Equilibria of Organic Reactions

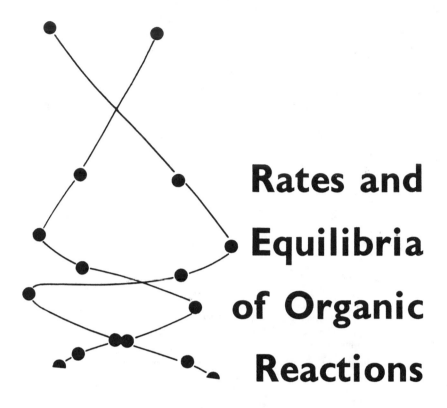

Rates and Equilibria of Organic Reactions

As Treated by Statistical, Thermodynamic, and
Extrathermodynamic Methods

John E. Leffler · FLORIDA STATE UNIVERSITY
AND
Ernest Grunwald · BRANDEIS UNIVERSITY

Dover Publications, Inc. · NEW YORK

Published in Canada by General Publishing Company, Ltd., 30 Lesmill Road, Don Mills, Toronto, Ontario.

Published in the United Kingdom by Constable and Company, Ltd., 10 Orange Street, London WC2H 7EG.

This Dover edition, first published in 1989, is an unabridged and unaltered republication of the work first published by John Wiley and Sons, Inc., New York, in 1963.

Manufactured in the United States of America
Dover Publications, Inc., 31 East 2nd Street, Mineola, N.Y. 11501

Library of Congress Cataloging-in-Publication Data

Leffler, John E.
    Rates and equilibria of organic reactions as treated by statistical, thermodynamic, and extrathermodynamic methods / John E. Leffler and Ernest Grunwald.
        p.   cm.
    Reprint. Originally published: New York : Wiley, c1963.
    Bibliography: p.
    Includes indexes.
    ISBN 0-486-66068-0
    1. Chemical equilibrium. 2. Chemical reaction, Rate of. 3. Chemistry, Organic. I. Grunwald, Ernest, 1923–   . II. Title.
QD503.L43   1989
547.1'392—dc20   89-11941
                                                                CIP

Apologia

Unless we can know something without knowing
everything, it is obvious that we can never know something.

Bertrand Russell

Preface

Organic reactions usually involve complex molecules reacting in the liquid phase. The theoretical analysis of their rates and equilibria is necessarily less exact than that of small molecules reacting in the gas phase, for it is rarely possible to establish the relevant sets of energy levels with sufficient accuracy or completeness by direct measurement or by quantum mechanical calculation. An approximate treatment can be made from either of two points of view. In one, the formal framework of the complete microscopic analysis is retained, but various simplifying assumptions concerning the microscopic fine structure of the system are made in order to render the calculations tractable. In the other approach, the starting point is a macroscopic or thermodynamic analysis to which is added just enough microscopic detail to allow explicitly for structural and medium effects.

The point of view emphasized in this book belongs very definitely in the latter category. We make extensive use of the concept of *subspecies* because this allows us to introduce into the formal thermodynamic analysis such ideas as conformational isomerism or solvation without having to go all the way to the ultimate treatment in terms of the individual energy levels. A major part of the book is devoted to extrathermodynamic relationships (relationships not directly resulting from the axioms of thermodynamics alone) among free energies or other thermodynamic quantities. The derivation of extrathermodynamic relationships requires the introduction of microscopic detail, but the amount of microscopic information actually needed is surprisingly little. For example, in the derivation of linear free energy relationships for substituent or medium effects on reactions, we need to postulate that microscopic interactions exist, *but we do not need to identify the mechanisms of the interactions*.

We devote so much space to extrathermodynamic relationships because we are ourselves greatly impressed not only by the usefulness of such relationships in predicting rates and equilibria but also by the generality of the underlying theory and by the ease and versatility of its application. The theory provides a logical framework that encompasses all of the known empirical extrathermodynamic relationships and classifies them naturally, thus tying together a lot of seemingly independent relationships into a single coherent and rational structure. The logical framework is sufficiently general to go beyond existing facts; it provides a useful basis for predicting new extrathermodynamic relationships that can then be tested by experiment; it also provides a useful starting point for the investigation of the actual mechanisms of substituent or medium effects.

In addition to the largely thermodynamic material, we have included chapters on very fast rate processes that occur as elementary steps in many reactions, and on solid state phenomena, two subjects we personally find exciting.

We would like to acknowledge the help and encouragement of our wives and to thank a number of people who have read and criticized our manuscript at various stages, especially George S. Hammond, Charles F. Jumper, Leo Mandelkern, Saul Meiboom, Robert W. Taft, Jr., and Edel Wasserman.

JOHN E. LEFFLER
ERNEST GRUNWALD

Tallahassee, Fla.
Murray Hill, N.J.
April 1963

Contents

Glossary of Symbols

1. Equilibrium from the Statistical Point of View, 1

2. Equilibrium and the Gibbs Free Energy, 15

3. Free Energy, Enthalpy, and Entropy, 40

4. Concerning Rates of Reaction, 57

5. Rates of Interconversion of Subspecies, 76

6. Theoretical Introduction to Extrathermodynamic Relationships, 128

7. Extrathermodynamic Free Energy Relationships.
   I. *Substituent Effects*, 171

8. Extrathermodynamic Free Energy Relationships.
   II. *Medium Effects*, 263

9. Extrathermodynamic Analysis of Enthalpy and Entropy Changes, 315

10. Some Mechanochemical Phenomena, 403

Author Index, 429

Subject Index, 443

Glossary of Symbols

Our symbols are based on the current journal literature and reflect that salutary lack of coordinated planning that is typical of active research fields. We have, however, tried to introduce a measure of order through the use of distinctive typography. Extrathermodynamic parameters and variables are represented by sans serifs or by SMALL CAPITALS. Thermodynamic quantities are represented by *italics*. Generic chemical symbols, such as RX or RBH^+, are in *italics*.

A	pre-exponential factor in the Boltzmann equation, the Arrhenius equation, and some other equations of analogous mathematical form.
a, b	coefficients in extrathermodynamic relationships by linear combination of two model processes.
a_i	activity of the ith component.
B, B_A, B_B	slope in the Brønsted relationship.
B_i	intrinsic term for the ith component in the power-series expansion of the free energy of a solution.
b	ionic radius.
C_p	heat capacity at constant pressure.
C_v	heat capacity at constant volume.
c_i	concentration of the ith component (in moles or formula weights per liter).
D	dielectric constant.
D_A	diffusion coefficient of the molecular species A.
E	energy.
E	activation energy.
$\bar{E}^\circ$	standard partial molar energy.
E_S	steric substituent constant of the acyl group in ester hydrolysis.
e	magnitude of the electric charge of the electron.
F	Gibbs free energy.
$\bar{F}_i$	partial molar free energy of the ith component.

$\bar{F}_i{}^\circ$	standard partial molar free energy of the ith component.
F_R, F_X	independent additive terms (per mole) in the formal representation of $\bar{F}_{RX}^\circ$.
f	force of retraction in a stretched fiber; tension.
$f(x)$	a function of x.
g_j	number of quantum levels having an energy value ϵ_j; degeneracy factor for ϵ_j.
H	enthalpy.
$\bar{H}_i{}^\circ$	standard partial molar enthalpy of the ith component.
H_0, H_0', H_-, H_R	acidity functions.
h	Planck's constant.
$\mathscr{I}$	ionization potential.
$I_{R,X}$	first-order interaction term for the interaction between R and X in the formal representation of $\bar{F}_{RX}^\circ$.
$I_{R,M}$	first-order interaction term for the interaction between R and the medium M in the formal representation of $\bar{F}_{RX}^\circ$ in M.
I_R, I_X, I_M	independent factors in the approximate representation of first-order interaction terms.
$II_{R,X,M}$	second-order interaction term in the formal representation of $\bar{F}_{RX}^\circ$ in the medium M.
II_R, II_X, II_M	independent factors in the approximate representation of $II_{R,X,M}$.
J	rate of application of shear energy.
j	rotational quantum number.
K	equilibrium constant.
K_c	equilibrium constant in terms of molar concentrations.
K_a	equilibrium constant in terms of activities.
K_0	equilibrium constant for the standard reference reaction.
K_A	acid dissociation constant.
K_B	base dissociation constant.
K_γ	$\gamma_{products}/\gamma_{reactants}$.
k	rate constant.
k_0	rate constant for the standard reference reaction.
$\mathbf{k}$	force constant of a bond.
k	Boltzmann's constant.
L	length of a fiber.
L_i	length of a fiber in the isotropic or unstrained state.
l'	length of a "statistical element" of a polymer chain.
M_i	molecular weight of the ith species.
M	interaction variable characteristic of the medium.

m	mass per molecule.
m	substrate parameter in the mY equation.
N	number of molecules.
N_i	number of molecules to be found in the ith energy level.
$N_A, N_B \ldots$	number of molecules belonging to the species or subspecies $A, B \ldots$.
N_0	Avogadro's number.
N$_i$	mole fraction of the ith component.
n	number of collisions in an encounter (Z/Z_E).
n_i	number of moles of the ith component.
n	nucleophilic parameter based on the reactions of methyl bromide.
n'	number of "statistical elements" in a polymer chain.
P	pressure.
P	probability factor in collision theory.
P	probability that an encounter between the reactant molecules will lead to reaction.
$\mathscr{P}$	time-dependent property of a solution.
$\mathscr{P}_{\parallel}$	component of $\mathscr{P}$ which is in phase with a changing variable of state.
$\mathscr{P}_{\perp}$	component of $\mathscr{P}$ which is 90° out of phase with a changing variable of state.
p	operator, $\equiv -\log_{10}$.
Q	partition function.
$\boldsymbol{R}$	gas-constant per mole.
R	interaction variable characteristic of the substituent.
r	temperature-independent interaction variable characteristic of the substituent.
R	mixing parameter in the Yukawa-Tsuno equation.
r	length of a polymer chain.
S	entropy.
$\bar{S}_i{}^{\circ}$	standard partial molar entropy of the ith component.
s	average distance a molecule diffuses in time t.
s	reaction parameter in the sE_S equation for steric effects.
s	substrate parameter in the ns equation for nucleophilic reactivity.
T	absolute temperature.
t	time
V	volume.
$\bar{V}_i$	partial molar volume of the ith component.
v	potential energy.

V_0 · minimum value of the potential energy function of a harmonic oscillator.

v · velocity.

v_2 · volume fraction of polymer in a network which is swollen or diluted by solvent.

W · work.

w · substrate parameter in the Bunnett equation.

X · interaction variable characteristic of the reaction zone.

x_i · concentration of the ith solute species or subspecies (in arbitrary units).

Y, Y_0, Y_- · solvent parameters in the mY equations.

Z · collision frequency at unit concentrations.

Z_E · encounter frequency at unit concentrations.

Z_{ij} · average number of collisions experienced by a molecule in a vibrational level ϵ_i before transition to ϵ_j takes place. The translational energies are thermally distributed.

Z · transition energy of the charge-transfer absorption band of 1-ethyl-4-carbomethoxypyridinium iodide ion pair.

z_i · charge number of the ith ionic species.

α · polarizability.

α · parameter in the rate-equilibrium relationship $(= \delta \Delta F^{\ddagger}/\delta \Delta F^{\circ})$.

α_i · fraction of the formal species which exists in the form of the ith subspecies.

β · isokinetic temperature $(= \delta \Delta H^{\ddagger}/\delta \Delta S^{\ddagger}$ for a single interaction mechanism).

$\left.\begin{array}{l}\beta_1, \beta_2 \\ \beta', \beta''\end{array}\right\}$ · values of β for individual interaction mechanisms.

β_{ij} · pairwise interaction term for components i and j in the power-series expansion of the free energy of a solution.

γ_i · activity coefficient of the ith component.

Δ · reaction operator; e.g., $\Delta F^{\circ} = F^{\circ}_{\text{product}} - F^{\circ}_{\text{reactant}}$.

$\left.\begin{array}{l}\Delta F^{\ddagger}, \Delta H^{\ddagger}, \\ \Delta S^{\ddagger}\end{array}\right\}$ · free energy, enthalpy, and entropy of activation.

δ_L · operator, for changing the length of a fiber of constant composition; e.g., $\delta_L F = F_L - F_{L_0}$.

δ_R · substituent stabilization operator; e.g.,

$$\delta_R F^{\circ} = F^{\circ}_{RX} - F^{\circ}_{R_0 X}.$$

δ_M solvent stabilization operator; e.g.,

$$\delta_M \bar{F}_i^\circ = \bar{F}_{i,\text{ in solvent M}}^\circ - \bar{F}_{i,\text{ in standard solvent}}^\circ.$$

δf half-width of a spectral line (in cycles/sec). See figure 5–6.

ϵ_i energy of the ith energy level.

$\zeta(x)$ a function of x.

η viscosity.

η_{ij} pairwise interaction term for components i and j in the power-series expansion of the entropy of a solution.

κ transmission coefficient in transition-state theory.

λ critical distance for bond-breaking in a polymer chain.

μ reduced mass of a two-particle system $[= m_1 m_2 / (m_1 + m_2)$, where m_1 and m_2 are the masses of the particles].

μ ionic strength $\left(= \dfrac{1}{2} \Sigma c_i z_i^2 \right).$

$\bar{\mu}$ dipole moment.

ν number of polymer chains.

ν frequency.

ν_{osc} characteristic frequency of a harmonic oscillator.

ν_{max} (in absorption spectroscopy) frequency of maximum absorption of energy.

ξ friction coefficient.

ρ reaction parameter in $\rho\sigma$ and $\rho\sigma^+$ relationships for aromatic reactions.

ρ', ρ'' reaction parameters for individual interaction mechanisms.

ρ' reaction parameter for side-chain reactions at the 1-position in the 4-X-bicyclo-[2.2.2] octane system.

ρ^* reaction parameter for polar effects in aliphatic reactions.

σ substituent parameter in the $\rho\sigma$ relationship for aromatic side-chain reactions.

σ', σ'' substituent parameters for individual interaction mechanisms.

σ^+ substituent parameter for reactions in which a formal positive charge is produced on, or adjacent to, an aromatic ring.

σ^- substituent parameter for reactions in which a non-bonding pair of electrons is produced on, or adjacent to, an aromatic ring.

σ°	"normal" aromatic substituent parameter.
σ'	substituent parameter for the 4-X-bicyclo [2.2.2] octane -1 series.
σ^*	polar substituent parameter for aliphatic reactions.
σ_{AB}	collision diameter; i.e., sum of molecular radii of A and B.
τ	transmission coefficient for aromatic substituent effects.
τ	relaxation time.
τ	mean time that a molecule can spend in a given state.
τ_{μ}	dielectric relaxation time per molecule.
ϕ_i	volume fraction of the ith component.
$\phi(x)$	a function of x.
χ_{ij}	pairwise interaction term for components i and j in the power-series expansion of the enthalpy of a solution.
$\psi(x)$	a function of x.

Rates and Equilibria of Organic Reactions

1

Equilibrium from the

Statistical Point of View

As far as I can see, the only foundation of the doctrine of probability, which (though not satisfactory for a mind devoted to the "absolute") seems at least not more mysterious than science as a whole, is the empirical attitude: The laws of probability are valid just as any other physical law in virtue of the agreement of their consequences with experience.

Max Born
Experiment and Theory in Physics

The course of any chemical system, the major and minor reactions that take place, and the composition of the system as a function of time are predictable if the values of the rate and equilibrium constants are predictable. Since rate constants are treated most conveniently by the transition-state theory, which is itself an extension of theories dealing with equilibrium, the study of equilibrium is the logical starting point for an understanding of chemical reactions. In this chapter we shall discuss equilibrium from the microscopic or molecular-statistical point of view. In the next chapter we shall discuss equilibrium from the thermodynamic point of view, and we shall show the relationships between the two treatments. In the statistical or microscopic treatment of equilibrium the fundamental concept is that of probability. In the thermodynamic treatment of equilibrium we again encounter probability, in the guise of entropy.

EQUILIBRIUM AS AN EXERCISE IN PROBABILITY

A system at equilibrium may be defined as being in a state of maximum probability. The various possible physical states, from which it is our task to select the most probable one, are defined in macroscopic terms. When each of these possible states is examined in microscopic detail, it is found that the same macroscopic properties (i.e., the same macroscopic state) can be produced by a large number of different microscopic arrangements of the molecules and the energy. The number of such different arrangements is the measure of the probability of a given macroscopic state.

The total probability of all the possible macroscopic states is of course unity. However, for systems containing a large number of particles the probability of the most probable state is so very close to unity that once the equilibrium has been achieved the macroscopic system is never observed to depart from it. That is to say, a state differing from the equilibrium state in its macroscopic properties by even a very small amount has already a negligibly small probability. A system on its way to the equilibrium state proceeds through macroscopic states of progressively higher probability.

In predicting the most probable state of the system, we must eliminate from consideration not only those states that are incompatible with the given physical conditions (such as constant volume and energy) but also those states that are merely inaccessible because the rate of attaining them is prohibitively slow. Most chemical problems concern states of meta-stable or quasi equilibrium, and the true or infinite-time equilibrium state may be excluded from consideration among the possible states because of the lack of a sufficiently fast mechanism for attaining it. For example, if we are interested in the equilibrium (1) between nitromethane and its *aci*-form, we will neglect the extremely slow reaction leading to carbon dioxide and ammonia.

$$CH_3\!-\!NO_2 \;\rightleftharpoons\; CH_2\!=\!N\!\!\begin{array}{c} \nearrow O \\[2pt] \searrow \\ OH \end{array} \tag{1}$$

$$CH_3NO_2 \;\rightleftharpoons\; CO_2 + NH_3$$

FACTORS ENTERING THE PROBABILITY CALCULATION

Besides the fundamental idea of probability, equilibrium calculations make use of the concept of quantized energy levels. An *energy level* is a complete description of the state of a molecule (or of a set of atoms) and makes use of *all* the quantum numbers. The value of the energy, ϵ, is an important *property* of the energy level but does not suffice to describe it completely. Since molecules in different energy levels differ by at least one quantum number, they also differ in at least one observable property and can therefore be distinguished even when the energies happen to be the same. A *chemical species* in the usual sense is a mixture of many different energy levels; these energy levels can be regarded as isomers constituting the chemical species.

The problem of calculating the equilibrium state of a system is essentially this: Given the possible energy levels of an isolated system, distribute the finite amount of matter and energy in the system among those energy levels and do so in the most probable way.

THE BOLTZMANN DISTRIBUTION

For simplicity let us first consider a system consisting of a single chemical species in which *no* chemical reaction occurs at an appreciable rate. Even though the molecules all belong to the same chemical species, they can be classified still further by assignment to quantized energy levels. The energies of all the molecules must add up to the total energy of the system, but there are many ways in which that amount of energy can be apportioned among the molecules. For example, Figure 1-1 shows some of the ways in which 30 molecules could be distributed among five levels equally spaced in energy so that the total energy remains constant. When the system contains a large number of molecules, one of the distributions is very much more probable than all the others, and this is to all intents and purposes the way the energy will be distributed at equilibrium. By counting permutations or

Energy of the Level (energy units)	Number of Molecules in Each Level					
5	0	1	3	0	0	0
4	0	3	1	3	2	1
3	0	5	5	6	7	9
2	30	7	5	9	10	9
1	0	14	16	12	11	11
Total molecules	30	30	30	30	30	30
Total energy units	60	60	60	60	60	60

Figure 1-1. Six of the possible distributions of 30 molecules having a total energy of 60 units among 5 equally spaced energy levels. Each distribution corresponds to a different macroscopic state of the system. Some distributions can be achieved in more ways than others. For example, the first distribution shown can be achieved in only one way, while the second distribution can be achieved in many ways: any one of the thirty molecules could occupy the fifth level, any three of the remaining twenty-nine molecules could occupy the fourth level, and so on.

the number of distinguishable ways of achieving a given distribution of the energy, it can be shown that the Boltzmann distribution (equation 2) is in fact the most probable or equilibrium distribution.

$$N_i = Ae^{-\epsilon_i/kT} \tag{2}$$

$$A = \frac{N}{\sum e^{-\epsilon_i/kT}} \tag{3}$$

In these equations N_i is the number of molecules to be found in the level of energy ϵ_i and N is the total number of molecules.

The choice of the zero point for the energy is arbitrary. As can be derived from equation 3, a change in the choice of zero point is canceled in its effect on the actual distribution because of its occurrence in the pre-exponential as well as in the exponential term. The value of the proportionality constant A is fixed by the fact that the sum of the molecules in all the energy levels must equal the total number, N, of molecules.

$$\sum N_i = N = A \sum e^{-\epsilon_i/kT} \tag{4}$$

The Boltzmann equation has two important qualitative properties. The first of these is that at constant temperature a given energy level will contain fewer molecules the higher the energy of the level. The reason for this is the finite constant amount of total energy at constant temperature. The other important property is that the difference in the relative populations of levels of high and low energy is less pronounced at higher temperatures. For example, at an infinitely high temperature the exponential term $e^{-\epsilon_i/kT}$ would become unity and each level would be populated by the same number of molecules as any other level, whereas at a very low temperature all of the molecules would have to be in the lowest levels.

Application of the Boltzmann Distribution to Chemical Reactions

When we consider systems in which chemical reactions occur, we find that the reactions do not alter the nature of the problem at all. Equation 2 continues to apply to the population of the ith energy level *regardless of the chemical species to which the ith level happens to belong.* This should not be surprising, since the notion of molecular or chemical species is completely arbitrary. A chemical species is a collection of energy levels, and we may include or exclude any energy levels we choose. In practice, the degree of difference we demand between two energy levels before we are willing to call them members of different molecular species depends on the time scale of the experiments that we propose to do. Thus 1,2-dichloroethane is a single chemical species to the person who is subjecting it to fractional distillation but a mixture of *trans* and *gauche* forms to the spectroscopist who is subjecting it to high-frequency radiation. At room temperature the

trans-1, 2-dichloroethane   gauche-1, 2-dichloroethane

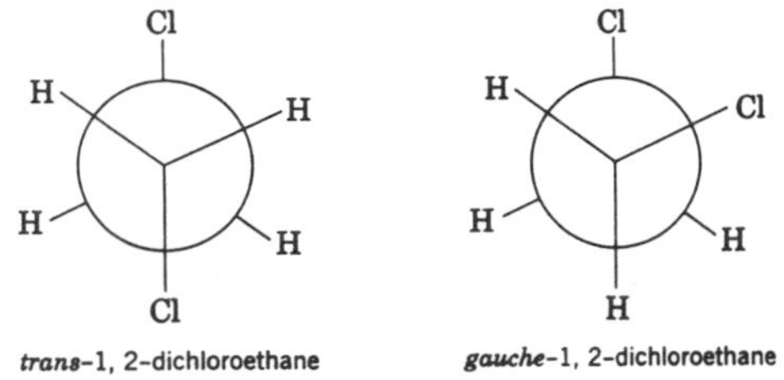

Figure 1-2. Subspecies of 1,2-dichloroethane.

average time between conversions from one form into the other is about 10^{-10} sec.

Since we usually wish to know how many molecules of a particular chemical species are present in a reaction mixture rather than how many molecules are in a particular energy level, we must specify which energy levels are to be counted as members of the chemical species in which we are interested. For example, if *gauche*-1,2-dichloroethane is the species in question, we count only those energy levels having the *gauche* configuration. But if we are interested simply in 1,2-dichloroethane, we must include not only *gauche* and *trans* energy levels but also higher energy levels which because of free rotation have no assignable configuration.[1]

For a simple system in which there are only two substances A and B in the equilibrium mixture, the equilibrium constant is the number of molecules in all the levels that are considered to belong to species B divided by the number of molecules in all the levels that are considered to belong to species A. The equilibrium constant is obtained by dividing the Boltzmann expression for the populations of the B species by that for the populations of the A species, as in equation 5.

$$N_A = A \sum e^{-\epsilon_{iA}/kT}$$

$$N_B = A \sum e^{-\epsilon_{iB}/kT}$$

$$K = \frac{N_B}{N_A} = \frac{\sum e^{-\epsilon_{iB}/kT}}{\sum e^{-\epsilon_{iA}/kT}} \tag{5}$$

It is convenient to recast equation 5 into an equivalent form which stresses the difference between the energy of the ith level and that of the lowest level for each species. The result is equation 6.

$$K = \frac{N_B}{N_A} = e^{-(\epsilon_{0B}-\epsilon_{0A})/kT} \times \frac{\sum e^{-(\epsilon_{iB}-\epsilon_{0B})/kT}}{\sum e^{-(\epsilon_{iA}-\epsilon_{0A})/kT}} \tag{6}$$

[1] The higher energy levels alluded to are by themselves a chemical species, namely the transition state for the interconversion of the *trans* and *gauche* isomers. Their population is quite small in comparison to that of either of the isomers.

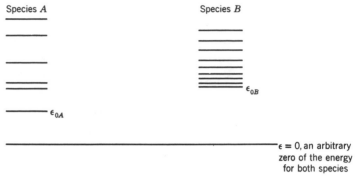

Figure 1-3. An energy level diagram for a hypothetical system showing differences in the lowest levels and also differences in the closeness of spacing of the levels.

If, as often happens, the energy of a molecule is not changed by changing one of its quantum numbers, it is usually the practice to combine these levels of equal energy into a single "degenerate" energy level. When this is done it is necessary to increase the importance of the degenerate energy level in the equation by multiplying the corresponding exponential term by its degeneracy factor, g, as in equation 7.

$$\frac{N_B}{N_A} = e^{-(\epsilon_{0B}-\epsilon_{0A})/kT} \times \frac{\sum g_j e^{-(\epsilon_{jB}-\epsilon_{0B})/kT}}{\sum g_j e^{-(\epsilon_{jA}-\epsilon_{0A})/kT}} \tag{7}$$

An example of a degenerate energy level is the ground state of the hydrogen atom. Neglecting interactions with the nuclear spin, this level has a degeneracy of two because, in the absence of a magnetic field, a change in the direction of the electron spin does not change the energy.

Figure 1-3 is an energy level diagram illustrating the fact that the lowest levels of the two species need not have the same energy and also that the levels of one species may be more densely packed on the energy scale than those of the other species.

THE PARTITION FUNCTION

For the sake of compactness the equilibrium equation is often written in terms of *partition functions*, Q, defined as in equation 8.

$$Q = \sum g_j e^{-(\epsilon_j - \epsilon_0)/kT} \tag{8}$$

By using the partition functions equation 9 is obtained.

$$\frac{N_B}{N_A} = e^{-(\epsilon_{0B}-\epsilon_{0A})/kT} \cdot \frac{Q_B}{Q_A} = e^{-\Delta\epsilon_0/kT} \cdot \frac{Q_B}{Q_A} \tag{9}$$

According to equation 9 the equilibrium constant depends on two factors, the difference between the energies of the lowest levels and the partition function ratio. The effect of the first factor is such that the species with the less energetic ground level is favored over the other. But the effect of this factor is modified, *and in some cases even reversed*, by the second factor, Q_B/Q_A. The partition function is a sum that includes a term for each energy level. The more energy levels there are, the greater the partition function. However, energy levels near the ground level count for more than do levels of higher energy because the negative exponential becomes a very small number as the value of ϵ_i is made large. Qualitatively the partition function is a measure of the density of spacing of energy levels within the range from the lowest level up to levels whose additional energy is still only a small multiple of kT. In Figure 1-3 species B obviously has the larger partition function. Even if species A has some high-energy but closely spaced energy levels (not shown), these would not offset the closely spaced energy levels of species B, for the latter are heavily weighted by virtue of being close to ϵ_{0B}.

Effect of Molecular Constraints on the Partition Function

A useful qualitative way of thinking about the partition function is to consider it to be a measure of the freedom of the molecule from constraint. For example, a rigid molecule will have a smaller partition function than an otherwise similar flexible molecule. In terms of energy levels a molecule is rigid because the lowest energy level corresponding to the internal motion is a high one and therefore one that can not contribute much to the partition function. Neither in molecules nor in the mechanics of macroscopic objects is there such a thing as an absolutely rigid body; rigidity is relative and is merely a matter of the energy required to produce a deformation. Another example of constraint is imprisonment of the molecule in a small volume; in the quantum mechanical model of the particle in a box the density of the translational energy levels is directly proportional to the volume of the box. An important consequence of this effect is that partition functions are smaller than they otherwise would be for solutes (partially confined by solvent cages) and for clathrate complexes.[2] Still another type of constraint is a restriction on rotation of the molecule as a whole. Almost any nonspherical molecule in a crystal is under such constraint, so is any dipolar molecule oriented by the presence of a nearby ion or another dipolar molecule.

We consider it important to emphasize these qualitative aspects because an actual calculation of the partition function is only rarely possible,

[2] An example of a clathrate complex is a crystal of hydroquinone containing hydrogen sulfide molecules in the regularly spaced cavities of the crystal lattice.

especially for reactions in solution. In any liquid the interactions of the molecules are so complex that, in fact, the energy levels can only be defined rigorously for the entire beaker rather than for individual molecules. For reactions in solution the concept of energy levels populated by individual molecules is an idealization.

Factoring of the Partition Function

For ideal gaseous systems the partition function can sometimes be calculated. It is worthwhile to examine the methods used in such calculations, because the ideas underlying them are useful in qualitative thinking about liquid systems. The basis of the calculation for ideal gases is the approximation that the total energy of a molecule can be separated into translational, rotational, vibrational, and electronic contributions, each of which can be treated independently. For example, the spacing of the vibrational energy levels is independent of the translational velocity of the molecule and almost independent of its rotational velocity. As the result of the additivity of the various kinds of energy, the partition function is factorable.

$$E = \epsilon_{\text{trans}} + \epsilon_{\text{rot}} + \epsilon_{\text{vib}} + \epsilon_{\text{elec}} \tag{10}$$

$$Q = Q_{\text{trans}} \times Q_{\text{rot}} \times Q_{\text{vib}} \times Q_{\text{elec}} \tag{11}$$

For any particle the translational partition function is given by equation 12, where m is the mass of the particle and V is the volume of the container.

$$Q_{\text{trans}} = \frac{(2\pi m k T)^{3/2} V}{h^3} \tag{12}$$

If the complete set of rotational and vibrational energy levels is known, perhaps from microwave, infrared and Raman spectra, it is possible to calculate Q_{vib} and Q_{rot}. When the energy levels are not known in detail, Q_{vib} and Q_{rot} can still be calculated at least approximately by assuming that the molecule behaves like a *rigid* rotator and that its vibrations are like those of a set of *harmonic* oscillators, as in equations 13 and 14.

$$Q_{\text{rot}} = \sum (2j + 1)e^{-B(j+1)/kT} \tag{13}$$

($B = h^2/8\pi^2\mathscr{I}$, where $\mathscr{I}$ is the moment of inertia; j is the rotational quantum number.)

$$Q_{\text{vib}} = \frac{1}{(1 - e^{-h\nu_{\text{osc}}/kT})} \tag{14}$$

Each mode of vibration will have its own characteristic frequency ν_{osc} and a corresponding set of energy levels $\epsilon_0 + h\nu_{\text{osc}}$, $\epsilon_0 + 2h\nu_{\text{osc}}$, etc. Equation 14 is mathematically equivalent to the more usual expression in terms of a summation, equation 15.

$$Q_{\text{vib}} = 1 + e^{-h\nu_{\text{osc}}/kT} + e^{-2h\nu_{\text{osc}}/kT} + \cdots \tag{15}$$

The spacing of energy levels depends on the type of energy level being considered as well as on the presence of constraints. Thus for translation the energy levels are usually so closely spaced that they may be regarded as a continuum. For example, if a molecule is constrained to remain in a space having a volume of one cubic centimeter, the energy levels may be as close together as 10^{-18} kcal/mole. Rotational levels are much further apart, but their spacing (about 10^{-2} kcal/mole) is still small compared to the average kinetic energy per mole at room temperature.

The spacing of vibrational levels varies widely, depending upon the stiffness of the bonds that are being distorted. For strong covalent bonds the vibrational energy levels may be as much as 10 kcal/mole apart; hence the levels above the first are not very important to the partition function at room temperature. For weak bonds, such as the hydrogen bond between a pair of alcohol molecules, the vibrational levels may be spaced as closely as 0.2 kcal/mole.

The spacing of electronic levels is also widely variable. However, most electronic transitions, corresponding as they do to absorption of light in the visible or ultraviolet, require 40 kcal/mole or more. Hence the electronic partition function is usually equal merely to the multiplicity of the lowest electronic level, the other electronic levels being so high that their populations are negligible. Chichibabin's hydrocarbon, in which a diamagnetic singlet ground state and a paramagnetic diradical coexist at room temperature, is exceptional.[3] A more typical example is nitrobenzene, in which the population of the electronically excited state at room temperature is about 10^{-85} %.[4]

Returning now to the consideration of the liquid phase, we find that the freedom of translation and rotation of the molecules is so reduced that it is no longer profitable to discuss this part of the problem in the same terms that were used for the gas phase. Translation tends to be replaced in a liquid, at least in part, by a quasi-crystalline vibration or oscillation about a position corresponding to a potential energy minimum. Rotation of the molecule tends to be replaced by a libration about a preferred orientation of minimum potential energy. On the other hand, the internal vibrational partition function for a molecule in the liquid must often be quite a bit like that for the same molecule in the gas, although the energy levels may be spaced somewhat differently. Complex formation between neighboring molecules gives rise to new molecular species with new vibrational modes and new vibrational energy levels corresponding to movements of the two

[3] C. A. Hutchison, Jr., A. Kowalsky, R. C. Pastor and G. W. Wheland, *J. Chem. Phys.* **20**, 1485 (1952).
[4] W. M. Schubert, J. Robins and J. L. Haun, *J. Am. Chem. Soc.* **79**, 910 (1957).

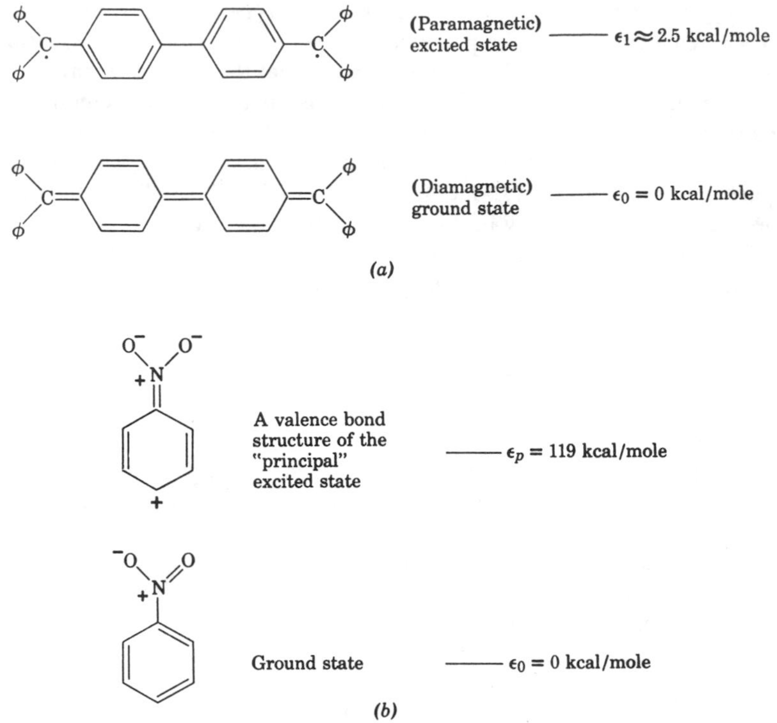

(Paramagnetic) excited state ———— $\epsilon_1 \approx 2.5$ kcal/mole

(Diamagnetic) ground state ———— $\epsilon_0 = 0$ kcal/mole

(a)

A valence bond structure of the "principal" excited state ———— $\epsilon_p = 119$ kcal/mole

Ground state ———— $\epsilon_0 = 0$ kcal/mole

(b)

Figure 1-4. (a) Two electronic levels of Chichibabin's hydrocarbon. At room temperature the population of the excited state is about 4%. (b) Two electronic levels of nitrobenzene. At room temperature the population of the excited state is about 10^{-85}%.

components with respect to each other. Moreover, the complex formation may modify the energy levels of the component molecules.

Zero-Point Vibrational Energy and the Vibrational Partition Function

For a classical oscillator, such as a pendulum, the total energy is constant, but there is an interconversion of potential and kinetic energy during each oscillation. The potential energy is a function of the configuration of the oscillator and has its minimum value at the equilibrium configuration, the latter being the configuration that the oscillator has when at rest. Quantum mechanical oscillators are never at rest, but in order to solve the wave equation it is necessary to express the potential energy just as in the classical case, as a function of the precise configuration. Such a potential energy function is shown in Figure 1-5.

In spite of our use of a potential energy function taken from classical mechanics, it should be recalled that the uncertainty principle will not permit us to know the momentum and thus the kinetic energy at the same time that we know the configuration of the oscillator. We therefore can not separate the energy of the quantum mechanical oscillator into potential and kinetic parts but are restricted to describing it only in terms of its total energy. The allowed values of the total energy are represented by the energy levels (horizontal lines) in Figure 1-5. The quantum mechanical counterpart of the concept of equilibrium configuration is simply the minimum in the potential energy function. The minimum value of the potential energy function, v_0, is the purely electronic part of the energy of the molecule. Part of the approximation inherent in the separation of the total energy into contributions from the electronic and vibrational parts (equation 10) is the assumption that the electronic energy is independent of the vibration of the nuclei. Calculations of such things as resonance energies or bond energies are therefore carried out as though there were no vibration and as though it were possible to have a fixed equilibrium bond distance, r_0.

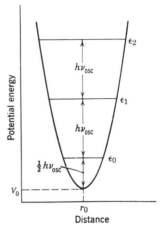

Figure 1-5. Energy levels and the potential function for a simple harmonic oscillator. v_0 is the classical minimum potential energy and r_0 the position of a classical oscillator when at rest. The lowest energy level possible for a quantum mechanical oscillator is ε_0, which is $\frac{1}{2}h\nu_{osc}$ above v_0.

Since actual molecules must have at least a minimum amount of vibrational energy, the energy of the ground-state level, labeled ϵ_0 in Figure 1-5, is a sum consisting of the electronic energy v_0 and the *zero-point vibrational energy*, $\epsilon_0 - v_0$.

The zero-point vibrational energy is related to the spacing of the vibrational levels and hence to the magnitude of the vibrational partition function. To show the nature of this relationship it is useful to assume that the vibration is like that of a simple harmonic oscillator, as in Figure 1-5. The difference in energy between the nth and 0th levels is then given by equation 16.

$$\epsilon_n - \epsilon_0 = nh\nu_{osc} \tag{16}$$

The zero-point vibrational energy is given by equation 17.

$$\epsilon_0 - v_0 = \tfrac{1}{2}h\nu_{osc} \tag{17}$$

From equations 16 and 17 it can be seen that the zero-point vibrational energy and the spacing of the energy levels are related by their dependence on the same parameter, the "oscillator frequency" ν_{osc}. The vibrational partition function computed from the energy levels of equation 16 (for a harmonic oscillator) is given by equation 14.

When the oscillation can be approximated as the relative motion of two rigid parts of the molecule, ν_{osc} is given by equation 18. In equation 18, k is the force constant of the bond and μ is the reduced mass, $m_1 m_2/(m_1 + m_2)$. Although

$$\nu_{osc} = \frac{1}{2\pi} \sqrt{k/\mu} \tag{18}$$

equation 18 is meant for a diatomic molecule or for the vibrations of two weakly bonded parts of an otherwise strongly bonded molecule, it is also a good approximation for the vibration of a light atom bonded to a heavy molecule. Examples of the latter application are the carbon-hydrogen and carbon-deuterium stretching vibrations. The ratio of the reduced masses, μ_H/μ_D, is very nearly equal to $\frac{1}{2}$. Since an isotopic substitution does not appreciably alter the electronic nature of the bond (force constants and v_0 are unchanged), the ratio of the oscillator frequencies, and hence of the zero-point vibrational energies, is merely the square root of 2.

$$\frac{\nu_{oscH}}{\nu_{oscD}} = \sqrt{\frac{\mu_D}{\mu_H}} = \sqrt{2} \tag{19}$$

As an illustration of the magnitude of this effect we may note that C—H stretching vibrations are in the neighborhood of 9×10^{13} sec^{-1}, and C—D stretching vibrations are in the neighborhood of 6.4×10^{13} sec^{-1}. The corresponding zero-point energies are 4.3 kcal for the C—H bond and 3.0 kcal for the C—D bond. At room temperature a change of 1.3 kcal in ϵ_0 could change an equilibrium constant by a factor of 9.

Since the value of ν_{osc}, and hence the spacing of the energy levels, is greater for a C—H bond than for a C—D bond, the value of Q_{vib} for the deuterium compound is greater than that for the hydrogen compound. The effects of the isotopic substitution are in such a direction as to favor the deuterium compound both through the change in ϵ_0 and the change in Q_{vib}. However, the magnitude of the isotope effect on Q_{vib} is negligible at room temperature and would not need to be taken into account in a calculation. This can be seen by putting the appropriate value of ν_{osc} in equation 14. For the C—H bond Q_{vib} is 1.0000005, whereas for the C—D bond it is 1.00004. The close approach to unity in both cases means that virtually all the bonds, whether C—H or C—D, are in their lowest vibrational state.

We can generalize the result for the carbon hydrogen bond and say that for the stretching vibrations of ordinary covalent bonds, the effect of changes in the zero-point vibrational energy on the equilibrium constant is much greater than the effect of the related changes in the partition function. In order for a given vibration to increase Q_{vib} by even 10%, ν_{osc} must be less than 1.5×10^{13} sec^{-1} (in wave numbers, less than 500 cm^{-1}). Table 1-1

Table 1-1. Relationship between Oscillator Frequency, Zero-point Energy, and Vibrational Partition Function for Harmonic Oscillators

THE VIBRATION		$\epsilon_0 - \nu_0$	$e^{+(\epsilon_0 - \nu_0)/kT}$	Q_{vib}
sec^{-1}	cm^{-1}	kcal/mole	(298.16°)	(298.16°)
8.99×10^{13}	3000	4.287	1380	1.0000005
3.00×10^{13}	1000	1.429	11.2	1.008
1.50×10^{13}	500	0.714	3.34	1.097
7.49×10^{12}	250	0.357	1.83	1.43
4.50×10^{12}	150	0.214	1.44	1.94
3.00×10^{12}	100	0.143	1.27	2.61
1.50×10^{12}	50	0.0714	1.13	4.66

shows the relationship between the characteristic frequency or the wave number of a vibration and its effect on the zero-point vibrational energy and on the partition function. It is clear from the table that the stretching and bending vibrations corresponding to the easily measurable bands in the infrared, which lie between 3500 cm^{-1} and 600 cm^{-1}, do not *individually* have an important effect on Q_{vib}. When the molecule is large and has a large number of such vibrations, the cumulative effect on the vibrational partition function may of course be considerable. However, in most reactions, even of large molecules, the number of vibrational modes that are significantly changed by the reaction is small. Hence even the cumulative effect of such vibrations is likely to cancel out in the ratio of partition functions.

The isomers *cis-* and *trans*-decalin furnish an example of the effect of ordinary bond vibrations and skeletal vibrations in a reaction involving fairly large molecules. Values of the fundamental frequencies of *cis*-decalin are given in Table 1-2.[5] Of the 78 vibrations, 67 are at frequencies above 600 cm^{-1}, and their cumulative contribution to Q_{vib} is a factor of 1.457. More important is the contribution of the frequencies below 600 cm^{-1}, which amount to a factor of 33.06. For the equilibrium between *cis-* and *trans*-decalin it is found that with perhaps just one exception the vibration frequencies are so nearly identical that their contribution to the ratio

[5] T. Miyazawa and K. S. Pitzer, *J. Am. Chem. Soc.*, **80**, 60 (1958).

Table 1-2. Probable Values of Fundamental Vibration Frequencies of cis-Decalin[5] (cm^{-1})

2890 (18)[a], 1444 (8)[b], 1384, 1350, 1342, 1327, 1312, 1300 (6)[c], 1292, 1270, 1258, 1241, 1216, 1210, 1171, 1165, 1143, 1125, 1114, 1089, 1073, 1069, 1041, 1011, 978, 969, 929, 888, 877, 855, 849, 835, 800, 793, 752, 742, 706, 661, 594, 538, 489, *446*, 375, 348, 317, 191, 150 (2)[c], 148.

Q_{vib} for cis-Decalin: 48.17

[a] The average of the frequencies for 18 C—H stretching vibrations.
[b] The average of the frequencies for 8 CH$_2$ bending vibrations.
[c] Average frequency for several unassigned vibrations.

$Q_{vib}(trans)/Q_{vib}(cis)$ cancels out. The exception is a skeletal vibration, perhaps the vibration at 446 cm^{-1} in the cis-isomer, which is changed to 350 cm^{-1} in the trans-isomer. The net value of $Q_{vib}(trans)/Q_{vib}(cis)$ at 298.16° is therefore

$$\frac{1 - e^{-446 \times 0.00482}}{1 - e^{-350 \times 0.00482}} = 1.085$$

Examination of molecular models shows that trans-decalin has a more open structure than cis-decalin, and the existence of a skeletal vibration of lower energy for the trans-form is plausible.

cis-Decalin   trans-Decalin

Large molecules like the decalins can have vibrations of low frequency which are individually of consequence to the partition function even though all the bonds in the molecule are fairly rigid. The reason for this is that it is possible to deform the skeleton of a large molecule as a whole without having to change the length or angle of any given bond by very much. Other types of vibration, which are important contributors to the partition functions even of small molecules, are internal rotations or torsions about single bonds, the vibrations of very weak bonds such as those found in hydrogen-bonded or π-bonded complexes, and the oscillation of entire molecules about orientations of minimum potential energy in condensed phases. Vibrations of this type correspond to fundamental frequencies in the far infrared and microwave regions. The bonds joining two reagents in a transition state may also be weak enough for certain motions of the reagents relative to each other to be important to the partition function of the transition state, although, as we shall see, any motion identifiable with progress along the reaction coordinate must be excluded on theoretical grounds.

2

Equilibrium and
the Gibbs Free Energy

Wenn Mund und Gaumen sich erlaben
Muss die Nase auch was haben.
                                    Villiger

INTRODUCTION

In Chapter 1 the state of equilibrium was defined in terms of probability, and the probability was related to the fine structure of the system. In this chapter we shall regard the problem of equilibrium from a more operational point of view, defining the equilibrium state in terms of macroscopic processes involving the system as a whole and not, at least at the outset, requiring any assumptions about the microscopic structure of the system. Eventually we shall find that the two criteria for equilibrium are equivalent, having common roots in the notion of probability.

If we are prevented from making a microscopic analysis of a system, we can only define the state of the system in terms of its macroscopic properties and in terms of what the system can be made to do. Our intuitive definition is that a system is at equilibrium under a given set of conditions if its properties remain constant indefinitely and are not affected by the introduction of any conceivable catalyst. To borrow a term from the biologists, a system at equilibrium is not *irritable*. This intuitive definition can be sharpened and made amenable to a mathematical formulation if we introduce the idea of *useful work*. (Useful work is defined to be the total work minus the expansion work, $\int P\,dV$.) We can then define a system at equilibrium by saying that the system cannot be made to do any useful work under the given set of conditions, even if irritated.

If the system is not in chemical equilibrium under the given set of conditions, it can be made to do useful work while moving towards equilibrium if the equilibration is harnessed by means of a suitable mechanism. For example, if the reaction takes place in an electrical cell, electrical work can be produced. The amount of useful work can be maximized if the mechanism by which the process is harnessed is allowed to operate reversibly.

The thermodynamic conditions for equilibrium are obtained most directly if we consider only those mechanisms that operate reversibly. For any infinitesimal reversible change, the useful work, dW_{rev}, can be related to other thermodynamic quantities by means of equation 1.

$$dW_{rev} = -dE - P\,dV + T\,dS. \tag{1}$$

Equation 1 can be simplified in a number of special cases. For example, if the change takes place at constant energy and volume, then equation 2 applies.

$$dE = 0, \qquad dV = 0;$$

therefore

$$dW_{rev} = T\,dS. \tag{2}$$

If the initial state of the system is already an equilibrium state, no useful work can be produced. Hence

$$dW_{rev} = 0 = T\,dS, \tag{3}$$

and

$$dS = 0. \tag{4}$$

According to equation 4, the condition for equilibrium in a system of constant energy and volume is that the entropy must be at one of its extreme values. Reflection will show that this extreme value is a maximum. The useful work to be derived from the system as it moves towards equilibrium is always a positive quantity; hence dS is always positive. The entropy therefore increases with every step that the system moves closer to equilibrium, reaching a maximum at equilibrium.

The special case of equilibrium at constant energy and volume is enlightening because of its close relationship to statistical theory. It will be recalled that in Chapter 1 the equilibrium composition was the one having the most probable distribution of the molecules among the energy levels. In arriving at the most probable distribution, the total energy of the system was assumed to be fixed, and constancy of volume was implied by constancy of the energy levels. The thermodynamic criterion of maximum entropy therefore applies under the same conditions as the statistical criterion of maximum probability, showing the close relationship between these two variables.

The application of equation 1 to processes taking place at constant temperature and pressure is of great practical importance. Chemical reactions are usually investigated at a constant, controlled temperature. Reactions in the gas phase are often studied also at constant pressure; and reactions in a condensed phase are rather insensitive to pressure and are usually studied at or near atmospheric pressure. As is shown in equation 5, the criterion for equilibrium at constant pressure and temperature is the attainment of an extreme value for the Gibbs free energy, in this case a minimum rather than a maximum.

$$dW_{rev} = 0 \begin{cases} = -dE - P\,dV + T\,dS \\ = -d(E + PV - TS) \\ = -d(H - TS) \\ = -dF \end{cases} \tag{5}$$

The measure of the displacement of the system from equilibrium at constant pressure and temperature is the decrease in free energy in going from the displaced state to the equilibrium state.

EQUILIBRIUM IN DILUTE SYSTEMS

In order to apply the criterion of minimum free energy to a *chemical* equilibrium, it is necessary to have some kind of an expression relating the free energy to the composition. Since the free energy at constant composition is proportional to the amount of matter, it can readily[1] be shown that the free energy is given by equation 6.

$$F = n_1 \left(\frac{\partial F}{\partial n_1}\right)_{n_2, n_3, \ldots, T, P} + n_2 \left(\frac{\partial F}{\partial n_2}\right)_{n_1, n_3, \ldots, T, P}$$
$$+ n_3 \left(\frac{\partial F}{\partial n_3}\right)_{n_1, n_2, \ldots, T, P} + \cdots \tag{6}$$

In equation 6, n_i is the number of moles of the ith component, and the sum extends over all components. A saving of space may be achieved by inventing quantities called *partial molar free energies* in such a way that equation 6 becomes simply equation 7.

$$F = n_1 \bar{F}_1 + n_2 \bar{F}_2 + n_3 \bar{F}_3 + \cdots \tag{7}$$

Physically the quantity $\bar{F}_i$ is equal to the limit of the change in the free energy of the mixture per mole of the added component i at constant temperature and pressure, as the amount of the added component is made to approach zero. Since the dimension of partial molar free energy is that of energy *per mole*, it is an *intensive* quantity, that is, a quantity independent of the extension of the system, the total number of moles. Although the partial molar free energy varies with the composition, it is convenient, in view of equation 7, to think of the total free energy as the sum of contributions from the individual components. Thus the partial molar free energies may be regarded as molar free energies of the corresponding components in solution.

[1] From Euler's theorem for a homogeneous function of the first degree.

For dilute solutions the dependence of F_i on the molar concentration c_i is delightfully simple:

$$F_i = F_i^\circ + RT \ln c_i \tag{8}$$

Although no attempt is made here to derive equation 8, the following remarks may help to give an insight into its meaning. According to one model, as the solute i becomes more dilute, the number of alternative sites among which the solute may be distributed in the liquid lattice increases; this is an example of a decrease in constraint and therefore entails an increase in probability, an increase in entropy, and an increase in the partition function.

Corresponding to the equation $F = H - TS$, we may write $\bar{F}_i = \bar{H}_i - T\bar{S}_i$. Since the quantity $R \ln c_i$ of equation 8 is obviously a negative entropy term, it might be tempting to equate the term $\bar{H}_i$ with the term F_i°. This would be incorrect, however. The partial molar entropy of a solute includes not only a contribution due to its concentration but also contributions due to such constraints as the orientation of molecules with respect to their neighbors. The quantities $R \ln c_i$ and $\bar{S}_i$ are not identical; neither are the quantities F_i° and $\bar{H}_i$.

Since a solute molecule in a dilute solution is surrounded only by solvent molecules, in such solutions the quantity F_i° is largely a measure of the free energy of the interaction between a solute molecule and *solvent* molecules, a measure of the stability of the solute in an environment composed of pure solvent. In view of this property of F_i°, equation 8 cannot in general be expected to hold at concentrations so high that other solute molecules constitute part of the immediate environment.

We are now in a position to derive an expression for the equilibrium constant of a reaction at constant temperature and pressure in *dilute* solution. Let us take as an example the familiar reaction (9).

$$A + B \rightleftharpoons 2C \tag{9}$$

The differential of the free energy, which is equal to zero at equilibrium, is given by equation 10.

$$dF = 0 \begin{cases} = \dfrac{\partial F}{\partial n_A} \, dn_A + \dfrac{\partial F}{\partial n_B} \, dn_B + \cdots \\[2mm] = F_A \, dn_A + F_B \, dn_B + \cdots \end{cases} \tag{10}$$

When C is formed from A and B, $-dn_A = -dn_B = \tfrac{1}{2}dn_C$, giving us equation 11 as the equilibrium condition.

$$dF = 0 = -dn_A(F_A + F_B - 2F_C)$$

or

$$F_A + F_B - 2F_C = 0 \tag{11}$$

When each partial molar free energy is written as a function of the concentration of the corresponding component as in equation 8, equation 12 is obtained.

$$F_A{}^\circ + RT \ln c_A + F_B{}^\circ + RT \ln c_B - 2F_C{}^\circ - 2RT \ln c_C = 0$$

or

$$-RT \ln \frac{c_C{}^2}{c_A c_B} = \Delta F^\circ \qquad \qquad (12)$$

The result obtained in equation 12 can be generalized to hold for any reaction as equation 13.

$$-RT \ln K_c = \Delta F^\circ \qquad (13)$$

K_c is the usual equilibrium constant expressed in terms of concentrations and ΔF° is the *standard free energy change* for the reaction.

As mentioned on page 16, the criterion for equilibrium in systems of constant energy and constant volume rather than constant temperature and pressure is the maximization of the entropy rather than the minimization of the free energy. It is equally possible to derive an expression for an equilibrium constant from the consequences of maximizing the entropy. The result is equation 14.

$$-RT \ln (K_c)_V = \Delta \bar{E}^\circ - T \Delta \bar{S}^\circ \qquad (14)$$

ACTIVITY COEFFICIENTS

The quantities $F_i{}^\circ$ in equation 8 and ΔF° in equation 13 are constants only if the reaction medium is constant. A constant reaction medium is a constant environment for each solute molecule and therefore restricts the application of equations 8 and 13 to dilute solutions. The deviation of the behavior of the actual solution from equation 8 at higher solute concentrations can be taken into account either by allowing the quantity $F_i{}^\circ$ to be a variable and using any given value of $F_i{}^\circ$ only for some narrow, specified range of concentrations or by replacing equation 8 by one containing a concentration-dependent correction term as in equation 15.

$$F_i = \underset{\text{(variable)}}{F_i{}^\circ} + RT \ln c_i \qquad (8)$$

$$F_i = \underset{\text{(constant)}}{F_i{}^\circ} + RT \ln c_i + \underset{\text{(variable)}}{RT \ln \gamma_i} \qquad (15)$$

Analogously, the expression for the equilibrium constant becomes equation 16.

$$-RT \ln K_c = \underset{\text{(variable)}}{\Delta F^\circ} \qquad (13)$$

$$-RT \ln K_c = \underset{\text{(constant)}}{\Delta F^\circ} + \underset{\text{(variable)}}{RT \ln \gamma_c{}^2/\gamma_A \gamma_B}$$

$$= \underset{\text{(constant)}}{\Delta F^\circ} + \underset{\text{(variable)}}{RT \ln K_\gamma} \qquad (16)$$

The variable γ is called an *activity coefficient*. The variable term $RT \ln \gamma$ is cast in that form because of the great convenience[2] of combining γ_i and c_i into an *activity*, a_i.

$$a_i = \gamma_i c_i \tag{17}$$

$$F_i = \underset{\text{(constant)}}{F_i^\circ} + RT \ln a_i \tag{18}$$

$$-RT \ln K_a = \underset{\text{(constant)}}{\Delta F^\circ} \tag{19}$$

The Reference State

As for any energy function, the absolute value of the free energy is subject to arbitrary choice, and only differences in the quantity have physical reality. This is not a difficulty, however, because in any actual calculation we use differences in free energy, like the quantity $F_i - F_i^\circ$. But even this quantity is subject to an arbitrary choice. If we rewrite equation 15 in the form (20), we see that $F_i - F_i^\circ$ is not defined completely until a numerical value is chosen for the free energy quantity $RT \ln \gamma_i$.

$$F_i - F_i^\circ = RT \ln c_i + RT \ln \gamma_i \tag{20}$$

The solution or state for which $RT \ln \gamma_i$ is equal to zero (so that the concentration of i is equal to its activity) is known as the *reference state*. The choice of the reference state is arbitrary. However, if we wish to obtain equation 8 for dilute solutions, the reference state must also be a dilute solution. Conversely, if in a series of experiments the solutions differ only negligibly from the reference state, then γ_i is unity and a_i may be replaced by c_i.

It is rarely possible to choose a reference state in such a way that in a series of experiments the coefficients γ_i are negligibly different from unity in *all* of the solutions, even if some average composition is chosen as the reference state. The actual choice of a reference state will depend on considerations of practical or theoretical convenience.

In order to simplify the theoretical treatment of the interaction between solute and medium, it is desirable for the reference state to consist of a solution in a medium of essentially one component, that is, the reference state should be a very dilute solution. The nominal solvent chosen for the dilute solution (By nominal solvent we mean the solvent in which the reagents are put into solution rather than the resulting solution.) should have a structure and properties that conform to the assumptions of the particular theoretical approach being used. For example, some theories

---

[2] For the reaction $A + B \rightleftharpoons 2C$, $K_\gamma = \gamma_C^2 / \gamma_A \gamma_B$, $K_a = a_C^2 / a_A a_B$.

require spherically symmetrical interactions among the solvent molecules. Other popular assumptions are that the solvent molecules are small compared to the solute molecules, or of equal size. However, water, even though it is much more sophisticated in its properties than are most theories, is often chosen as a reference solvent because of its practical importance and because of the large amount of data already available.

In many experiments practical considerations dictate the choice of rather complicated reference states. For example, solubility or the rate of attainment of equilibrium might force the choice of a mixed solvent as the reference state. It might not be possible to do enough experiments to allow extrapolation (always a risky process) to the behavior of the reaction in one of the pure solvent components. Because of limitation in the accuracy of chemical analysis, it is often desirable to choose as a reference state a solution that is not even particularly dilute. In such cases the reference state will be chosen in such a way as to make most of the activity coefficients involved as near unity as possible. For example, if a series of experiments is to be done in a rather narrow range of concentration in a single solvent, that solvent and the midpoint of the concentration range might be chosen as the reference state. This is essentially what is done when K_c rather than K_a is reported for the reactions of a series of related compounds in a single solvent and in approximately the same range of concentration, not necessarily dilute. Of course, with such a reference state the activity coefficients will probably be very different from unity in the very dilute solution.

Often, for reactions involving ions, a rather concentrated solution of some salt other than those involved in the reaction is chosen as the nominal solvent. The reason for the use of the swamping salt technique is that a large part of the medium effect on an ionic reaction is electrical in nature and a concentrated salt solution provides a medium in which the electrical effects are more nearly constant. This stratagem is most successful when the concentrations of the reagent salts are very small compared to that of the inert salt. A less successful variant of this technique, used when the two concentrations are of the same order of magnitude, is to keep the ionic strength, μ, rather than the medium constant.

$$\mu = \tfrac{1}{2} \sum_i c_i z_i^2 \tag{21}$$

The ionic strength is defined by equation 21 in terms of the concentrations and charge numbers, z_i, of the ionic species. At *low* ionic strengths in water the activity coefficient may be given approximately by equation 22, the Debye-Hückel limiting law.

$$\log \gamma_i = -a z_i^2 \sqrt{\mu} \tag{22}$$

The coefficient, a, is known as the Debye-Hückel limiting slope. It is calculated from the equation,

$$a = \frac{1.825 \times 10^6}{(DT)^{3/2}},$$

and has the value 0.5091 for water at 25°.

In using the constant ionic strength method of overcoming complications due to medium effects, the amount of inert salt added is varied at the same time as the concentration of the reagent salts. The specific interactions among the ions that always accompany the electrical effects cause variations in the activity coefficients which may be fairly serious. One ion is not always a good model for another one.

SOME CHEMICAL OPERATORS

The Solvent Stabilization Operator

There are two formalisms for describing the effect of a change in the nominal solvent. One of these is to use a different reference state for each different nominal solvent. The reference state for each solvent will be the dilute solution in that solvent. The other device is to keep the reference state constant and to add a variable correction term $RT \ln \gamma$ as in equation 15. However, the activity coefficients needed to compensate for changes in the nominal solvent often differ from unity by several orders of magnitude. Since most chemists prefer to think of the activity coefficient as a small correction term, it is customary, although not logically necessary, to vary the reference state and with it F°_{solute}.

A convenient quantity for discussing effects due to changes in the nominal solvent is the *difference* between the standard partial molar free energies of a solute in two solvents. This quantity may be symbolized by $\delta_M \bar{F}_i^\circ$. In discussing a *series* of solvents, it is convenient to choose one of the solvents as an arbitrary standard.

$$\delta_M \bar{F}_i^\circ = \bar{F}^\circ_{i,\,\text{medium } M} - \bar{F}^\circ_{i,\,\text{standard solvent}} \tag{23}$$

The quantities $\delta_M \bar{F}_i^\circ$, whether they happen to be positive or negative, will be referred to as *solvent stabilizations*.

It has been customary in the chemical literature to use the symbol Δ indiscriminately for increments due to reaction, changes of structure, or changes of solvent. To avoid possible confusion we will restrict our use of Δ to represent changes associated only with chemical reactions, and we will use δ_M and δ_R, respectively, for medium and substituent effects.

The effect of changes in the solvent on a chemical equilibrium can be treated analogously. Using a reference state that is changed whenever the nominal solvent is changed, $K_a = K_c$ for the dilute solution in each solvent, and the standard free energy change for a reaction, ΔF°, is a variable depending on the nominal solvent. The variation in ΔF° can be described by the symbol $\delta_M \Delta F^\circ$. In general, δ_M is an *operator* that describes the effect of a solvent change on whatever quantity follows it, as in $\delta_M \bar{F}_i^\circ$ or $\delta_M \Delta \bar{H}^\circ$. The symbol Δ is also an operator and indicates the effect of a chemical reaction on whatever quantity follows it. Thus in a reaction $A + B \rightleftharpoons 2C$, we have

$$\Delta F^\circ = 2F_C^\circ - F_A^\circ - F_B^\circ$$
$$\Delta \bar{H}^\circ = 2\bar{H}_C^\circ - \bar{H}_A^\circ - \bar{H}_B^\circ$$
$$\Delta \delta_M F^\circ = 2\delta_M F_C^\circ - \delta_M F_A^\circ - \delta_M F_B^\circ$$

and so on.

It is useful to note that the operators δ_M and Δ commute:

$$\delta_M \Delta F^\circ = \Delta \delta_M F^\circ$$

Figure 2-1 shows the standard partial molar free energies of a reacting system consisting of a simple isomerization reaction in two solvents,

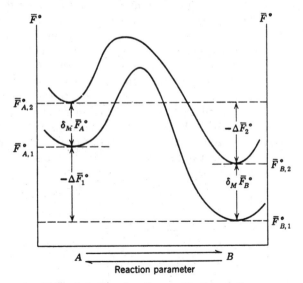

Figure 2-1. Reaction diagram for the reactions

$A \rightleftharpoons B$ in solvent 1
$A \rightleftharpoons B$ in solvent 2

designated 1 and 2. It can be seen by inspection of the figure that equation 24 holds and hence that the operators commute (equation 25).

$$-\Delta F_1^{\circ} + \delta_M F_A^{\circ} = \delta_M F_B^{\circ} - \Delta F_2^{\circ} \tag{24}$$

Rearranging equation 24,

$$\Delta F_2^{\circ} - \Delta F_1^{\circ} = \delta_M \Delta F^{\circ}$$

$$= \delta_M F_B^{\circ} - \delta_M F_A^{\circ} = \Delta \delta_M F^{\circ} \tag{25}$$

The practical consequence of the commutability of δ_M and Δ is that there are two alternative ways of measuring $\delta_M \Delta F^{\circ}$: the measurement of ΔF° for the reaction in two solvents or the measurement of the difference in the individual solvent stabilizations, $\delta_M F^{\circ}$, for the products and the reagents. The latter possibility illustrates the usefulness and the physical reality of standard partial molar free energies for individual substances, for the same value can be used in all reactions in which the given substance is involved. The use of solvent stabilizations for individual reagents is analogous to the use of heats of formation to compute heats of reaction.

It is difficult to find a modern illustration of the relationship between solvent effects on equilibrium constants and solvent stabilizations of the substances involved in the equilibrium because the thermodynamic theory is not regarded as being in need of testing. Numerous examples of the measurement of solvent stabilizations can be found, but no one bothers to determine equilibrium constants in a series of solvents directly unless the data that would permit their calculation are missing. We are therefore forced to use a rather old and well-known example to show how the relationship of Figure 2-1 and equation 25 works in practice.[3,4] Table 2-1 shows that the variation in the equilibrium constants for the keto-enol tautomerization of benzoyl camphor in a series of solvents is consistent with the solubilities of the two reagents, from which the solvent stabilizations can be estimated.

keto form   enol form

[3] O. Dimroth, *Ann.*, **399,** 91 (1913).
[4] L. P. Hammett, *Physical Organic Chemistry*, McGraw-Hill, 1940, p. 89.

Formally, the dissolving of a solute in a saturated solution may be regarded as an equilibrium, with the solubility or, in the case of gases, the Henry's law constant taking the place of the equilibrium constant.

Solvent + Crystals $\rightleftharpoons$ Dilute, but saturated solution

Solvent + Gas $\rightleftharpoons$ Dilute solution saturated at partial pressure P

$$\delta_M(-RT \ln \text{solubility}) = \delta_M \Delta F^\circ_{\text{solubility}} \equiv \delta_M \bar{F}^\circ_i \qquad (26)$$

The agreement between the observed and computed decrements in ΔF° (Table 2-1) is quite satisfactory in view of the fact that the concentrations of

Table 2-1. Keto-Enol Equilibrium of Benzoyl Camphor

SOLVENT	K_c at 0°	enol Solubility / keto Solubility	OBSERVED CHANGE (kcal/mole) $\delta_M \Delta F^\circ$	$\Delta\delta_M \bar{F}^\circ$
Ether (standard)	6.81	6.39	(0)	(0)
Ethyl acetate	1.98	1.81	−0.67	−0.68
Ethanol	1.67	1.57	−0.76	−0.76
Methanol	0.87	0.75	−1.12	−1.16
Acetone	0.85	0.80	−1.13	−1.13

some of the saturated solutions obtained in the solubility measurements are undoubtedly rather high. At high solute concentrations the solubility is only partly determined by the solvent stabilization of the solute, for solute-solute interactions also begin to be important.

The accurate measurement of solvent stabilizations is possible only in certain favorable cases. For a solid substance, the solubility must be small and the solid phase in contact with the solution must be the same in all experiments. For example, the result would be vitiated by the formation of a solid solvate or of a new (and more stable) crystalline modification. The measurement of Henry's law constants is limited to substances that have sufficient vapor pressure. For ionic substances other equilibria can sometimes be used, for example, an emf measurement. At times one is hard put to find a suitable method, especially for nonvolatile liquids. One stratagem is to measure the equilibrium constant for a chemical equilibrium involving the substance and for which the solvent stabilizations of the other reagents and products are already known.

It will be noted that the variation with the solvent of the equilibrium constant in the keto-enol equilibrium of benzoyl camphor is very small compared to that found in a typical equilibrium between ionic reagents. On the other hand, K_c does range over a factor of eight, and this is enough variation to be of qualitative, practical importance. The same thing is true of nonionic reactions in general: The solvent effects on K_c are smaller than those encountered in ionic reactions but certainly not small enough to justify their being ignored either in practical or theoretical contexts.

The Substituent Stabilization Operator

A substituent group in the reagent molecules can change the partial molar free energies of the reagents and hence shift the equilibrium. By choosing an arbitrary reference substituent such as the hydrogen atom, substituent effects can be treated in a way that is formally analogous to the use of a solvent-stabilization operator in the treatment of medium effects. In fact, the analogy is partly mechanistic as well as formal. A solvent molecule attached to a solute differs from a substituent only in that its point of attachment is less rigorously defined. Conversely, the effect of a substituent is sometimes computed in terms of electrostatic interactions, a procedure that resembles the electrostatic interpretation of solvent effects.

The substituent operator δ_R denotes the change that takes place in the expression following the operator when the standard substituent R_0 (usually the hydrogen atom) is replaced by a new one, R.

$$\delta_R \Delta F° = \Delta F_R° - \Delta F_{R_0}°$$

As in the case of the solvent stabilization operator, the substituent stabilization operator commutes with the reaction operator. This can be derived from a reaction diagram, Figure 2-2, in just the same way as was

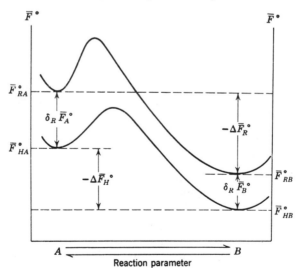

Figure 2-2. Reaction diagram for the reactions

$$RA \rightleftharpoons RB$$
$$HA \rightleftharpoons HB$$

done for the solvent operator. In fact, all three operators Δ, δ_M, and δ_R commute and may be permuted in any of six possible ways. For example, $\delta_M \Delta \delta_R F°$ is equal to $\Delta \delta_R \delta_M F°$. One such permutation is shown by the data in Table 2-2, solvent and substituent effects on the "principal" ultraviolet transition of substituted nitrobenzenes.[5] The quantity $h\nu_{max}$ is the

transition energy at the wavelength of maximum transition probability and is analogous to a thermodynamic Δ-quantity. The commutability of the

Table 2-2. "Principal" Transition Energies for para-Substituted Nitrobenzenes[5]

para-SUBSTITUENT	$h\nu_{max}$, cal/mole		$\delta_M h\nu_{max}$
	GAS PHASE	WATER	
CH_3CH_2-	113,843	100,070	+3773
$(CH_3)_3C-$	113,614	100,270	+3344
$\delta_R h\nu_{max}$	+229	−200	
$\delta_M \delta_R h\nu_{max}$ =	229 −	(−200) =	429
$\delta_R \delta_M h\nu_{max}$ =	3773 −	3344 =	429

substituent and solvent operators must not be taken to mean that the substituent and solvent effects are _independent_. Thus $\delta_R h\nu_{max}$ in water not only has a different value from that in the gas phase, it even differs in sign.

THE ANALYSIS OF ACTIVITY COEFFICIENTS AND SOLVENT STABILIZATIONS IN TERMS OF INTERACTION COEFFICIENTS

The standard free energy of a solute species depends on the immediate environment of its molecules. This immediate or microscopic environment changes as the macroscopic composition of the mixture is changed, and there is nothing but a conventional difference between the effect of a change

[5] W. M. Schubert, J. Robins, and J. L. Haun, _J. Am. Chem. Soc._ **79**, 910 (1957).

in the number of solute molecules in the immediate neighborhood of a given molecule and the effect of a change in the nominal solvent. Formally, the total free energy of a mixture of nonelectrolytes can be expressed as a power series function of the concentrations of all of the components, solute and solvent being treated exactly alike. Equation 27 is a modification of a series expression developed by Scatchard and Prentiss.[5a] The modification consists in the use of molar concentrations rather than mole fractions in order that the expression which will eventually be obtained for the *partial molar* free energy will reduce to equation 8 for dilute solutes.

$$\frac{F}{RT} = \sum_i n_i B_i + \sum_i n_i \ln c_i + \left(\frac{1}{V}\right) \sum_i \sum_j n_i n_j \beta_{ij} + \left(\frac{1}{V^2}\right) \sum_i \sum_j \sum_k n_i n_j n_k \gamma_{ijk} + \cdots$$

$$= \begin{array}{c} \text{intrinsic} \\ \text{terms} \end{array} + \begin{array}{c} \text{concentra-} \\ \text{tion} \\ \text{terms} \end{array} + \begin{array}{c} \text{pairwise} \\ \text{interaction} \\ \text{terms} \end{array} + \begin{array}{c} \text{higher order interaction} \\ \text{terms} \end{array} \qquad (27)$$

Equation 27 represents the total free energy, plausibly enough, as the resultant of intrinsic contributions from each component proportional to the number of moles present, contributions due to the concentration of each component and contributions due to interactions between molecules.

In the intrinsic terms the n_i are the number of moles of the corresponding components and the B_i are constants characteristic only of the nature of the components and independent of the concentration or the environment. Concentration and environmental contributions are taken care of by other terms.

The concentration terms will be seen to correspond to the $\ln c_i$ term of equation 8. The pairwise interaction terms contain coefficients β_{ij}, which, although empirical, are roughly a measure of the interaction between the pairs of molecules i and j. Since F/RT is an extensive quantity, proportional to the number of moles of mixture, and since the coefficients β_{ij} are intensive quantities, each of the terms $n_i n_j \beta_{ij}$ must be divided by a suitable extensive quantity to keep the dimensions consistent. In equation 27 the $n_i n_j \beta_{ij}$ are divided by the volume of the mixture, V, in order that subsequent expressions for the partial molar free energies will involve molar concentrations. Because of this stratagem, the β_{ij} have the dimensions of a molar volume.

Just as the terms involving the β_{ij} may be interpreted, roughly, as pairwise interactions, the next higher terms in the series may be interpreted, even more roughly, as the effect of a third molecule on the interaction of two other molecules. Still higher terms may be included in the series as needed. Even though the physical interpretation of β_{ij} and γ_{ijk} in terms of molecular interactions is plausible and scientifically fruitful, it will be

[5a] G. Scatchard and S. S. Prentiss, *J. Am. Chem. Soc.*, **56**, 1486, 2314 (1934).

realized that there is no mathematical necessity to associate any single term in a power series with a simplified physical model, and in fact we will refrain from doing so in the case of the higher terms.

If we accept equation 27, then we must accept for the *partial molar* free energy an expression obtained by differentiating equation 27. For example, F_3 for the third component in a three-component system can be shown to be given by equation 28.

$$F_3 \equiv \left(\frac{\partial F}{\partial n_3}\right)_{n_1, n_2, T, P} = RT B_3 + RT \ln c_3 + RT[1 + (2\beta_{33} - \bar{V}_3)c_3$$
$$+ (2\beta_{23} - \bar{V}_3)c_2 + (2\beta_{13} - \bar{V}_3)c_1 - \bar{V}_3(\beta_{11}c_1^2 + \beta_{22}c_2^2 + \beta_{33}c_3^2$$
$$+ 2\beta_{12}c_1c_2 + 2\beta_{13}c_1c_3 + 2\beta_{23}c_2c_3) + \cdots] \tag{28}$$

The quantity $\bar{V}_3$ is the partial molar volume, $\partial V/\partial n_3$, of component 3. In the approximation shown in equation 28, terms involving the coefficients γ_{ijk} are neglected, and $\bar{V}_3$ is treated as independent of the composition. Since the concentrations c_i are in units of moles/liter, $\bar{V}_3$ and the β_{ij} must be expressed in liters/mole.

The significance of equation 28 can be appreciated most readily by examining the simplified forms that are obtained in certain special cases. One such application allows a mechanistic interpretation of activity coefficients.

Interpretation of Activity Coefficients

A special case previously treated by means of activity coefficients is that of a solution in a solvent of nominally constant composition in which the solute concentration c_3 is allowed to become too high for the complete neglect of solute-solute interactions. In order to make the problem more interesting, we shall let the nominal solvent have two components whose concentrations in the absence of solute are c_1 and c_2. The derivation of equation 30 makes certain simplifying assumptions but does allow for the change in concentration of the solvent components upon the addition of solute. (This gives rise to the term in $\bar{V}_3^2$.) The constant terms *not* involving the solute concentration have been grouped together under the heading of F_3°.

$$F_3 = F_3^\circ + RT \ln c_3 + RT \ln \gamma_3 \tag{29}$$
$$F_3 = F_3^\circ + RT \ln c_3 + RT c_3[2\beta_{33} - \bar{V}_3(1 + 4\beta_{13}c_1 + 4\beta_{23}c_2)]$$
$$+ RT c_3 \bar{V}_3^2[c_1 + c_2 + 2\beta_{11}c_1^2 + 4\beta_{12}c_1c_2 + 2\beta_{22}c_2^2]$$
$$+ \text{terms proportional to } c_3^2 + \cdots \tag{30}$$

Although the correction term $RT \ln \gamma_3$ of equation 29 is completely formal and Newtonian in the sense of *hypotheses non fingo*, the corresponding parts of equation 30 can be discussed in mechanistic terms. Thus it can be seen by inspection of the quantities within the brackets that deviations from ideality are due largely to three effects: solute-solute interactions (the 33 subscript), solute-solvent interactions (the 13 and 23 subscripts), and solvent-solvent interactions (the 11, 12, and 22 subscripts).

The solution just discussed is nondilute, in the sense that interactions of the type $\beta_{33}c_3$ between solute molecules could not be ignored. However, a solution containing *two* solute species can be nondilute even in the range of concentrations in which either solute, present without the other, would be considered dilute. Let component 1 be the nominal solvent and components 2 and 3 be the solutes. Even if the self-interaction terms $\beta_{22}c_2$ and $\beta_{33}c_3$ are negligible, the solution might still not be properly considered dilute, for the interaction terms $\beta_{23}c_2$ and $\beta_{23}c_3$ may be large. This possibility might be realized, for example, by the existence of stable complexes formed by a reaction between the two solute species. Under these conditions we obtain equations 31 and 32 by differentiating equation 27 and neglecting small terms.

$$\bar{F}_2 = \bar{F}_2{}^\circ + RT \ln c_2 + 2RT\beta_{23}c_3 + \cdots \tag{31}$$

$$\bar{F}_3 = \bar{F}_3{}^\circ + RT \ln c_3 + 2RT\beta_{23}c_2 + \cdots \tag{32}$$

Equations 31 and 32 differ from the corresponding equations for dilute solutes only because of the presence of an added term consisting of an interaction coefficient multiplied by the concentration of the other dilute solute. Since both equations involve the same interaction coefficient β_{23} (which can also be written β_{32}), the effect of the two solution components on each other is symmetrical. It follows that any theory designed to account for the interaction of the two solutes must also be a symmetrical one. That is, the theory must be invariant with respect to the operation of interchanging the subscripts used for the two solutes. The equivalence of the two interaction coefficients is fortunate because in practice it may be easier to evaluate β_{23} from an experiment corresponding to one of the equations 31 and 32 than from an experiment corresponding to the other.

Another way of expressing this relationship is that, in dilute solution,

$$(\partial \bar{F}_3 / \partial c_2)_{c_3} = (\partial \bar{F}_2 / \partial c_3)_{c_2}.$$

Although this relationship becomes inexact at higher concentrations, the analogous expression,

$$(\partial \bar{F}_3 / \partial m_2)_{m_3} = (\partial \bar{F}_2 / \partial m_3)_{m_2},$$

which employs *molal* rather than *molar* concentrations, is correct at all concentrations.

Interpretation of Solvent Stabilizations

Another simple application of the formal expansion equations for the free energy of solutions is to the dilute solution of a single solute in a variable solvent. Consider, for example, a system consisting of a dilute solute, component 3, in a binary solvent consisting of components 1 and 2. Again we define the dilute solution as one in which solute-solute interactions can be neglected. Hence the terms involving β_{33} as well as other terms proportional to c_3 are omitted, and the concentrations of the solvent components are taken as those in the absence of solute. With these approximations we are able to transform equation 28 into 33.

$$F_3 = RT[B_3 + 1 + (2\beta_{23} - \bar{V}_3)c_2 + (2\beta_{13} - \bar{V}_3)c_1 - \bar{V}_3(\beta_{11}c_1{}^2$$
$$+ 2\beta_{12}c_1c_2 + \beta_{22}c_2{}^2) + \cdots] + RT \ln c_3 \quad (33)$$

Comparing equation 33 with our previous expression for dilute solutions we see that the term $F_3{}^\circ$ is a function of the concentrations and the nature of the solvent components and that it is divisible into contributions due to specific molecular interactions. It is also worth noting that the solvent effect depends in part on solvent-solvent interactions as well as on solvent-solute interactions.

CHEMICAL MODELS AND THE FORMAL NATURE OF THE SOLUTE SPECIES

In principle it is possible to explain interactions between species in solution purely in physical terms, but in practice it is frequently more convenient to use a chemical explanation. The reason for this is that the physical explanation must often be made very complicated in order to deal adequately with the experimental facts.

When the molecules and their interactions with the surrounding medium are spherically symmetrical, that is, when the medium is in effect a continuum, the use of physical models is sufficiently simple and accurate to make the spoils worthy of the stratagems. On the other hand, when the interactions between adjacent molecules are strong only for particular orientations of the molecules it is logical to regard such oriented pairs as molecular complexes, for example, hydrogen-bonded or π-complexes. When the formation of such complexes is not sufficient to account fully for the properties of the solution, the chemical explanation can be supplemented by a physical treatment in which the complexes interact with the rest of the solution, usually treated as a continuum. In general, the chemical approach is more successful in dealing with the highly specific interactions between adjacent molecules, whereas the physical approach is more convenient for dealing with the less specific interactions of molecules at a distance.

In some cases it is possible to give a molecular or chemical interpretation to the interaction between substituent and solvent effects. A solvent molecule might be considered to stabilize a solute molecule by interacting specifically with a particular substituent rather than with the solvent molecule as a whole. The substituent, thus modified by solvation, will have an effect on the rest of the molecule different from that of an unsolvated substituent. This can be appreciated most readily in terms of a rather drastic example: the effect of a dimethylamino group on nitrobenzene in the two solvents benzene and sulfuric acid.

dimethylaminonitrobenzene in benzene

dimethylaminonitrobenzene in sulfuric acid

Hydrogen bonding between a substituent and the solvent can also be expected to modify the effect of a substituent, although to a lesser extent than actual protonation. Presumably the formation of a π-complex between an aromatic substituent and an aromatic solvent component would also modify the effect of the aromatic substituent.[6]

Interdependence of solvent and substituent effects can also often be interpreted in terms of a less microscopic treatment which makes use of the electrical properties of the solvent. For example, an increase in the dielectric constant of the medium will increase the relative stability of polar resonance structures involving the substituent. Thus in predicting the effect of a *para*-nitro group on the base strength of aniline, it is desirable to take into account the increase in the importance of the dipolar resonance structures involving the nitro group in the more polar solvents.[7]

Another example of the desirability of supplementing the purely physical approach with a more chemical one is provided by the behavior of ions in concentrated solutions. Thus the ionic strength is a convenient parameter

[6] J. E. Leffler and R. A. Hubbard, II, *J. Org. Chem.*, **19**, 1089 (1954); M. G. Alder and J. E. Leffler, *J. Am. Chem. Soc.*, **76**, 1425 (1954); B. B. Smith and J. E. Leffler, *J. Am. Chem. Soc.*, **77**, 2509 (1955).
[7] B. Gutbezahl and E. Grunwald, *J. Am. Chem. Soc.*, **75**, 559 (1953).

for predicting the interaction of an ion with other ions at the rather large distances characteristic of dilute solutions, but it does not account for the fact that lithium and potassium ions are not interchangeable solvent components in concentrated solutions. The specific chemical effects have still greater relative importance in reactions that neither create nor destroy ions. Medium effects in such reactions are only rarely describable with any accuracy in terms of macroscopic solvent parameters, such as the dielectric constant, but require descriptions on the molecular level.

INTERPRETATION OF FREE ENERGY QUANTITIES WHEN THE FORMAL SOLUTE CONSISTS OF SUBSPECIES

If medium effects are to be analyzed in terms of complex formation, it is necessary to investigate the effect of such subsidiary equilibria on the partial molar free energy of the solute. The solute is represented by its usual molecular formula only by virtue of a convention according to which the presence of isomers and solvation complexes is ignored. However, in spite of the complexity of the actual solution, the experimentally determined partial molar free energy can be shown to have a surprisingly simple meaning: it is equal to the partial molar free energy of the unsolvated, uncomplexed monomeric solute. It also turns out that the partial molar free energy is equal to that of *any* unsolvated monomeric *subspecies* of the solute, even including species resembling those suggested by the conventional molecular formula.

The Partial Molar Free Energies of Solvation Species

Consider, for example, the partial molar free energy of the solute "acetic acid" in the solvent "benzene." The formal thermodynamic expression for the free energy of such a solution is shown in equation 34, in which $n_{``A"}$ and $n_{``B"}$ are the total numbers of formula weights of "acetic acid" and "benzene."

$$F = n_{``A"} \bar{F}_{``A"} + n_{``B"} \bar{F}_{``B"} \tag{34}$$

Now let us consider that the actual solution consists of n_1 moles of acetic acid monomer, n_2 moles of acetic acid dimer, and n_c moles of a one-to-one hydrogen-bonded acetic acid-benzene complex. We then get equation 35 as an alternative expression for the free energy in terms of these simple, and presumably actual, species rather than in terms of the all-inclusive formal species.

$$F = n_1 \bar{F}_1 + n_2 \bar{F}_2 + n_c \bar{F}_c + (n_{``B"} - n_c)\bar{F}_{C_6 H_6} \tag{35}$$

Because of the equilibrium among the complexes, we also have the relationships (36).

$$2F_1 = F_2$$
$$F_c = F_1 + F_{C_6H_6} \tag{36}$$

Hence equation 35 reduces to equation 37:

$$F = (n_1 + 2n_2 + n_c)F_1 + n_{``B"} F_{C_6H_6} \tag{37}$$

But $n_1 + 2n_2 + n_c$ is equal to $n_{``A"}$, and therefore the free energy is given just as well by equation 38 as by the formal equation 34.

$$F = n_{``A"} F_1 + n_{``B"} F_{C_6H_6} \tag{38}$$

The formal quantity $F_{``A"}$ must therefore be interchangeable with and equal to the quantity $F_1$; the formal quantity $F_{``B"}$ must be interchangeable with and equal to the quantity $F_{C_6H_6}$. *In general, the partial molar free energy of the formal species is always equal to that of the monomeric, unsolvated species.*

The partial molar free energy of a *solvated* species can be obtained from equations analogous to (36). Thus the partial molar free energy of the acetic acid-benzene complex is given by equation 39.

$$F_c = F_1 + F_{C_6H_6} = F_{``A"} + F_{``B"} \tag{39}$$

The *Standard* Partial Molar Free Energy of Solvation Species

In order to predict the effects of solvation and other subsidiary equilibria on the equilibrium constants of reactions involving the formal solute, it is necessary to know not the partial molar free energy of the solute but its *standard* partial molar free energy. The latter can be obtained as follows:

$$F_{\text{formal}} = F^\circ_{\text{formal}} + RT \ln c_{\text{formal}} \tag{40}$$

$$F_{\substack{\text{unsolvated}\\\text{monomer}}} = F^\circ_{\substack{\text{unsolvated}\\\text{monomer}}} + RT \ln c_{\substack{\text{unsolvated}\\\text{monomer}}} \tag{41}$$

$$F_{\text{formal}} = F_{\substack{\text{unsolvated}\\\text{monomer}}} \tag{42}$$

Letting α denote the fraction of the solute in the form of unsolvated monomer, equations 41 and 42 give rise to equation 43.

$$F_{\text{formal}} = F^\circ_{\substack{\text{unsolvated}\\\text{monomer}}} + RT \ln c_{\text{formal}} + RT \ln \alpha \tag{43}$$

Hence (44) is the desired expression for the *standard* partial molar free energy.

$$F^\circ_{\text{formal}} = F^\circ_{\substack{\text{unsolvated}\\\text{monomer}}} + RT \ln \alpha \tag{44}$$

Since the term $RT \ln \alpha$ in equation 44 is negative, the formal standard partial molar free energy is less than it would have been in the absence of complex formation. Thus complex formation *always* has the effect of reducing the standard partial molar free energy of the solute. When the solvation complexes are highly stable, the reduction is considerable, for α is then very small. But it should be noted that *some* reduction results even when the solvation complexes are of higher energy, and therefore intrinsically of lower stability, than the unsolvated monomer.

Subsidiary Equilibria Involving Covalent Bond Formation

Although the types of chemical interactions most frequently responsible for stabilization by the solvent are hydrogen-bonding, π-complexing, and dipole-dipole association, the interaction sometimes involves the formation of covalent bonds. A well-known example is the dissociation of acids in water. The formal species is the unsolvated proton, and the activity of "hydrogen ion" is equal to that of the unsolvated proton. Yet it is doubtful whether a beaker of dilute aqueous acid actually contains even one free proton, the predominant species being the various kinds of solvated oxonium ion. A large part of the medium effect on acid dissociation is due to the inherent differences in free energy between different covalently protonated solvent molecules.

Another example of the effect of a subsidiary equilibrium on solute activity is the behavior of carbonic acid in water. If the nominal solute is carbonic acid, then the partial molar free energy is that of the unsolvated species H_2CO_3. However, the subsidiary equilibrium between H_2CO_3 and carbon dioxide causes the actual concentration of H_2CO_3 to be much lower than the nominal concentration. Hence the strength of the acid "H_2CO_3," the value of K_A given in the handbooks, is much lower than would be expected by analogy with other carboxylic acids.

$$K_A(\text{formal "}H_2CO_3\text{"}) = K_A(H_2CO_3) \cdot \alpha_{H_2CO_3}$$

Correcting for the subsidiary equilibrium with carbon dioxide, we find that $K_A(H_2CO_3)$ is actually greater than K_A for benzoic acid.[8]

[8] D. Berg and A. Pattersen, Jr., *J. Am. Chem. Soc.*, **75**, 5197 (1953).

THE PARTIAL MOLAR FREE ENERGIES OF ISOMERIC SPECIES

We have seen that the problem of the solvated molecule is complicated by the existence of various solvation complexes and that the partial molar free energy of the all-inclusive formal species is equal to that of the uncomplexed monomer. We shall now go one step further and treat the uncomplexed monomer itself as a formal species, treating it as a mixture of isomeric subspecies. This is always possible, for a molecular species can always be regarded as a mixture of rapidly interconverting subspecies. For example, the subspecies might be conformational isomers, or they might be defined by some other criterion involving molecular geometry. Or, indeed, the definition might not involve molecular geometry at all, for in principle a subspecies can be generated by taking the population of any arbitrary subset of the set of energy levels comprising the formal species.

Because of the free energy criterion for equilibrium, it is necessary to equate the partial molar free energy of the unsolvated monomer with the partial molar free energy of any of its isomeric subspecies. It is important, however, to remember that the subspecies must be *isomeric*. Thus the partial molar free energy of "acetic acid" is equal to that of 1,1-dihydroxyethylene (if that equilibrium is established), and to that of CH_3COOH, and to that of any rotameric conformation of CH_3COOH, and to that of the ion

pair $H^+ {}^-O{-}\overset{\overset{\displaystyle O}{\|}}{C}CH_3$, and even to that of the ion pair $CH_3CO^+ {}^,OH^-$, but *not* to that of any of the solvated or dimeric species, for example, the solvated ion pair.

The *Standard* Partial Molar Free Energy of Isomeric Subspecies

Just as a subsidiary equilibrium involving solvation stabilizes the formal species, subsidiary equilibria involving isomerism of the monomer also have a stabilizing effect. Let us consider a formal monomeric species A and treat it as an equilibrium mixture of the isomeric subspecies $A_1, A_2, \dots$. To derive a relationship between the standard partial molar free energy of the formal species A and that of the subspecies A_1, we begin with the definitions shown in equations 45.

$$F_A = F_A{}^\circ + RT \ln c_A \tag{45a}$$

$$F_{A_1} = F_{A_1}^\circ + RT \ln c_{A_1} \tag{45b}$$

$$c_{A_1} = \alpha_1 c_A \tag{45c}$$

Since $F_A = F_{A_1}$, equation 46 can be derived at once.

$$F_A = F_{A_1}^\circ + RT \ln c_{A_1}$$
$$= F_{A_1}^\circ + RT \ln \alpha_1 + RT \ln c_A \qquad (46)$$

On comparing equations 46 and 45a we obtain the desired relationship, equation 47.

$$F_A^\circ = F_{A_1}^\circ + RT \ln \alpha_1 \qquad (47)$$

It can be seen that the *standard* partial molar free energy of the formal species A is equal to that of the isomeric subspecies A_1, *plus* an inherently

Table 2-3. Equilibrium Constants for the Trans-Gauche Isomerism of 1,2-Dibromoethane at 25°

MEDIUM	DIELECTRIC CONSTANT	N_{gauche}/N_{trans}
Gas	1	0.164
n-Hexane	1.9	0.251
Cyclohexane	2.0	0.260
CCl$_4$, dilute	2.2	0.307
CCl$_4$, $N_A = 0.22$[a]	2.5	0.439
CS$_2$	2.7	0.453
CCl$_4$, $N_A = 0.44$[a]	2.9	0.439
1,2-Dibromoethane, pure liquid	4.8	0.554
Methanol	33	0.892

[a] The formal mole fraction of 1,2-dibromoethane

negative term involving the mole fraction, always less than unity, of that subspecies. An analogous result would of course have been obtained for any other isomeric subspecies. The fact that the formal quantity F_A° is smaller than the standard partial molar free energy of any isomer is a reflection of the fact that, other things being equal, the existence of subspecies produces stabilization.

There are not very many circumstances in which it is particularly useful to subdivide the reagent into subspecies. Occasionally, however, a simple theory can be developed that can easily predict the behavior of one of the subspecies but not of the others. To get the standard partial molar free energy change, the theoretical value of $\Delta F_{subspecies}^\circ$ is then supplemented by the experimentally determined term $\Delta RT \ln \alpha_{subspecies}$. For example, a quantum-mechanical theory or the calculation of an electrostatic interaction might be applicable with sufficient accuracy only to a single geometrical form of the molecule. A familiar qualitative example of this kind of reasoning is the explanation of the reactions of cyclohexane derivatives in terms of the amount of polar or equatorial isomer.[9]

Factors that might modify the subsidiary equilibria among the isomers include such things as the steric energies of certain conformations or even a solvent effect on torsional isomerism. The latter is illustrated by the considerable solvent effect on the geometrical equilibrium between *trans-* and *gauche-*1,2-dibromoethanes, which would be a factor in any chemical equilibrium involving this substance.[10] The approximate parallelism of the equilibrium constants in Table 2-3 with the dielectric constant of the medium can be attributed to the fact that the *gauche-*isomer has a substantial dipole moment.

THE CONNECTION BETWEEN THE THERMODYNAMIC AND STATISTICAL MECHANICAL TREATMENTS OF EQUILIBRIUM

In the analysis of the formal solute into subspecies, there is a type of subspecies that is not further analyzable into still simpler components. Such a subspecies is the population of a single energy level, and indeed this is perhaps the best way of defining an energy level.

If in the analysis of the standard partial molar free energy according to equation 47 the lowest energy level of A is taken as the subspecies A_1, then the standard partial molar free energy of A is given by equation 48.

$$F_A^\circ = F_{A,\text{lowest}}^\circ + RT \ln \alpha_{A,\text{lowest}} \tag{48}$$
$$\underset{\text{energy level}}{} \qquad \underset{\text{energy level}}{}$$

If a similar choice of subspecies is made for a reaction product B, then the equilibrium constant for the interconversion of A and B is given by equation 49.

$$-RT \ln K = \Delta F^\circ = \Delta F_0^\circ + RT \ln \frac{\alpha_{B,\text{lowest energy level}}}{\alpha_{A,\text{lowest energy level}}} \tag{49}$$

or

$$K = e^{-\Delta F_0^\circ / RT} \times \frac{\alpha_{A,\text{lowest energy level}}}{\alpha_{B,\text{lowest energy level}}} \tag{50}$$

According to equation 50 the problem of equilibrium reduces to the determination of the ground level free energy differences and to the determination of the fraction of each reagent in its ground energy level. But the

[9] S. Winstein and J. Takahashi, *Tetrahedron*, **2**, 316 (1958); E. L. Eliel and C. A. Lukach, *J. Am. Chem. Soc.*, **79**, 5986 (1957); E. L. Eliel and R. S. Ro, *J. Am. Chem. Soc.*, **79**, 5995 (1957); S. Winstein and N. J. Holness, *J. Am. Chem. Soc.*, **77**, 5562 (1955).

[10] J. A. A. Ketelaar and N. van Meurs, *Rec. trav. chim.*, **76**, 495 (1957).

standard *free* energy of a molecular species *consisting of molecules in a single energy level* can be shown by a purely thermodynamic argument to be equal to the standard energy of that species,[11] which in turn is simply equal to Avogadro's number times the energy of the level.

$$\Delta F_0^\circ = N_0(\epsilon_{0B} - \epsilon_{0A}) \tag{51}$$

By means of the Boltzmann distribution law it can be shown that

$$\frac{\alpha_{A,\,\text{lowest energy level}}}{\alpha_{B,\,\text{lowest energy level}}} = \frac{\sum_B e^{-(\epsilon_i - \epsilon_0)_B/kT}}{\sum_A e^{-(\epsilon_i - \epsilon_0)_A/kT}} \tag{52}$$

Equation 50 is therefore equivalent to the statistical mechanical equation 6 of Chapter 1.

[11] We specify merely that $C_v^\circ = (\partial E^\circ/\partial T)_v = 0$ for each subspecies at all temperatures; thus $\Delta C_v^\circ = 0$. On integrating $(\partial \Delta S^\circ/\partial T)_v = \Delta C_v^\circ/T$, we then find that ΔS° has a constant value independent of the temperature. Finally, it follows from the third law of thermodynamics that this constant value must be zero.

3

Free Energy, Enthalpy, and Entropy

No, we are in a jungle and find our way by trial and error, building our road behind us as we proceed. We do not find signposts at crossroads, but our own scouts erect them, to help the rest.

Max Born
Experiment and Theory in Physics

The variation of equilibrium constants with temperature is conveniently analyzed in terms of the variation of the standard free energy change with temperature. The basic thermodynamic equations are (1) through (3).

$$\Delta \bar{F}^\circ = \Delta \bar{H}^\circ - T \Delta \bar{S}^\circ \tag{1}$$

$$\left(\frac{\partial(\Delta \bar{F}^\circ)}{\partial T}\right)_P = -\Delta \bar{S}^\circ \tag{2}$$

$$\left(\frac{\partial(\Delta \bar{F}^\circ/T)}{\partial(1/T)}\right)_P = \Delta \bar{H}^\circ \tag{3}$$

The corresponding equations in terms of equilibrium constants are (4) through (6).

$$-RT \ln K = \Delta \bar{F}^\circ \tag{4}$$

$$R \ln K + RT \left(\frac{\partial \ln K}{\partial T}\right)_P = \Delta \bar{S}^\circ \tag{5}$$

$$RT^2 \left(\frac{\partial \ln K}{\partial T}\right)_P = \Delta \bar{H}^\circ \tag{6}$$

ENTHALPY

The partial molar enthalpy or heat content is related to the partial molar energy by equation 7.

$$\bar{H} = \bar{E} + P\bar{V} \tag{7}$$

For condensed systems at one atmosphere the magnitude of the term $P\bar{V}$ is of the order of 0.01 kcal. The quantity $P \Delta \bar{V}$ is then even smaller, and the reaction increments $\Delta \bar{H}$ and $\Delta \bar{E}$ can therefore be considered equal for practical purposes.

$$\Delta \bar{F}^\circ \cong \Delta \bar{E}^\circ - T \Delta \bar{S}^\circ \tag{8}$$

The relationship between $\bar{H}$ and $\bar{H}^\circ$ can be obtained from the differential equation which defines the enthalpy.

$$\bar{H} = -T^2\left(\frac{\partial(\bar{F}/T)}{\partial T}\right)_{P,c} = -T^2\frac{\partial}{\partial T}\left[\frac{\bar{F}^\circ}{T} + R\ln\gamma + R\ln c\right]_{P,c} \tag{9}$$

$$\bar{H} = \bar{H}^\circ - RT^2\left(\frac{\partial\ln\gamma}{\partial T}\right)_{P,c} \tag{10}$$

It will be noted that equation 10 no longer contains a term proportional to $\ln c$. As will be seen later, the partial molar entropy accounts entirely for the $-RT\ln c$ term in the expression for the partial molar free energy, even though the term $-RT\ln c$ accounts for only a part of the partial molar entropy.

In dilute solutions the second term on the right in equation 10 vanishes, and $\bar{H}$ and $\bar{H}^\circ$ become equivalent.

Enthalpy and the Zero-Point Energy

In equation 9 of Chapter 1, the equilibrium constant was related to the zero-point energy difference of the reagent and product molecules and to the ratio of their partition functions. The corresponding equation in terms of the standard free energy change is (11).

$$\Delta\bar{F}^\circ = N_0\,\Delta\epsilon_0 - RT\ln(Q_B/Q_A) \tag{11}$$
$$= N_0\,\Delta\epsilon_0 - N_0kT\ln(Q_B/Q_A)$$

In spite of the superficial resemblance, the term $N_0\,\Delta\epsilon_0$ is not equal to $\Delta\bar{E}^\circ$, nor is $RT\ln(Q_B/Q_A)$ equal to $T\,\Delta\bar{S}^\circ$. The actual relationships are given by equations 12 and 13.

$$\Delta\bar{E}^\circ = N_0\,\Delta\epsilon_0 + N_0kT^2\left[\frac{\partial\ln(Q_B/Q_A)}{\partial T}\right] \tag{12}$$

$$\Delta\bar{S}^\circ = N_0k\ln(Q_B/Q_A) + N_0kT\left[\frac{\partial\ln(Q_B/Q_A)}{\partial T}\right] \tag{13}$$

The nature of the additional terms in equations 12 and 13 can be most readily apprehended by reference to an energy level diagram, Figure 3-1. The *thermodynamic* energy per molecule, $\bar{E}^\circ/N_0$, is an average quantity dependent on the energies of all the levels weighted according to their respective populations. On the other hand, the quantity $\Delta\epsilon_0$ is merely the difference in the zero-point energies.

Since $N_0\,\Delta\epsilon_0$ contains bond energies, resonance energies, and other forms of potential energy, it is a quantity of great theoretical interest, but unfortunately it is only $\Delta\bar{E}^\circ$ that can be determined by the actual measurement of the dependence of the equilibrium constant on the temperature.

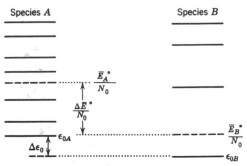

Figure 3-1. Energy level diagram showing the relationship between the quantities $\Delta\epsilon_0$ and $\Delta\bar{E}°/N_0$ for a hypothetical reaction $A \rightleftharpoons B$.

It is worthwhile to enquire into the error that would be made by neglecting the difference between the two quantities. The difference, which we shall call the kinetic energy contribution to the average energy, is given by equation 14.

$$\Delta\bar{E}° - N_0\,\Delta\epsilon_0 = RT^2\left[\frac{\partial \ln (Q_B/Q_A)}{\partial T}\right] \tag{14}$$

The importance of the kinetic energy contribution will depend on the extent to which the excited energy levels are actually populated. The populations will depend on the temperature (which we shall assume to be in the neighborhood of 300°K) and the spacing of the energy levels. At ordinary temperatures it is possible to neglect the population of even the lowest excited levels corresponding to the vibration of a strong bond, but, on the other hand, the lesser energy required for the excitation of such motions as translation, rotation, and internal rotation about single bonds makes for a considerable population of those levels. Therefore we should expect differences between $\Delta\bar{E}°$ and $N_0\,\Delta\epsilon_0$ whenever the reaction changes the number of freely translating molecules or interferes with an internal rotation or torsion. An example of the latter would be a reaction in which a double bond or ring is created.

Although it is not easy to estimate the kinetic energy contribution, its *maximum possible* value can be predicted by assuming the kinetic energy to be classical, that is, that the energy levels are infinitely close together. For this limiting case the kinetic energy is the sum of contributions $\frac{1}{2}RT$, one for each *squared term*. A *squared term* is a term in the classical equation for the total energy that contains either the square of a velocity or the square of a space coordinate. There is one squared term for each degree of freedom of motion. Thus if the reaction creates an additional molecule that is free to move independently, there is a contribution of $3 \times \frac{1}{2}RT$ for the translation of that molecule along the three spatial axes and a contribution of $3 \times \frac{1}{2}RT$ for its rotation (if it is nonlinear) about three axes. For each possibility of unhindered internal rotation that is present in the reaction products but not in the reagents there is a contribution of $\frac{1}{2}RT$. For each new *hindered* rotation or for a vibration there are two squared terms, one involving a spatial coordinate and the other involving a velocity, and hence a contribution of $2 \times \frac{1}{2}RT$.

For an isomerization reaction in which there is no change in the possibility of torsional motion, for example, one in which there is no change in the number of rings or double bonds or in the amount of steric hindrance, the kinetic energy contribution will be close to zero. For a reaction like the recombination of a pair of nonlinear polyatomic radicals, the only appreciable change in the kinetic energy term is due to the loss of three degrees of translational and three degrees of rotational freedom and the gain of one (usually hindered) torsional degree of freedom. The net loss in kinetic energy for such a reaction is therefore no more than $3 \times \frac{1}{2}RT + 3 \times \frac{1}{2}RT - 2 \times \frac{1}{2}RT$ or about 1.2 kcal/mole at room temperature.

The reader should note, however, that the above discussion of the maximum (classical) contribution to be expected from kinetic energy has entirely neglected the effect of solute-solvent interactions, both on the magnitude and on the number of energy levels. Even a simple isomerization reaction can have an appreciable kinetic energy change in solution, since the reagent and product molecules will usually be solvated and will constrain the motions of solvent molecules to different extents.

The Variation of Enthalpy with Temperature

Closely connected with the kinetic energy contribution to the enthalpy is the extent to which the enthalpy depends on the temperature, for it is only that part of the enthalpy which is a function of the temperature. The thermodynamic expressions for the dependence of the enthalpy on the temperature are equations 15 and 16.

$$\left(\frac{\partial \bar{H}^\circ}{\partial T}\right)_P = \bar{C}_P{}^\circ \tag{15}$$

$$\left(\frac{\partial \Delta \bar{H}^\circ}{\partial T}\right)_P = \Delta \bar{C}_P{}^\circ \tag{16}$$

Although the heat capacity at constant pressure, $\bar{C}_P{}^\circ$, is by no means negligible, the reaction increment $\Delta \bar{C}_P{}^\circ$ is often so small that experiments within a range of 100° or so will not reveal the change in $\Delta \bar{H}^\circ$. This will be especially true for gas phase reactions not involving much change in

Table 3-1. Standard Enthalpy and Entropy Changes for the Ionization of Cyanoacetic Acid in Water

Temperature, °C	5	15	25	35	45
$\Delta \bar{H}°$ cal/mole	−32	−502	−888	−1255	−1705
$\Delta \bar{S}°$ cal/mole deg.	−11.30	−12.96	−14.28	−15.49	−16.92

freedom of molecular motion. But even for reactions in solution, the known examples of an experimentally detectable change in $\Delta \bar{H}°$ within a moderate temperature range all seem to be for reactions that entail the creation or neutralization of electrical charges.

The solvation of ions is particularly strong.[1] A major change in solvation changes the freedom of motion of quite a few entire molecules and has an effect on $\Delta \bar{C}_P°$ which can be large compared to the heat capacity change that would be associated with the reaction in the absence of solvent,

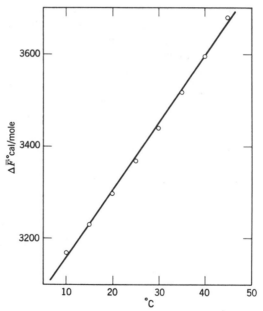

Figure 3-2. The nearly linear dependence of the free energy on the temperature for the ionization of cyanoacetic acid in water.

[1] By means of microwave spectroscopy it is possible to estimate the number of solvent molecules whose attachment to the solute molecule is firm enough to make the aggregate rotate as a unit. For example, the number of water molecules fastened to this extent to the ions of sodium chloride is about six. G. H. Haggis, J. B. Hasted, and T. J. Buchanan, *J. Chem. Phys.*, **20**, 1452 (1952).

especially if the reaction is one that changes only one or two bonds. For example, the standard enthalpies (and entropies) of ionization of cyano-acetic acid in water vary considerably with temperature, as can be seen from the data in Table 3-1. The $\Delta \bar{C}_P^\circ$ for this reaction is unusually large, about -50 cal/mole deg.[2] It should be noted that ΔF° as predicted by equation 1, using *constant* values of $\Delta \bar{H}^\circ$ and ΔS°, is quite accurate in spite of the variation of these quantities, since the temperature-induced changes in ΔS° are in such a direction as to compensate for the corresponding changes in $\Delta \bar{H}^\circ$. This effect is illustrated by Figure 3-2 in which ΔF° for the ionization of cyanoacetic acid is plotted against the temperature. In spite of the known 1.67 kcal/mole change in $\Delta \bar{H}^\circ$ within the 40° temperature range, the data are approximated by a straight line with a maximum deviation of only 0.015 kcal. The worst error in the equilibrium constant that would be caused by using a value of ΔF° from the straight line rather than from the curve is about 3%.

We shall see in the next section that the freedom of motion, which is the important factor in the kinetic energy and in the heat capacity, is also the important factor in the entropy.

ENTROPY

As a background for an intuitive and qualitative understanding of the $-T \Delta S^\circ$ contribution to the standard free energy changes of reactions, it is useful to analyze the entropy in terms of partition functions. The desired equations can be obtained by differentiating the expression for the free energy.

$$F^\circ = N_0 \epsilon_0 - N_0 kT \ln Q + P\bar{V} \tag{17}$$

$$-S^\circ = \left(\frac{\partial (F^\circ - P\bar{V})}{\partial T} \right)_V = -N_0 k \ln Q - N_0 kT \left(\frac{\partial \ln Q}{\partial T} \right)_V \tag{18}$$

The second term in equation 18 is simply the kinetic energy contribution to the total energy divided by the absolute temperature. Since a structural or environmental change which increases the partition function always increases the kinetic energy, it also increases the entropy. Hence constraints on translational, rotational, or internal molecular motions which decrease the partition function (Chapter 1) also decrease the entropy. If a molecule is constrained to occupy a small volume, either by a container or by a solvent cage, it will have a smaller entropy. If a molecule is hindered in its internal or over-all rotation, it will also have a smaller entropy. Examples

[2] F. S. Feates and D. J. G. Ives, *J. Chem. Soc.*, p. 2798 (1956).

are the effects of *ortho*-substituents or ring closure on an otherwise freely rotating group and the effect of an orienting electrical field on an otherwise freely rotating dipole. A constraint on the vibrational motion of a molecule will also decrease the entropy, but this effect will be noticeable only if the vibration interfered with is one that would otherwise have had energy levels low enough to be populated appreciably in the experimental temperature range. This will not ordinarily be the case for the stretching and bending vibrations of covalent bonds (corresponding to bands in the infrared above 600 cm^{-1}), but vibrations of very weak bonds (such as hydrogen bonds) and also torsional vibrations about a single bond axis can make significant contributions. Excited electronic levels can be excluded from consideration in most cases, for the energy required to populate them is too high. However, it is interesting to note that the principle that constraints tend to decrease the entropy is completely general. Thus, in theory, even a constraint on the motion of electrons decreases the entropy, although the effect may be numerically insignificant. For example, the electron delocalization found in conjugated molecules (a relaxation of constraint) has the effect of splitting the electronic energy levels and hence of increasing the partition function. If a molecule is substituted in such a way as to prevent a subsidiary equilibrium, this also decreases the entropy of the formal species.

The relative importance to the entropy of motions corresponding to different degrees of freedom is in the order opposite to that of the energy required to activate the molecule to the first excited state. Thus the contribution to the entropy decreases in the order translation > free rotation > hindered rotation > vibration > electronic transition. The relative importance of various factors in their effect on ΔS° is also in opposition to their importance for $N_0 \Delta \epsilon_0$. In determining $\Delta \epsilon_0$, the electronic state of the molecules is of foremost importance and the energy of excitation to the first vibrational level is also important because it determines the magnitude $(\frac{1}{2} h \nu_{osc})$ of the zero-point vibrational energy. These factors make only a small contribution to the entropy. On the other hand, those factors that are dominant in the entropy and in the kinetic energy are unimportant to the zero-point energy.

Entropy as a Function of the Composition

The dependence of the partial molar entropy on the concentration can be derived from the corresponding expression for the free energy by differentiation.

$$\bar{F} = \bar{F}^\circ + RT \ln c$$

$$\bar{S} = -\left(\frac{\partial \bar{F}}{\partial T}\right)_P = -\left(\frac{\partial \bar{F}^\circ}{\partial T}\right)_P - R \ln c = \bar{S}^\circ - R \ln c \qquad (19)$$

The quantity $\bar{S}^\circ$ is simply the partial derivative of the standard partial molar free energy; the standard state is not given an independent definition for the entropy. For concentrated solutions the differentiation of the free energy expression gives equation 20 for the entropy.

$$\bar{S} = -(\partial \bar{F}/\partial T)_P = \bar{S}^\circ - R \ln c - R \ln \gamma - RT(\partial \ln \gamma/\partial T)_P \quad (20)$$

The terms $-R \ln \gamma$ and $-RT(\partial \ln \gamma/\partial T)_P$, which describe the deviation of the entropy from its dilute solution value, are derived from the corresponding term in the equation for the free energy.

The $-R \ln c$ term in the expression for the partial molar entropy causes the entropy to decrease with increasing concentration. The effect is analogous to the effect of a volume constraint in decreasing the entropy of a gas. The terms containing the activity coefficient include the effect of solute-solute interactions. Solute-solute interactions can bring about a decrease in the entropy if the interacting solute particles suffer a loss of translational or rotational freedom. This might be the case, for example, if the molecules are mutually held and oriented by their dipole fields. However, it is quite possible for the net effect of the interaction to be a net *increase* in entropy, since the solute might be more constrained by its interactions with solvent molecules than by its interaction with other solute molecules. The standard partial molar entropy, of course, depends very much on the nature of the interaction of an isolated solute molecule with its surrounding solvent molecules.

The Variation of Entropy with Temperature

The data of Table 3-1 for the ionization of cyanoacetic acid show that not only $\Delta \bar{H}^\circ$ but also $\Delta \bar{S}^\circ$ can vary with the temperature. Just as in the case of the enthalpy, the variation of $\Delta \bar{S}^\circ$ is seldom observed, and for the same reasons. Formally the variation of the entropy is given by equations 21.

$$\left(\frac{\partial \bar{S}^\circ}{\partial T}\right)_P = \frac{\bar{C}_P{}^\circ}{T}$$
$$\left(\frac{\partial \Delta \bar{S}^\circ}{\partial T}\right)_P = \frac{\Delta \bar{C}_P{}^\circ}{T} \quad (21)$$

Since $(\partial \Delta \bar{H}^\circ/\partial T)_P = \Delta \bar{C}_P{}^\circ$, the variation of entropy with temperature, over a narrow temperature range, is proportional to that of the enthalpy, and the slope of the relationship is equal to the average value of $1/T$. This is the reason for the cancellation of errors in simultaneously neglecting the variation of both enthalpy and entropy with temperature in computing the free energy according to equation 1.

MEDIUM AND SUBSTITUENT EFFECTS ON THE PARTIAL MOLAR ENTHALPY AND ENTROPY

The Medium and Substituent Delta Operators

The operators δ_M and δ_R which were applied to free energy quantities are also applicable to the corresponding enthalpy and entropy quantities. Thus $\delta_M \bar{H}^\circ$ is the change in partial molar enthalpy caused by a change in the reference medium. The free energy increments are functions of the enthalpy and entropy increments, that is, any of the medium or structural effects on the free energy can be separated into effects on the enthalpy and the entropy.

$$\delta_M F^\circ = \delta_M \bar{H}^\circ - T \delta_M \bar{S}^\circ \tag{22}$$

$$\delta_M \Delta F^\circ = \delta_M \Delta \bar{H}^\circ - T \delta_M \Delta \bar{S}^\circ \tag{23}$$

$$\delta_R \Delta F^\circ = \delta_R \Delta \bar{H}^\circ - T \delta_R \Delta \bar{S}^\circ \tag{24}$$

The Relative Importance of Enthalpy and Entropy Effects

Equations 22 through 24 show the free energy increments as the resultants of contributions from the enthalpy and the entropy increments. The enthalpy and entropy increments are sufficiently independent so that they sometimes have the same sign and sometimes opposite signs. When they have the same sign, sometimes $\delta \bar{H}^\circ$ is the larger quantity and sometimes $T \delta \bar{S}^\circ$. We therefore have four possible combinations, all of which have been observed in practice.

1. Cooperative effects in which $\delta \bar{H}^\circ$ and $T \delta \bar{S}^\circ$ are of opposite sign.
2. Enthalpy-controlled effects in which the two terms oppose each other, but the enthalpy term is the larger.
3. Entropy-controlled effects in which the two terms oppose each other, but the $T \delta \bar{S}^\circ$ term is the larger.
4. Compensating effects in which the two terms oppose each other and are almost equal at the given temperature. It is sometimes possible, by a moderate change of temperature, to convert an effect that is slightly enthalpy-controlled into one that is slightly entropy-controlled.

The above categories are of course also found for quantities involving the $\delta\Delta$ operators.

Because entropy effects due to the solvent are so difficult to calculate *a priori*, some theoretical calculations of medium effects have treated the free energy increments in solution simply as potential energy quantities. Implicit in this procedure is the assumption that any solvation entropy effects are either negligibly small or are cancelled by the concomitant solvation enthalpy effects. We believe that this assumption is rather unsafe, especially when the experimentally observed effects are entropy-controlled. For example, a popular model process for theoretical calculation is the ionization of carboxylic acids in water. Yet the experimental enthalpy-entropy relationship for this reaction is extremely complex, as shown by

Table 3-2. Relative Thermodynamic Quantities for the Ionization of Some Carboxylic Acids in Water at 25°, cal/mole

ACIDS $(A-B)$	$\delta_R \Delta \bar{F}^{\circ a}$	$\delta_R \Delta \bar{H}^{\circ a}$	$-T\delta_R \Delta \bar{S}^{\circ a}$	CATEGORY
Formic-Acetic[b]	−1369	+69	−1438	Entropy-controlled
Iodoacetic-Acetic[c]	−2157	−1304	−853	Cooperative
Cyanoacetic- Iodoacetic[c]	−962	+528	−1490	Entropy-controlled
o-Nitrobenzoic- Benzoic[d]	−2690	−2960	+270	Enthalpy-controlled
m-Nitrobenzoic- Benzoic[e]	−1040	+240	−1280	Entropy-controlled
p-Methoxybenzoic- Benzoic[f]	+350	+480	−130	Enthalpy-controlled

[a] $\Delta \bar{F}_A{}^\circ - \Delta \bar{F}_B{}^\circ$; $\Delta \bar{H}_A{}^\circ - \Delta \bar{H}_B{}^\circ$; $-T(\Delta \bar{S}_A{}^\circ - \Delta \bar{S}_B{}^\circ)$.
[b] H. S. Harned and B. B. Owen, *The Physical Chemistry of Electrolytic Solutions*, Reinhold, 1943, p. 514.
[c] F. S. Feates and D. J. G. Ives, *J. Chem. Soc.*, 2798 (1956).
[d] L. P. Hammett, *Physical Organic Chemistry*, McGraw-Hill, 1940, p. 84.
[e] G. Briegleb and A. Bieber, *Z. Elektrochem.*, **55**, 250 (1951).
[f] T. W. Zawidzki, H. M. Papée, and K. J. Laidler, *Trans. Faraday Soc.*, **55**, 1743 (1959).

the data in Table 3-2 where the substituent effects fall into three different categories.

The comparison of the ionization free energies of formic and acetic acids is particularly instructive. The methyl group is almost certainly electron-releasing relative to hydrogen, and its electrical effect within the molecule must therefore be to change the potential energies in such a way as to make acetic the weaker acid. But even though the change in the free energy of reaction is in the expected direction in this instance, it is found that the change in the *enthalpy* of reaction is actually in the contrary direction. Experimentally, the overwhelming contribution to the value of $\delta_R \Delta F^\circ$ results from the large increase in the entropy of reaction.

Enthalpy and Entropy Predictions as Tests of the Validity of a Theoretical Model

It is frequently found that two or more quite different and mutually incompatible theoretical models will fit a set of observed free energy quantities equally well. Although this situation may appear to be fortunate from an empirical point of view, it means that the free energy is not a sensitive tool for judging the validity of theoretical models and mechanisms. On the other hand, the enthalpy and entropy are usually more sensitive to changes in the model from which they are calculated. The failure of free energy measurements to detect flaws in the model is analogous to the phenomenon mentioned in the section on the temperature variation of the enthalpy, where it was pointed out that a rather bad value of the enthalpy can lead to a satisfactory fit of the free energy provided that a compensatingly erroneous value is used for the entropy. Compensating enthalpy and entropy effects are frequent characteristics of theoretical models. To take a simple example, suppose that the decrease in enthalpy obtained by permitting one solvent molecule to become hydrogen-bonded to a solute molecule is 6 kcal/mole and the corresponding decrease in $T \Delta S°$ is 5.9 kcal/mole. Theories differing in the *number* of solvent molecules bonded to the solute molecule will differ appreciably in their predictions of the solvation enthalpy and entropy but will differ only slightly in their predictions of the free energy of solvation.

Another example of the tendency for the enthalpy and entropy to compensate in a theoretical model is provided by the application of the theory of resonance to the effect of an aromatic nitro substituent. It is well known that the resonance energy is at a maximum when the nitro group is coplanar with the benzene ring, and, insofar as the contribution from resonance energy is concerned, the enthalpy for the substance is a minimum for the coplanar structure. However, the coplanar model suffers a decrease in entropy which partially compensates for the stabilization due to the resonance energy. Thus the planar and the freely rotating models do not differ in free energy by as much as might have been expected. Because of the geometrical requirements for resonance, the resonance energy is invariably compensated in part by a negative *resonance entropy*, which reduces the net resonance stabilization or resonance free energy.

SOME SOLVENT EFFECT MODELS

Just as in the case of solvent effects on the partial molar free energy of a solute, it is sometimes convenient to treat solvent effects on the enthalpy and entropy in terms of complexing between the solute and one or more solvent molecules. Actual solutions may be more complex than this model suggests, for the "solvated species" is really an equilibrium mixture of complexes involving different numbers of solvent molecules. A further complication of the model arises from the fact that the number of solvent molecules whose motion is affected by interaction with a given solute molecule is indefinite, since even solvent molecules at a considerable distance from the solute may be constrained to some extent. However, the species in solution can nevertheless be represented by formulas such as $A \cdot \bar{n} M$ if it is kept in mind that $\bar{n}$ is an average rather than an actual solvation number and that even the actual solvation numbers are somewhat indefinite.

The relationship (25) between the formal partial molar enthalpy or entropy and that of the representative solute-solvent complex can be derived by means of a thermodynamic cycle.

$$\bar{H}^{\circ}_{A, \text{formal}} = \bar{H}^{\circ}_{A \cdot \bar{n} M} - \bar{n} \bar{H}_{M}{}^{\circ}$$
$$\bar{S}^{\circ}_{A, \text{formal}} = \bar{S}^{\circ}_{A \cdot \bar{n} M} - \bar{n} \bar{S}_{M}{}^{\circ} \tag{25}$$

In equations 25 the quantities $\bar{H}^{\circ}_{A \cdot \bar{n} M}$ and $\bar{S}^{\circ}_{A \cdot \bar{n} M}$ are the weighted averages of the enthalpies and entropies of all the solute species, the weighting factors being the number of molecules of each species. The quantities $\bar{H}_{M}{}^{\circ}$ and $\bar{S}_{M}{}^{\circ}$ are the enthalpy and entropy of the uncomplexed solvent molecules.

Although the solute-solvent complex model involves certain difficulties, it does show the special importance of the difference between the gas-phase quantity $N_0 \Delta \epsilon_0$ and the quantity $\Delta \bar{H}^{\circ}$ in solution. These quantities differ not only because of the kinetic energy of the unsolvated molecules, but also because of the much more important solvent contributions which take the form $\Delta(\bar{n} \bar{H}_{M}{}^{\circ})$.

It should be noted that the relationship $\bar{F}_{A, \text{formal}} = \bar{F}_{A, \text{uncomplexed}}$ has no parallel in terms of enthalpies and entropies, since its derivation depends on the fact that the *free energy* change for complex formation at equilibrium must be zero. Although $\bar{H}_{A, \text{formal}} - T\bar{S}_{A, \text{formal}} = \bar{H}_{A, \text{uncomplexed}} - T\bar{S}_{A, \text{uncomplexed}}$, it is not the case that the enthalpy and entropy terms in this equation are separately equal.

For neutral molecules, the mechanisms of solvation include both hydrogen-bonding and the formation of π-complexes. The equilibrium constants for the formation of such complexes can be measured in inert solvents and are often near unity. The corresponding standard free energy changes are frequently within a kilocalorie of zero. Conversely, the enthalpies, which are exothermic, are usually more than one kilocalorie and the values of $-T \Delta \bar{S}^{\circ}$ are of similar magnitude. Examples are given in Table 3-3. If the results may be extrapolated to complexing in which one

Table 3-3. Complex Formation at 25°

REACTANTS	SOLVENT	K (liters/mole)	$\Delta \bar{F}^\circ$ (kcal/mole)	$\Delta \bar{H}^\circ$ (kcal/mole)	$\Delta \bar{S}^\circ$ (cal/mole deg)	Ref.
ϕOH + Me$_3$N	Cyclohexane	86	−2.6	−5.7	−10.4	a
EtOH + EtOH	CCl$_4$	0.64	0.26	−2.66	−9.80	b
ϕOH + ϕOH	C$_6$H$_6$	0.57	0.33	−2.4	−9.2	c
ϕOH + Me$_2$CO	CCl$_4$	8.5	−1.27			d
HOAc + HOAc	C$_6$H$_6$	151	−2.97	−8.89	−19.82	e
HOAc + HOAc	H$_2$O	0.037	1.95	0.0	−6.55	e
Picric acid + Naphthalene	CHCl$_3$	1.01	0.006	−1.4	−4.7	f
Iodine +						
Benzene	CCl$_4$	0.151	1.12	−1.32	−8.19	g
p-Xylene	CCl$_4$	0.315	0.68	−2.18	−9.60	g
Mesitylene	CCl$_4$	0.580	0.32	−2.86	−10.67	g
sym-Triethylbenzene	CCl$_4$	0.509	0.40	−2.64	−10.20	g
sym-Tri-t-butylbenzene	CCl$_4$	0.279	0.75	−2.18	−9.83	g
Durene	CCl$_4$	0.630	0.27	−2.78	−10.24	g
Hexamethylbenzene	CCl$_4$	1.52	−0.25	−3.73	−11.68	g
Hexaethylbenzene	CCl$_4$	0.367	0.59	−1.79	−7.98	g
t-Butyl alcohol	CCl$_4$	1.08	−0.05	−3.4	−11.2	g

sym-Trinitrobenzene +					
Acenaphthene	$Cl_2CHCHCl_2$	2.43	−0.53	−2.45	h
Benzene	CCl_4			ca. −0.6	i
Naphthalene	CCl_4			−3.4	i
Phenanthrene	CCl_4			−4.0	i
Anthracene	CCl_4			−4.4	i
Styrene	CCl_4			−1.81	i
m-Dinitrobenzene +					
Acenaphthene	$Cl_2CHCHCl_2$	0.32	0.68	−1.35	h
Naphthalene	CCl_4			−1.6	i
Nitrobenzene +					
Naphthalene	CCl_4			very small	i
Picric acid +					
Acenaphthene	$Cl_2CHCHCl_2$	2.02	−0.42	−0.95	h

[a] R. L. Denyer, A. Gilchrist, J. A. Pegg, J. Smith, T. E. Tomlinson, and L. E. Sutton, J. Chem. Soc., p. 3889 (1955).
[b] W. C. Coburn, Jr., and E. Grunwald, J. Am. Chem. Soc., 80, 1318 (1958).
[c] E. N. Lassettre and R. G. Dickinson, ibid., 61, 54 (1939).
[d] J. M. Widom, R. J. Philippe, and M. E. Hobbs, ibid., 79, 1383 (1957).
[e] M. Davies and D. M. L. Griffiths, Z. physik. Chem. N. F., 2, 353 (1954).
[f] S. D. Ross and I. Kuntz, J. Am. Chem. Soc., 76, 74 (1954).
[g] R. M. Keefer and L. J. Andrews, ibid., 77, 2164 (1955).
[h] H. v. Halban and E. Zimpelmann, Z. physik. Chem., A117, 461 (1925).
[i] G. Briegleb, "Zwischenmolekulare Kräfte," F. Enke, Stuttgart (1937).

partner is the solvent, it can be remarked that an investigation of enthalpies and entropies is more likely to reveal the existence of solute-solvent interactions than is an investigation of free energies.

A striking example of the effect of π-complexing by a solvent component on a reaction rate is the four-fold acceleration of the acetolysis of 2,4,7-trinitrofluorenyl-9-tosylate at 56° by the addition of 0.06 molar phenanthrene to the reaction medium. The transition state for this reaction forms a more stable π-complex with phenanthrene than does the ground state because of the partial positive charge developed on the trinitrofluorenyl group. The added aromatic hydrocarbon has an even larger effect on the enthalpy and entropy of activation. The formal value of $\Delta H^{\ddagger}$ drops from 25.8 to 22.7 kcal/mole, that of $\Delta S^{\ddagger}$ from -11.6 to -18.2 cal/mole degree.[3]

The thermodynamic functions for the dimerization of acetic acid have been given in Table 3-3 for the solvent water as well as for the more nearly inert solvent benzene. The enthalpies and entropies for the dimerization in benzene are much more negative than those for the dimerization in water. The reason for this difference is that in water the monomeric acetic acid is hydrogen-bonded to water molecules to a greater extent than is the dimer, and the formation of new hydrogen bonds is therefore accompanied by the rupture of old ones.

The major advantage of the solute-solvent complex model is that it accounts for the considerable specificity of solvent effects on free energies and the very great specificity of solvent effects on enthalpies and entropies.

In the theories of medium effects that treat the solvent as a continuum with certain electrical properties, the electrical work required to convert the charge distribution of the reagents into that of the products is a free energy quantity. Since the electrical work is expressed as some function of the dielectric constant, there is an entropy contribution to the free energy whenever the temperature derivative of the dielectric constant differs from zero. Because the temperature derivative of the dielectric constant is negative for almost all liquids, the electrostatic part of the standard partial molar entropy of an ion is usually negative. This can be seen by differentiating the free energy term proportional to $(1 - D)/D$.

When the electrostatic treatment of medium effects does not fit the data, it is fairly certain that specific short range interactions, such as those dealt with by the complexing model, are to blame. On the other hand, when the electrostatic treatment does fit the data, it is likely that the fit is assisted by parallelisms between the dielectric constants and the molecular complexing abilities of the series of media in question. It also helps if the solvents of the series all belong to the same chemical type.

[3] A. K. Colter, private communication.

Fundamentally the electrostatic calculation of the thermodynamic solution quantities rests on an analogy. Just as solvent molecules are oriented by the microscopically complex electrical field in the neighborhood of solute molecules, they are also oriented by the external field applied during the measurement of the dielectric constant. In both cases the solvent molecules undergo a change in constraint, and a parallelism between the corresponding thermodynamic quantities for the two processes should not be surprising. However, the electrostatic calculation of the thermodynamic quantities completely *a priori* and without the use of adjustable parameters is rarely successful.

SOME SUBSTITUENT EFFECT MODELS

The free energy increments $\delta_R F^\circ$ and $\delta_R \Delta F^\circ$ depend both on the nature of the molecule in which the substituent finds itself and on the nature of its enveloping solvent; the corresponding enthalpy and entropy increments are even more sensitive to these factors.

It is convenient to divide substituents into two classes, *rigid substituents* in which no internal motions are excited at ordinary temperatures, and *mobile substituents* in which internal rotations or rotations of the substituent about the bond connecting it to the host molecule are excited.

An example of a rigid substituent would be a halogen atom or a cyano group. If such a substituent is introduced in a position where it is crowded by other groups, the length and angles of its bond to the host molecule and the contribution of that bond to the zero-point potential energy of the molecule may be altered. However, even if the crowding is not sufficiently severe to cause an abnormality in the contribution of the substituent to the potential energy of the molecule *in vacuo*, it may still interfere with the normal solvation of the substituent. This will affect both the enthalpy and the entropy. Another factor that can change the effect of a substituent is a change in the electron demand or polarization of the rest of the molecule. Such a change may also alter the nature of the bond to the substituent and hence its contribution to the potential energy of the molecule. However, any such change in the nature of the bond to a rigid substituent will have an appreciable effect on the entropy only if the resulting change in charge distribution alters the solvation.

In mobile substituents such as alkyl, alkoxyl, and nitro groups, both crowding and electron-demand can have a direct effect on the internal mobility of the molecule and therefore affect the contribution of the substituent to the entropy. For example an *n*-butyl group flanked by another group in the benzene ring will suffer a constraint of its internal rotation. A nitro group in a molecule in which it has an unusually strong resonance interaction will not only lower the potential energy but will also suffer a negative resonance entropy because of the increased barrier to rotation associated with the resonance. The mobile substituent presents a problem intermediate in difficulty between that of the rigid substituent and that of a complexed solvent molecule. A complexed solvent molecule resembles a mobile substituent except for the greater uncertainty in its point of attachment.

The reader will by now have gathered that the *a priori* calculation of medium effects and of substituent effects in solution is very difficult, if feasible at all. We do not want to leave him to any such gloomy conclusion without mentioning an alleviating circumstance. In many cases the effect of solvents and substituents can be calculated by using the behavior of chemical analogs as the primary data. Such semiempirical or extrathermodynamic relationships among thermodynamic quantities can be almost as useful and enlightening as more rigorous calculations from first principles. They will be taken up in later chapters.

4

Concerning Rates of Reaction

No single thing abides, but all things flow.
Fragment to fragment clings and thus they grow
Until we know and name them.
Then by degrees they change and are no more
The things we know.

Lucretius

There are two principal theories in use for dealing with the problem of reaction rates. The collision theory is based largely on the kinetic theory of gases and uses a mechanical model; the transition state theory is based largely on thermodynamics and uses a three-dimensional surface as a model, the vertical coordinate being the energy. Either theory is simple enough so that a useful qualitative insight into the nature of rate processes can be gained by visualizing the model for a given reaction. Although the transition state theory is the more generally useful of the two, particularly for organic reactions, the collision theory is nevertheless convenient for certain special purposes and is of historical importance because of the influence it has had on modes of thinking about reaction mechanisms.

THE COLLISION THEORY

The basic assumption of the collision theory is that reaction is the result of a collision but that the collision is ineffective unless the kinetic energy of the colliding molecules along their line of centers equals or exceeds a critical value called the *activation energy*. The necessity for proper orientations and other requirements independent of the energy of the colliding molecules gives rise to an additional parameter in the theory, the probability factor, P. The rate constant or rate at unit concentrations of the reagents is given by equation 1.

$$k = PZe^{-E/RT} \tag{1}$$

The factor Z is the frequency at which collisions occur between the reagent molecules when they are present at unit concentration. According to the kinetic theory of gases, the frequency of collisions between molecules of two species A and B, in units of liter/mole sec, is given by equation 2.

$$Z = \left[\frac{(M_A + M_B)8\pi RT}{M_A M_B} \right]^{1/2} \frac{\sigma_{AB}^2 N_0}{1000} \tag{2}$$

The factor σ_{AB} is the sum of the molecular radii of A and B. Since Z is proportional only to the square root of the temperature, its variation within the range of temperature normally employed in evaluating the parameters of equation 1 is usually imperceptible. Equation 1 is therefore very much like the Arrhenius equation,

$$k = Ae^{-E/RT},$$

in which the temperature occurs only in the exponential factor.

Insertion of typical values for the molecular weights and radii gives collision frequencies of the order of 10^{11} or 10^{12} liters moles^{-1} sec^{-1}. For a very simple reaction in the gas phase, for which P perhaps actually represents merely the fraction of the collisions in which the reagent molecules are properly oriented, P should be a fraction between unity and about 0.01. On the other hand, the transition state theory predicts that a "normal" PZ factor will be found only for reactions in which one of the reagents is a single atom. As we shall see, the additional complications in solution make the calculation of PZ factors for reactions in solution even more unreliable. Both theory and experiment agree that the value of the PZ factor varies considerably for reactions in solution and that no single value can be regarded as "normal," in spite of the persistent use of the term in the chemical literature.

Although the most common way of forcing agreement between the simplest form of the collision theory and experiment is to use the probability factor P as a disposable parameter, an alternative is to dispense with P entirely and to use an effective rather than an actual collision diameter. By using a nomenclature consistent with the ballistic tendencies of nuclear chemists, the *effective cross section* for a reaction is defined by equation 3.

$$B_\sigma = \pi(\sigma_{\text{effective}})^2 \tag{3}$$

Although it is not customary to use the effective cross section terminology in ordinary thermal reactions, the rate constant would be given as a function of the effective cross section by equation 4.

$$k = \left[\frac{(M_A + M_B)8RT}{\pi M_A M_B} \right]^{1} \frac{B_\sigma N_0}{1000} e^{-E/RT} \tag{4}$$

Because of its geometrical connotation, the use of the effective cross section is especially popular for reactions not requiring thermal activation. The most frequent usage is in connection with the reactions of "hot" molecules or ions produced in radiochemical experiments or by the electron bombardment of molecular beams. In ordinary thermal reactions, collisions in which the colliding particles have energies appreciably in *excess* of the minimum critical value necessary for reaction are exceedingly rare. This is the justification for using a single value for B_σ in equation 4. On the other hand, in hot particle chemistry the effective collision cross section depends on the energy of the particle, normally being less when the energy is excessive.

The Cage Effect

Analogy with the behavior of a model consisting of a tray containing a densely but not rigidly packed layer of agitated marbles indicates that collisions of molecules in solution should occur in sets.[1] Because of the confinement by surrounding solvent molecules, any two molecules that once collide are likely to collide several or even many times before they are separated by the escape of one of them from the solvent cage. The rate of a collisional process in solution will therefore be dependent on two quantities. One of these is the number, Z_E, of first collisions, or sets, and is called the *encounter number*. The encounter number will tend to decline with increasing viscosity, η, since the molecules must diffuse together before an encounter can take place.

$$Z_E \propto \eta^{-1}$$

The other important quantity is the number of collisions, n, in a single set of collisions or encounter. This number tends to increase with η, since the same factors that increase the viscosity also increase the resistance of the cage wall to the escape of the reagents.

$$n \propto \eta$$

Combination of these equations indicates that the total collision number $n \times Z_E$ will be independent of the viscosity, and any reaction rate *that is proportional to the number of collisions* should also be independent of the viscosity.

In liquids of ordinary viscosity the actual frequency of collisions between solute molecules at one molar concentration is of the order of 10^{12} or 10^{13} and the number of collisions in a set is of the order of 10 or 10^2. The collision frequency in solution is one or two orders of magnitude greater than in the gas phase at the same concentration because the volume actually available to the solute molecules is less than the volume of the solution. Thus for liquids of ordinary viscosity the *encounter* frequency is of the same order of magnitude as the gas phase collision frequency.

[1] E. Rabinowitch and W. C. Wood, *Trans. Faraday Soc.*, **32**, 1381 (1936); **33**, 1225 (1937).

If the scope of application of the theory of Brownian motion is enlarged to include the motion of molecules, then Z_E, the number of encounters at unit concentration, can be calculated from the experimental values of the diffusion coefficients D_A and D_B.[2] The diffusion coefficient is the number of moles of material diffusing across unit area in unit time under the influence of a concentration gradient of unity. For encounters between two ions, the encounter number is given by equation 5.

$$Z_E = \frac{8\pi N_0 q(D_A + D_B)}{1000(1 - e^{-2q/\sigma_{AB}})} \qquad (5)$$

where

$$q = \frac{-z_A z_B e^2}{2DkT}$$

If at least one of the reagents is a nonelectrolyte, the encounter number is given by equation 6.

$$Z_E = \frac{4\pi N_0(D_A + D_B)\sigma_{AB}}{1000} \qquad (6)$$

Under certain circumstances the rate of a reaction will not be proportional to the number of collisions. Suppose, for example, that reaction is always achieved at an early collision of each set; then the rest of the collisions of the encounter are wasted, and the rate is proportional to the number of encounters rather than the number of collisions. If an encounter includes enough collisions to ensure reaction, it does not matter how many additional collisions it includes. The rate of such a process will therefore tend to be inversely proportional to the viscosity rather than independent of it.

There are two conditions under which reaction is likely to take place at a relatively early stage of the encounter. One is that the total probability of reaction at one collision, $Pe^{-E/RT}$, is close to unity; the other is that the encounters consist, on the average, of unusually long sequences of collisions. For ordinary liquids it is calculated that the transition from encounter-controlled to collision-controlled rates should occur at an activation energy of about 2 kcal/mole if the probability factor is unity. For activation energies in the more common range of 10 to 30 kcal/mole, the effect of viscosity should not appear until the liquid is almost a glass.

An example of a reaction that is viscosity-sensitive because of a low activation energy is the quenching of fluorescence. A metastable, electronically excited molecule can lose its excitation energy instead of fluorescing if it encounters a quencher, that is, a molecule that either

[2] M. v. Smoluchowski, *Physik. Z.*, **17**, 557, 585 (1916); J. A. Christiansen, *Z. physik. Chem.*, **113**, 35 (1924); P. Debye, *Trans. Electrochem. Soc.*, **82**, 265 (1942).

accepts the electronic energy or catalyzes its conversion to vibrational energy. Because of the high efficiency of quenchers, the measured rate of the quenching process is actually that of the diffusion of the quencher to the excited molecule and the measured activation energy is that associated with the process of diffusion.[3] Diffusion-controlled processes are also common in very fast proton-transfer reactions.[4]

An example of a reaction that is viscosity-controlled partly because of its low activation energy and partly because the reaction medium is unusually viscous is the recombination reaction of growing chain radicals in vinyl polymerization.

$$2RCH_2CHX\cdot \rightarrow RCH_2CHX\text{---}CHXCH_2R$$

During the polymerization the viscosity is increased because of the formation of the polymer. If at the end of the reaction the reaction mixture attains a glassy state, it is not unusual for the radical recombination reaction to be slowed to such an extent that radicals are actually left over, stabilized by their isolation from one another by the rigid medium.[5] When the radicals in such a polymer sample are freed by dissolving the polymer in a less viscous solvent, they recombine immediately. Samples of polymer containing stuck free radicals will grow when brought into contact with fresh monomer because the molecules of the monomer, being much smaller than those of the radical, are still able to diffuse into the glassy medium.[6] Another example of the same phenomenon is the marked acceleration that takes place in the polymerization of methyl methacrylate in the absence of a solvent.[7] This reaction is a typical chain reaction and is therefore very sensitive to any factor that removes the chain-carrying radicals. In the last stages of the reaction, the high viscosity of the medium interferes with the combination of the radicals with one another but does not interfere to the same extent with the diffusion of the small monomer radicals to the sites of the comparatively immobilized radicals. Therefore the amount of polymer produced by each radical before its chain is terminated, and hence the rate of the polymerization, tends to increase.

The cage effect is also a factor to be reckoned with in any dissociation reaction that produces reactive molecules, such as free radicals or ions, in pairs. Since the dissociation necessarily takes place inside a solvent cage, the reactive fragments of the original molecule remain in contact for some

[3] B. Williamson and V. K. La Mer, *J. Am. Chem. Soc.*, **70**, 717 (1948).

[4] M. T. Emerson, E. Grunwald, and R. A. Kromhout, *J. Chem. Phys.*, **33**, 547 (1960).

[5] S. S. Medvedev, *Acta Physicochimica U.R.S.S.*, **19**, 457 (1944).

[6] G. H. Miller and A. K. Bakhtiar, *Can. J. Chem.*, **35**, 584 (1957); D. J. E. Ingram, M. C. R. Symons, and M. G. Townsend, *Trans. Faraday Soc.*, **54**, 409 (1958); H. W. Melville, *Proc. Roy. Soc.*, A **163**, 511 (1937).

[7] G. V. Schulz and G. Harborth, *Makromol. Chem.*, **1**, 106 (1947).

time and may recombine rather than separate by diffusion. Recombination of the dissociation products before they have escaped from the cage is called *primary recombination*.[8]

Even after escape of the original partners from their cage, but while they are still fairly close together, the probability that they will again encounter each other and recombine is somewhat greater than that for their combination with similar molecules originating in other cages. Such recombination of the original partners outside of the cage in which they were produced is called *secondary recombination*. The inclusive term for either primary or secondary recombination is *geminate* recombination (L. *gemini*, twins).

THE TRANSITION STATE THEORY

The transition state theory rests on three fundamental assumptions:[9]

1. In the process of reaction the molecules must traverse certain states of potential energy higher than the average potential energies of either the reactant or product states.
2. The populations of these higher potential energy levels that must be traversed by the reacting molecules are in statistical equilibrium with the ground states from which the reacting molecules came.
3. The rate of reaction is proportional to the concentration of molecules in these higher energy levels.

In order for the transition state theory to apply, the reaction must be a thermal one, and not diffusion-controlled. The limitation to thermal reactions is necessary because of the assumption of statistical equilibrium, which would of course not be valid for "hot particle" reactions. The limitation to reactions that are not diffusion-controlled is merely a matter of convenience. In principle the process of diffusion can itself be treated by means of transition state theory. However, we will want to avoid comparing reactions that are diffusion-controlled with reactions that are not, because the factors that determine the potential energy barriers to the two types of reaction are quite different.

Graphical Representation of Reacting Systems

For each reaction it is possible to define a *reaction coordinate*, motion along which corresponds to the transformation from reactants through transition state to products. Points along the reaction coordinate correspond to possible values of some function that describes the conformation

[8] R. M. Noyes, *J. Am. Chem. Soc.*, **77**, 2042 (1955).
[9] S. Glasstone, K. J. Laidler, and H. Eyring, *The Theory of Rate Processes*, McGraw-Hill, 1941.

of the reacting molecule. For example, in the *cis-trans* isomerization of an olefin, the function might be the angle of twist. For reactions in which some bond is broken, the function might be an internuclear distance, as it might also be in reactions in which a bond is being formed. If the system had only this one degree of freedom, it would be possible to represent the reaction of any given molecule by means of a two-dimensional diagram in which the potential energy is plotted as a function of the reaction co-ordinate, as in Figure 4-1.

The motion along the reaction coordinate *may* correspond to some one of the normal modes of vibration of the activated complex, but it need not be that simple. (E. Thiele and D. J. Wilson, *J. Phys. Chem.*, **64**, 473 (1960); N. B. Slater, *ibid.*, **64**, 476 (1960).) In some reactions a concerted motion of a rather complex kind, involving changes in the lengths or angles of several bonds, is needed to achieve reaction. For example, in the *cis-trans* isomerization of olefins it appears that as the double bond twists it also changes its length. (See, for example, B. S. Rabinovitch and K. W. Michel, *J. Am. Chem. Soc.*, **81**, 5065 (1959).) A complicated motion might have to be represented as a combination of two or even more normal modes.

In an actual molecule there are usually additional motions along other coordinates which are not *necessary* for reaction but which nevertheless can and do occur to various extents in the individual transition states. Such motions contribute to the energy of the transition state species and must be considered in any complete theory. The transition state species, like any

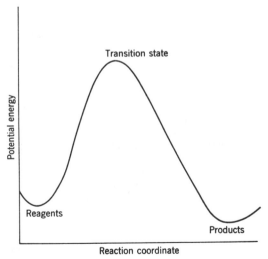

Figure 4-1. Graphical representation of a reacting system.

molecule, can consist of more than the population of just one energy level. We might call the motions not included in the reaction coordinate *irrelevant* motions; examples of such motions might be the change in the length of some bond or in the conformation of some part of the molecule remote from the site of the reaction.

Strictly speaking, the complete graphical representation of the potential energy of a reacting system requires a separate coordinate for each kind of motion. However, to simplify the polydimensional problem so that we can visualize the surface and use our geometrical intuition, it is convenient to suppose that the motions other than those along the reaction coordinate can be represented by a single additional coordinate. The three dimensions will then be a z-dimension representing the energy, a y-dimension representing motion along the reaction coordinate, and an x-dimension representing the other degrees of freedom. On such a three-dimensional surface there is a valley representing the reagents, separated from a valley representing the products by a ridge of high energy. The ridge, or energy barrier, is not of uniform height, however, but contains a pass within which there is a saddle point, or to use mountaineering terminology, a col, dividing the two sides of the mountain range. The *transition state* is the region in the neighborhood of the col or saddle point (Figure 4-2).

In becoming products the reacting molecules *must* surmount the higher energy levels represented by the col; but there are still higher energy levels that most of the reacting molecules will avoid, just as a conservative traveler crossing a mountain range will choose a path that crosses the range by the lowest possible route. Actually the reagent molecules will explore *all* the possible paths, but because of their tendency to obey the Boltzmann distribution law, most of them will be found on the low road, that is, the lower energy levels, and to a good approximation the reaction can be regarded as following a set of reaction paths very close to the lowest possible one.

Because the most traveled paths may tend to converge only near the transition state (where they are crowded together by the steep side walls of the pass), it is usually not meaningful to discuss *the* mechanism for the activation process. This situation is analogous to the impossibility of determining reaction mechanisms from equilibrium constants alone. The transition state treatment of reaction rates allows us to deduce from the rates something about the ground states and something about the transition states, but nothing about the states that the molecules presumably must traverse in becoming transition states.

For any given reacting molecule, the transition state is the highest point on its path over the barrier. However, since a collection of reacting molecules will follow more than one path, the set of energy levels that we call the transition-state *species* consists of the transition states for all the

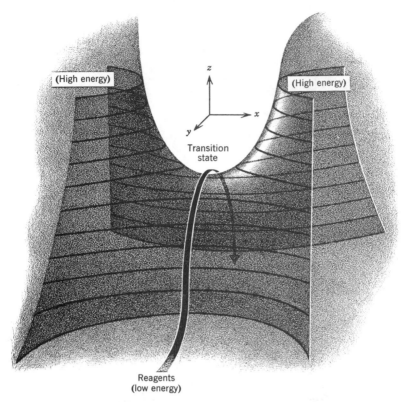

Figure 4-2. Front view of a simplified potential energy surface. The z-dimension represents the potential energy, the y-dimension represents the reaction coordinate, and the x-dimension represents the other degrees of freedom. Note the shapes of the potential energy contour lines. Readers unacquainted with the delights of contour maps, or for that matter anyone who merely wants to read a very enjoyable book, should consult C. E. Montague's *The Right Place*, Chatto and Windus, 1924, especially chapters 3 and 4.

individual paths and must be represented by a region in the col rather than by a single point.

In order to represent a statistical collection of reacting molecules by means of a simple two-dimensional diagram we must replace the energy of the individual molecule by a new ordinate, usually the standard partial molar free energy, a statistical quantity. The point at the maximum in such a diagram (Figure 4-3) gives the standard partial molar free energy of the transition state species; the points at the minima give the standard partial molar free energies of the reagents and products. If there is more than one

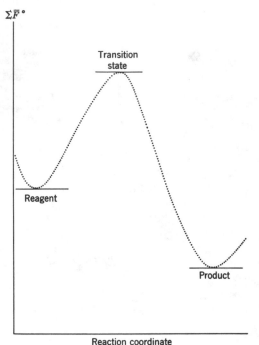

Figure 4-3. Graphical representation of the standard free energy in a reacting system.

reagent, the ordinate can be taken to represent the sum of the standard partial molar free energies of the reagents. The meaning of intermediate points, those that are neither maxima nor minima, is less obvious. Such points represent partially activated reagent or product molecules in which the reaction coordinate has a value different from that of the average value for the formal reagent or product species. The intermediate points represent subspecies of the reagent and product species. Subspecies (Chapter 2) have higher standard partial molar free energies than do the formal species of which they constitute a small part. Free energy versus reaction coordinate diagrams are extremely useful in any discussion having to do with reaction rates or equilibria.

The Thermodynamic Properties of Transition States

One of the important ideas of transition state theory is that the transition state is in equilibrium with the reagent and that the population of the transition state energy levels may be approximated by means of the Boltzmann distribution law. The correctness of this assumption is not immediately obvious because we must presume that the lifetime of molecules in the transition state is too short to permit them to interchange their energy with one another. The reason for the Boltzmann-like distribution in the transition state is not collisional equilibration of the levels within that species, but rather that transition state molecules are produced from reagent molecules which themselves have a Boltzmann distribution of energies.

When we consider a system in which equilibrium between reagents and products has already been established, there is of course no doubt about the applicability of the Boltzmann distribution law to the population of the transition state energy levels. However, there is something peculiar about the dynamics of the transition state population. The molecules in the transition state may be divided into two subspecies, those having a momentum along the reaction coordinate in the reagent direction, and those having a momentum along the reaction coordinate in the product direction. It is an additional assumption of the theory that all, or almost all, of the transition state molecules which are about to decompose into product were formed from reagent molecules, and that the transition state molecules which are about to decompose into reagent were formed from product molecules. This assumption is equivalent to saying that whenever a transition state molecule is transformed into one of the high energy levels of the product, it is almost invariably deactivated into one of the lower energy levels of the product rather than "reflected" back into the transition state.

When the reagents and products are not in equilibrium, the concentration of the part of the transition state corresponding to reaction in the forward direction is still governed by the Boltzmann distribution and the concentration of the *reagent*. This is true because virtually all of the forward-moving transition state molecules are formed from reagent molecules, and it does not matter whether product is present or not. In other words, there is always a quasi-equilibrium between a reagent and the subspecies of the transition state leading away from the reagent, even when the reagent and product are not in equilibrium.

The assumption that a reagent molecule, after entering the transition state, almost always moves forward to yield product is a corollary of the assumption that there is a quasi-equilibrium between the reagent and its transition state subspecies. Consider an energy level of the reagent near, but not quite at, the transition state. At thermodynamic equilibrium, the rates at which molecules enter and leave the level are equal. Some of the molecules will be entering from lower levels, and an equal number will leave for lower levels. Some of the molecules will be entering from product by way of the transition state, and an equal number will leave by the same route. When there is no product, everything is the same except that there are no longer any molecules entering by way of the transition state. This change will have a negligible effect on the concentration of molecules in the level only if the rate at which molecules enter from, or leave for, the transition state is very slow compared to the rate at which they enter from, or leave for, the lower levels. Hence the postulate of quasi-equilibrium is equivalent to the postulate that the rate of deactivation is large compared to the rate of further activation to the transition state, and hence that a molecule coming from the transition state is almost invariably handed down to lower levels rather than being reflected back to the transition state.

For the special case in which the reagent and product *are* in equilibrium, the two transition state subspecies are present in equal concentration. Equal concentrations of the two transition state subspecies of course correspond to equal rates in the forward and reverse directions.

As a result of these qualitative considerations, the transition state can be dealt with by means of the same mathematical devices that we are in the habit of applying to the description of ordinary molecular species. Thus the statistical-mechanical expression for the equilibrium constant between the reagents and the forward-moving transition state subspecies is given by equation 7. The thermodynamic expression for this equilibrium constant is given by equation 8.

$$K_{act} = \frac{Q_{trans}}{Q_{reag}} \exp\left(- \frac{\epsilon_{0\,trans} - \epsilon_{0\,reag}}{kT}\right) \tag{7}$$

$$-RT \ln K_{act} = \Delta F_{act}^{\circ} = \Delta \bar{H}_{act}^{\circ} - T \Delta \bar{S}_{act}^{\circ} \tag{8}$$

The statistical-mechanical and thermodynamic quantities in equations 7 and 8 need not be discussed at length, since they are similar in principle to

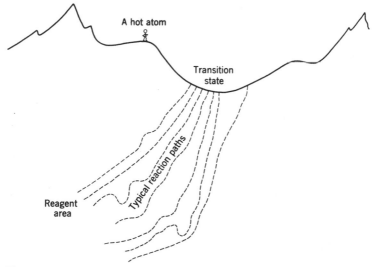

Figure 4-4. The profile of the barrier for a reaction of relatively positive activation entropy.

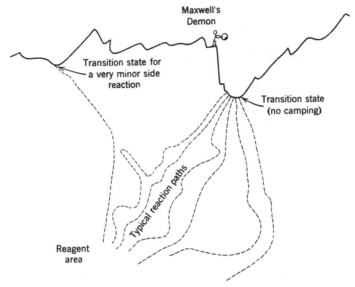

Figure 4-5. The profile of the barrier for a reaction of relatively negative activation entropy due to unusual constraints in the transition state.

those for equilibria, which have been treated in earlier chapters. However, it may be helpful to visualize the relationship between these quantities and the shape of the potential energy surface for the reactions. If there is a large number of paths crossing the barrier whose highest points have approximately the same energy, the surface must have a relatively broad pass with gently sloping side walls, as shown in the sketch of Figure 4-4. Such a transition state has a relatively positive entropy and a relatively large partition function. On the other hand, if the transition state is constrained by steep side walls to lie in a narrow defile, as in Figure 4-5, there are relatively few low energy paths over the barrier, the entropy is less positive, and the partition function is smaller. A transition state within such a narrow defile is one for which very few motions other than that along the reaction coordinate are possible without a large increase in energy.

The Transition State Rate Equation

The transition state theory considers that the rate of the forward reaction is proportional to the concentration of the forward-moving transition state molecules. The motion that leads to the decay of a transition state molecule into products corresponds to what in an ordinary molecule would be one of the bond vibrations or some special combination of several bond vibrations, that is, one of the vibrational degrees of freedom in the transition state is unstable, having a potential energy maximum rather than a minimum. It can be shown in various ways[9] that the partition function for this unstable degree of freedom, Q_u, multiplied by the velocity of motion in the product direction, $v^{\ddagger}$, is equal to $\kappa kT/h$.

$$Q_u v^{\ddagger} = \frac{\kappa kT}{h} \tag{9}$$

In equation 9, k is Boltzmann's constant, h is Planck's constant, and κ is the *transmission coefficient*, that is, the fraction of the forward-moving transition state molecules that actually become product and are not reflected. The transmission coefficient will usually be close to unity.

The reaction rate is equal to the concentration, $c^{\ddagger}$, of the forward-moving transition state species multiplied by the velocity of motion in the product direction.

$$\text{Reaction rate} = c^{\ddagger} v^{\ddagger}$$

Equation 10 gives the transition state concentration $c^{\ddagger}$ in terms of the product of the reagent concentrations and the equilibrium constant for activation. Equation 11 gives it in terms of the partition functions.

$$c^{\ddagger} = K_{\text{act}} \prod c_r \tag{10}$$

$$c^{\ddagger} = \frac{Q_{\text{trans}}}{Q_{\text{reag}}} \exp\left(-\frac{\epsilon_{0\,\text{trans}} - \epsilon_{0\,\text{reag}}}{kT}\right) \cdot \prod c_r \tag{11}$$

In view of equation 9, it is desirable to factor the partition function for the transition state, Q_{trans}, into one term for the unstable degree of freedom and another term, $Q^{\ddagger}$, for all the other degrees of freedom.

$$Q_{\text{trans}} = Q_u \times Q^{\ddagger} \tag{12}$$

The reaction rate is then given by equation 13.

$$\text{Reaction rate} = c^{\ddagger} v^{\ddagger}$$

$$= v^{\ddagger} Q_u \frac{Q^{\ddagger}}{Q_{\text{reag}}} \exp\left(-\frac{\epsilon_{0\,\text{trans}} - \epsilon_{0\,\text{reag}}}{kT}\right) \cdot \prod c_r$$

$$= \kappa \frac{kT}{h} \cdot \frac{Q^{\ddagger}}{Q_{\text{reag}}} \exp\left(-\frac{\epsilon_{0\,\text{trans}} - \epsilon_{0\,\text{reag}}}{kT}\right) \cdot \prod c_r \tag{13}$$

Equation 13 can be simplified by defining a constant $K^{\ddagger}$.

$$K^{\ddagger} = \frac{Q^{\ddagger}}{Q_{\text{reag}}} \exp\left(-\frac{\epsilon_{0\,\text{trans}} - \epsilon_{0\,\text{reag}}}{kT}\right) \tag{14}$$

Since $K^{\ddagger}$ differs from an ordinary equilibrium constant only because of the lack of one factor in the partition function, it is logical to use it to define quantities $\Delta F^{\ddagger}$, $\Delta H^{\ddagger}$, and $\Delta S^{\ddagger}$, which we shall call simply the free energy, enthalpy, and entropy of activation. They are analogous to the usual standard partial molar quantities of reaction.

$$\Delta F^{\ddagger} = -RT \ln K^{\ddagger} \tag{15}$$

$$\Delta H^{\ddagger} = RT^2 \left(\frac{\partial \ln K^{\ddagger}}{\partial T} \right)_P \tag{16}$$

$$\Delta S^{\ddagger} = -\left(\frac{\partial \Delta F^{\ddagger}}{\partial T} \right)_P \tag{17}$$

The rate constant, which is equal to the reaction rate divided by the concentrations of the reagents, is then given by equation 18.

$$k = \frac{\kappa kT}{h} K^{\ddagger}$$

$$= \frac{\kappa kT}{h} e^{-\Delta F^{\ddagger}/RT} = \frac{\kappa kT}{h} e^{-\Delta H^{\ddagger}/RT} e^{\Delta S^{\ddagger}/R} \tag{18}$$

On comparing equation 18 with the Arrhenius equation, $k = Ae^{-E/RT}$, it can be shown that the activation energy E is related to the enthalpy of activation by equation 19.

$$E = \Delta H^{\ddagger} + RT \tag{19}$$

The enthalpy of activation can be divided into a potential energy term and a kinetic energy term. Neither of these bears any simple relationship to the Arrhenius activation energy.

The activation entropy is related to the A factor of the Arrhenius equation. Since experimental results are expressed in the literature about equally often in terms of A (or of PZ of the collision theory) and in terms of $\Delta S^{\ddagger}$, it is convenient to have an equation for converting them. Equation 20 is for use when A is given for the second as the unit of time and for temperatures in the neighborhood of 300°K. The error introduced by using it at slightly different temperatures is usually negligible.

$$\Delta S^{\ddagger} = 4.575 \log_{10} A - 60.53 \tag{20}$$

The Activity of the Transition State

The reaction rate according to transition state theory is proportional to the concentration of the transition state rather than to its activity. The *activity* of the transition state enters the rate problem only because the *concentration* of the transition state is given by a thermodynamic equilibrium constant. The resulting expression for the transition state concentration contains the ratio of activity coefficients $\gamma_A \gamma_B / \gamma^{\ddagger}$. Because $\gamma^{\ddagger}$ is not constant, the rate can be in no constant proportion to the activities of the reagents. If reaction rates were proportional to the *activity* of the transition state, $\gamma^{\ddagger}$ would cancel out, and the rates *would* be proportional to the reagent activities. The failure of rates to be proportional to reagent activities is one of the main reasons for the development of transition state theory.

The Comparison of Transition States

In later chapters we shall discuss extrathermodynamic relationships among various quantities such as $\delta_M \Delta F^{\ddagger}$ or $\delta_R \Delta F^{\ddagger}$. We shall even have occasion to discuss the relations between quantities involving $\Delta F^{\ddagger}$ and quantities involving ΔF°, using the existence of empirical relationships between such quantities to deduce something about the structure or the solvation of transition states. In this chapter we will confine ourselves to a brief discussion of the general features that must be taken into account in comparing one transition state with another and in comparing a transition state with some ground state species used as a model.

In equations 9, 13, and 18 there occurs a so-called transmission factor κ which allows for the possibility that the barrier-crossing process is in some cases reversed. It is generally assumed that κ is close to unity, that is, that most molecules passing the barrier are immediately deactivated to low energy levels of the product. If κ is indeed different from unity, it will change the absolute value of $\Delta S^{\ddagger}$ and hence that of $\Delta F^{\ddagger}$. If, however, κ is constant in a series of related reactions, then the *relative* values of $\Delta S^{\ddagger}$ in the series will have a simple meaning.

We will also wish to compare changes in rate constants with changes in equilibrium constants, that is, $\delta \Delta F^{\ddagger}$ with $\delta \Delta F^{\circ}$. This seems to be justified pragmatically to the extent that the resulting relationships (such as the Hammett relationship, Chapter 7) are fairly simple rather than highly complex functions. The procedure is not rigidly logical because of the fact that "free energy of activation" is not rigorously a free energy and is therefore not the same kind of quantity as ΔF°. The reason for the qualitative difference between the equilibrium and kinetic parameters is the omission of one partition function factor in the definition of "free energy of activation" and "entropy of activation" and, also, the possible error in the assumption that the transmission coefficient is unity.

We will also wish to compare solvent stabilizations of transition states not only with other quantities of exactly the same kind but also with solvent stabilizations of ordinary molecules. Because of the considerable success in the interpretation of kinetic solvent effects by comparison with such equilibrium phenomena as the orientation of solvent molecules in electrical condensers or the effect of solvents on equilibrium constants, we are confident that the process is justified. There is, however, one property of the transition state that must be kept in mind in thinking about its solvation, and that is the short lifetime of the transition state as compared to an ordinary molecule. Solvent molecules may not always be able to attain quite the same orientation about a transition state as about an ordinary molecule of comparable polarity—the relaxation time for the reorientation of solvent molecules is too long. This is analogous to the much more drastic rule for photochemical processes, for which the Frank-Condon principle requires that the solvation shell of the newly photo-activated molecule have essentially the same composition and configuration as that of the ground state molecule. We should therefore consider that if the solvent effect on a rate constant suggests a certain degree of polarity for the transition state, the actual transition state may be slightly more polar than might be thought. This difficulty will be less serious, of course, if we are comparing solvent effects on two rate constants than if we are comparing solvent effects on a rate constant with solvent effects on an equilibrium constant.

DYNAMIC REVERSIBILITY AND REFLECTABILITY

Both the mechanics of the special theory of relativity and classical mechanics predict that the equations of motion for a system of particles will remain valid for either a change in the sign of the time variable (dynamic reversibility) or for a change in the sign of one of the space variables (reflectability).[10] These properties of the equations of motion have two consequences. One is that any isolated mechanical system can with equal probability be run backward, with reversed velocities for all the particles. The other is that the mirror image of any isolated system will behave in all other respects just like the original.

Dynamic reversibility implies that any chemical reaction mechanism must also exist in reverse. It is not possible to have one mechanism, that is, one transition-state configuration, exclusively for the forward process and an entirely different one exclusively for the reverse process. Thus if the mechanism of a reaction in one direction can be ascertained, the mechanism for the reverse direction is known immediately. This principle is known as the principle of *microscopic reversibility* because the statistical behavior of *macroscopic* collections of particles is effectively irreversible in consequence of the second law of thermodynamics.[11] Presumably a system that violated

---

[10] R. C. Tolman, *The Principles of Statistical Mechanics*, Oxford University Press, 1938.
[11] In fact, the principle of microscopic reversibility is the basis of the derivation of the reciprocal relations in Onsager's system of irreversible thermodynamics. The experimental verification of the Onsager reciprocal relations, and hence of the principle of microscopic reversibility, has been reviewed by D. G. Miller, *Chem. Revs.*, **60**, 15 (1960).

the principle of microscopic reversibility would also be capable of violating the second law of thermodynamics. If there were separate paths for the forward and reverse processes, it should be possible to block one without blocking the other, in the fashion of Maxwell's Demon. The reaction could then be made to run uphill past the point of equilibrium.

From the point of view of transition state theory, the principle of microscopic reversibility is either a geometrical or a grammatical truism: a potential energy barrier can not have more than one lowest pass. This pass will be the route preferred by the majority of the molecules for reaction in either direction. It is possible for a reaction to proceed through two passes, but both passes must be used in both directions.

The reflectability principle explains the observation of equal rates (i.e., equal probability) of formation of the enantiomorphs from optically inactive starting materials. The failure of this principle on the nuclear level in the experiments of Wu et al.[12] should not be taken to mean that it is no longer applicable on the molecular level, nor that the principle of microscopic reversibility is invalid.

THE TUNNELING EFFECT

The hydrogen atom is subject to a considerable quantum mechanical uncertainty in its position since its momentum, and hence the uncertainty in the momentum, are relatively small at thermal velocities. Although the uncertainty is much less than that for less massive particles, such as the electron, it is considerably greater than that for other atoms and may in some instances cause detectable effects on reaction rates. The thickness of the potential energy barrier near the top of the col need only be of the same order of magnitude as the de Broglie wavelength of the hydrogen atom for the effect to become great enough to cause a breakdown of the simple transition state treatment, which neglects quantum jumps through the barrier relative to the classical process of surmounting the barrier. For protons or hydrogen atoms moving at the thermal velocities characteristic of ordinary temperatures, the de Broglie wavelength of hydrogen is about 10^{-8} to 10^{-9} cm, not very small compared to the changes in internuclear distances involved in reactions. At temperatures well below room temperature the de Broglie wavelength will be still larger. Hydrogen atom abstraction by methyl radicals[13] and the proton transfers involved in at least some

[12] C. S. Wu, E. Ambler, R. W. Hayward, D. D. Hoppes, and R. P. Hudson, *Phys. Rev.*, **105**, 1413 (1957).
[13] H. S. Johnston, *Advances in Chemical Physics*, vol. 3, Academic Press, 1960, p. 131; H. S. Johnston and D. Rapp, *J. Am. Chem. Soc.*, **83**, 1 (1961).

acid- or base-catalyzed reactions[14] appear to be examples of the tunneling effect.

Because of the greater masses of deuterium or tritium atoms, these atoms will be much less subject to tunneling than the hydrogen atom. Hence the tunneling effect should appear as a deuterium or tritium isotope effect inexplicably large on the basis of simple considerations of zero-point vibrational energies alone.[13,14] Because the de Broglie wavelength of the hydrogen atom is greater at lower temperatures, the anomaly should be especially marked for experiments at low temperatures. In the base-catalyzed bromination of ethyl cyclopentanone-2-carboxylate the activation energy difference for the bromination of the deuterium and protium compounds is as much as 1.2 kcal/mole greater than would be expected in the absence of tunneling.[15] The activation entropies for the deuterium compound are also markedly higher than those for the hydrogen compound, as is to be expected if some factor is making the rate for the hydrogen compound abnormally high at the lower extremity of the range of temperatures over which the rates are measured.

[14] R. P. Bell, *The Proton in Chemistry*, Cornell University Press, 1959.
[15] R. P. Bell, J. A. Fendley, and J. R. Hulett, *Proc. Roy. Soc.*, **A235,** 453 (1956); J. R. Hullet, *ibid.*, **A251,** 274 (1959).

5

Rates of Interconversion

of Subspecies

In discussions of the effects of solvent or structure on reactivity it is often desirable to analyze the usual chemical species into subspecies, such as conformational isomers, solvation complexes, or even the populations of individual energy levels. We have already discussed the thermodynamic consequences of this approach. In this chapter we shall consider the rates of interconversion of such subspecies. These rates are usually very fast in comparison with other chemical reactions but not so fast as to prevent their being measured. It is thus possible to test experimentally any assumptions concerning the equilibration of the subspecies. For example, it can be determined experimentally whether the rate of interconversion of vibrational and translational energy is high enough to produce a Boltzmann distribution of the energy in a chemically reacting system.

RELAXATION PROCESSES

Most of the processes concerned in subsidiary equilibria or in the equilibria between reagents and transition states are fast enough to make it necessary to apply special methods in estimating their rates. One technique is to observe the relaxation, or return to equilibrium, of a system that has been slightly perturbed by the imposition of a change in one of the variables of state. The imposed change in the variables of state can be in the form of an oscillation about a fixed value. For example, if a sound wave is passed

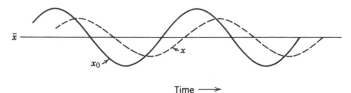

$\bar{x}$

x_0 x

Time $\longrightarrow$

Figure 5-1. The solid curve represents the equilibrium value, x_0, as a function of time; the dotted curve represents the actual value, x, as a function of time.

through a system, there will be periodic changes in temperature and pressure. If the rates of interconversion among the subspecies are high compared to the rate of change in the variables of state, there will be a new equilibrium composition for each value of the variable of state, for the equilibrium will keep up with the change in the variable. If, on the other hand, the oscillations in the variables of state are very fast compared to the rates of subspecies interconversion, the fluctuations of the populations of the subspecies will be negligible.

For *very* slow oscillations of the variables of state, any changes in the system take place reversibly: the system has at all times the particular equilibrium composition corresponding to the values of the variables of state at that time. For *very* fast oscillations, no chemical changes take place. Of interest to us is the behavior of the system when oscillations of intermediate frequency are imposed on it. During such oscillations of the variables of state, the detailed composition of the system changes, but with a phase lag. Let x represent the actual concentration of some subspecies, and x_0 the concentration as it would be if the system actually kept up with the disturbance, that is, x_0 is the equilibrium concentration that would actually correspond to the instantaneous values of the variables of state. A system like that represented by Figure 5-1 is continually in some state of disequilibrium. To maintain such a state, work must be done on the system, which therefore absorbs energy. The energy, of course, is supplied by the external mechanism that is producing the disturbance. If the oscillation in the variable of state is extremely rapid, the system in effect experiences only a time-average value of the variable of state and behaves as it would under static conditions; therefore no energy is absorbed. If the oscillation is extremely slow, the system is in some state of equilibrium at all times, and again no work needs to be done on the system.

Mathematical Description of Relaxation Processes

If a system reacts by a single step process and is *only slightly displaced from equilibrium*, it will relax (i.e., return to equilibrium) according to a first-order rate law. The rate law will be first order even if the reaction step is polymolecular. First let us consider the simple unimolecular process $A \rightleftharpoons B$, with which we are already quite familiar. Let x_{A_0} and x_{B_0} be the equilibrium concentrations and x_A and x_B the actual concentrations.

$$\Delta x = x_{A_0} - x_A = x_B - x_{B_0} \tag{1}$$

The rate of the process is given by equation 2.

$$\frac{d\,\Delta x}{dt} = -\frac{dx_A}{dt} = k_1 x_A - k_{-1} x_B$$

$$= k_1 x_{A_0} - k_{-1} x_{B_0} - (k_1 + k_{-1})\,\Delta x \tag{2}$$

Since x_{A_0} and x_{B_0} are the equilibrium concentrations, $k_1 x_{A_0} = k_{-1} x_{B_0}$, and the rate is given by the first-order equation 3.

$$\frac{d\,\Delta x}{dt} = -(k_1 + k_{-1})\,\Delta x \tag{3}$$

In general, if $A + B + \cdots$ react reversibly to form $C + D + \cdots$, the rate of attainment of equilibrium may be written as in equation 4.

$$\frac{d\,\Delta x}{dt} = k_+ f_+(\Delta x) - k_- f_-(\Delta x) \tag{4}$$

In equation 4, $k_+ f_+$ and $k_- f_-$ are the appropriate rate laws, functions of the equilibrium concentrations and of Δx. The functions f_+ and f_- may be expanded in a Taylor series about the point $\Delta x = 0$.

$$f_+(\Delta x) = f_+(0) + \left(\frac{df_+}{d(\Delta x)}\right)_{\Delta x = 0} \cdot \Delta x + \binom{\text{higher order}}{\text{terms}}$$

$$f_-(\Delta x) = f_-(0) + \left(\frac{df_-}{d(\Delta x)}\right)_{\Delta x = 0} \cdot \Delta x + \binom{\text{higher order}}{\text{terms}} \tag{5}$$

For systems of chemical interest, the first-derivative terms in equations 5 are different from zero. The higher-order terms can be neglected since Δx is small. Furthermore, $k_+ f_+(0) = k_- f_-(0)$. Equation 4 therefore reduces to equation 6.

$$\frac{d\,\Delta x}{dt} = \left[k_+ \left(\frac{df_+}{d\,\Delta x}\right)_{\Delta x = 0} - k_- \left(\frac{df_-}{d\,\Delta x}\right)_{\Delta x = 0} \right] \cdot \Delta x \tag{6}$$

Equation 6 shows that the relaxation of Δx is a first-order process, although the term in brackets depends on the equilibrium concentrations x_{A_0}, x_{B_0}, x_{C_0}, $\ldots$.

Because the field of fast reaction kinetics has been infiltrated by physicists, there is a tendency in the literature to express the rates in terms of relaxation times, τ, rather than rate constants.

$$\frac{d \ln \Delta x}{dt} = -\frac{1}{\tau} \tag{7}$$

On comparing equations 6 and 7, it is seen that $1/\tau$ is related to the rate constants of the given relaxation process by equation 8.

$$\frac{1}{\tau} = -k_+ \left(\frac{df_+}{d\,\Delta x}\right)_{\Delta x=0} + k_- \left(\frac{df_-}{d\,\Delta x}\right)_{\Delta x=0} \tag{8}$$

In general, the terms $k_+ (df_+/d\,\Delta x)_{\Delta x=0}$ and $k_-(df_-/d\,\Delta x)_{\Delta x=0}$ are pseudo-first-order rate constants, dependent on the equilibrium concentrations. However, if the reactions are truly first order, as in equation 2, then $1/\tau$ is independent of the equilibrium concentrations, as seen from equation 9.

$$\frac{1}{\tau} = k_1 + k_{-1} \tag{9}$$

To calculate k_1 and k_{-1} from an observed relaxation time, it is necessary to know the equilibrium constant $K = k_1/k_{-1}$. If K is large, $k_1 \gg k_{-1}$, and $1/\tau \approx k_1$.

To see how equation 8 might be applied to other than simple unimolecular processes, consider the reaction (10).

$$NH_4OH \underset{k_-}{\overset{k_+}{\rightleftharpoons}} NH_4^+ + OH^- \tag{10}$$

$$f_+(\Delta x) = [NH_4OH]_0 - \Delta x$$

$$f_-(\Delta x) = ([NH_4^+]_0 + \Delta x) \cdot ([OH^-]_0 + \Delta x)$$

$$\left(\frac{df_+}{d\,\Delta x}\right)_{\Delta x=0} = -1$$

$$\left(\frac{df_-}{d\,\Delta x}\right)_{\Delta x=0} = [NH_4^+]_0 + [OH^-]_0$$

Hence

$$\frac{1}{\tau} = k_+ + k_-([NH_4^+]_0 + [OH^-]_0) \tag{11}$$

It is seen that $1/\tau$ depends on the sum of the concentrations of the reaction products, so it should be possible to get values of k_+ and k_- by measuring $1/\tau$ at two different concentrations. In practice, however, it often happens that either the term k_+ or the term $k_-([NH_4^+]_0 + [OH^-]_0)$ is so much greater than the other that only one of the rate constants can be obtained accurately in this way. The other is calculated from the equilibrium constant. Table 5-1 shows some data obtained for this reaction by use of the dissociation field method.

In using a relaxation method for determining the rate of a reaction we measure not Δx but some physical property $\mathscr{P}$ of the system that varies in phase with Δx. Since the property being measured is in phase with Δx, it is *not* in phase with the oscillatory disturbance used for excitation. The

Table 5-1. *Relaxation Times in the System* $NH_4OH—H_2O$ *at* 20°

[NH_4OH]	[NH_4^+] + [OH^-]	τ(sec)[a]	k_-(l/mole sec)[b] (equation 11)
66.6×10^{-4}	6.8×10^{-4}	Too short to measure	
6.0×10^{-4}	2.03×10^{-4}	1.5×10^{-7}	3.0×10^{10}
1.5×10^{-4}	1.02×10^{-4}	3.7×10^{-7}	2.3×10^{10}
0.5×10^{-4}	0.59×10^{-4}	6.1×10^{-7}	2.2×10^{10}
			Average 2.5×10^{10}

[a] M. Eigen and J. Schoen, *Z. Elektrochem.*, **59**, 483 (1955).
[b] $K_B = k_+/k_- = 1.71 \times 10^{-5}$ at 20°.

measured property $\mathscr{P}$ is customarily analyzed into a component $\mathscr{P}_{\parallel}$ that is in phase with the original exciting disturbance and a component $\mathscr{P}_{\perp}$ that is 90° out of phase.

When the in-phase component, $\mathscr{P}_{\parallel}$, is plotted against the frequency ν of the exciting sonic or electrical disturbance, a sigmoid curve like that of Figure 5-2 and equation 12 is obtained. The reciprocal of the relaxation

$$\mathscr{P}_{\parallel} = \frac{(\mathscr{P}_{\parallel})_0 - (\mathscr{P}_{\parallel})_\infty}{1 + 4\pi^2\nu^2\tau^2} + (\mathscr{P}_{\parallel})_\infty \tag{12}$$

$$\tau^{-1} = 2\pi\nu_{\frac{1}{2}} \tag{13}$$

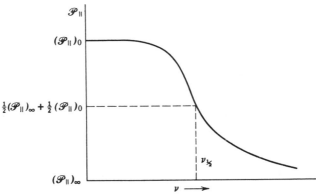

Figure 5-2. Dispersion with frequency of the in-phase component of a physical property when there is a relaxation process.

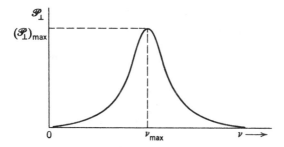

Figure 5-3. The out-of-phase component of a physical property plotted as a function of the frequency when there is a relaxation process.

time is equal to 2π times the frequency, $\nu_{1/2}$, at which the measured value of $\mathscr{P}_{\parallel}$ has dropped to $\frac{1}{2}$ the difference between its static (or zero-frequency) value and its high-frequency value (equation 13).

When the component $\mathscr{P}_{\perp}$ (which is 90° out of phase with the incident disturbance) is plotted against the frequency of the latter, the relationship is that of Figure 5-3 and equation 14.

$$\mathscr{P}_{\perp} = \frac{4\pi(\mathscr{P}_{\perp})_{max}\nu\tau}{1 + 4\pi^2\nu^2\tau^2} \tag{14}$$

$$2\pi\nu_{max} = \frac{1}{\tau} \tag{15}$$

The reciprocal of the relaxation time is then equal to 2π times the frequency at which the measured value of $\mathscr{P}_{\perp}$ is at its maximum. Incidentally, $\mathscr{P}_{\perp}$ is analogous to an extinction coefficient in ordinary spectrophotometry. Both quantities are proportional to the amount of energy absorbed by the system. Thus the curve shown in Figure 5-3 is sometimes referred to as the absorption curve.

Types of Relaxation Experiments

The incident disturbance in a relaxation experiment can be electrical, electromagnetic, mechanical, thermal, or chemical. The disturbance is often, but not necessarily, sinusoidal; it sometimes consists of square waves or even of single pulses.[1]

Dissociation Field Method. The dissociation field method makes use of the fact that the position of the equilibrium of a reaction in which charge is generated depends on the strength of an applied electrical field. The

[1] M. Eigen, *Disc. Faraday, Soc.*, **17**, 194 (1954).

property $\mathscr{P}$ in this case is the electrolytic conductivity. High fields are used because at low fields the effect is difficult to measure.[2,3] In the method used by Eigen and Schoen[3] a potential of 200,000 volts is applied in a single pulse lasting for 10^{-5} to 10^{-7} sec. The electrolyte can be considered to be a nonohmic conductor whose conductivity is a function of the field. The change of conductivity can be displayed on an oscilloscope.

The Temperature Jump Method. It is possible to raise the temperature of a solution containing a strong electrolyte by several degrees in less than 10^{-8} sec, by discharging a large condenser through the solution.[1] The rate of establishment of the new equilibrium at the higher temperature can then be measured by following the conductance of the solution, or by means of some other convenient property, such as an optical absorption or rotation.

Flash Photolysis. The flash photolysis method produces free radicals or electronically excited molecules by exposure to intense square pulses of visible or ultraviolet light, usually generated by means of a rotating sector that periodically occludes the light source. During the period following the pulse, the decay in concentration of radicals or electronically excited states is followed either by the absorption or emission spectra of the species or by some secondary chemical process such as radical-induced polymerization.[4,5]

A closely related method is the use of a phosphorimeter, in which the excited species produces phosphorescence radiation which is measured during the dark periods.[6,7] This measures the relaxation of electronically excited states.

The Spectrophone. A method rather similar to flash photolysis or phosphorimetry is the excitation of molecular vibration by infrared. A pulse of infrared radiation produces vibrational excitation whose relaxation by conversion into translational energy of the gas increases the temperature and pressure of the system. Pulses of infrared radiation applied at regular intervals to a system of constant volume produce a periodic fluctuation of temperature and pressure. The device is called the spectrophone and the result is called the optic-acoustic effect. The optic-acoustic effect was discovered in 1880 by Alexander Graham Bell, who reported that a cigar irradiated in pulses produced sound waves.[8]

[2] R. G. Pearson, *Disc. Faraday Soc.*, **17**, 187 (1954).

[3] M. Eigen and J. Schoen, *Z. Elektrochem.*, **59**, 483 (1955).

[4] R. G. W. Norrish, *Z. Elektrochem.*, **56**, 705 (1952); R. G. W. Norrish and G. Porter, *Disc. Faraday Soc.*, **17**, 40 (1954).

[5] H. Linschitz and K. Sarkanen, *J. Am. Chem. Soc.*, **80**, 4826 (1958).

[6] G. N. Lewis and M. Kasha, *J. Am. Chem. Soc.*, **66**, 2100 (1944); M. Kasha, *Chem. Reviews*, **41**, 401 (1947).

[7] R. J. Keirs, R. D. Britt, Jr., and W. E. Wentworth, *Anal. Chem.*, **29**, 202 (1957).

[8] A. G. Bell, *Proc. Amer. Assoc. Adv. Sci.*, **29**, 115 (1880). Scientific enthusiasm for tobacco has a long and honorable history. Lavoisier even lost his head over it.

Dielectric Relaxation. The dielectric relaxation method measures the dielectric constant and the dielectric loss (absorption of energy by the dielectric) as a function of the frequency of the applied alternating voltage.[9] The change in dielectric constant and the dielectric loss are due to the orientation of dipoles associated with molecules or with parts of molecules under the influence of the applied electrical field. The relaxation measured is that of dipoles partly oriented in the field to dipoles randomly oriented. The mechanism can be rotation of the entire molecule, or rotation of a polar group within the molecule, or both simultaneously.

Shock Waves. The rapid passage of a compression wave through a system is associated with changes in temperature as well as pressure, since the compression is nearly adiabatic. Thus the passage of a sound wave or a shock wave through a medium displaces the system from equilibrium through the combined action of pressure and temperature changes.

In a shock wave experiment[10] the shock front is the site of conversion of the mechanical energy of mass flow into random molecular kinetic energy of translation and rotation. The time required for this conversion is very short compared to that required for the further conversion of the translational and rotational energy into vibrational energy. Immediately behind the shock front is a region in which the vibrational degrees of freedom are being excited at the expense of the excess (over the equilibrium amount) energy of the translational and rotational degrees of freedom. This relaxation of the motion of the molecule as a whole into internal motions causes a drop in temperature and an increase in density. The density gradient and its accompanying refractive index gradient cause *schlieren*, which can be photographed. From the thickness of the region in which the transition of energy between different modes is taking place, the rate of the process or its relaxation time can be computed. The advantage of this method is that translational temperatures up to 6000° can readily be obtained. By a translational temperature of 6000° is meant a velocity distribution among the molecules corresponding to that which they would have at equilibrium at 6000°. In contrast, the vibrational temperature is initially room temperature. A special advantage of the shock tube technique over other experiments in which the temperature change is produced by conduction is that the rise in the translational temperature takes place in about 10^{-10} to 10^{-9} sec, so that the starting time for the relaxation process is quite precisely known and relaxation times as short as 10^{-8} sec are measurable. The translational-vibrational relaxation is sometimes followed

[9] P. Debye, *Polar Molecules*, Chemical Catalog Co., New York, 1929.
[10] W. Payman and W. C. F. Shepherd, *Proc. Roy. Soc.*, A **186**, 293 (1946); W. Bleakney, D. K. Weimer, and C. H. Fletcher, *Rev. Sci. Inst.*, **20**, 807 (1949); E. L. Resler, S. C. Lin, and A. Kantrowitz, *J. Appl. Phys.*, **23**, 1390 (1952).

by other relaxations, which can also be measured. These might include the dissociation of molecules into atoms or free radicals.[11]

Ultrasonics.   In the ultrasonic method of putting mechanical energy into a system, the observed properties are the sound velocity and the sound absorption.[12]   For an ideal gas the sound velocity is a simple function of C_v, the molar heat capacity at constant volume (equation 16).

$$v_{\text{sound}} = \sqrt{\left(1 + \frac{R}{C_v}\right)\frac{RT}{M}} \qquad (16)$$

The heat capacity C_v is in turn a function of the sound frequency, since it depends on how many of the various modes of molecular motion are actually excited.

At very low sound frequencies, all modes are excited and C_v has its static value.   The static value of the heat capacity is the sum of contributions from translational ($\frac{3}{2}R$), rotational ($\frac{3}{2}R$ for nonlinear molecules), and vibrational motions.   At higher frequencies C_v falls below the static value and levels off at a value corresponding to excitation of translation and rotation without any excitation of vibration.   For nonlinear molecules the limiting value is $3R$, half of which is the contribution from translation and the other half from rotation.   We may therefore assume that during the compression phase of the sound cycle, energy is initially put into translational and rotational motions of the molecules, and, if that phase lasts long enough, vibrational modes also become excited.   During the decompression phase, these processes are reversed.   If the frequency of the sound is too high, the vibrational equilibrium lags, leading to a deficit of vibrationally excited molecules in the compression phase of the cycle and an excess in the decompression phase.

On the basis of this theory, C_v for a gas in which the excitation of vibrational motions can be characterized by a single relaxation time, τ, is given as a function of ν by equation 17.

$$C_v = 3R + C_{v,\text{ internal}}\left(\frac{1}{1 + 4\pi^2\nu^2\tau^2}\right) \qquad (17)$$

If the relaxation process is complicated, the accurate description of C_v as a function of ν may require several relaxation times.   These might correspond to the excitation of different vibrational modes or to the displacement of chemical equilibria.   Figure 5-4 shows an example of a simple and of a more complicated relaxation.

[11] D. F. Hornig, *J. Phys. Chem.*, **61**, 856 (1957).
[12] K. F. Herzfeld and V. Griffing, *J. Phys. Chem.*, **61**, 844 (1957);   M. Eigen, *Disc. Faraday Soc.*, **17**, 194 (1954).

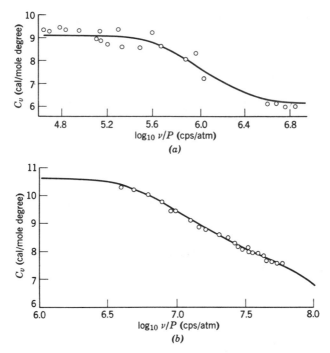

Figure 5-4. (*a*) Experimental results (circles) for methyl chloride at 100°. Solid line is the theoretical curve for $\tau = 1.8 \times 10^{-7}$ sec at 1 atm. The static value of C_v, as deduced from spectroscopic data, is 9.06 cal/mole deg. After P. G. T. Fogg, P. A. Hanks and J. D. Lambert, *Proc. Roy. Soc.* (*London*), **A219**, 493 (1953). (*b*) Experimental results (circles) for ethane at 25°. Solid line is the theoretical curve for a double relaxation process in which the torsional mode is excited with a relaxation time, τ_1, of 0.28×10^{-8} sec, and all other modes are being excited with a single τ_2 of 1.79×10^{-8} sec. After J. D. Lambert and R. Salter, *Proc. Roy. Soc.* (*London*), **A253**, 280 (1959).

For gases at low-to-moderate pressures, τ is inversely proportional to the pressure; a plot of C_v versus ν/P for various pressures is a single curve. This indicates that the translational-vibrational relaxation process takes place almost exclusively by bimolecular rather than higher order collisions. For example, if termolecular collisions were involved, $1/\tau$ would be proportional to P^2 rather than P.

Because of the phase lag at suitable frequencies, the system through which the sound is passing is not at equilibrium, and energy is absorbed. This energy is in addition to that which would be expected on the basis merely of the frictional losses, which produce a background absorption. The frequency of maximum absorption obeys equation 18.

$$\nu_{max} = \frac{1}{2\pi\tau} \tag{18}$$

The Heterogeneous Flash Initiation Method. Irradiation of a medium containing suspended particles, such as charcoal or carbon-black, by means of a brief and intense flash of light produces very rapid, localized heating to temperatures up to 2000° in the neighborhood of the solid particles.[13] In principle the method is like that employing shock waves, but the reaction medium is inherently more complicated.

The Chemical Methods. Introduction of a very reactive species, for example, atomic hydrogen, into a substrate that is also very reactive, such as ozone, gives reaction products that initially exist in excited vibrational states.[14,15] The relaxation of these species can then be studied by means of their infrared absorption, infrared emission, or the vibrational fine structure of their electronic absorption bands.

WIDTHS OF SPECTRAL LINES

Another approach to the problem of measuring fast interconversion reactions among subspecies uses the fact that the widths of the lines in their absorption or emission spectra depend on the lifetimes of the species. This principle has been most often applied to nuclear magnetic and electron spin resonance spectra.[16,17]

Consider, for example, the problem of rotameric isomerism in acetamides, which arises because the C—N bond has partial double bond character. The rate of such a *cis-trans* interconversion is equal to the rate of rotation about the C—N bond. For N,N-dimethyl acetamide, the rate of this rotation can be estimated from the proton magnetic resonance spectrum.[18] If the rotation in N,N-dimethyl acetamide were slow, the proton magnetic resonance spectrum would consist of three lines, one for the acetyl protons and one each for the two nonequivalent N-methyl groups. This is the

[13] L. S. Nelson and J. L. Lundberg, *J. Phys. Chem.*, **63**, 433 (1959).

[14] J. C. Polanyi, *J. Chem. Phys.*, **31**, 1338 (1959).

[15] F. J. Lipscomb, R. G. W. Norrish, and B. A. Thrush, *Proc. Roy. Soc.*, A **233**, 455 (1956).

[16] J. A. Pople, W. G. Schneider, and H. J. Bernstein, *High Resolution Nuclear Magnetic Resonance*, McGraw-Hill, 1959.

[17] R. L. Ward and S. I. Weissman, *J. Am. Chem. Soc.*, **79**, 2086, 6579 (1957).

[18] H. S. Gutowsky and C. H. Holm, *J. Chem. Phys.*, **25**, 1228 (1956).

$$CH_3 \quad O$$

(structure: acetyl group CH₃ and O double-bonded to C; C bonded to N; N bonded to two CH₃ groups labeled a and b)

$$CH_3 \qquad CH_3$$
$$\quad a \qquad\quad b$$

spectrum actually found at $-24°$. As the temperature is raised, the line due to the acetyl protons does not change, but the two lines due to the N-methyl groups first broaden and, at about $52°$, coalesce into a single broad line that gradually becomes a single narrow line as the temperature is still further increased (Figure 5-5).

These phenomena are due to the increasing rate at which the methyl groups interchange between the two nonequivalent positions. At the highest temperatures, the rate is so great that the differences between the two positions are averaged out, and only a single line corresponding to an average species is seen. When we say that the interconversion of the two positions is fast, we mean fast with reference to the time scale of the particular experiment that is being carried out.

If the property being measured is a pair of absorption lines, then a convenient unit for expressing the time scale of the experiment is the difference between the frequencies of the two lines, that is, the beat frequency, $\delta\nu = |\nu_1 - \nu_2|$. Of course, the reason that the beat frequency is an appropriate measure of the experimental time scale is that it is necessary to count oscillations for at least a time equivalent to one beat to know that there are indeed two frequencies rather than one. If the frequency of exchange between the two states, or, in the case of the amide rotation, the frequency of the rotation, is fast compared to the beat frequency, the separate states will be undetectable; instead of beats one observes a wave train consisting of a single, average frequency, with periodic fluctuations in phase. This accounts for the observation of only a single line at very high interconversion rates.

We would now like to consider why it is that before the two lines coalesce into a single line they first broaden. This can be most easily explained in terms of the uncertainty principle, which may be written as equation 19.

$$\delta E \times \delta t \approx \frac{h}{2\pi} \tag{19}$$

The quantity δt in equation 19 is the time taken to measure the energy of a given state. This, of course, cannot exceed the lifetime of the state. As the rate of exchange with other states becomes more rapid, δt necessarily

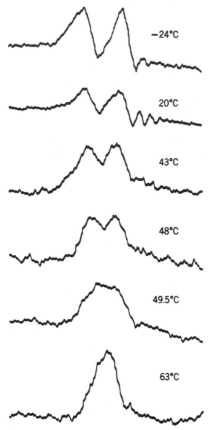

−24°C

20°C

43°C

48°C

49.5°C

63°C

Figure 5-5. The proton magnetic resonance spectrum of the N-methyl groups in N,N-dimethylacetamide at various temperatures. From H. S. Gutowsky and C. H. Holm, *J. Chem. Phys.*, **25**, 1233 (1956).

decreases. This means that the uncertainty in the energy, δE, increases, which in turn means that spectral lines involving the state become broader.

In order to extract actual rates of exchange from the spectroscopic data, it is convenient to define the *line width* as the width at half-height (Figure 5-6) and to redefine τ as the mean time that a molecule can spend in a given state without exchanging. That is,

$$\frac{1}{\tau_i} = \frac{\text{Rate of removal of molecules from the } i\text{th state by exchange}}{\text{Number of molecules in the } i\text{th state}}$$

In general, τ is different for different states.

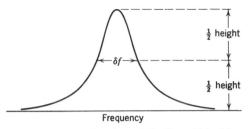

Figure 5-6. Explanation of the line width, δf.

The relationship between the line width, δf, and τ is simple for two special cases. These special cases have one feature in common, namely, that τ must be the same for the upper and lower states between which the given absorption or emission of energy takes place. The first special case is called *lifetime broadening* and arises when $1/\tau$ is small compared to the beat frequency $\delta\nu$. The value of τ for the species responsible for a given line in the spectrum is then related to the width of that line by equation 20.

$$\frac{1}{\tau} = \pi(\delta f - \delta f_0) \tag{20}$$

The quantity δf_0 in equation 20 is the width in the absence of exchange; however, this will not necessarily be an experimentally accessible quantity for all systems. The lifetime broadening method has the advantage that the species responsible for the broadened spectral lines are almost always known. Moreover, the results can be confirmed by using lines from different parts of the spectrum.[19] In contrast, the relaxation methods may give us accurate relaxation times, but quite often leave us in doubt as to the nature of the species that is relaxing.

The second special case is called *exchange narrowing* and arises when $1/\tau$ is large compared to the beat frequency. In nuclear magnetic resonance, if the two lines whose coalescence is being measured result from a difference in chemical shift, equation 21 is precise, although for other applications, such as spin-spin interaction, it may be only a first approximation.[19a]

$$\pi(\delta f - \delta f_0) = P_A{}^2 P_B{}^2 (2\pi\delta\nu)^2(\tau_A + \tau_B) \tag{21}$$

In equation 21, P_A is the population of the species A divided by the sum of the populations of the exchanging species A and B. For example, in *proton* magnetic resonance P_A and P_B are the fractions of the total number of exchangeable protons in the two species.

The quantity δf_0 in equation 21 is the line width in the limit of extremely rapid exchange. The sum of the relaxation times $\tau_A + \tau_B$, which is

[19] E. Grunwald, A. Loewenstein, and S. Meiboom, *J. Chem. Phys.*, **27**, 630 (1957).
[19a] J. Kaplan, *J. Chem. Phys.*, **28**, 278 (1958).

obtained by a single experiment corresponding to equation 21, can, in all cases for which the equation holds, be resolved into its components by means of equation 22, which expresses the fact that the two species A and B are in dynamic equilibrium with each other.

$$\frac{P_A}{\tau_A} = \frac{P_B}{\tau_B} \tag{22}$$

It is possible that exchange narrowing does not occur even though $1/\tau$ is much greater than the beat frequency $\delta\nu$. For convenience of discussion, let us represent our dynamic system of molecules by a mechanical system of oscillators, such that the frequency of each oscillator can switch back and forth between two characteristic values, ν_1 and ν_2, with an average time delay of τ seconds. If τ is short compared to $\delta\nu$, the oscillator will in effect have a single average frequency $\bar{\nu}$ only if the phase angle of the oscillation varies continuously at the time of switching. That is, the oscillations just before switching must be coherent with those just after.

Returning to our dynamic system of molecules, the requirement of coherence can be satisfied only if the actual process of exchange takes place in a time that is short compared both to τ and to $1/\delta\nu$. All transition state complexes and reaction intermediates must therefore be suitably short-lived. Furthermore, the characteristic frequencies $\nu^\ddagger$ and ν_i of all transition state complexes and reaction intermediates must not be vastly different from ν_1 and ν_2.

Although line width methods have been most widely used in electron spin and nuclear magnetic resonance, they are also sometimes of semi-quantitative usefulness for the estimation of rate constants from infrared, Raman, and electronic spectra. For example, the broadness of the band corresponding to OH stretching vibrations in the Raman spectra of concentrated aqueous acids leads to the conclusion that the relaxation time for the exchange of protons between H_3O^+ and adjacent H_2O molecules is about 10^{-13} seconds.[20] The first overtone of the OH stretching vibration of ethanol, dilute in carbon tetrachloride, consists of a partially resolved doublet at 1.4081 and 1.4130 microns.[21] If this doublet is due to the existence of *trans* and *gauche* rotamers, the fact that it can be resolved sets an upper limit of about 10^{12} for $1/\tau$ for the interconversion.

trans   gauche

[20] E. Wicke, M. Eigen, and T. Ackermann, *Z. physik. Chem.* N.F, **1**, 340 (1954).

[21] R. Piccolini and S. Winstein, *Tetrahedron Letters*, No. 13, p. 4 (1959).

THE EFFECTIVENESS OF A COLLISION

In the gas phase the energies of the individual molecules are quantized. In order to transfer energy between molecules during a collision, the quantum donated by one molecule must be of a magnitude acceptable to the other. This means that energy transfer is more probable the more energy levels there are available to the molecules and the more closely the levels are spaced, since the probability of matching two energy intervals is then greater. Because of the requirement for matching, energy is only rarely transferred between vibrational levels of two molecules, or between a vibrational level and a rotational level, or between rotational levels of two molecules. On the other hand, translational energy, being essentially continuous, is readily transferred in a collision. Hence the most common mechanism for the excitation or de-excitation of any internal mode of motion is conversion from or to *translational* energy in a collision. It is customary to express such internal-to-translational energy transfers in terms of the number of collisions that must occur, on the average, before one of them is successful. The reciprocal of this number is the probability that any single collision will be successful.

The number of collisions required to excite motion in a given degree of freedom is easily calculated if the number required for de-excitation is known. In order to derive the relationship needed, let us consider temporarily that the excited and unexcited states are in equilibrium. The equilibrium condition expressed by equation 23 can also be expressed

$$\frac{N_1}{N_0} = e^{-(\epsilon_1 - \epsilon_0)/kT} \tag{23}$$

dynamically by equating the rates of excitation and de-excitation. Let Z_{01} and Z_{10} be the average number of collisions needed for the transfer of the required amount of energy, and let Z be the average number of collisions experienced by one molecule in one second. Then $1/Z_{01}$ and $1/Z_{10}$ are the probabilities that a single collision will be successful in transferring the required energy, and the dynamic equilibrium may be written as equation 24.

$$\frac{1}{Z_{01}} ZN_0 = \frac{1}{Z_{10}} ZN_1 \tag{24}$$

From equations 23 and 24 we get the desired result (25).

$$Z_{01} = Z_{10} e^{(\epsilon_1 - \epsilon_0)/kT} \tag{25}$$

Some Theoretical Models[22]

The number of collisions required to effect a transfer of energy between internal and translational degrees of freedom depends on the average energy of the collisions (and hence the temperature) and on the nature of the colliding molecules. A complete theory is yet to be developed, but

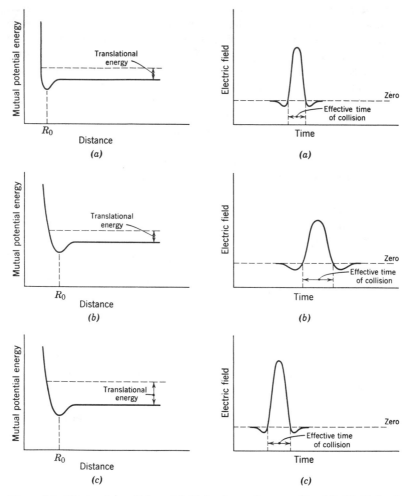

Figure 5-7. Effects of a collision. (a) Molecular surfaces are "hard." (b) Molecular surfaces are "soft." (c) Same as (b), but total energy is greater.

[22] For a somewhat different and more quantitative approach, see M. E. Jacox and S. H. Bauer, *J. Phys. Chem.*, **61,** 833 (1957).

some insight into the mechanism of this process may be gained if we consider some simplified models. According to one model, the collision may be regarded as the vibration of a composite molecule. As shown in Figure 5-7, the mutual potential energy of a pair of approaching molecules varies with distance in much the same way as, say, the potential energy of a pair of approaching chlorine atoms. In both cases the potential energy passes through a minimum, indicating that a stable composite unit can exist. Of course, for the collision complex, the potential well is much more shallow and the corresponding vibration much more anharmonic than for the Cl_2 molecule. Since the translation is represented as a very anharmonic vibration, the problem of energy transfer between translation and any given internal motion is thus reduced to that for a pair of coupled oscillators.

From this point of view the anharmonicity of the intermolecular vibration is very important, for energy cannot be transferred from one quantum oscillator to another if both are perfectly harmonic. To obtain a rough idea of the factors that influence the probability of energy transfer, it is useful to remember that each of the molecules in a collision is exposed to the strong electric field produced by the charge distribution of the other. Let us temporarily ignore the presence of this other molecule and think of the collision merely as a single pulse of electrical disturbance acting on the given molecule. To simplify the problem further, let us suppose also that the given molecule is excited and that it is to be de-excited under the influence of the electrical pulse.

The height and shape of the electrical pulse associated with the collision depend on the nature of the potential energy curve and on the translational energy, as shown qualitatively in Figure 5-7. An increase in the translational energy or a steepening in the repulsion branch of the potential energy curve will reduce the time during which the pulse is effective. We may assume that the pulse is effective only at very short distances of approach, say at distances less than R_0. This part of the pulse is conveniently represented as a Fourier sum of components of various frequencies and amplitudes. Because of the anharmonicity of the pulse, there will be a large number of such Fourier components, and one of them will match exactly the frequency of the internal motion whose energy is to be converted into translational energy.

One advantage of letting an electrical pulse represent the actual collision is that this procedure emphasizes the analogy between collisional de-excitation and radiation induced de-excitation. According to the Einstein theory, emission of a quantum of radiation from an excited molecule is catalyzed by the presence of radiation of matching frequency, even though the catalyzing radiation is not absorbed. Let us surmise that in a collision, the Fourier component of matching frequency can similarly catalyze de-excitation. If so, the probability that the collision will result in de-excitation is proportional to the intensity of the Fourier component of matching frequency. The intensity of this component may be expected to be high if the time during which the pulse is effective is of the order of $1/\nu_{osc}$; it is low for pulses that are either much longer or much shorter than this optimum. We thus arrive at a picture in which collisions are inefficient at producing de-excitation if the translational energy is either too low or too high. The majority of the effective collisions is located in a fairly narrow band of translational energies such that the frequency matching is good.

Although specific effects depending on the nature of the colliding molecules are undoubtedly important, we may guess that the catalytically effective band of collisions consists of translational energies that are similar to, or at least of the same order of magnitude as, the quantum to be de-excited. Hence it should be relatively easy to remove excitation energy in small quanta or in quanta of a size comparable to the average translational energy per molecule. On the other hand, if the mechanism of de-excitation requires the transfer of quanta that are large compared to the average translational energy, only a small fraction of all collisions, in which the molecules have the required large catalytic amount of energy, will be effective. This fraction will be given by a Boltzmann expression even though the process being considered is de-excitation and does not actually *consume* any of the energy of the collision. A seemingly paradoxical corollary is that a "hot" molecule in a gas is protected from collisional de-excitation by the very size of its excitation energy, unless the excitation energy can be removed in small quanta.

In the remainder of this section we shall mainly discuss the process of de-excitation. The number of collisions required for the reverse process of excitation is readily derived from that for de-excitation by means of equation 25.

Rotational-Translational Interconversion in the Gas Phase

Except for light molecules, such as hydrogen, the energy spacing of the rotational levels is an order of magnitude smaller than the average translational energy per molecule. The size of the quantum transferred in a rotational de-excitation is therefore small, and we may expect that a substantial fraction of all collisions is catalytically effective. As a matter of fact, it can be deduced from the results of shock tube experiments that this kind of energy transfer requires only about five collisions for all but very light nonpolar molecules, such as hydrogen.[11] It may therefore be assumed in most experiments that if translational equilibrium has been achieved, so has the equilibrium involving rotation of the entire molecule.

Vibrational-Translational Interconversion in the Gas Phase

The energy spacing of vibrational levels is comparable to or greater than the average translational energy per molecule at ordinary temperatures. We therefore expect that vibrational de-excitation requires more collisions than does rotational de-excitation, at least for some substances. As a matter of fact, the vibrational-translational interconversion is slow enough for diatomic or simple polyatomic molecules to produce a measurable relaxation effect in the ultrasonic range. Analysis of the resulting relaxation times leads to a model of the relaxation mechanism.

To begin with, let us consider a simple set of vibrational energy levels, such as that shown in Figure 5-8. The energy levels are divided into separate columns for the different modes. The excitation energy is distributed over all the levels. It is clear that a molecule in any level above ϵ_1 can be de-excited by a number of alternative routes. For example, a molecule in level ϵ_1' could be de-excited by direct transition to ϵ_0, or it could first move to ϵ_1 and then to ϵ_0. Since each of the possible transitions proceeds with its own characteristic relaxation time, one would expect to find an entire spectrum of vibrational relaxation times in the ultrasonic range.

Most substances, however, exhibit only a single vibrational relaxation time within the range of measurement. Of course, two or more relaxation times close together are not easily resolved experimentally, but it is highly unlikely that *all* of the physically important relaxation times should be nearly identical. The idea therefore suggests itself that de-excitation from *any* of the excited levels to the ground state proceeds with the same rate-determining step. Since the only process that could possibly be common to all de-excitation mechanisms is the transition from ϵ_1 to ϵ_0, the observed relaxation time is assigned to this transition. According to this theory, molecules in higher levels are de-activated by first cascading rapidly down to ϵ_1 and then proceeding more slowly to ϵ_0. The relaxation times for the steps in the rapid initial cascade are presumed to be too short for ultrasonic measurement. Direct transitions from higher levels to the ground levels, such as the process $\epsilon_1' \to \epsilon_0$, are presumed to be too infrequent to affect the relaxation spectrum. Thus the process $\epsilon_1 \to \epsilon_0$ is the only one observed.

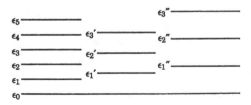

Figure 5-8. A set of vibrational energy levels.

Table 5-2. *Transfer of Energy from the First Excited Vibrational Level to Translation in the Gas Phase (from ultrasonic relaxation data)*

MOLECULE LOSING VIBRATIONAL ENERGY	MOLECULE WITH WHICH IT COLLIDES	TEMP. °C	Z_{10}	Ref.
O_2	O_2	15	2.1×10^7	a
		1100	36,000	a
		2640	1,800	a
Cl_2	Cl_2	15	34,000	a
		1100	200	a
	He	20	900	a
N_2O	N_2O	25	7,500	b
	He	25	14,000	b
	A	25	33,000	b
	H_2O	25	105	b
CO_2	CO_2	25	57,000	b
	A	25	47,000	b
	H_2O	25	105	b
	CO	20	6,200	c
CH_4	CH_4	20	8,400	c
	H_2	20	2,700	c
	CO_2	20	460	c
SCO	SCO	20	17,700	c
	O_2	20	3,200	c
	CO	20	2,000	c
	H_2	20	200	c
	He	20	1,250	c
	A	20	7,750	c
C_2H_6	C_2H_6	25	74^g	d
			20^g	
C_2F_6	C_2F_6	25	7	d
CH_3CHF_2	CH_3CHF_2	25	<2	d
Propane	Propane	25	5	d
n-Butane	n-Butane	25	<2	d
Isobutane	Isobutane	25	<2	d
Neopentane	Neopentane	25	<2	d
CH_3CH_2Cl	CH_3CH_2Cl	25	9(?)	d
CH_3CH_2F	CH_3CH_2F	25	7(?)	d
CH_3OH	CH_3OH	25	71(?), 10(?)	d, e
CH_3F	CH_3F	19	4,800	f
		100	5,890	f
CH_3Cl	CH_3Cl	21	300	f
		100	330	f
		188	160	f

[a] Compiled by K. F. Herzfeld and V. Griffing, *J. Phys. Chem.*, **61**, 844 (1957).
[b] A. Eucken and E. Nümann, *Z. physik. Chem.*, **B36**, 163 (1937).
[c] A. Eucken and S. Aybar, *Z. physik. Chem.*, **B46**, 195 (1940).
[d] J. D. Lambert and R. Salter, *Proc. Roy. Soc.*, **A253**, 277 (1959).
[e] C. Ener, A. Busala, and J. C. Hubbard, *J. Chem. Phys.*, **23**, 155 (1955).
[f] P. G. Corran, J. D. Lambert, R. Salter, and B. Warburton, *Proc. Roy. Soc.*, **A244**, 212 (1958).
[g] Two relaxation times (Figure 5-4b).

For CO_2 there is direct evidence that the single relaxation time observed by the ultrasonic method is probably not illusory. The various modes of vibration can be excited selectively simply by using the appropriate infrared wavelength, and their relaxation times can then be measured directly in a spectrophone. All wavelengths give the same relaxation time.[22]

The results of ultrasonic measurements for some representative vibrational de-excitation processes are summarized in Table 5-2. Collision numbers, Z_{10}, have been computed from the relaxation data on the assumption that the transition from ϵ_1 to ϵ_0 is rate determining. It should be noted, however, that very similar collision numbers would have been computed on the assumption of any other reasonable de-excitation mechanism.

In most cases the values of Z_{10} decrease with increasing temperature, indicating that activation energy is required for de-excitation. Further analysis shows that for diatomic and simple polyatomic molecules of related structure, the activation energy is approximately proportional to the size of the quantum to be de-excited. The results are represented by the empirical equation 26,

$$Z_{10} = A_{10}e^{qh\nu_1/kT} \tag{26}$$

where $h\nu_1 = \epsilon_1 - \epsilon_0$, ν_1 being the lowest of the fundamental vibration frequencies of the given molecule. A_{10} and q are parameters characteristic of the colliding molecules.

In Figure 5-9 values of $\log(Z_{10}/A_{10})$ are plotted versus ν_1 for a number of polyatomic molecules.[23] The figure shows the expected linear trend, but it also shows a dispersion into two families of different molecular structure. One of these consists of substances not having any hydrogen atoms, and for these q is about 4. For substances having peripheral hydrogen or deuterium atoms q is about 2. A possible reason for the greater effectiveness of collisions involving hydrogen compounds is that the relatively small and unpolarizable hydrogen atom steepens the repulsion branch of the intermolecular potential energy curve.

The correlation of Z_{10} with the *lowest* fundamental vibration frequency for each substance is consistent with the assumed mechanism that de-excitation from the lowest excited state ($\epsilon_1 \rightarrow \epsilon_0$) is rate determining. It is now of interest to re-examine the rates of the other de-excitation steps. Let us assume that the number of collisions required for de-excitation from level ϵ_j to level ϵ_i is given by equation 27, which is analogous to the empirical equation 26. The parameter A_{ji}

$$Z_{ji} = A_{ji}e^{qh\nu_{ji}/kT} \tag{27}$$

[23] J. D. Lambert and R. Salter, *Proc. Roy. Soc.*, **A253**, 277 (1959).

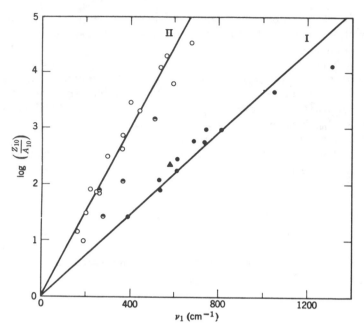

Figure. 5-9. Relation between collision life of vibrational energy at 300°K and lowest fundamental vibration frequency for molecules showing a single relaxation process. Values listed in order from the bottom upward.

$\bigcirc$, molecules containing no hydrogen atom; C_2F_4, CF_2Br_2, CF_2BrCl, CF_2Cl_2, $CFCl_3$, CCl_4, CF_3Br, CF_3Cl, SF_6, CF_4, CS_2, N_2O, COS, Cl_2, CO_2.
$\ominus$, molecules containing one hydrogen atom; $CHCl_2F$, $CHCl_3$, $CHClF_2$, CHF_3.
$\bullet$, molecules containing two or more hydrogen atoms; CH_2ClF, CH_3I, CH_2F_2, CH_3Br, C_2H_2, CH_3Cl, C_2H_4O, C_2H_4, cyclo-C_3H_6, CH_3F, CH_4.
$\blacktriangle$, deuterated molecules; CD_3Br.

is the reciprocal of a transition probability, and selection rules are therefore of some importance. As a first approximation let us assume that the selection rules are similar to those for radiation-induced transitions.[22] Thus crossovers from levels belonging to one mode to those belonging to another (such as $\epsilon_1' \to \epsilon_1$) should be relatively improbable, as should be transitions involving the transfer of two or more quanta within a given mode ($\epsilon_2 \to \epsilon_0$). We may expect, however, that these selection rules apply less rigorously to collision complexes than to isolated molecules, since the vibrations of collision complexes are more anharmonic. Indeed, in really violent collisions, such as those involving "hot" molecules in a shock tube

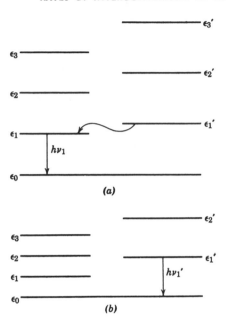

Figure 5-10. (a) Two-step and (b) one-step
mechanisms for de-excitation.

or fragments recoiling from nuclear reactions, the molecules are deformed
to such an extent that all selection rules probably break down.

The result of the remaining degree of forbiddenness is that transitions
with relatively low probabilities take place, but only if the exponential
factors are not too unfavorable. Thus if two levels ϵ_1' and ϵ_1 of different
modes are close together, as in Figure 5-10a, a step-by-step transfer of
energy will occur. If two levels of different modes also differ considerably
in energy, a larger quantum jump between like modes will become competitive, as in Figure 5-10b. In the latter case, the molecule can exhibit two
relaxation times instead of one. An example is ethane, for which Z_{10} is
equal to 20 and $Z_{1',0}$ is equal to 74.[23]

Some molecular species are *more* efficient at transferring energy from
vibrational to translational modes than would be expected solely on the
basis of the spacing of their vibrational energy levels. For example,
1,1-difluoroethane has a gap between the ground state and the lowest
vibrationally excited state of 220 cm^{-1}, yet Z_{10} is less than 2. In contrast,
hexafluoroethane with a gap of only 100 cm^{-1} has a Z_{10} of 7. The important factor here may be dipole-dipole forces, which are exerted between

the polar 1,1-difluoroethane molecules. A similar example is the unusual efficiency of water in deactivating vibrationally excited CO_2 molecules.[24] For collisions between two CO_2 molecules at 20°, Z_{10} is 57,000; for collisions between a CO_2 molecule and a water molecule, Z_{10} is 105. On the other hand, for CO_2 and argon, Z_{10} is 47,000.

The role of dipole-dipole forces in collisions is illustrated also by the abnormal temperature coefficients of Z_{10} for polar gases. Whereas for non-polar molecules the values of Z_{10} generally *decrease* with increasing temperature, the values for methyl chloride and for methyl fluoride *increase* between 20° and 100° (Table 5-2). This apparent abnormality may be explained if we make two hypotheses. First, that there is a partial alignment of molecular dipoles during the approach of the colliding molecules. This alignment is more nearly complete at the lower temperatures, where, on the average, the molecules come together more slowly. Second, that the extent of this alignment of dipoles in the collision complex has a pronounced effect on the transition probability for de-excitation, the transition probability passing through a sharp maximum when the two dipoles are optimally aligned. The combined effect of these two factors could more than offset a less favorable exponential factor.

Collisions are also very efficient between large flexible molecules. For example, the Z_{10} for *n*-butane, iso-butane and neo-pentane are all less than 2.[23] A possible reason for this is that large flexible molecules engage in "wrestling collisions,"[24a] in which various parts of the colliding molecules come into contact at different times.[25] This is analogous to a whole series of collisions, like an encounter in the enforced propinquity of a solvent cage.

In the Liquid Phase

Translation-Rotation. Rotation of a molecule in a liquid may be a reorientation to a new potential minimum within the solvent cage, or it may be a geared rotation for which a cooperative motion of solvent molecules is required. For the rotation of polar molecules, dielectric relaxation is a convenient measurement. The quantity measured is the relaxation of a nonequilibrium population of the various potential-minimum orientations. The relaxation time is therefore of the same order of magnitude as the fastest of the various reorientations involved. Table 5-3 (pp. 102–103) gives some representative times.

[24] A. Euken and E. Nümann, *Z. phys. Chem.*, **B36,** 163 (1937).
[24a] J. C. McCoubrey, J. B. Parke, and A. R. Ubbelohde, *Proc. Roy. Soc.*, **A223,** 155 (1954).
[25] In the French literature, "collision amoureuse."

For a given liquid the temperature dependence of the relaxation time very closely parallels that of the viscosity.[26] The Debye equation (28) for a dipolar sphere rotating in a viscous continuum gives the relaxation time as a function of the molecular radius, a.

$$\tau = \frac{4\pi\eta a^3}{kT} \tag{28}$$

According to this model the relaxation time should be proportional to η/T, a relationship that has been found to hold fairly well for liquid alcohols and alkyl halides.[26] For example, the relaxation time of isopropyl bromide between room temperature and $-170°$ changes by a factor of 10^8, while η/T changes by a factor of 10^9. For the most part, the relaxation time increases with increasing size of the molecule, and the values of a needed to obtain a good fit with equation 28 are reasonable.

There are, however, some serious discrepancies. For molecules conforming most nearly to the model of a dipolar sphere, the macroscopic viscosity of the medium appears to be a poor measure of the viscous drag actually encountered by the rotating molecule. For example, the relaxation time of t-butyl chloride in dilute solution at $20°$ is almost the same in n-heptane as it is in Nujol, in spite of the 200-fold difference in macroscopic viscosity (Table 5-3). In fact, molecular rotation proceeds with low activation energy even in solid t-butyl chloride.[27] One reason for the failure of the viscous continuum model in the case of small, relatively nonpolar spherical molecules is that such molecules can rotate on their lattice sites without having to push adjacent molecules out of the way.

For substances with nonspherical molecules, the viscous continuum model is a better approximation, yet even here deviations due to the discrete structure of the solvent appear. The interpretation of these deviations is particularly simple for some substances for which the dielectric relaxation time in a series of nonpolar solvents does parallel the macroscopic viscosity. When the relaxation of such a substance is studied in a series of polar solvents, the measured relaxation times are generally longer than they would be in a nonpolar solvent of equal viscosity. This suggests that the actual process in the polar solvents is the geared or cooperative rotation of a cluster of molecules. For example, if the actual relaxing species were a cluster of four identical molecules, then the concentration of the actual relaxing species would be only one fourth that of the stoichiometric

[26] D. J. Denney, *J. Chem. Phys.*, **30**, 159 (1959); D. W. Davidson and R. H. Cole, *J. Chem. Phys.*, **19**, 1484 (1951).
[27] C. P. Smyth, *J. Phys. Chem.*, **58**, 580 (1954).

Table 5-3. Representative Dielectric Relaxation Times for Molecular Rotation in Liquids

ROTATING SPECIES	SOLVENT AND TEMP.	$\eta_{solvent}$ (centipoise)	τ (sec)	REF.
Benzyl chloride	Benzene, 20°	0.65	2.1×10^{-11}	a
bis(chloromethyl)durene	Benzene, 20°	0.65	4.35×10^{-11}	a
Anthrone[1]	Benzene, 20°	0.65	2.5×10^{-11}	b
Fluorenone[1]	Benzene, 20°	0.65	2.0×10^{-11}	b
Phenanthrenequinone[1]	Benzene, 20°	0.65	3.1×10^{-11}	b
(thianthrene structure)	C_2Cl_4, 20°	0.9	3.0×10^{-11}	c
(dimethylthianthrene structure)	C_2Cl_4, 20°	0.9	7.6×10^{-11}	c
Picric acid	Benzene, 20°	0.65	5.0×10^{-11}	d
2,4,6-Tribromophenol	Benzene, 20°	0.65	2.2×10^{-11}	d
α-Chloronaphthalene	Pure liquid, 20°	3.33	5.8×10^{-11}	e
	n-Heptane, 20°	0.42	1.1×10^{-11}	e
	Nujol, 20°	108	6.4×10^{-11}	e

t-Butyl chloride	Pure liquid, 20°	0.53	0.48×10^{-11}	e
	CCl_4, 20°	0.97	0.35×10^{-11}	e
	n-Heptane, 20°	0.42	0.2×10^{-11}	e
	Nujol, 20°	108	0.3×10^{-11}	e
C_2H_5Br	Pure liquid, 25°	0.38	0.9×10^{-11}	f
$n\text{-}C_4H_9Br$	Pure liquid, 25°	0.60	2.2×10^{-11}	f
$n\text{-}C_6H_{13}Br$	Pure liquid, 25°	0.94	8.1×10^{-11}	f
$n\text{-}C_{10}H_{21}Br$	Pure liquid, 25°	2.29	32.5×10^{-11}	f
$n\text{-}C_{16}H_{33}Br$	Pure liquid, 25°	6.73	76×10^{-11}	f
Toluene	Pure liquid, 20°	0.59	0.75×10^{-11}	g
γ-Picoline	Pure liquid, 20°	1.07	1.33×10^{-11}	g
Pyridine	Pure liquid, 20°	1.12	0.71×10^{-11}	g
Diethylamine	C_2Cl_4, 20°	0.9	3.4×10^{-11}	c
Tri-isoamylamine	C_2Cl_4, 21°	0.9	4.0×10^{-11}	c
2,4,6-Tri-t-butylphenol	trans-Decalin, 20°	2.13	50×10^{-11}	h

[a] W. P. Purcell, K. Fish, and C. P. Smyth, J. Am. Chem. Soc., 82, 6299 (1960).

[b] D. A. Pitt and C. P. Smyth, ibid., 80, 1061 (1958).

[c] A. H. Price, J. Phys. Chem., 62, 773 (1958).

[d] A. H. Price, ibid., 64, 1442 (1960).

[e] A. J. Curtis, P. L. McGeer, G. B. Rathmann, and C. P. Smyth, J. Am. Chem. Soc., 74, 644 (1952).

[f] K. Higasi, K. Bergmann, and C. P. Smyth, J. Phys. Chem., 64, 880 (1960).

[g] R. S. Holland and C. P. Smyth, ibid., 59, 1088 (1955).

[h] M. Davies and R. J. Meakins, J. Chem. Phys., 26, 1584 (1957).

[i] Activation enthalpies for molecular rotation in benzene: anthrone, 2.1 kcal; fluorenone, 2.0 kcal; phenanthrenequinone, 2.6 kcal Ref. b.

concentration of the individual molecules. Therefore the relaxation time *per molecule* would be less than the observed relaxation time for the cluster by a factor of one fourth. For a model consisting of a polar molecule in a solvent treated as a continuum, the relationship between the observed relaxation time of the geared rotation and the molecular relaxation time, τ_μ, of the polar molecule is given approximately by equation 29.[28]

$$\tau_\mu = \frac{2D_0 + D_\infty}{3D_0} \tau \tag{29}$$

In equation 29, D_0 is the static value of the dielectric constant and D_∞ is the high frequency value. For nonpolar solvents D_0 is equal to D_∞, which in turn is equal to the square of the refractive index. When D_0 and D_∞ are equal, it is seen from equation 29 that τ_μ is simply equal to τ.

Sometimes the dielectric relaxation is slower than anticipated even in nonpolar solvents because of solute-solvent complex formation. For example, picric acid in benzene has a relaxation time of 50×10^{-12} seconds at $20°$, about three times the anticipated value, whereas 2,4,6-tribromophenol behaves normally.[29] The benzene-picric acid complex responds as a large single molecule, which rotates more slowly.

In the rotation of flat molecules there is a hydrodynamic effect. Because the efficiency of the molecule as a propeller[30] depends on the axis about which the molecule is rotating, the relaxation times for rotation about different axes are quite different. Consider, for example, the rotation of a flat rectangular molecule. Rotation about an axis perpendicular to the plane will meet the least resistance from the solvent, and rotation about the long axis within the plane will be somewhat easier than rotation about the short axis within the plane. In dielectric relaxation the axis of rotation is perpendicular to the molecular dipole. Therefore, if we examine two approximately planar molecules of about the same size and shape, one having its dipole vector in the plane and the other having its dipole vector perpendicular to the plane, we find a great difference in the dielectric relaxation times. A good example is the behavior of metal-free heptaphenylchlorophenylporphyrazine, which has its dipole in the molecular plane, in contrast to the behavior of ferric octaphenylporphyrazine chloride, the dipole of which is perpendicular to the plane.[31]

[28] J. G. Powles, *J. Chem. Phys.*, **21**, 633 (1953); R. S. Holland and C. P. Smyth, *J. Phys. Chem.*, **59**, 1088 (1955).

[29] A. H. Price, *J. Phys. Chem.*, **64**, 1442 (1960).

[30] The propeller form of triphenylmethyl should rotate about its shortest axis as it moves through the solution, that is, the rotational and translational motions of a molecule of this sort should not be separable.

[31] D. A. Pitt and C. P. Smyth, *J. Phys. Chem.*, **63**, 582 (1959).

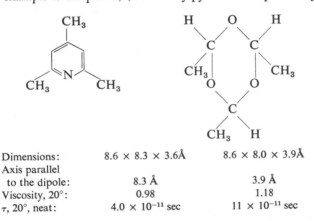

Dimensions:	$20.3 \times 19.0 \times 6.25$Å	$19.0 \times 19.0 \times 7.15$Å
Axis parallel to the dipole:	20.3 Å	7.15 Å
τ in benzene at 20°:	30.2×10^{-11} sec	70.6×10^{-11} sec
$\Delta H_{rot}^{\ddagger}$:	2.5 kcal/mole	2.6 kcal/mole
$\Delta S_{rot}^{\ddagger}$:	−6.4 e.u.	−7.9 e.u.

Another example is the pair 2,4,6-trimethylpyridine and paraldehyde.[32]

Dimensions:	$8.6 \times 8.3 \times 3.6$Å	$8.6 \times 8.0 \times 3.9$Å
Axis parallel to the dipole:	8.3 Å	3.9 Å
Viscosity, 20°:	0.98	1.18
τ, 20°, neat:	4.0×10^{-11} sec	11×10^{-11} sec

Using 10^{-10} to 10^{-11} seconds as typical of the relaxation time for molecular rotation in liquids and 10^{-13} seconds for the time between successive collisions, it is clear that about 10^2 to 10^3 collisions would be required, on the average, to effect rotational relaxation. This is a much greater number than we previously found to be required in the gas phase for translational-rotational relaxation. The reason for the apparent discrepancy is that "rotation" at low energy levels in a liquid is not really a rotation but rather a libration about orientations of minimum potential

32 R. C. Miller and C. P. Smyth, *J. Phys. Chem.*, **60**, 1354 (1956).

energy. The lower energy levels are therefore analogous to those for torsions within a molecule. The energy of a molecule freely rotating in a liquid must be at least sufficient to keep the molecule from becoming trapped in one of the potential energy troughs corresponding to libration. The activation energies found for molecular rotations are about 3 kcal/mole. Since the Boltzmann factor $e^{-3000/RT}$ is between 10^{-2} and 10^{-3} at ordinary temperatures, the transfer of a molecule from a librating to a freely rotating state must in fact occur at nearly every collision having the requisite energy.

Vibration-Translation. In the gas phase we found that for all but diatomic and small polyatomic molecules, the conversion of vibrational into translational energy takes place at nearly every collision. The requirement for the reverse process is merely that the translational energy of the colliding molecules be at least equal to that of the resulting vibration. Collisions in the liquid phase should be just as efficient, if not more so. This is consistent with the fact that the fluorescence radiation emitted by an electronically excited molecule in solution is always that to be expected from the ground vibrational state of the electronically excited species. Since the absorption spectrum shows that the electronically excited species first produced consists mostly of states that are also vibrationally excited, the relaxation of the vibrational part of the energy must be faster than the relaxation of the electronic part of the energy. A typical lifetime for a fluorescent species in solution is 10^{-8} sec.

THE EFFECT OF AN ENCOUNTER

As was pointed out in Chapter 4, collisions in liquids occur in sets. The number of collisions in one encounter is of the order of 10^2 for liquids of ordinary viscosity at room temperature. We now want to estimate the probability that various processes will take place during a single encounter as opposed to a single collision. One way of estimating the duration of an encounter is to calculate the average time taken by a molecule in diffusing a distance equal to one molecular diameter. From the theory of Brownian motion, the time required for a molecule to move a distance s in an arbitrary direction is equal to $s^2/6D$, where D is the diffusion coefficient of the substance. If the origin is taken as the position of a second molecule that is itself in motion and s is allowed to be one molecular diameter, the mean lifetime of an encounter is given by equation 30.

$$t_e = \frac{s^2}{6(D_1 + D_2)} \tag{30}$$

Since a diffusion coefficient might plausibly be of the order of 10^{-5} cm²/sec for a molecule whose diameter is 5 Å, the mean lifetime of an encounter should be about 10^{-11} sec.

The rotational relaxation time for such a molecule would be in the range 10^{-10} to 10^{-11} sec. For small- or medium-sized molecules, the factors influencing the rate of rotation and the rate of escape from solvent cages tend to cancel, so that it may be said quite generally that escape from the solvent cage will be somewhat faster than rotation. Nevertheless, the fact that rotation can *sometimes* take place during the lifetime of a single encounter has important chemical consequences. One of these is that the primary recombination of radicals or ions produced in a solvent cage can be accompanied by racemization, although the racemization is not likely to be complete. For example, in the decomposition of optically active β-phenylisobutyryl peroxide, about 50 % of the ester in the product is believed to arise from primary recombination of the alkyl and acyloxy radicals within the cage. If the formation of the ester from alkyl and acyloxy radicals which have escaped from the cage is suppressed by using radical scavengers, the alkyl moiety of the ester is found to be 15 % racemized.[33]

If we could be sure that the products that seem to have been formed within the cage were entirely the result of the primary recombination of radicals, we could estimate the rate constant for that process. The yield of the relevant products in the presence of scavengers is reduced by 50 %. The fact that about half the reactive fragments appear to escape from the cage indicates that the mean time before recombination of a pair of radicals is about equal to the mean lifetime of an encounter. The unimolecular rate constant for the conversion of radical pairs to molecules should then be of the order of 10^{11} sec⁻¹.

If the caged pair consists of ions in which the charges are at least partially localized on a functional group, the likelihood of relative rotation of the partners before recombination takes place is even less than that for caged pairs of radicals. For example, the observed structural and solvent effects on the polar decomposition of decalyl perbenzoate (31) and of p-methoxy-p'-nitrobenzoyl peroxide (32) strongly suggest an intimate ion-pair mechanism.[34,35]

However, the decomposition of O^{18}-labeled decalyl perbenzoate shows that the two oxygens of the benzoate ion do not become equivalent during the reaction.[36] If there is an ion pair intermediate, its collapse to rearranged

---

[33] D. F. DeTar and R. Hunt, private communication.
[34] R. Criegee and R. Kaspar, *Ann.*, **560**, 127 (1948); P. D. Bartlett and J. L. Kice, *J. Am. Chem. Soc.*, **75**, 5591 (1953); H. L. Goering and A. C. Olson, *J. Am. Chem. Soc.*, **75**, 5853 (1953).
[35] J. E. Leffler, *J. Am. Chem. Soc.*, **72**, 67 (1950).
[36] D. B. Denney, *J. Am. Chem. Soc.*, **77**, 1706 (1955).

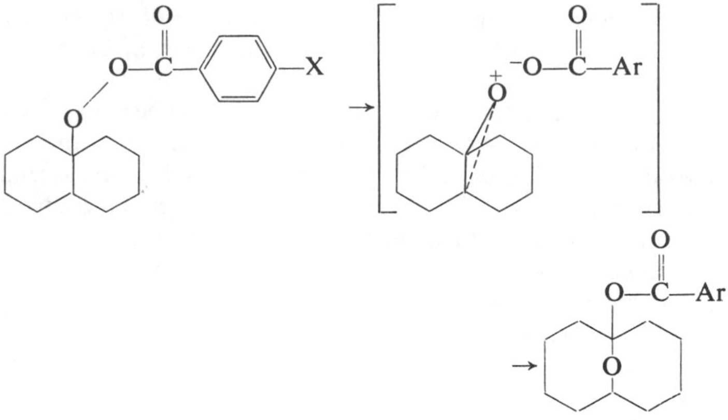

(31)

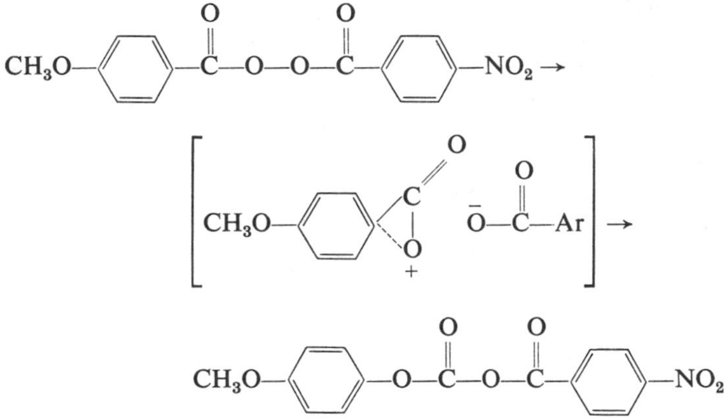

(32)

product is at least fifty times as fast as even a partial reorientation of the benzoate ion. Similarly, the two oxygens of the anisyloxy cation in the carboxy-inversion reaction of the *p*-methoxy-*p'*-nitrobenzoyl peroxide never become equivalent.[37] These results prove that the ion pairs involved cannot be long-lived; they might indeed be transition states rather than intermediates.

The rearrangement of decalyl perbenzoate in aqueous methanol is accompanied by solvolysis; if these reactions have a common ion-pair intermediate, the rate of collapse of this intermediate to rearranged product must be of the same order of magnitude as the separation of the ion pair by diffusion. Separation by diffusion is slow enough so that it is likely that the benzoate would have time to do some rotating during the lifetime of the ion pair. It may be, however, that the benzoate ion is so polarized by the

[37] D. B. Denney, *J. Am. Chem. Soc.*, **78**, 590 (1956).

adjacent positive charge of the cation that it should really be written as

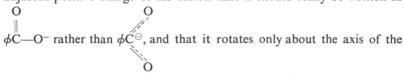

ϕC—O^- rather than $\phi C^{\ominus}$, and that it rotates only about the axis of the electrostatic bond between O^- and the cation. This possibility is rendered more plausible by the fact that the infrared spectrum of triethylammonium acetate in carbon tetrachloride shows the characteristic carbonyl group absorption frequency rather than that characteristic of symmetrical carboxylate ions.[38]

In the gas phase, molecules formed by the collision of two polyatomic fragments have a good chance of decomposing before the excess vibrational energy of the newly formed bond is lost to some other mode of motion. In solution such redissociation is not likely, for an encounter need consist of only a few collisions to make de-excitation of a particular mode of vibration virtually certain.

RE-EVALUATION OF THE TRANSITION-STATE THEORY

The transition-state theory as presented in Chapter 4 assumed the existence of a quasi-equilibrium between the reagent and the forward-moving subspecies of the transition state. The theory also assumed that the population of this subspecies is relatively independent of the amount of product present. It has been shown (page 67) that these assumptions imply, and are implied by, the assumption that a molecule once past the barrier rapidly cascades to low-lying energy levels of the product. We now have direct experimental evidence that deactivation of vibrationally excited energy levels of organic molecules in the liquid phase is indeed rapid. The only assumption still needed for the application of the transition-state theory to liquid-phase organic reactions is that the band of energy levels near the transition state behaves in the same way and is also rapidly deactivated. Although this is still an assumption, it is less drastic than the one it replaces.

THE INTERCONVERSION OF ROTAMERS

The rates of interconversion of rotational or conformational isomers can vary all the way from rates low enough to permit separation of the mixtures

[38] G. M. Barrow and E. A. Yerger, *J. Am. Chem. Soc.*, **76**, 5211 (1954); E. A. Yerger and G. M. Barrow, *ibid.*, **77**, 4474 (1955).

Table 5-4. Some Energy Barriers for Rotation about Covalent Bonds in Gas Molecules

MOLECULE	BARRIER HEIGHT (kcal)	REF.
H_3C—CH_3	2.9	a
H_3C—CH_2Cl	3.4	b
H_3C—CH_2Br	2.8	c
H_3C—CH_2CN	3.3	d
H_3C—$COOH$	0.48	e
H_3C—$COCH_3$	0.76	f
H_3C—$COBr$	1.31	g
H_3C—CHO	1.15	h
H_3C—$COCN$	1.21	i
H_3C—$CH{=}CH_2$	1.98	j
H_3C—$CH{=}CHF$	2.2	k
H_3C—$\overset{\displaystyle O}{\overset{\displaystyle \triangle}{CH}}CH_2$	2.56	l
H_3C—SiH_3	1.67	m
H_3C—SiH_2F	1.56	n
H_3C—$SiHF_2$	1.26	o
H_3C—$CH(CH_3)_2$	3.9	p
H_3C—$CF(CH_3)_2$	4.3	p
H_3C—$P(CH_3)_2$	2.6	p
H_3C—$N(CH_3)_2$	4.4	q
HO—OH (cis → trans)	1.29	r
HO—OH (trans → cis)	0.59	r
$H\overset{\displaystyle O}{\underset{\displaystyle \|}{C}}$—$OH$ (cis → trans)	10.9	s
$H\overset{\displaystyle O}{\underset{\displaystyle \|}{C}}$—$OH$ (trans → cis)	8.9	s
CH_3BF_2	0.014	t
(pseudorotation of pucker)	0	u
(pseudorotation of pucker)	2.8	v

(a) D. R. Lide, Jr., *J. Chem. Phys.*, **29**, 1426 (1958).   (b) R. S. Wagner and B. P. Dailey, *ibid.*, **26**, 1588 (1957).   (c) R. S. Wagner, B. P. Dailey, and N. Solimene, *ibid.*, **26**, 1593 (1957).   (d) R. G. Lerner and B. P. Dailey, *ibid.*, **26**, 678 (1957).   (e) W. Tabor, *ibid.*, **27**, 974 (1957).   (f) J. D. Swalen and C. C. Costain, *ibid.*, **31**, 1562 (1959).   (g) L. C. Krisher, *ibid.*, **33**, 1237 (1960).   (h) R. W. Kilb, C. C. Lin, and E. B. Wilson, Jr., *ibid.*, **26**, 1695 (1957).   (i) L. C. Krisher and E. B. Wilson, Jr., *ibid.*, **31**, 882 (1959); **33**, 304 (1960).   (j) D. R. Lide, Jr., and D. E. Mann, *ibid.*, **27**, 868 (1957); D. R. Herschbach and L. C. Krisher, *ibid.*, **28**, 728 (1958).   (k) S. Siegel, *ibid.*, **27**, 989 (1957).   (l) D. R. Herschbach and J. D. Swalen, *ibid.*, **29**, 761 (1958).   (m) R. W. Kilb and L. Pierce, *ibid.*, **27**, 108 (1957).   (n) L. Pierce, *ibid.*, **29**, 383 (1958).   (o) J. D. Swalen and B. P. Stoicheff, *ibid.*, **28**, 671 (1958).   (p) D. R. Lide, Jr., and D. E. Mann, *ibid.*, **29**, 914 (1958).   (q) D. R. Lide, Jr., and D. E. Mann, *ibid.*, **28**, 572 (1958).   (r) E. Hirota, *ibid.*, **28**, 839 (1958).   (s) T. Miyazawa and K. S. Pitzer, *ibid.*, **30**, 1076 (1959).   (t) R. E. Naylor, Jr., and E. B. Wilson, Jr., *ibid.*, **26**, 1057 (1957).   (u) J. E. Kilpatrick, K. S. Pitzer and R. Spitzer, *J. Am. Chem. Soc.*, **69**, 2483 (1947); J. P. McCullough, *J. Chem. Phys.*, **29**, 966 (1958).   (v) W. N. Hubbard et al., *J. Am. Chem. Soc.*, **74**, 6025 (1952).

by ordinary chemical techniques to rates even greater than those for rotation of molecules as a whole. Beginning at one extreme we have double bonds, then single bonds of varying degrees of double bond character or steric hindrance to rotation, and finally we have free rotation. Even in ethane, however, the rotation is not completely free. The extreme of free rotation is more nearly realized in systems like nitromethane or toluene[39] in which all the rotational positions about the single bond axis have approximately the same energy. In ethane, the eclipsed conformation is 2.9 kcal higher in energy than the staggered conformation.[40]

Table 5-4 contains some values that have been observed for torsional energy barriers for various single bonds. Table 5-5 contains some typical rate constants or relaxation times for torsions and for the interconversion of conformational isomers. The rate measurements may be precise only to within 50%, although the techniques are constantly being improved. For the faster processes there is often some doubt as to whether the process responsible for an observed relaxation time has been correctly identified.

In some instances the relaxation spectrum shows two characteristic frequencies. One of these is normally in the range expected for rotation of the molecule as a whole. The other may be assigned to an internal rotation. Following the pioneering work of E. Fischer,[41] the internal rotation responsible for the second relaxation time is identified by the amount of energy absorbed, for this energy is proportional, in first approximation, to the square of the effective dipole moment of the rotating group.[41,42] For example, 2,4,6-tri-t-butylphenol in decalin has two relaxation times, 5×10^{-10} and 5×10^{-12} sec at 20°.[43] From the amounts of energy absorbed in the two relaxations, the corresponding dipole moments are calculated to be 0.4 and 1.3 Debye units. The bond moment of the O—H bond is about 1.6 Debye units and the component of the *molecular dipole moment* in the plane of the ring is about 0.55 Debye units. Since the axis of rotation is perpendicular to the dipole axis, it is clear that the absorption at the higher frequency is due to the rotation of the OH group about the C—O bond axis, whereas that at the lower frequency is due to the tumbling of the entire molecule.

Another method of assigning the relaxation time is to change the structure of the molecule so as to interfere with the suspected internal rotation. For example, diphenyl ether in benzene solution has the

[39] E. Tannenbaum, R. J. Myers, and W. D. Gwinn, *J. Chem. Phys.*, **25**, 42 (1956).

[40] K. S. Pitzer, *Disc. Faraday Soc.*, **10**, 66 (1951).

[41] E. Fischer, *Z. für Naturforschung*, **4A**, 707 (1949).

[42] C. F. J. Böttcher, *Theory of Electric Polarisation*, Elsevier Publishing Co., New York, 1952, pp. 323, 349.

[43] M. Davies and R. J. Meakins, *J. Chem. Phys.*, **26**, 1584 (1957).

Table 5-5. Rate Constants for Internal Rotation about Covalent Bonds

INTERNAL ROTATION	EXPERIMENTAL CONDITIONS	$\frac{1}{\tau}$ (sec^{-1})	REF.
H_3C—CH_2Br	Pure liquid, 25°	7×10^{11}	a
H_7C_3—CH_2Br	Pure liquid, 25°	3×10^{11}	a
$H_{11}C_5$—CH_2Br	Pure liquid, 25°	3×10^{11}	a
$H_{19}C_9$—CH_2Br	Pure liquid, 25°	2×10^{11}	a
$H_{31}C_{15}$—CH_2Br	Pure liquid, 25°	3×10^{11}	a
C_6H_5—CH_2Cl	Benzene, 20°	4×10^{11}	b
$(CH_3)_2HC$—$CH(CH_3)_2$	Pure liquid, 25°		
$trans \rightarrow gauche$		ca. $1 \times 10^{10}(?)$p	c
$gauche \rightarrow trans$		ca. $5 \times 10^{10}(?)$p	c
$-\overset{\displaystyle \mid}{\underset{\displaystyle \mid}{C}}-OH$	Decalin, 0°		
2,4,6-tri-t-Butylphenol		7×10^{10}	d, q
4-bromo-2,6-di-t-Butylphenol		4×10^{10}	d
Tricyclohexylcarbinol		6×10^{10}	d
CH_3—$\dot{C}HCO_2^-$	Irradiated crystalline alanine, $-100°C$	2.9×10^8	k

Cyclohexane, chair → boat	CS$_2$, 25°C	2.0 × 10^5	l, m
Perfluorocyclohexane, chair → boat	Pure liquid, 25.5°C	3.5 × 10^4	m, n
CH$_3$O—NO, *trans* → *cis*	Pure liquid, −38°	200	e
	25°	2 × 10^4	e, q
$\overset{\text{O}}{\underset{\|\|}{}}$ HC—OC$_2$H$_5$, *trans* → *cis*	Pure liquid, 25°	8.5 × 10^5	f
	n-Hexane, 25°	13 × 10^5	f
(CH$_3$)$_2$N—NO	Pure liquid, 25°	1 × 10^{-4}	g, q
	180°	110	g
	CCl$_4$ or CH$_3$OH, 25°	2.0	o
O=HC—NH$_2$	Acetone, 25°	0.5	h, r
O=HC—N(CH$_3$)$_2$	Pure liquid, 25°	0.7	i, p

Table 5-5 (continued)

INTERNAL ROTATION	EXPERIMENTAL CONDITIONS	$\dfrac{1}{\tau}$ (sec^{-1})	REF.
$CH_3\overset{O}{\overset{\|}{C}}-N(CH_3)_2$	Pure liquid, 25°	0.5	i, p
(biphenyl with OCH$_3$, COOH, COOH, OCH$_3$ substituents)	Methanol, 100° Dioxane, 100°	16.6×10^{-6} 5×10^{-6}	j j

(a) K. Higasi, K. Bergmann, and C. P. Smyth, *J. Phys. Chem.*, **64**, 880 (1960). (b) W. P. Purcell, K. Fish, and C. P. Smyth, *J. Am. Chem. Soc.*, **82**, 6299 (1960). (c) J. H. Chen and A. A. Petrauskas, *J. Chem. Phys.*, **30**, 304 (1959). (d) M. Davies and R. J. Meakins, *J. Chem. Phys.*, **26**, 1584 (1957). (e) P. Gray and L. W. Reeves, *J. Chem. Phys.*, **32**, 1878 (1960). (f) D. N. Hall and J. Lamb, *Trans. Faraday Soc.*, **55**, 784 (1959). (g) C. E. Looney, W. D. Phillips, and E. L. Reilly, *J. Am. Chem. Soc.*, **79**, 6136 (1957). (h) B. Sunners, L. H. Piette, and W. G. Schneider, *Can. J. Chem.*, **38**, 681 (1960). (i) H. S. Gutowsky and C. H. Holm, *J. Chem. Phys.*, **25**, 1228 (1956). (j) B. M. Graybill and J. E. Leffler, *J. Phys. Chem.*, **63**, 1461 (1959). (k) I. Miyagawa and K. Itoh, *J. Chem. Phys.*, **36**, 2157 (1962). (l) S. Meiboom, private communication. The value of $1/\tau$ is extrapolated from data at somewhat lower temperatures. (m) Experimentally one observes a molecular process in which the axial protons (or fluorine atoms) become equatorial, while the equatorial protons (or fluorine atoms) simultaneously become axial. The rate-determining step for this exchange is believed to be the formation of a flexible boat conformation. (n) G. V. D. Tiers, *Proc. Chem. Soc.*, 389 (1960). (o) A. Loewenstein, J. F. Neumer, and J. D. Roberts, *J. Am. Chem. Soc.*, **82**, 3599 (1960). (p) Units employed by original authors are not quite clear. (q) Calculated from Arrhenius parameters. (r) Estimated by us from published curves.

unusually low reduced relaxation time, τ_μ/η at 60°, of 0.23 $\times$ 10^{-9}, in contrast to a value of 1.19 $\times$ 10^{-9} for benzophenone.[44] Since the relaxation time for dibenzofuran is much increased over that for diphenyl ether,[45] it

$10^9\, \tau_\mu/\eta$ 0.9   0.23

$10^9\, \tau_\mu/\eta$ 1.19   0.70

follows that the unusually low relaxation time of the latter is due to an internal rotation as represented by equation 33.[44,46] The reason that this

$$\text{(33)}$$

rotation is able to produce a change in dipole moment is that the two forms have dipole resonance structures, as shown in equation 34.

$$\text{(34)}$$

If this explanation is correct, the fact that benzophenone exhibits only a normal molecular relaxation time implies that a similar resonance effect in benzophenone is too small for detectable energy absorption. The rather low relaxation times for diphenylamine[47] and triphenylamine[48] have also been ascribed to internal rotations.

[44] K. Higasi and C. P. Smyth, *J. Am. Chem. Soc.*, **82**, 4759 (1960).
[45] G. W. Nederbragt, oral communication to D. M. Roberti, O. F. Kalman, and C. P. Smyth, *J. Am. Chem. Soc.*, **82**, 3525 (1960).
[46] H. Shimizu, S. Fujiwara, and Y. Morino, *J. Chem. Phys.*, **34**, 1467 (1961).
[47] E. Fischer, *Z. für Elektrochemie*, **53**, 16 (1949).
[48] B. B. Howard, unpublished results.

Table 5-6. Activation Parameters for Rotation About Covalent Bonds

COMPOUND AND BOND	SOLVENT	METHOD	$\Delta H^{\ddagger}$ (kcal)	$\Delta S^{\ddagger}$ (e.u.)*	REF.
$H_{2n+1}C_n$—CH_2Br (average for n = 3 to 7)	Pure liquid	Dielectric	1.15	−2.2	a
C—OH in 2,4,6-tri-*t*-Butylphenol	Decalin	Dielectric	2.2	−0.5	b
CH_3—$\dot{C}HCO_2^-$	Irradiated crystalline alanine	EPR	3.3	0.4	h
Cyclohexane, chair → boat	CS_2	NMR	11.4	4.0	i
Perfluorocyclohexane, chair → boat	Pure liquid	NMR	7.5	−10.7	j
CH_3O—NO (*trans* → *cis*)	Pure liquid	NMR	9.4 ± 2	−7.6 ± 8	c
$(CH_3)_2N$—NO	Pure liquid	NMR	22.1	−2.5	d
	CCl_4 CH_3OH	NMR NMR	7.2 6.2	−33(?) −36(?)	k k
O=HC—N(CH_3)$_2$	Pure liquid	NMR	7 ± 3	−36 ± 11(?)	e

$CH_3C(=O)—N(CH_3)_2$	Pure liquid	NMR	12 ± 2	−20 ± 8(?)	e
(biphenyl: $N(CH_3)_2$ / $\oplus N(CH_3)_3$)	Water	Polarimetric	26.71 ± 0.06	−7.11 ± 0.15	f
	Dimethylsulfoxide	Polarimetric	24.66 ± 0.06	−10.62 ± 0.16	f
(biphenyl: OCH_3 $COOH$ / $COOH$ OCH_3)	Methanol	Polarimetric	23.24 ± 0.06	−18.58 ± 0.17	g
	Dioxane	Polarimetric	25.15 ± 0.07	−15.85 ± 0.20	g

(*) Transmission coefficient has been assumed to be unity. (a) K. Higasi, K. Bergmann, and C. P. Smyth, *J. Phys. Chem.*, **64**, 880 (1960). (b) M. Davies and R. J. Meakins, *J. Chem. Phys.*, **26**, 1584 (1957). (c) P. Gray and L. W. Reeves, *J. Chem. Phys.*, **32**, 1878 (1960); L. H. Piette and W. A. Anderson, *J. Chem. Phys.*, **30**, 899 (1959). (d) C. E. Looney, W. D. Phillips, and E. L. Reilly, *J. Am. Chem. Soc.*, **79**, 6136 (1957). (e) H. S. Gutowsky and C. H. Holm, *J. Chem. Phys.*, **25**, 1228 (1956). (f) J. E. Leffler and W. H. Graham, *J. Phys. Chem.*, **63**, 687 (1959). (g) B. M. Graybill and J. E. Leffler, *J. Phys. Chem.*, **63**, 1461 (1959). (h) I. Miyagawa and K. Itoh, *J. Chem. Phys.*, **36**, 2157 (1962). (i) S. Meiboom, private communication. See also note (m) in Table 5-5. (j) G. V. D. Tiers, *Proc. Chem. Soc.*, 389 (1960). (k) A. Loewenstein, J. F. Neumer, and J. D. Roberts, *J. Am. Chem. Soc.*, **82**, 3599 (1960).

Benzyl chloride in benzene solution shows two relaxation times.[49] The shorter of these vanishes in molecules in which flanking methyl groups prevent the rotation of the chloromethyl group about the carbon-carbon bond axis. In n-alkyl bromides the flexibility of the molecule might lead us to expect not just two but several relaxation times. These probably exist but have not been resolvable. However, it is possible to analyze the dielectric data for the n-alkyl bromides in terms of an assumed distribution of relaxation times. For each n-alkyl bromide the shortest of the component relaxation times is that of rotation of the smallest polar unit, the CH_2Br group. Its value is independent of the length of the alkyl group.[50]

Relaxation times for internal rotation have in some cases been measured as a function of temperature (Table 5-6).

$$\frac{1}{\tau} = \frac{\kappa kT}{h} e^{\Delta S^{\ddagger}/R} e^{-\Delta H^{\ddagger}/RT} \tag{35}$$

Theoretically, two categories can be foreseen. Going from a group trapped in a potential well to a transition state corresponding to free rotation, we expect a positive $\Delta S^{\ddagger}$ because of the decrease in constraint. There may even be a further positive contribution to $\Delta S^{\ddagger}$ because of desolvation, since it should be difficult to solvate a spinning top. The transmission coefficient, κ, for such a process may be a fraction depending on the number of potential wells, rather than unity. The freely rotating transition state may drop into any of the potential wells, including the one from which it was excited. If there were two such potential wells, the effect would be equivalent to -1.37 units of entropy of activation.

We expect somewhat different behavior if the mechanism of interconversion of two stable rotamers involves a static rather than a freely rotating transition state. Such a process might even have an inherently negative entropy of activation and might have a further negative contribution to the entropy because of solvation of the transition state. For example, in a biphenyl the transition state might be a coplanar configuration with a partially double bond between the rings and with a considerable polarity, especially if the structures contributing to the partial double bond are zwitterionic.

[49] W. P. Purcell, K. Fish, and C. P. Smyth, *J. Am. Chem. Soc.*, **82**, 6299 (1960).
[50] K. Higasi, K. Bergmann, and C. P. Smyth, *J. Phys. Chem.*, **64**, 880 (1960).

So far as we are aware, there are as yet no relaxation data bearing on changes in the conformation of solvation complexes, although it should be possible for one partner of such a complex to rotate about the axis connecting the two partners. This would be analogous to rotation about a single bond. For example, it is probable that in a π-complex like that between picric acid and phenanthrene, there is more than one orientation corresponding to a potential minimum.

RATE CONSTANTS FOR MIXTURES OF ISOMERS

The analysis of reaction rate constants for equilibrium mixtures of conformational isomers has recently received considerable attention.[51] For example, in the cyclohexane system the reactive group may occupy either an axial or an equatorial position. Judging by the data for cyclic compounds in Table 5-5, the equilibration of axial and equatorial subspecies is fast compared to the reaction of the functional group in either. For mobile systems of this type, the rate constant for reaction of the functional group has been written as in equation 36,

$$k = k_a \alpha_a + k_e(1 - \alpha_a) \tag{36}$$

in which α_a is the fraction of the axial isomer in the reagent. The constants k_a and k_e are discussed as though they were rate constants for the reaction of their respective species under conditions where the interconversion is impossible. Although this treatment is formally correct, we would like to re-examine it to remove any lingering impression that the successful application of equation 36 has anything to do with the mechanism of activation. The potential energy surface for a process of this type might look like Figure 5-11a or 5-11b. In Figure 5-11b there is only one transition state conformation, and the two subspecies merge as they approach the transition state. In Figure 5-11a the conformational distinction persists, and it would be quite logical to label one transition state equatorial and the other axial. Each transition state is connected with its respective subspecies by a ravine, all points of which are separated from the other ravine by a small barrier. At all levels molecules jump back and forth across this barrier, and at any given level the rates of crossing in the two directions are equal. It follows, therefore, that a molecule arriving at the equatorial transition state *cannot* be said to have originated from the equatorial subspecies. Nevertheless, equation 36 is correct, for there is

[51] D. Y. Curtin, *Record Chem. Progress*, **15**, 111 (1954); S. Winstein and N. J. Holness, *J. Am. Chem. Soc.*, **77**, 5562 (1955); E. L. Eliel and C. A. Lukach, *J. Am. Chem. Soc.*, **79**, 5986, 6583 (1957); E. L. Eliel and R. S. Ro, *ibid.*, **79**, 5992, 5995, 6583 (1957).

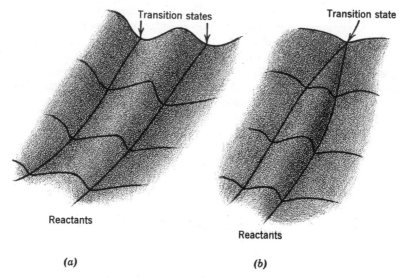

Figure 5-11. Simplified potential surfaces for the reaction of a mixture of conformational isomers. (*a*) Reaction coordinates remain separate. (*b*) Reaction coordinates intersect at transition-state level.

detailed balancing at all levels. If the barrier between the subspecies were suddenly to become impermeable, the rates at which the two transition states are traversed would remain unchanged. Similar things may be said about any analysis of rate constants into contributions from separate but rapidly equilibrating subspecies.

THE LIFETIMES OF SOLVATION COMPLEXES

Any molecule in a liquid phase is solvated in the sense that it interacts with its neighbors. These interactions are often detectable by their effect on the thermodynamic properties of the solution. For example, the low compressibility of aqueous glucose and maltose solutions has been taken to indicate that each sugar molecule is associated with four to five tightly bound and relatively incompressible molecules of water.[52] However, even when there is good evidence for strong solute-solvent interaction, it is by no means certain that anything other than a transient association is involved. The solvation complex model is at its most appropriate when the solute and one or several solvent molecules diffuse and rotate as a single entity rather

[52] H. Shiio, T. Ogawa, and H. Yoshihashi, *J. Am. Chem. Soc.*, **77**, 4980 (1955).

than separately. This implies that the lifetime of the complex must be long compared to relaxation times for diffusion and rotation.

The measurement of the lifetime of the solvation complex is possible only in a few cases. For example, the mean lifetimes of the OH · · · · N hydrogen bonds between water molecules and ammonia, methylamine, and trimethylamine are estimated from proton-deuteron exchange experiments in strong acid to be 2.0×10^{-12}, 1.2×10^{-11} and 1.0×10^{-10} sec.[53] If the partners in such a complex were diffusing and rotating as independent units, the mean lifetimes of the hydrogen bonds would be expected to fall between 2×10^{-12} and 4×10^{-12} sec. Thus ammonia (2.0×10^{-12}) in aqueous solution appears to rotate and diffuse by breaking and remaking its hydrogen bonds, whereas methylamine and trimethylamine rotate and diffuse as solvated units.

In most instances no direct measurement of the lifetime of the complex is possible, and one must be satisfied merely with the establishment of upper or lower bounds. One such technique uses the relaxation time for diffusion. If the mobility of the solute is abnormally low, it can be inferred that the solute has combined with a number of solvent molecules to form a slower-moving complex whose lifetime is long compared to the relaxation time for its diffusion. A limitation of this approach is that the mobility to be expected for the unsolvated solute cannot usually be predicted with great accuracy. Hence the method yields definitive results only when the effect of solvation is large, which in turn implies that the method is probably limited to instances where the solvation complex contains at least two or three solvent molecules or where the solvent molecules are large.

Another technique uses the relaxation time for molecular rotation. For polar solutes in *nonpolar* solvents, the rotational relaxation time of the solute is measured directly. The result is then analyzed to see if rotation is abnormally slow. For example, the dielectric relaxation time of picric acid in benzene is appreciably longer than that of other substances with similar structures (Table 5-3), suggesting that a benzene-picric acid complex is formed. If this is so, the lifetime of the complex must be longer than 5×10^{-11} sec, the dielectric relaxation time.

In *polar* solvents it is usually not possible to measure the rotational relaxation time of the *solute* unless it differs by at least one order of magnitude from that of the solvent. The reason for this is that the dielectric effect of the solute cannot be resolved from the much greater effect of the solvent. In such a case we can only analyze the dielectric data to see if any appreciable number of *solvent* molecules are prevented from rotating as a result of the addition of the solute. This approach has serious limitations, however.

[53] M. T. Emerson, E. Grunwald, M. L. Kaplan, and R. A. Kromhout, *J. Am. Chem. Soc.*, **82**, 6307 (1960).

The analysis requires a fairly detailed model of the dielectric behavior of the pure solvent and hence an undesirably large number of *ad hoc* assumptions. Even if these are granted, a conclusion can be drawn only if it turns out that solvent molecules have indeed been immobilized by the solute. A negative result does not prove the absence of solvation complexes, for a solvent molecule in a long-lived complex may yet be free to rotate about the bond joining it to the solute. For example, dielectric data for aqueous solutions of such polar substances as dioxane, *n*-propyl alcohol, phenol, propionic acid, and ethylamine indicate that water is not "irrotationally bound" to these solutes.[54] This could mean that the lifetimes of the solvation complexes are, at most, of the same order of magnitude as the relaxation times for molecular rotation, 10^{-10} to 10^{-11} sec. But it could also mean simply that the activation required for the rotation of a water molecule in a solvation complex is similar to that for the rotation of a water molecule within the quasi-crystalline solvent lattice (Figure 5-12).

The solvation of *ions* in polar solvents has been one of the most intensively studied problems in physical chemistry. From ionic mobilities of small ions in water it is quite clear that the kinetic unit includes several

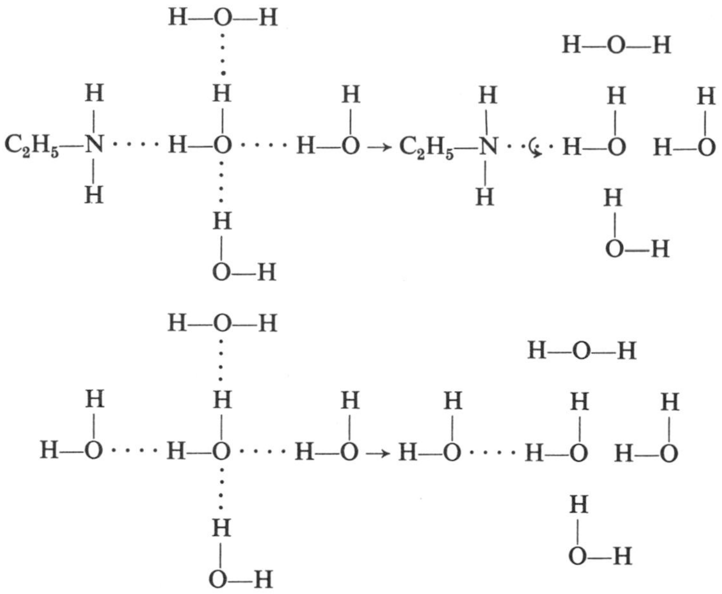

Figure 5-12. Models of molecular rotation of a water molecule. In these models, three hydrogen bonds are broken before rotation takes place.

[54] G. H. Haggis, J. B. Hasted, and T. J. Buchanan, *J. Chem. Phys.*, **20**, 1452 (1952).

water molecules.[55] The conductivity data[55] and the dielectric relaxation data[54] indicate that small cations have from four to six bound water molecules but that large ions with well-shielded charges, such as tetraethylammonium ion, do not form kinetically stable hydration complexes. Halide anions and perhaps propionate anions appear to form hydrates stable enough to diffuse as such, but without restricting the ability of their component water molecules to rotate independently of the larger unit.[54,55]

If the shapes and sizes of cations and solvent molecules associated in a solvate are suitable, the solvate may be quite a stable chemical species. This seems to be the case for the complex between dioxane and potassium ion[56] (formula 37) and for hexahydrate chromic ion.[57] As the thermo-

(37)

dynamic stability of the solvation complex increases, so does its lifetime. In extreme cases the lifetime is great enough to be measurable by the rate of isotopic exchange between the solvation complex and the bulk of the solvent. For example, the lifetime of a water molecule in O^{18}-labeled hexahydrate chromic ion can be as long as several hours.[57]

Something can be learned about the lifetimes of solvation complexes in hydroxylic solvents by measuring the lifetimes for some model systems consisting of solutions of the hydroxylic compounds in inert solvents. The model is an imperfect one, for processes in hydroxylic solvents, in comparison to processes in dilute solutions of such compounds in inert solvents, are more likely to involve the concerted formation of new hydrogen bonds as the old ones are broken.

The association of t-butyl alcohol in cyclohexane solution has been studied by the ultrasonic relaxation method.[58] The major species appear to be the monomer and tetramer. The rate constant for the dissociation of the tetramer at 27° is 1.1×10^8 sec^{-1}. This rate is lower by a factor of 10^2 than that for the dissociation of the trimethylamine-water complex in water. Either the t-butyl alcohol tetramer is cyclic, and two hydrogen bonds must be broken to prevent the ring from being immediately

[55] E. R. Nightingale, Jr., *J. Phys. Chem.*, **63**, 1381 (1959); R. A. Robinson and R. H. Stokes, *Electrolyte Solutions*, Academic Press, New York, 1955, ch. 6.
[56] E. Grunwald, G. Baughman, and G. Kohnstam, *J. Am. Chem. Soc.*, **82**, 5801 (1960).
[57] R. A. Plane and H. Taube, *J. Phys. Chem.*, **56**, 33 (1952).
[58] R. S. Musa and M. Eisner, *J. Chem. Phys.*, **30**, 227 (1959).

reformed, or the dissociation of the amine-water complex is unusually fast because the transition state for that process is stabilized by new hydrogen bonds already being formed. It is interesting to note in this connection that the dissociation of carboxylic acid dimers appears to proceed with about the same rate constant in nonpolar solvents as in the glacial acid. The rate constant for the dissociation of benzoic acid dimers in carbon tetrachloride at 25° is 7.4 × 10⁵ sec⁻¹, and in toluene, 3.7 × 10⁶ sec⁻¹.[59] The rate constant for dissociation of acetic acid dimers in glacial acetic acid at 25°, also measured by ultrasonic absorption, is 1.6 × 10⁵ sec⁻¹, not so very different.[60]

REACTIONS LEADING TO PRODUCTS IN EXCITED STATES

The initial product of the reaction of an atom with a molecule or the initial product of a photolysis reaction is likely to be an excited state rather than a ground state. If de-excitation is slow enough, these intermediates can be detected by their physical properties or by their distinctive chemical reactions. The chemical detection of excited products usually requires either that the reaction be carried out in the gas phase or that the reagent used as the detector be the solvent.

The excess energy of the activated complex usually appears initially as vibrational rather than translational or rotational energy. Such a vibrationally excited state will persist for some time if the molecules are small and the process is in the gas phase at low pressures (see, for example, Table 5-2). An interesting feature of such processes is that the excess vibrational energy is often found almost entirely in one of the products of a reaction to the exclusion of the other. For example, in reaction (38),

$$H\cdot + O_3 \rightarrow HO\cdot + O_2 \tag{38}$$

most of the energy produced is found as vibrational energy of the hydroxyl radicals, not the oxygen molecules.[61] Polanyi[62] has shown for the special case of an atom reacting with a diatomic molecule that lack of repulsion between the two product fragments leads to vibrational excitation of the molecular product at the expense of translational and rotational excitation.

[59] W. Maier, L. Borucki, B. Dischler, P. Manogg, and H. Rieseberg, *Z. physik. Chem.* (*Frankfurt*), **26,** 27 (1960).

[60] J. Lamb and J. M. M. Pinkerton, *Proc. Roy. Soc.*, **A199,** 114 (1949); E. Freedman, *J. Chem. Phys.*, **21,** 1784 (1953).

[61] J. D. McKinley, Jr., D. Garvin, and M. J. Boudart, *J. Chem. Phys.*, **23,** 784 (1955); W. D. McGrath and R. G. W. Norrish, *Z. physik. Chem.* (*Frankfurt*), **15,** 245 (1958).

[62] M. G. Evans and M. Polanyi, *Trans. Faraday Soc.*, **35,** 178 (1939).

The flash photolysis of ClO_2 [63] gives oxygen molecules with up to eight vibrational energy quanta but nevertheless in the ground electronic state.

$$O + ClO_2 \rightarrow ClO + O_2^*$$

Because of the tendency to conserve electron spin angular momentum, the photochemically excited state usually has the same electronic multiplicity as the ground state; so does any *rapidly* formed decomposition product of the excited state. The thermal decomposition of N_2O gives oxygen atoms in their triplet state, but the reaction has a low PZ factor.[64]

In the photolysis of diazomethane to nitrogen and methylene, the electron spin conservation rule predicts that the methylene will be in a singlet state rather than its electronic ground state, which is believed to be a triplet.[65,66] Conversion of singlet methylene to triplet methylene can, however, take place if the methylene undergoes enough collisions with an unreactive diluent.

If the triplet methylene can be regarded as analogous to an ordinary free radical, and singlet methylene as analogous to a carbonium ion, some differences in chemical reactivity might be expected. The products from the photolysis of diazomethane in the presence of various substrates do in fact depend on whether the reaction is carried out in the liquid or the gas phase and, if in the gas phase, whether at high or low pressures.

The photolysis of diazomethane in solution in 2,2,4-, 2,2,3-, and 2,3,4-trimethylpentane gives a random or statistical mixture of products corresponding to the single-step reaction (39).[67]

$$-\overset{|}{\underset{|}{C}}-H + CH_2: \rightarrow -\overset{|}{\underset{|}{C}}-CH_3 \qquad (39)$$

In contrast, the photolysis in the gas phase in the presence of inert gases gives a methylene that reacts somewhat selectively with tertiary or secondary hydrogens in preference to primary hydrogens, the ratio being 1.5 to 1.2 to 1.0.[68]

It has been suggested that the random methylene insertion reaction is

[63] F. J. Lipscomb, R. G. W. Norrish, and B. A. Thrush, *Proc. Roy. Soc.*, A233, 455 (1956).

[64] E. K. Gill and K. J. Laidler, *Can. J. Chem.*, 36, 1570 (1958).

[65] D. B. Richardson, M. C. Simmons, and I. Dvoretsky, *J. Am. Chem. Soc.*, 82, 5001 (1960).

[66] G. Herzberg and J. Shoosmith, *Nature*, 183, 1801 (1959).

[67] W. von E. Doering, R. G. Buttery, R. G. Laughlin, and N. Chaudhuri, *J. Am. Chem. Soc.*, 78, 3224 (1956).

[68] H. M. Frey. *J. Am. Chem. Soc.*, 80, 5005 (1958).

characteristic of the singlet, whereas the nonrandom, selective reaction is characteristic of the triplet ground state.[65] The triplet state is the product of collisional deactivation.

A difference in product under different reaction conditions is also noted for the reaction of methylene with alkenes. However, in this case the effect of a solvent or of gas at a pressure sufficient to produce collisional deactivation is more difficult to interpret, for there might be an open-chain intermediate in the addition of either kind of methylene to the double bond, and the fate of such an intermediate might itself depend on deactivation by

Table 5-7. Major Products of the Photolytic Reaction of Diazomethane with Cis-2-butene[a]

AT HIGH ARGON PRESSURES	AT LOW ARGON PRESSURES
cis- and trans-1,2-Dimethyl-cyclopropane in nearly equal amounts	Mostly the cis isomer
cis- and trans-2-Pentene in nearly equal amounts	Mostly the cis isomer
Mostly 3-methyl-1-butene	Mostly 2-methyl-2-butene

[a] From H. M. Frey, *J. Am. Chem. Soc.*, **82**, 5947 (1960).

collisions. A further difficulty is that vibrationally activated dimethyl cyclopropanes can undergo *cis-trans* geometrical isomerization or rearrangement to pentenes.[69]

Photolysis of diazomethane in the gas phase in the presence of 2-butene, with argon as an added inert gas to catalyze singlet-triplet transitions, gives both of the geometrically isomeric dimethyl cyclopropanes, both of the geometrically isomeric 2-pentenes, and 3-methyl-1-butene.[70] The latter is not produced in the isomerization of vibrationally excited 1,2-dimethylcyclopropane; that reaction gives only 2-methyl-2-butene and 2-methyl-1-butene.

The product composition in the photolysis reaction depends on the argon pressure, as shown in Table 5-7. At low argon pressures the methylene apparently reacts stereospecifically as the singlet, as shown in equation 40. At high argon pressures the singlet methylene is collisionally deactivated to the triplet before reacting, and the results are those indicated by equation 41.

[69] B. S. Rabinovitch and D. W. Setser, *J. Am. Chem. Soc.*, **83**, 750 (1961).
[70] H. M. Frey, *J. Am. Chem. Soc.*, **82**, 5947 (1960).

$$CH_2: + \quad \begin{array}{c} H \quad H \\ | \quad | \\ C{=}C \\ / \qquad \backslash \\ CH_3 \qquad CH_3 \end{array} \quad \rightarrow \quad H{-}\overset{\displaystyle CH_2}{\underset{\textstyle CH_3}{\overset{/\quad\backslash}{C\text{———}C}}}{-}H \text{ plus } \begin{array}{c} CH_3 \qquad H \\ \backslash \qquad / \\ C{=}C \\ / \qquad \backslash \\ CH_3 \qquad CH_3 \end{array}$$

(singlet)

$$\text{and} \quad \begin{array}{c} H \quad H \\ | \quad | \\ C{=}C \\ / \qquad \backslash \\ CH_3 \qquad CH_2CH_3 \end{array} \qquad (40)$$

Oxygen prevents the formation of trans-1,2-dimethylcyclopropane, 3-methyl-1-butene and trans-2-pentene almost completely even at high pressures, either by reacting with triplet methylene or with its triplet reaction product. The formation of trans-2-pentene can be attributed to the migration of a methyl group in the diradical.

The photolysis of diazomethane in cis- or trans-2-butene as *solvents* is also stereospecific, giving the appropriate cis- or trans-dimethylcyclopropanes and 2-pentenes.[71]

Divalent carbon compounds containing heavy atoms, CBr_2 for example, might be expected to exist as a rapidly established equilibrium mixture of the singlet and triplet states. Spin-orbital coupling is enhanced by heavy atoms[72] and collisional deactivation would also be expected to be more efficient. Whenever there is such a rapid equilibration, of course, our earlier remarks concerning the reactions of equilibrating subspecies would apply, and the nature of the product could not be taken to imply anything about the mechanism by which the transition state leading to that product is formed.

$$\overset{\displaystyle \cdot}{CH_2} \cdot + \quad \begin{array}{c} H \quad H \\ | \quad | \\ C{=}C \\ / \qquad \backslash \\ CH_3 \qquad CH_3 \end{array} \quad \rightarrow \quad \begin{array}{c} H \quad H \\ | \quad | \\ C{-}\underset{\cdot}{C}{-}CH_3 \\ / \qquad \backslash \\ CH_3 \qquad CH_2 \\ \qquad \cdot \end{array}$$

(triplet)

(41)

$$\xrightarrow[\text{equilibration}]{\text{rotameric}} CH_3{-}\overset{\displaystyle CH_2}{\overset{/\quad\backslash}{CH\text{———}CH}}{-}CH_3 \text{ plus } CH_2{=}CH{-}\overset{\displaystyle CH_3}{\underset{\textstyle CH_3}{\overset{/}{\underset{\backslash}{CH}}}}$$

(cis and trans)

[71] R. C. Woodworth and P. S. Skell, *J. Am. Chem. Soc.*, **81**, 3383 (1959).
[72] M. Kasha, *J. Chem. Phys.*, **20**, 71 (1952).

6

Theoretical Introduction to
Extrathermodynamic Relationships

For even the most sober scientific investigator in science, the most thoroughgoing positivist, cannot dispense with fiction; he must at least make use of categories, and they are already fictions, analogical fictions, or labels, which give us the same pleasure as children receive when they are told the "name" of a thing.

Havelock Ellis
The Dance of Life

Any explanation of the effect of structural or medium changes on the rates or equilibria of organic reactions must in some way reflect the inherent complexity of the problem. All explanations involve some kind of comparison with a suitable model or set of models. The comparison may involve simple models, such as the interaction of a pair of charged spheres, or it may involve more complex models, such as an actual substituent effect in some actual chemical reaction. In the first case the explanation is complex in its mathematical structure because of the large number of simple models that must be considered simultaneously. In the second case the mathematical structure may be simple and the number of models small because the models are themselves complex.

When the explanation of substituent or medium effects is given in terms of similar effects in a model reaction, the quantities compared are thermodynamic, usually free energies, enthalpies, or entropies. The simple relationships often found among such quantities are not, however, part of the formal structure of thermodynamics. Hence they are called *extrathermodynamic relationships*. Although the relationships themselves are outside of thermodynamics, the *approach* resembles that of thermodynamics in the sense that the detailed microscopic mechanisms need not be explicitly identified. For example, the effect on the ionization constant of an *alpha*-bromine atom in propionic acid is a suitable model for the effect of an *alpha*-bromine atom in butyric acid, even if the mechanisms by which the effect is transmitted are not precisely known. On the other hand, the mathematical form of an extrathermodynamic relationship, and particularly the reagent structures for which it fails, do give valuable information about the microscopic mechanisms. Our discussion would therefore be incomplete if we did not occasionally attempt to identify a particular mechanism. In doing so, we run the risk that our identification might be in error. But it should be noted that even when we are wrong, the mistake in the identification will in no way affect the validity of the extrathermodynamic relationship that suggested the identification.

In some cases where the identification of the microscopic mechanisms is particularly clear-cut, the quantitative aspects of the extrathermodynamic relationships seem to necessitate revisions in current theories of these mechanisms. One example is that resonance interactions even in rigid molecules often involve appreciable changes in entropy as well as in enthalpy. Another is that steric effects are often found to be simply superimposable on polar effects without distorting the latter. This is hard to understand if nonbonded repulsions are relatively hard, as suggested by the Lennard-Jones r^{-12} potential function. The observed steric effects are much more compatible with softer potential functions, such as are often derived from scattering in molecular beams.[1]

Useful extrathermodynamic relationships are usually simple in form. For example, substituent effects $\delta_R \Delta F_1^\circ$ in the ionization of a series of carboxylic acids in one solvent are simply proportional to the same substituent effects, $\delta_R \Delta F_2^\circ$, in a second solvent.[2]

$$\delta_R \Delta F_1^\circ \propto \delta_R \Delta F_2^\circ \tag{1}$$

The fit is illustrated in Figure 6-1. Another example is the simple proportionality that exists between $\delta_R \Delta \bar{H}^\circ$ and $\delta_R \Delta \bar{S}^\circ$ for the rearrangement of aminotriazoles.[3]

$$\delta_R \Delta \bar{H}^\circ = \beta\, \delta_R \Delta \bar{S}^\circ \tag{2}$$

$$\tag{3}$$

The fit of equation 2 is shown in Figure 6-2.

[1] I. Amdur and R. R. Bertrand, *J. Chem. Phys.*, **36**, 1078, (1962); I. Amdur, J. E. Jordan, and S. O. Colgate, *ibid.*, **34**, 1525 (1961); I. Amdur, E. A. Mason, and J. E. Jordan, *ibid.*, **27**, 527 (1957).
[2] E. Grunwald and B. J. Berkowitz, *J. Am. Chem. Soc.*, **73**, 4939 (1951).
[3] E. Lieber, C. N. Ramachandra Rao, and T. S. Chao, *J. Am. Chem. Soc.*, **79**, 5962 (1957).

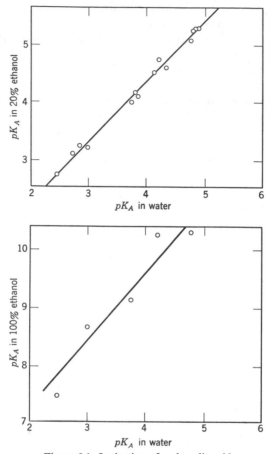

Figure 6-1. Ionization of carboxylic acids.

The precision of fit of a simple extrathermodynamic relationship between two processes improves with the degree of resemblance of the two processes. Thus in Figure 6-1 the fit is better when substituent effects on the ionization of carboxylic acids are compared in water and in 20% alcohol rather than in water and in absolute alcohol. Similarly, in the base-catalyzed decomposition of nitramide in anisole, the fit of the rate constants for a series of bases to a simple relationship involving the base dissociation constants is improved if the base dissociation constants are measured in m-cresol rather than in water.[4]

[4] G. C. Fettes, J. A. Kerr, A. McClure, J. S. Slater, C. Steel, and A. F. Trotman-Dickenson, *J. Chem. Soc.*, 2811 (1957).

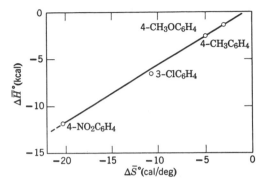

Figure 6-2. An extrathermodynamic relationship between thermodynamic quantities for the rearrangement of substituted aminotriazoles.

In the imidazole-catalyzed hydrolysis of various esters, the logarithm of the rate constant is not a linear function of the logarithm of the acid ionization constant of the alcohol being displaced, although there is a trend. A linear relationship is obtained when the logarithm of the rate constant for the imidazole-catalyzed reaction is plotted against the logarithm of the rate constant for the hydroxide ion-catalyzed reaction, although even this relationship breaks down for methyl p-nitrobenzoate.[5] Similarly, a plot of the logarithm of the rate constant for the base-catalyzed hydrolysis of ethyl esters of aliphatic acids against the pK_A of the acid gives only a scatter diagram, whereas if the hydrolysis of the benzyl esters is used as the reference reaction, at least a roughly linear correlation is obtained.[6] However, the closeness of resemblance of the processes becomes less critical if the variation in the structure of the reagent is kept relatively small. Thus in the hydrolysis of ethyl esters of *meta*- and *para*-substituted benzoic acids, the ionization of the acid serves just as well for a reference reaction as does the hydrolysis of another series of esters.

For some substituent or medium effects, no simple direct comparison with any single model reaction gives an adequate fit, and the technique of comparison by linear combination is used.[7] For example, substituent effects in the epoxidation of substituted *cis*-dimethylstilbenes are not well correlated with substituent effects in the ionization of benzoic acids, nor

[5] M. L. Bender and B. W. Turnquest, *J. Am. Chem. Soc.*, **79**, 1656 (1957).

[6] D. P. Evans, J. J. Gordon, and H. B. Watson, *J. Chem. Soc.*, 1439 (1938); E. Tommila, *Ann. Acad. Sci. Fennicae*, **A57**, No. 4, p. 3 (1941); *C.A.* **38**, 6173 (1944).

[7] Curve fitting by linear combination of two models is a device frequently used in engineering, as well as in the valence bond and LCAO methods in quantum mechanics. The device has the virtue that it almost always improves the fit but cannot in any case make it worse.

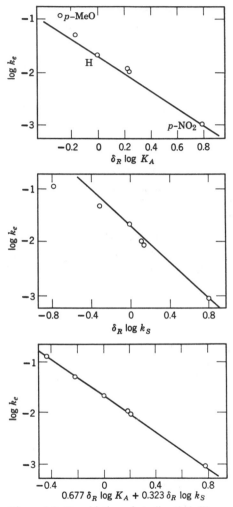

Figure 6-3. Epoxidation of *cis*-dimethylstilbenes.

are they well correlated with substituent effects in the solvolysis of 2-phenyl-2-propyl chloride. They are, however, well fitted by equation 4, where **a** and **b** are constants.[8]

$$\underset{\text{stilbene epoxidation}}{\delta_R \log k_e} = \underset{\text{acid ionization}}{\mathbf{a}\, \delta_R \log K_A} + \underset{\text{solvolysis}}{\mathbf{b}\, \delta_R \log k_s} \tag{4}$$

Figure 6-3 shows the improvement in fit by using the linear combination of models rather than the single models. In the solvolysis reaction a charge is

[8] Y. Yukawa and Y. Tsuno, *Bull. Chem. Soc. Japan*, **32**, 971 (1959).

generated in a position that permits direct resonance interaction with the ring; in the acid ionization the charge is somewhat insulated from the ring. The epoxidation reaction apparently resembles both models in some respect.

Statistical Corrections

In order to compare the chemical effects of substituents in related reactions, it is necessary first to correct the reactivities for those differences that result merely from a difference in the number of reactive sites. For example, in ethylene diamine there are two equivalent NH_2-groups, both of which can function either as the reactive site or as the substituent. We are however interested in the reactivity of a hypothetical molecule in which only one of the NH_2-groups functions as the reactive site, while the other functions as the substituent. Hence we must divide the observed rate constant by a statistical factor of two. In general, if in a rate process a reagent has p equivalent reactive sites, the rate constant must be divided by a statistical factor of p. If there are two reagents with p_1 and p_2 equivalent reactive sites, respectively, the rate constant must be divided by a statistical factor of $p_1 \cdot p_2$.

Similar considerations apply to chemical equilibria. For example, if a reagent has p equivalent reactive sites and a product has q equivalent reactive sites, the equilibrium constant must be divided by a statistical factor of p/q.

EXTRATHERMODYNAMIC RELATIONSHIPS IN THE GAS PHASE

Reactions in the gas phase should be comparatively easy to interpret because their substituent effects are not complicated, or even overwhelmed,[9] by solvation phenomena. They are therefore of considerable theoretical interest. For this reason we have made a diligent search for extrathermodynamic relationships involving gas phase reactions. The results of this survey have been disappointingly meager and are summarized in Tables 6-1 and 6-2.

The failure to find extensive extrathermodynamic relationships for gas phase reactions is in sharp contrast to the ease with which such relationships can be found for liquid phase reactions. We doubt, however, that it

[9] See, for example, R. P. Bell, *The Proton in Chemistry*, Cornell University Press, 1959, for the effect of hydration on the acidities of the halogen acids, and E. F. Caldin, *J. Chem. Soc.*, p. 3345 (1959) for the contribution of solvation energies to activation energies.

Table 6-1. Some Extrathermodynamic Relationships in the Gas Phase

1. $\Delta F^{\ddagger}$ for the reaction of sodium atoms with ArBr is linearly related to $\Delta F^{\ddagger}$ for the reaction of sodium atoms with ArCl at 520°K. F. Riding, J. Scanlan, and E. Warhurst, *Trans. Faraday Soc.*, **52**, 1354 (1956).

2. $\Delta F^{\ddagger}$'s for the reactions of methyl and trifluoromethyl free radicals with a series of hydrocarbons at 254°C are linearly related. A. F. Trotman-Dickenson, *Free Radicals*, Methuen, 1959.

3. $\Delta F^{\ddagger}_O$'s for the reaction of oxygen atoms with alkenes are linearly related to the heats of hydrogenation of the same alkenes. The plot of the $\Delta F^{\ddagger}_O$'s against $\Delta F^{\ddagger}_H$ for the reaction of hydrogen atoms with the alkenes is a scatter diagram, however. R. J. Cvetanović, *J. Chem. Phys.*, **30**, 19 (1959); P. E. M. Allen, H. W. Melville, and J. C. Robb, *Proc. Roy. Soc.* (*London*), **A218**, 311 (1953).

4. $\Delta F^{\ddagger}_{Br}$ for the elimination of hydrogen bromide from ethyl, *i*-propyl and *t*-butyl bromide is linearly related to $\Delta F^{\ddagger}_X$ for the elimination of HX, where X is Cl, acetate, and formate. E. Gordon, S. J. W. Price, and A. F. Trotman-Dickenson, *J. Chem. Soc.*, 2813 (1957).

5. $\Delta F^{\ddagger}$ for the pyrolysis of *alkyl* iodides into alkyl radicals and iodine atoms is linearly related to $\Delta F^{\ddagger}$ for the reaction of sodium atoms with the corresponding alkyl chlorides. E. T. Butler and M. Polanyi, *Trans. Faraday Soc.*, **39**, 19 (1943). A similar plot for *aryl* groups is a scatter diagram. M. Szwarc and D. Williams, *Proc. Roy. Soc.*, **A219**, 353 (1953).

6. *Scatter diagrams* are found for $\Delta \bar{H}^{\circ}$ versus $\Delta \bar{S}^{\circ}$ in the complexing of trimethyl boron with a series of amines, and for $\Delta H^{\ddagger}$ versus $\Delta S^{\ddagger}$ in the homolysis of alkyl bromides and in the pyrolysis of esters to alkenes and carboxylic acids. H. C. Brown and G. K. Barbaras, *J. Am. Chem. Soc.*, **75**, 6 (1953); H. C. Brown, M. D. Taylor, and S. Sujishi, *ibid.*, **73**, 2464 (1951); M. Szwarc and D. Williams, *Nature*, **170**, 290 (1952); A. T. Blades, *Can. J. Chem.*, **32**, 366 (1954).

7. A linear correlation is observed between $\Delta H^{\ddagger}$ and $\Delta S^{\ddagger}$ for the elimination of hydrogen chloride from a series of four alkyl chlorides. K. E. Howlett, *J. Chem. Soc.*, 945 (1953); D. H. R. Barton, A. J. Head, and R. J. Williams, *J. Chem. Soc.*, 2039 (1951).

Table 6-2. Some Relationships between Processes in the Gas and Liquid Phases

1. $\Delta F^{\ddagger}$ for elimination of hydrogen bromide from alkyl bromides in the gas phase is linearly related to the polar substituent constant σ^* for liquid phase reactions (see Chapter 7). A plot of E_{act} versus log A for the same reaction is a scatter diagram. J. H. S. Green, G. D. Harden, A. Maccoll, and P. J. Thomas, *J. Chem. Phys.*, **21**, 178 (1953).

2. $\Delta \bar{H}^{\circ}$ for hydrogenation of *trans*-disubstituted alkenes in the gas phase is correlated with σ^* as shown below:

$$\Delta \bar{H}^{\circ} = -2.2 \sum \sigma^* + 0.50 n_h + 0.20 n_c + a \; constant$$

Table 6-2 (continued)

The quantities n_h and n_c are the numbers of C—H and C—C single bonds conjugated with the double bond. R. W. Taft, Jr. and I. C. Lewis, *Tetrahedron*, **5**, 210 (1959).

3. The ionization potentials of *meta-* and *para-*substituted benzyl radicals ($ArCH_2 \cdot \rightarrow ArCH_2^+$) in the gas phase are linearly related to the polar substituent constants σ^+ for the substituents in liquid-phase reactions (see Chapter 7). A. G. Harrison, P. Kebarle, and F. P. Lossing, *J. Am. Chem. Soc.*, **83**, 777 (1961).

4. $\Delta F^{\ddagger}$ for the reaction of oxygen atoms with alkenes in the gas phase is linearly related to $\Delta F^{\ddagger}$ for the reaction of the alkenes with peracetic acid in the liquid phase and to $\Delta F^{\ddagger}$ for the reactions of the alkenes with bromine in the liquid phase. R. J. Cvetanović, *J. Chem. Phys.*, **30**, 19 (1959).

stems from any fundamental difference between matter in these two phases. For one thing, difficulties with wall effects and limited volatility make the gas-phase data scarce. For another thing, molecules that can conveniently be studied in the gas phase are usually so small that almost any structural change is a major one. This means that few reaction series in the gas phase are sufficiently alike so that one series could plausibly serve as a model for the other.

ADDITIVITY RULES

The mathematical simplicity of many extrathermodynamic relationships results in large part from the tendency of such quantities as the standard free energy to be additive functions of molecular structure. When the chemical change affects only a relatively small reaction zone within the molecule, the contribution to the free energy from the part of the molecule outside the reaction zone tends to cancel out in the quantity ΔF° or $\Delta F^{\ddagger}$. Thus if the reaction zones and the changes occurring in them during a reaction are very similar for a series of reagents, we expect such quantities as ΔF° and $\Delta \bar{F}^{\ddagger}$ also to be very similar.

Let us therefore digress to consider the representation of molar properties as additive-constitutive functions of molecular structure. To the lowest approximation, the molar values of the property are assumed to be *precisely* additive functions of independent contributions assignable to part-structures of the molecule. The nature of these part-structures is subject to arbitrary choice; they may be complex groupings of atoms or simply single atoms. If the lowest approximation is not adequate, one may achieve any desired higher degree of accuracy by using an appropriate higher and more complicated approximation. For example, the molar refraction is, to the lowest or zeroth approximation, the sum of independent contributions from the constituent atoms. In the next higher, or first approximation, note is taken of the immediate environment of the atom. Thus different numbers are assigned to the contribution from ether as compared to carbonyl oxygen atoms. In still higher approximations, the contribution from the atoms in question is regarded also as a function of more remote structural features of the molecule.

When a molar property is represented as an additive-constitutive function, certain representations are much to be preferred over others. The preferred representations are those in which all part-structures are treated in a logically consistent way. For example, if a distinction is made between the contribution to the property from ether and carbonyl oxygen atoms, a distinction should also be made between the contribution from amine and imine nitrogen atoms, at least in principle. One reason for insisting on logical consistency is that the accuracy of the molar value of the property is no greater than that of the least accurate additive term. It would therefore be pointless to improve the accuracy of the representation for some part-structures of the molecule but not for others.

When all part-structures are treated in a logically consistent way, it is possible to devise objective procedures for finding the functional form of the next higher approximation. One such procedure has been described by Benson and Buss.[10] In their procedure, the contributions from any two groups in the molecule are assumed to be *exactly* additive if the two groups are far enough apart. Thus the contribution from the two terminal methyl groups of an unbranched alkane approaches a constant or limiting value as the length of the alkane is increased.

Let P and Q denote terminal groups and B the intervening structure. As B becomes sufficiently long, the equilibrium constant, K, for reaction (5)

$$PBP + QBQ \rightleftharpoons PBQ + QBP \qquad (5)$$

approaches a limiting value of unity if B is unsymmetrical and of four if it is symmetrical. The value, four, is of course simply the statistical factor for the reaction of two molecules, each of which has two equivalent reactive sites.

The length of B is measured by the number of skeletal atoms, that is, by the number of structural units, such as $-CH_2-$ or $-O-$, in the chain. The successive approximations are generated by assuming that the limiting value of K is reached when B consists of zero, one, two, . . . structural units. Thus in the zeroth approximation, K assumes the limiting value even in the absence of any intervening structure. The first approximation

---

[10] S. W. Benson and J. H. Buss, *J. Chem. Phys.*, **29**, 546 (1958).

assumes that K reaches its limiting value if the intervening structure consists of at least one atom or CH_2 group. The second approximation requires at least two intervening atoms or CH_2 groups, and so on for still higher approximations.

In the zeroth approximation the thermodynamic property is seen to be simply the sum of atomic contributions (modified by the appropriate statistical correction), for conversion of a PP and a QQ bond into two PQ bonds has no effect. In the first approximation the nature of the bond becomes important, and in fact this approximation is a rule of simple bond additivity rather than atomic additivity. This is seen most readily if B is a single atom, for example, oxygen. On both sides of equation 5 we then have two PO bonds and two QO bonds.

The second approximation takes into account also the nature of the next-nearest neighbors. Thus nonbonded interactions of next-nearest neighbors are specifically included, but those of more remote parts of the molecule are not.

Table 6-3. Mean Errors of Fit[a]

ORDER OF APPROXIMATION	ΔH_f° (kcal/mole)	C_p° (cal/mole deg)	S° (cal/mole deg)
0	Useless	±1.5	±2
1	±2.5	±1.0	±0.7
2	±0.4	±0.3	±0.3

[a] For a sample of about 100 compounds of various types in the gas phase at 1 atm and 25°. Taken from S. W. Benson and J. H. Buss, *J. Chem. Phys.*, **29**, 546 (1958).

Successively higher approximations improve the accuracy with which the data are fitted, but only at the cost of using more empirical parameters. The rate of accretion of parameters as the order of the approximation is increased is extremely rapid. Thus a table of atomic parameters (zeroth approximation) requires N entries to deal with N different atoms. A table of bond parameters (the first approximation), on the other hand, will require $N!$ entries just for single bonds and still more entries if there are also multiple bonds. The second approximation requires multiple entries for each different kind of bond. A table for the third approximation would be extremely cumbersome.

In spite of the rapid increase in the number of disposable parameters, the improvement in the fit is not as great as might perhaps be expected. The reason for this is that the lower approximation already contains an average correction for the interactions that are treated specifically in the next higher approximation. Table 6-3 shows the degree of improvement in the

Table 6-4. *Bond Contributions to* ΔH_f°, C_p° *and* S° *of Gaseous Compounds at 25° and 1 atm*[a]

BOND	CONTRIBUTIONS TO		
	ΔH_f° (kcal)	C_p° (cal/deg)	S° (cal/deg)
C—H	−3.83	1.74	12.90
C—D	−4.73	2.06	13.60
C—C	2.73	1.98	−16.40
C—F	...	3.34	16.90
C—Cl	−7.4	4.64	19.70
C—Br	2.2	5.14	22.65
C—I	15.0	5.54	24.65
C—O	−12.0	2.7	−4.0
O—H	−27.0	2.7	24.0
O—D	−27.9	3.1	24.8
O—O	21.5	4.9	9.1
O—Cl	9.1	5.5	32.5
C—N	9.3	2.1	−12.8
N—H	−2.6	2.3	17.7
C—S	6.7	3.4	−1.5
S—H	−0.8	3.2	27.0
S—S	...	5.4	11.6
C_{vi}—C[b]	6.7	2.6	−14.3
C_{vi}—H[b]	3.2	2.6	13.8
C_{vi}—F[b]	...	4.6	18.6
C_{vi}—Cl[b]	−0.7	5.7	21.2
C_{vi}—Br[b]	9.7	6.3	24.1
C_{vi}—I[b]	...	6.7	26.1
····CO—H[c]	−13.9	4.2	26.8
····CO—C[c]	−14.4	3.7	−0.6
····CO—O[c]	−50.5	2.2	9.8
····CO—F[c]	...	5.7	31.6
····CO—Cl[c]	−27.0	7.2	35.2
Φ—H[d]	3.25	3.0	11.7
Φ—C[d]	7.25	4.5	−17.4
(NO₂)—O[d]	−3.0	...	43.1
(NO)—O[d]	9.0	...	35.5

[a] S. W. Benson and J. H. Buss, *J. Chem. Phys.*, **29**, 550 (1958).

[b] C_{vi}— represents a bond to the carbon atom of a vinyl group; the unit $\rangle C = C\langle$ is treated as a single tetravalent unit.

[c] ····CO— represents a bond to the carbon atom of a carbonyl group; the unit $\rangle CO$ is treated as a single bivalent unit.

[d] NO and NO₂ are here considered as univalent terminal groups; Φ denotes the hexavalent carbon skeleton of the benzene ring,

For example, ΔH_f° for C_6H_6 is 6 × 3.25 kcal/mole.

fit of standard enthalpies of formation, molar heat capacities, and molar entropies for a sample consisting of about 100 compounds of various types in the gas phase.

Table 6-4 shows the parameters for use in the first, or bond-additivity, approximation.

When the standard free energy increments for reactions are represented as additive-constitutive functions, the various degrees of approximation give greatly differing results. The zeroth, or atomic additivity, approximation makes the completely useless prediction that ΔF° is in all cases simply equal to the statistical value. That is, all equilibrium constants would simply be equal to the statistical factor for the given reaction and hence would usually be unity. The first, or bond-additivity, approximation predicts different ΔF° for different reactions but only if the reactions differ in the type of bonding. For example, ΔF° for the reaction of p-nitrobenzyl bromide with cyanide ion would be the same as that for neopentyl bromide with cyanide ion. This is still inadequate for quantitative purposes but does express the basic organizing principle of organic chemistry, namely, that reactions can be usefully discussed as members of general classes rather than individually.

Except for statistical corrections, it is only in the second approximation, which takes account of next-nearest neighbor interactions, that equilibrium constants come to depend on parts of the molecule outside of the immediate reaction zone. It is therefore the second approximation which begins to predict that there will be substituent effects. Most extrathermodynamic relationships of simple form involve substituents quite remote from the reaction zone and could therefore be analyzed only by using an approximation of still higher order.

The existence of highly successful additivity rules for molecules in the gas phase in contrast with the virtual absence of linear extrathermodynamic relationships calls for some remark. Evidently an additivity principle alone is not sufficient to generate linear extrathermodynamic relationships. Some additional postulate is also needed.

For the purpose of generating linear extrathermodynamic relationships, additivity schemes of the type just discussed are not convenient. They involve too many parameters and place an unnecessary emphasis on the distance of the substituent from the reaction site.

A FORMAL METHOD FOR GENERATING EXTRATHERMO-DYNAMIC RELATIONSHIPS

In order to treat substituent effects in a manner that does not unduly emphasize the role of the distance of separation between the substituent and the reaction site, we use the concept of additivity in a modified way.

The molecule is divided into two zones, one of which contains the substituent and the other the region of primary importance in the reaction. Each of these zones will be regarded as contributing an additive term to the free energy and also as interacting with the other zone. We shall represent the zones in a molecule R—X by R and X. The standard free energy of a substance, RX, is then expressed as in equation 6.

$$F_{RX}^{\circ} = F_R + F_X + I_{R,X} \tag{6}$$

The quantities F_R and F_X are the independent additive terms (per mole), and $I_{R,X}$ is the term resulting from the interaction between R and X.

In a chemical reaction, the R region remains intact, and the additive term F_R cancels out in the quantities ΔF° or $\Delta F^{\ddagger}$. Similarly, the X region remains intact upon introduction of a substituent, and the additive term F_X cancels out in the quantities $\delta_R F^{\circ}$. To illustrate these relationships, consider the ionization of a carboxylic acid $R_0 COOH$. ΔF° for this reaction may be written as in equation 7.

$$\Delta F_{R_0 COOH}^{\circ} = F_{R_0 COO^-}^{\circ} + F_{H^+}^{\circ} - F_{R_0 COOH}^{\circ}$$

$$= F_{H^+}^{\circ} + F_{COO^-} - F_{COOH} + I_{R_0, COO^-} - I_{R_0, COOH} \tag{7}$$

The substituent effect on this reaction, if analyzed according to equation 7, will consist only of the interaction terms.

$$\delta_R \Delta F_{COOH}^{\circ} = \Delta F_{RCOOH}^{\circ} - \Delta F_{R_0 COOH}^{\circ}$$

$$= (I_{R,COO^-} - I_{R,COOH}) - (I_{R_0,COO^-} - I_{R_0,COOH}) \tag{8}$$

A typical extrathermodynamic relationship will involve substituent effects on the standard free energy changes or equilibrium constants of two related reactions. Thus the Hammett $\rho\sigma$ relationship (Chapter 7) is essentially a proportionality between $\delta_R \Delta F^{\circ}$ for a particular benzene side-chain reaction and $\delta_R \Delta F^{\circ}$ for the ionization of the corresponding benzoic acids. To analyze this relationship in our terminology, let R_0 represent phenyl and R the *meta*- or *para*-substituted phenyl group. Substituent effects on the ionization of benzoic acid are then given by equation 8. Similarly, substituent effects on the ionization of phenol are given by equation 9.

$$\delta_R \Delta F_{OH}^{\circ} = (F_{RO^-}^{\circ} - F_{ROH}^{\circ}) - (F_{R_0 O^-}^{\circ} - F_{R_0 OH}^{\circ})$$

$$= (I_{R,O^-} - I_{R,OH}) - (I_{R_0,O^-} - I_{R_0,OH}) \tag{9}$$

From equations 8 and 9 alone it is not possible to generate a linear relationship between $\delta_R \Delta F_{COOH}^{\circ}$ and $\delta_R \Delta F_{OH}^{\circ}$. The additional postulate needed is the assumption that the interaction terms are factorable. We will call this assumption the separability postulate.[11]

$$I_{R,X} = I_R \cdot I_X \tag{10}$$

Application of the separability postulate to equations 8 and 9 for benzoic acids and phenols leads to equations 11 and 12.

$$\delta_R \Delta F^\circ_{COOH} = (I_R - I_{R_0})(I_{COO^-} - I_{COOH}) \tag{11}$$

$$\delta_R \Delta F^\circ_{OH} = (I_R - I_{R_0})(I_{O^-} - I_{OH}) \tag{12}$$

A proportionality relationship follows immediately.

$$\delta_R \Delta F^\circ_{OH} = \frac{(I_{O^-} - I_{OH})}{(I_{COO^-} - I_{COOH})} \cdot \delta_R \Delta F^\circ_{COOH} \tag{13}$$

The fit to the linear relationship predicted in equation 13 is shown in Figure 6-4.

An important property of equations 11 and 12 is their symmetry, that is, the equal status of the interaction term for the reaction zone and that for the rest of the molecule. This is as it should be, because the designation of a given part of the molecule as the *reaction zone* is an arbitrary choice rather than a logical necessity. The operators δ_R and Δ, which we have earlier shown to be commutable, are also formally equivalent. The difference between them is that reactions like $RCOOH \rightarrow RCOO^-$ are readily carried out, whereas reactions like $RCOOH \rightarrow R_0COOH$ are not always convenient.

Implicit in the derivation of equation 13 are two assumptions. One of these is that structures R and X interact by only a single physical mechanism. By a *single interaction mechanism* we mean that the free energy of the interaction depends only on a *single pair* of independent variables, one characteristic of R and the other characteristic of X.[12] These might be such quantities as group dipole moment, charge, polarizability, or size.

[11] The separability postulate is a mathematical device of a type that is used frequently. Familiar examples are the factoring of wave functions and partition functions. However, we shall show later in this section that separability arises spontaneously from a less drastic postulate, namely, that the structural change is small.

[12] It would appear, in fact, that there is no useful definition of "single interaction mechanism" except in terms of independent variables. Consider, for example, a dipole-dipole interaction and a charge-dipole interaction; the two are physically distinct and are *independent* functions of any structural change. On the other hand, if two quantities are *necessarily* (as opposed to accidentally) proportional to one another, they actually represent only a single independent variable, and the interaction of this variable with some other independent variable corresponds to just one physically real phenomenon and not two. An example of two variables, one of which must be physically superfluous, would be an "inductive" charge displacement that had the property of always (and necessarily) being proportional to a "resonance" charge displacement. One interaction mechanism, the inductive effect, has recently been sharpened in concept and application by the work of R. P. Iczkowski and J. L. Margrave [*J. Am. Chem. Soc.*, **83**, 3547 (1961)] and W. F. Sager and C. D. Ritchie [*ibid.*, **83**, 3498 (1961)].

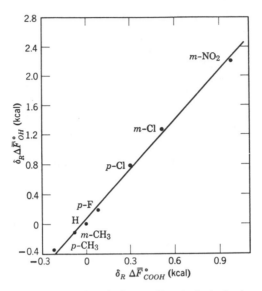

Figure 6-4. Substituent effects in the ionization of phenol (ordinate) and benzoic acid (abscissa) in water at 25°. The points for p-NO$_2$ and p-CN (not shown) deviate from the linear relationship, presumably because of the presence of a second interaction mechanism. Data from C. M. Judson and M. Kilpatrick, *J. Am. Chem. Soc.*, **71**, 3115 (1949); M. Kilpatrick and W. H. Mears, *ibid.*, **62**, 3047 (1940); D. T. Y. Chen and K. J. Laidler, *Trans. Faraday Soc.*, **58**, 480 (1962).

The second assumption implicit in the derivation of equation 13 concerns the scalar nature of the interacting variables. Even when the interaction proceeds by a single mechanism, we may hope to separate $I_{R,X}$ into independent factors only in special cases. To simplify the problem, we define $I_{R,X}$ so that its numerical value is equal to $I_{R,X} - I_{R_0,X_0}$, where R_0X_0 is a suitable reference compound. That is, I_{R_0,X_0} for the reference compound is arbitrarily assigned a value of zero.[13] We shall then be able to show that $I_{R,X}$ (or rather, $I_{R,X} - I_{R_0,X_0}$) can *always* be separated into independent factors if the variables are scalar quantities and if their changes are small. However, if the variables are vector or tensor quantities, $I_{R,X}$ can be separated only if there are suitable constraints on the vector

[13] It is always permissible to select the zero-point of free energy quantities so as to suit the convenience of the problem.

components or matrix elements. For example, in dipole-dipole interaction the angle between the dipole vectors must be reasonably constant, and the angle that either dipole makes with the radius vector joining it to the other dipole must be reasonably independent of the nature of the other dipole. Geometrical restrictions such as these are more likely to be satisfied by molecules whose structure is rigid than by highly flexible molecules. This is perhaps one of the reasons why precise extrathermodynamic relationships involving substituent effects are more numerous in the aromatic series than in the aliphatic series, although they are by no means uncommon in the latter.

If there is more than one mechanism of interaction between the parts of the molecule, these should be represented by separate interaction terms as in equation 14.

$$F_{RX}^{\circ} = F_R + F_X + I_{R,X} + I'_{R,X} + \cdots + \text{higher order}$$
$$\text{interaction terms} \quad (14)$$

In addition to the familiar terms of the type $I_{R,X}$, equation 14 contains interaction terms of higher order to take care of any tendency of the $I_{R,X}$ terms not to be additive. For example, a resonance effect might be sensitive to steric inhibition. If it is assumed that each interaction mechanism gives rise to an independent, factorable term, then an equation analogous to (11) can be written.

$$\delta_R \Delta F_{COOH}^{\circ} = (I_R - I_{R_0})(I_{COO^-} - I_{COOH})$$
$$+ (I_R' - I_{R_0}')(I_{COO^-}' - I_{COOH}') + \cdots \quad (15)$$

However, we now find that two $\delta_R \Delta F^{\circ}$ quantities will not, in general, be proportional. In Chapter 7 we will encounter examples of relationships involving several interaction mechanisms. In one of these a linear polar free energy relationship is obeyed only if a correction for the steric effect is first subtracted from each $\delta_R \Delta F^{\circ}$.

When a $\delta_R \Delta F^{\circ}$ quantity is dependent on two interaction mechanisms, it cannot, in general, be simply proportional to another $\delta_R \Delta F^{\circ}$ quantity, as we have seen. However, it can, in general, be represented as a linear combination of *two* such quantities, associated with two model processes A and B.

$$\delta_R \Delta F^{\circ} = a \, \delta_R \Delta F_A^{\circ} + b \, \delta_R \Delta F_B^{\circ} \quad (16)$$

Reaction A may involve only one of the interaction mechanisms and reaction B the other, or both reactions may involve both types of interaction. In the latter case, however, the relative importance of the two interaction mechanisms must be different in the two reactions taken as models. An example has been shown in equation 4 and Figure 6-3.

By analogy with equation 15, the quantities $\delta_R \Delta F^\circ$, $\delta_R \Delta F_A^\circ$, and $\delta_R \Delta F_B^\circ$ can be represented as in equations 17.

$$\delta_R \Delta F^\circ = (I_R - I_{R_0})\rho + (I'_R - I'_{R_0})\rho'$$
$$\delta_R \Delta F_A^\circ = (I_R - I_{R_0})\rho_A + (I'_R - I'_{R_0})\rho'_A \qquad (17)$$
$$\delta_R \Delta F_B^\circ = (I_R - I_{R_0})\rho_B + (I'_R - I'_{R_0})\rho'_B$$

Eliminating the quantities $(I_R - I_{R_0})$ and $(I_R' - I'_{R_0})$ from this set of equations, we obtain equation 18 for the desired linear combination.

$$\delta_R \Delta F^\circ = \frac{\rho\rho'_B - \rho'\rho_B}{\rho_A\rho'_B - \rho'_A\rho_B} \delta_R \Delta F_A^\circ + \frac{\rho_A\rho' - \rho'_A\rho}{\rho_A\rho'_B - \rho'_A\rho_B} \delta_R \Delta F_B^\circ \qquad (18)$$

On comparing equation 18 with equation 16 we find that the coefficients a and b are not necessarily positive.

A special case of the effect of two simultaneously operating interaction mechanisms, which is apparently quite common, results in dispersion of extrathermodynamic relationships into pairs of parallel lines. For example, in the hydrolysis of acetic anhydride catalyzed by nitrogen heterocycles, the plot of $\log k$ versus $\log K_B$, the base dissociation constant of the amine, consists of two parallel lines as shown in Figure 6-5. The upper line consists of points representing sterically unhindered amines. The lower line consists of points for sterically hindered amines.[14] A similar dispersion into parallel lines for different classes of amines, this time primary amines

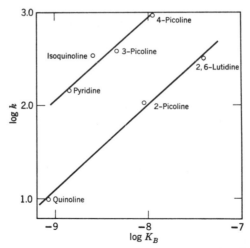

Figure 6-5. Amine-catalyzed hydrolysis of acetic anhydride.

[14] V. Gold and E. G. Jefferson, *J. Chem. Soc.*, p. 1409 (1953).

and secondary amines, is found for association constants of the amines with Ag^+ as compared with K_B.[15]

To interpret the dispersion, let us suppose that $l'_{R,X}$, a second interaction mechanism, is a discontinuous function that always has one or the other of two values. In δ_R operations, the contribution of this interaction mechanism either vanishes or is constant. If the contribution is zero, we have in effect a single interaction mechanism and a single line. If it is constant, we simply have a second line parallel to the first, since the first interaction mechanism is unaffected by the changing value of the second. There will usually be some obvious qualitative difference between the two classes of substituent corresponding to the two values of the second interaction mechanism. The most plausible explanation of the sudden discontinuous change in interaction mechanism is that it is occasioned by a minor change in reaction mechanism. Perhaps a certain degree of crowding causes the transition state to abandon the conformation electronically most favored and to take up a second-best conformation.

Experimentally it is known that extrathermodynamic relationships break down if too large a structural change is made. There are two possible reasons for this. The individual interaction terms may no longer be separable or additional interaction terms may become significant. We shall now demonstrate that the separability postulate is likely to apply only if the changes made in the independent variables are small. For convenience we shall assume a single interaction mechanism, and use the symbols R and X to denote the independent variables which are responsible for the interaction between the groups R and X. Furthermore, we shall define the interaction term $l_{R,X}$ in such a way that l_{R_0,X_0} (for the standard reference compound) is equal to zero.[13]

We may now write two alternative expressions for F°_{RX}: Equation 19, which is obtained directly from (6); and equation 20, which is a Taylor's series expansion of F°_{RX} about $F^\circ_{R_0 X_0}$.

$$
\begin{aligned}
F^\circ_{RX} &= F^\circ_{R_0 X_0} + (F_R - F_{R_0}) + (F_X - F_{X_0}) + l(_{R,X} - l_{R_0,X_0}) \\
&= F^\circ_{R_0 X_0} + (F_R - F_{R_0}) + (F_X - F_{X_0}) + l_{R,X}
\end{aligned}
\tag{19}
$$

$$
\begin{aligned}
F^\circ_{RX} &= F^\circ_{R_0 X_0} + \left(\frac{\partial F^\circ}{\partial R}\right)_{R_0 X_0} (R - R_0) + \left(\frac{\partial F^\circ}{\partial X}\right)_{R_0 X_0} (X - X_0) \\
&+ \left(\frac{\partial^2 F^\circ}{\partial R^2}\right)_{R_0 X_0} \frac{(R - R_0)^2}{2} + \left(\frac{\partial^2 F^\circ}{\partial X^2}\right)_{R_0 X_0} \frac{(X - X_0)^2}{2} \\
&+ \left(\frac{\partial^2 F^\circ}{\partial R \partial X}\right)_{R_0 X_0} (R - R_0)(X - X_0) + \cdots
\end{aligned}
\tag{20}
$$

[15] R. J. Bruehlman and F. H. Verhoek, *J. Am. Chem. Soc.*, **70**, 1401 (1948).

Strictly speaking, equation 20 should also contain additional series of terms involving independent variables other than R and X. Although such terms make significant contributions to the standard free energy, they have not been written explicitly, for they are irrelevant to the assumed single interaction mechanism and will vanish during the $\delta_R \Delta$ operation.

The terms in equation 20 may be sorted according to their independent variables. Those depending only on the change in R are, in sum, equal to $(F_R - F_{R_0})$; those depending only on the change in X are, in sum, equal to $(F_X - F_{X_0})$; and those depending on both changes are, in sum, equal to $I_{R,X}$. Therefore $I_{R,X}$ is given by equation 21.

$$I_{R,X} = \left(\frac{\partial^2 F^\circ}{\partial R \, \partial X}\right)_{R_0 X_0} (R - R_0)(X - X_0) + \cdots \tag{21}$$

Since the partial derivative in equation 21 is a constant characteristic only of $R_0 X_0$, the interaction term is in first approximation separable into independent factors, one characteristic of the change in R and the other characteristic of the change in X. Equation 21 is of course applicable without its higher order terms only if the changes in R and X are small. Hence, although the separability postulate has now been proven, the proof is rigorous only if the structural changes are limited to small ones.

As the changes in R and X become greater, the contribution of higher order terms may become significant. In the next higher approximation, the Taylor's series expansion is carried to the third derivatives, and $I_{R,X}$ is given by equation 22.

$$I_{R,X} = (R - R_0)(X - X_0)\left[\left(\frac{\partial^2 F^\circ}{\partial R \, \partial X}\right)_{R_0 X_0} + \left(\frac{\partial^3 F^\circ}{\partial R^2 \partial X}\right)_{R_0 X_0} \frac{(R - R_0)}{2}\right.$$
$$\left. + \left(\frac{\partial^3 F^\circ}{\partial R \, \partial X^2}\right)_{R_0 X_0} \frac{(X - X_0)}{2} + \cdots\right] \tag{22}$$

The coefficient of $(R - R_0)(X - X_0)$ is now a function rather than a constant and cannot be separated into independent factors. Hence we must expect that the separability of $I_{R,X}$ will cease when both of the terms containing the third derivatives are appreciable.

A FORMAL EXTRATHERMODYNAMIC TREATMENT OF SOLVATION EFFECTS

The standard partial molar free energy of a substance in solution is modified by solvation. The first approximation is that the effect of solvation on the intramolecular part of the free energy is negligible and that the effect due to interaction of the solute molecule with the surrounding medium can be expressed as an additive function of independent terms ascribable to the solvation of part-structures of the solute molecule. To this approximation the free energy is expressed by equation 23 in which the term $I_{R,X}$ is the same as before, and $I_{R,M}$ and $I_{X,M}$ are contributions from the *independent* solvation of R and X.

$$F^\circ_{RX} = F_R + F_X + I_{R,X} + I_{R,M} + I_{X,M} \tag{23}$$

In this equation all the interaction terms employ the same symbol, I, to suggest that they are analogous quantities. In the second approximation, equation 24, the failure of the interactions to be truly independent is acknowledged and represented by a term using the symbol $II_{M,R,X}$.

$$F^\circ_{RX} = F_R + F_X + I_{R,X} + I_{R,M} + I_{X,M} + II_{M,R,X} \tag{24}$$

The physical significance of the terms in equation 24 can be seen more clearly if we temporarily adopt an alternative method, equation 25, for representing the standard free energy.

$$F^\circ_{RX,\, in\, M} = F_{R,\, in\, M} + F_{X,\, in\, M} + I_{R,X,\, in\, M} \tag{25}$$

Equation 25 is analogous to (6) except that the terms on the right are functions of the medium rather than gas-phase quantities. On comparing (25) with (24), we see that

$$\begin{aligned} F_{R,\, in\, M} &= F_R + I_{R,M} \\ F_{X,\, in\, M} &= F_X + I_{X,M} \\ I_{R,X,\, in\, M} &= I_{R,X} + II_{R,X,M} \end{aligned} \tag{26}$$

In other words, each of the last three terms in equation 24 expresses the effect of solvation on the respective term without the M subscript.

The First Approximation

A consequence of expressing the free energy in solution only to the first approximation (equation 23) is that $\delta_M F^\circ_{RX}$ is an additive function of independent terms assignable to part-structures.

$$\delta_M F^\circ_{RX} = \delta_M I_{R,M} + \delta_M I_{X,M} \tag{27}$$

The first approximation is often adequate. For example, the free energy of hydrocarbon mixtures,[16] or of an alcohol in carbon tetrachloride,[17] can be fitted with fair accuracy if one assumes that the intermolecular interactions result from independent contributions due to suitably chosen part-structures. The first approximation is also made implicitly whenever the values of $\delta_M F^\circ_{RX}$ for a series of compounds are correlated linearly with an

[16] O. Redlich, E. L. Derr, and G. J. Pierotti, *J. Am. Chem. Soc.*, **81**, 2283, 2285 (1959).
[17] J. A. Barker, I. Brown, and F. Smith, *Disc. Faraday Soc.*, **15**, 141 (1953).

additive molar property of these compounds. Thus for the change of medium shown in equation 28, the standard partial molar free energy

$$RX \text{ (pure liquid)} \rightleftharpoons RX \text{ (solute in the solvent } M) \qquad (28)$$

change is given by the empirical equation 29.[18,19] In this equation, $\bar{P}$ is the

$$\delta_M \bar{F}_{RX}^{\circ} = a_{M,X} + k_M \bar{P}_{RX} \qquad (29)$$

parachor (an additive-constitutive function),[20] k_M is a parameter characteristic of the solvent medium, and $a_{M,X}$ is a parameter characteristic both of the solvent medium and of the polar part of the solute molecule. Thus $a_{M,X}$ is zero for saturated hydrocarbons but assumes nonzero values for compounds with polar functional groups. Some numerical values for the parameters are listed in Table 6-5. The fit of the equation is usually within 0.2 kcal/mole.

Table 6-5. Parameters for Predicting the Standard Partial Molar Free Energy Change According to Equation 29[a]

SOLVENT	TEMP.	k_M (kcal/parachor unit)	CLASS OF SOLUTES	$a_{M,X}$ (kcal/mole)
Water	25°	0.0178	Hydrocarbons and their halogen derivatives	0.0
			Ethers	−2.1
			Ketones	−2.6
			Aldehydes	−1.6
			Aliphatic esters[b]	−2.3
			Formate esters	−1.8
			Aromatic esters	−2.1
			Nitriles	−2.3
			Nitroalkanes	−1.2
			Nitroaromatics	−0.9
SO_2	−29°	0.0033_5	Alkanes	0.0
			Aromatic hydrocarbons	−1.5
NH_3	−33.3°	0.0120	Alkanes	0.0
	25°	0.0060	Alkanes	0.0

[a] From N. C. Deno and H. E. Berkheimer, *J. Chem. Eng. Data*, **5**, No. 1, 1–5 (1960).

[b] Except formate esters.

[18] The correlation seems to be best if one treats RX in the pure liquid state as if it were a solute, that is, $\bar{F}_{RX}^{\circ} = F^{\circ}(\text{liquid } RX) - RT \ln c_{RX}(\text{in the pure liquid})$.

[19] N. C. Deno and H. E. Berkheimer, *J. Chem. Eng. Data*, **5**, No. 1, 1–5 (1960).

[20] O. R. Quayle, *Chem. Revs.*, **53**, 484 (1953).

Extensive and highly accurate data are available for the solubility of α-amino acids and their derivatives in water and in ethanol.[21,22] For compounds of the type

$$NH_3^+ \cdot CHR \cdot CO_2^-, \quad HO \cdot CHR \cdot CO_2H,$$

$$H_2NCONH \cdot CHR \cdot CO_2H \text{ and } HCONH \cdot CHR \cdot CO_2H,$$

where R is an alkyl group, each additional CH_2-group in R has been found to increase the solubility in ethanol relative to that in water by a constant factor of 3.2 ± 0.3 at 25°.[22] Since $\delta_M \bar{F}° \approx RT \ln$ (solubility ratio), each CH_2-group therefore makes an additive contribution of $RT \ln 3.2$ (or 0.69 kcal) to the standard free energy change. This additivity rule breaks down, however, when R contains a polar substituent, the effect of an additional CH_2-group near the polar substituent being much smaller.[22]

Application of the first approximation to reactions gives equation 30.

$$RX + Y \rightleftharpoons X + RY$$
$$\delta_M \Delta \bar{F}° = \delta_M(\bar{F}_X° + \bar{F}_{RY}° - \bar{F}_{RX}° - \bar{F}_Y°)$$
$$= \delta_M(I_{Y,M} - I_{X,M}) - \delta_M(\bar{F}_Y° - \bar{F}_X°) \tag{30}$$

If the interaction of Y and X with solvent is the same whether or not the group is bonded to R, then $\delta_M I_{Y,M} = \delta_M \bar{F}_Y°$, $\delta_M I_{X,M} = \delta_M \bar{F}_X°$, and $\delta_M \Delta \bar{F}° = 0$. In other words, there is no effect of solvent on the equilibrium constant. If the attachment or nonattachment of the group R is important to the solvation, then $\delta_M \Delta \bar{F}°$ is not equal to zero, but it is at least independent of the *nature* of R. That is, $\delta_M \log K$ (or $\delta_M \log k$) is the same for all substituents. In such a case $\delta_R \Delta \bar{F}°$ will be independent of the solvent; a plot of $\log K$ for a series of compounds in one solvent will be a linear function of $\log K$ for the same series in a second solvent *and will have unit slope.*

The solvolysis of substituted 2-phenyl-2-propyl chlorides[23] provides a nearly exact example of the expected equality of $\delta_R \Delta \bar{F}^{\ddagger}$ in various solvents. The slopes of plots of $\delta_R \Delta F^{\ddagger}$ versus the same quantity in 90% acetone are 1.063 for methanol, 1.029 for ethanol, and 0.975 for 2-propanol.

A somewhat more complicated example is the dissociation of pyridine-silver ion complexes as the solvent is changed from acetonitrile to ethanol.[24] For complexes between silver ion and substituted pyridines we have the relationship (31) in which the slope is almost exactly unity, in accord with

[21] E. J. Cohn, T. L. McMeekin, J. T. Edsall, and J. H. Weare, *J. Am. Chem. Soc.*, **56**, 2270 (1934).

[22] E. J. Cohn and J. T. Edsall, *Proteins, Amino Acids and Peptides*, Reinhold Publishing Corp., 1943, ch. 9.

[23] Y. Okamoto, T. Inukai, and H. C. Brown, *J. Am. Chem. Soc.*, **80**, 4972 (1958).

[24] W. J. Peard and R. T. Pflaum, *J. Am. Chem. Soc.*, **80**, 1593 (1958).

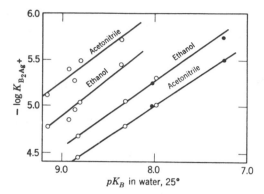

Figure 6-6. Stability constants of silver complexes with quinolines (upper pair of lines) and with pyridines (lower pair of lines). The solid points are for 2-picoline and 2,6-lutidine.

the first approximation. However, for complexes between silver ion and substituted quinolines we have the relationship (32), in which the slope is significantly different from unity.

$$(\log K_{B_2Ag^+})_{\text{acetonitrile}} = 0.98(\log K_{B_2Ag^+})_{\text{ethanol}} + 0.12 \tag{31}$$

$$(\log K_{B_2Ag^+})_{\text{acetonitrile}} = 1.3(\log K_{B_2Ag^+})_{\text{ethanol}} + 1.2 \tag{32}$$

The structural change from pyridines to quinolines also causes dispersion in the plot of $\log K_{B_2Ag^+}$ versus pK_B, as shown in Figure 6-6. It is clear from the figure that this small structural change is not only accompanied by a change in the slope of equations 31 and 32, but it even reverses the *sign* of the quantities $\delta_M \log K_{B_2Ag^+}$. The example is typical in showing that there are few reactions for which the first approximation is adequate in the treatment of the medium effect.

Factorability of Interaction Terms

Before proceeding to the second approximation, we wish to examine the factorability of the interaction terms in equations 23 and 24. For simplicity we assume a single interaction mechanism and denote the independent variables by R, X, and M. We shall find it instructive to analyze the effect of a series of δ-operations, first in terms of equation 24 and then by means of a Taylor's series. The initial system consists of the compound R_0X_0 in the medium M_0. The operator δ_X changes the X-variable of this system from X_0 to X but has no effect on the variables R and M. Similarly δ_M changes only the M-variable, from M_0 to M; and δ_R changes only the R-variable, from R_0 to R.

Beginning with equation 24, we obtain equations 33.

$$\delta_X F^\circ = \mathsf{F}_X - \mathsf{F}_{X_0} + (\mathsf{I}_{R_0,X} - \mathsf{I}_{R_0,X_0}) + (\mathsf{I}_{M_0,X} - \mathsf{I}_{M_0,X_0})$$
$$+ (\mathsf{II}_{R_0,X,M_0} - \mathsf{II}_{R_0,X_0,M_0})$$
$$= \delta_X \mathsf{F}_{X_0} + \delta_X \mathsf{I}_{R_0,X_0} + \delta_X \mathsf{I}_{M_0,X_0} + \delta_X \mathsf{II}_{R_0,X_0,M_0} \qquad (33a)$$

$$\delta_M \delta_X F^\circ = (\delta_X \mathsf{I}_{M,X_0} - \delta_X \mathsf{I}_{M_0,X_0}) + (\delta_X \mathsf{II}_{R_0,X_0,M} - \delta_X \mathsf{II}_{R_0,X_0,M_0})$$
$$= \delta_M \delta_X \mathsf{I}_{M_0,X_0} + \delta_M \delta_X \mathsf{II}_{R_0,X_0,M_0} \qquad (33b)$$

$$\delta_R \delta_M \delta_X F^\circ = \delta_M \delta_X \mathsf{II}_{R,X_0,M_0} - \delta_M \delta_X \mathsf{II}_{R_0,X_0,M_0}$$
$$= \delta_R \delta_M \delta_X \mathsf{II}_{R_0,X_0,M_0} \qquad (33c)$$

It is seen that each successive δ-operation results in a simpler expression than the preceding one as more and more interaction terms cancel out. Thus the initial δ_X-operation eliminates all terms without an X-subscript. The double operation $\delta_M \delta_X$ leaves only those terms that have both an M and an X subscript. And the final triple operation leaves only the $\mathsf{II}_{M,R,X}$ terms.

Moreover, if the changes in the variables are small, each successive δ-operation also reduces the magnitude of the resultant free-energy quantity. This is because each δ-operation is then the difference of two nearly equal numbers. Thus $|\delta_X F^\circ| \gg |\delta_M \delta_X F^\circ| \gg |\delta_R \delta_M \delta_X F^\circ|$. Since $\delta_R \delta_M \delta_X F^\circ$ involves only the $\mathsf{II}_{M,R,X}$ quantities, which vanish in the first approximation, we may describe the first approximation alternatively by saying that all $\delta_R \delta_M \delta_X F^\circ$ quantities are negligibly small.

The effect of the same series of δ-operations may be expressed in the form of a Taylor's series, as in equations 34. In these equations, all partial derivatives are evaluated at the point R_0, X_0, M_0.

$$\delta_X F^\circ = \frac{\partial F^\circ}{\partial X}(X - X_0) + \frac{1}{2}\frac{\partial^2 F^\circ}{\partial X^2}(X - X_0)^2 + \cdots \qquad (34a)$$

$$\delta_M \delta_X F^\circ = \left(\frac{\partial}{\partial M}\delta_X F^\circ\right)(M - M_0) + \frac{1}{2}\left(\frac{\partial^2}{\partial M^2}\delta_X F^\circ\right)(M - M_0)^2 + \cdots$$

$$= \frac{\partial^2 F^\circ}{\partial M\,\partial X}(M - M_0)(X - X_0) + \frac{1}{2}\frac{\partial^3 F^\circ}{\partial M\,\partial X^2}(M - M_0)(X - X_0)^2$$

$$+ \frac{1}{2}\frac{\partial^3 F^\circ}{\partial M^2\,\partial X}(M - M_0)^2(X - X_0) + \cdots \qquad (34b)$$

$$\delta_R \delta_M \delta_X F^\circ = \left(\frac{\partial}{\partial R}\delta_M \delta_X F^\circ\right)(R - R_0) + \cdots$$

$$= \frac{\partial^3 F^\circ}{\partial R\,\partial M\,\partial X}(R - R_0)(M - M_0)(X - X_0) + \cdots \qquad (34c)$$

In an actual application, the series expansions (34) are terminated at some suitable partial derivative, depending on the accuracy required. However, in order to achieve a consistent degree of accuracy, it is necessary that the highest terms retained in the several expansions be of the same order. Thus if the series (34a) is carried only to the term involving the second derivative $\partial^2 F^\circ/\partial X^2$, then the series (34b) should be represented by the single term proportional to $\partial^2 F^\circ/\partial X \, \partial M$, and the series (34c) is zero, since it starts off with a third derivative. The reader will recognize that an approximation which terminates the Taylor's series at the second derivatives is identical with the first approximation represented in equation 23. Hence in the first approximation we may write equations 35.

$$\delta_R \delta_X |_{R_0, X_0} = \left(\frac{\partial^2 F^\circ}{\partial R \, \partial M}\right)_{R_0, X_0, M_0} (R - R_0)(X - X_0) \tag{35a}$$

$$\delta_R \delta_M |_{R_0, M_0} = \left(\frac{\partial^2 F^\circ}{\partial R \, \partial M}\right)_{R_0, X_0, M_0} (R - R_0)(M - M_0) \tag{35b}$$

$$\delta_X \delta_M |_{X_0, M_0} = \left(\frac{\partial^2 F^\circ}{\partial X \, \partial M}\right)_{R_0, X_0, M_0} (X - X_0)(M - M_0) \tag{35c}$$

Since in equations 35 the results of the double δ-operations are separable into independent factors, the interaction coefficients themselves may also be represented as separable quantities. For example, $I_{R,M}$ may in this approximation be written as a product $I_R \cdot I_M$ (plus an arbitrary constant[13]), where I_R is proportional to R and I_M is proportional to M.

In the second approximation represented by equation 24, $\delta_R \delta_M \delta_X F^\circ$ does not vanish, and the Taylor's series expansions (34) must be carried at least to the third derivatives. If the series is carried exactly to the third derivatives, comparison of (33c) with (34c) leads directly to equation 36.

$$\delta_R \delta_M \delta_X II |_{R_0, X_0, M_0} = \frac{\partial^3 F^\circ}{\partial M \, \partial R \, \partial X}(M - M_0)(R - R_0)(X - X_0) \tag{36}$$

The simplest form of $II_{R,X,M}$ that will satisfy equation 36 is one for which $II_{R,X,M}$ is factorable, as shown in equation 37.

$$II_{R,X,M} = II_R \cdot II_X \cdot II_M \tag{37}$$

In (37), each of the independent factors is proportional to its respective interaction variable. Although equation 24 in no way limits the functional form that $II_{R,X,M}$ may take, equation 37 will be the highest approximation that we shall consider.

Since the lowest approximation to the Taylor's series in which $II_{R,X,M}$ does not vanish involves a third derivative, the use of equation 24 implies that all of equations 34 must be carried at least to the third derivatives. It then follows from equations 33b, 34b and 37 that the quantity $\delta_M\,\delta_X|_{X_0,M_0}$ is no longer simply proportional to $(X - X_0)(M - M_0)$ but also contains at least two additional terms, one proportional to $(X - X_0)^2(M - M_0)$ and the other proportional to $(X - X_0)(M - M_0)^2$. Hence the interaction terms $I_{X,M}$ can no longer be represented as factorable products $I_X \cdot I_M$. Since an analogous argument can be made for $I_{R,X}$ and $I_{R,M}$, *we may characterize the second approximation by the statement that the first-order interaction terms are not factorable, but that the second-order interaction terms may be represented as products of three independent factors.*

The Second Approximation

When the second approximation, equation 24, is applied to an actual problem, $\delta_M\,\Delta F^\circ$ is a function of the part-structure R. For simplicity let us consider a unimolecular process, which might be either an isomerization equilibrium or a first-order rate process.

$$RX_0 \rightleftharpoons RX$$
$$R_0X_0 \rightleftharpoons R_0X$$

The process is formulated so that the operator δ_X of the preceding section becomes the reaction operator Δ. The dependence of the medium effect on R enters through the second-order interaction term, as in equations 38.

$$\delta_M\,\Delta \bar{F}_R^\circ = \delta_M(I_{X,M} - I_{X_0,M}) + \delta_M(II_{M,R,X} - II_{M,R,X_0})$$
$$\delta_M\,\Delta \bar{F}_{R_0}^\circ = \delta_M(I_{X,M} - I_{X_0,M}) + \delta_M(II_{M,R_0,X} - II_{M,R_0,X_0})$$
(38)

The conclusion that in equation 38 only the second-order interaction terms are separable leads to rather complicated extrathermodynamic relationships, in which there is a great deal of cancellation of chemical effects. These are most simply treated as triple δ-operations, as in equation 39.

$$\delta_R\,\delta_M\,\Delta \bar{F}^\circ = \left(\frac{\partial^3 \bar{F}^\circ}{\partial M\,\partial R\,\partial X}\right)_{R_0X_0,M_0}(M - M_0)(R - R_0)(X - X_0) \quad (39)$$

Equation 39 also applies to bimolecular processes,

$$X + RX_0 \rightleftharpoons RX + X_0$$

since the standard free energies of X and X_0 cancel out in the δ_R-operation.

Equation 39 makes it possible to generate extrathermodynamic relationships of three types, depending on which of the three quantities: structure, medium, or nature of the reaction, is held constant. One of these is illustrated by the acid dissociation of three carboxylic acids in a long series of ethanol-water mixtures.[25] Let the standard acid be acetic and the other

[25] E. Grunwald and B. J. Berkowitz, *J. Am. Chem. Soc.*, **73**, 4939 (1951).

two benzoic and formic. The standard solvent is water. The three acids are represented by the symbols CH_3, Φ, and H. The result of applying the second approximation to this system is equation 40.

$$\delta_M[pK_A(\Phi) - pK_A(CH_3)] \propto \delta_M[pK_A(H) - pK_A(CH_3)] \tag{40}$$

Equation 40 may be written without the δ_M operator as (40a).

$$\{[pK_A(\Phi) - pK_A(CH_3)]_{\text{in } M} - [pK_A(\Phi) - pK_A(CH_3)]_{\text{in } H_2O}\}$$
$$\propto \{[pK_A(H) - pK_A(CH_3)]_{\text{in } M} - [pK_A(H) - pK_A(CH_3)]_{\text{in } H_2O}\} \tag{40a}$$

Detailed examples of the use of the second approximation for medium effects will be found in Chapter 8.

It sometimes happens that the first-order interaction terms are separable, as predicted by the first approximation, even though the second-order interaction term $\text{II}_{M,R,X}$ does not actually vanish. To derive the form of the resulting extrathermodynamic relationships, consider again an isomerization equilibrium or first-order rate process. If we assume a single solvent-solute interaction mechanism[26] and separability of the interaction terms, we may write

$$\text{I}_{X,M} = \text{I}_X \cdot \text{I}_M, \text{ etc.}$$
$$\text{II}_{M,R,X} = \text{II}_M \cdot \text{II}_R \cdot \text{II}_X, \text{ etc.}$$

Hence equations 38 may be cast into the form (41).

$$\delta_M \Delta \bar{F}_R^\circ = \delta_M \text{I}_M(\text{I}_X - \text{I}_{X_0}) + \delta_M \text{II}_M[\text{II}_R(\text{II}_X - \text{II}_{X_0})]$$
$$\delta_M \Delta \bar{F}_{R_0}^\circ = \delta_M \text{I}_M(\text{I}_X - \text{I}_{X_0}) + \delta_M \text{II}_M[\text{II}_{R_0}(\text{II}_X - \text{II}_{X_0})] \tag{41}$$

Since we are assuming a single solvent-solute interaction mechanism, there is only a single independent variable characteristic of the solvent. For small changes in this variable, the factors $\delta_M \text{I}_M$ and $\delta_M \text{II}_M$ are both proportional to the change in this variable and hence to each other. We will therefore replace equations 41 by 42, where λ is the proportionality constant relating $\delta_M \text{I}_M$ to $\delta_M \text{II}_M$.

$$\delta_M \Delta \bar{F}_R^\circ = \delta_M \text{II}_M[\lambda(\text{I}_X - \text{I}_{X_0}) + \text{II}_R(\text{II}_X - \text{II}_{X_0})]$$
$$\delta_M \Delta \bar{F}_{R_0}^\circ = \delta_M \text{II}_M[\lambda(\text{I}_X - \text{I}_{X_0}) + \text{II}_{R_0}(\text{II}_X - \text{II}_{X_0})] \tag{42}$$

Equations 42 lead to the extrathermodynamic relationship that $\delta_M \Delta \bar{F}_R^\circ$ is proportional to $\delta_M \Delta \bar{F}_{R_0}^\circ$

$$\delta_M \Delta \bar{F}_R^\circ = \left[\frac{\lambda(\text{I}_X - \text{I}_{X_0}) + \text{II}_R(\text{II}_X - \text{II}_{X_0})}{\lambda(\text{I}_X - \text{I}_{X_0}) + \text{II}_{R_0}(\text{II}_X - \text{II}_{X_0})}\right] \cdot \delta_M \Delta \bar{F}_{R_0}^\circ \tag{43}$$

[26] It is not necessary to assume that the interaction of R with X involves only a single mechanism. Thus $\text{II}_{M,R,X}$ might more generally be written in the form $\text{II}_M(\text{II}_R\text{II}_X + \text{II}'_R\text{II}'_X + \cdots)$, where the products II_RII_X, $\text{II}'_R\text{II}'_X$, ... result from the various interaction mechanisms between R and X.

Thus in this approximation, values of $\log K$ (or $\log k$) for RX_0 in a series of solvents vary linearly with those for the reference compound R_0X_0. However, *in contrast to the first approximation, the slope is now no longer unity.*

THE TEMPERATURE DEPENDENCE OF EXTRATHERMO-DYNAMIC RELATIONSHIPS

If a precise linear free energy relationship exists over a range of temperatures, the existence of relationships constraining the enthalpy and entropy is implied. Let us suppose that the substituent effects on the equilibrium constants for two reactions are given accurately by equations 44 and 45.

$$\text{At temperature } T_1: \quad \delta_R \log K = \lambda_1 \, \delta_R \log K' \tag{44}$$

$$\text{At temperature } T_2: \quad \delta_R \log K = \lambda_2 \, \delta_R \log K' \tag{45}$$

Moreover, each value of $\log K$ is given as a function of temperature by an equation of the form (46).

$$2.303 \log K = \frac{\Delta S^\circ}{R} - \frac{\Delta H^\circ}{RT} \tag{46}$$

It is usually safe to assume that the quantities ΔH° and ΔS° are independent of the temperature. Thus we can transform equations 44 and 45 into a form containing the four variables $\delta_R \Delta H^\circ$, $\delta_R \Delta S^\circ$, $\delta_R \Delta H^{\circ\prime}$, and $\delta_R \Delta S^{\circ\prime}$. In view of the constraints implied by equations 44 and 45, only two of these variables can be independent. Let the independent variables be $\delta_R \Delta H^\circ$ and $\delta_R \Delta H^{\circ\prime}$. To find the expressions for $\delta_R \Delta S^\circ$ and $\delta_R \Delta S^{\circ\prime}$, it is instructive to consider equation 47, which is derived from (44) through (46).

$$\frac{T_1 \delta_R \Delta S^\circ - \delta_R \Delta H^\circ}{T_2 \delta_R \Delta S^\circ - \delta_R \Delta H^\circ} = \frac{\lambda_1}{\lambda_2} \cdot \frac{T_1 \delta_R \Delta S^{\circ\prime} - \delta_R \Delta H^{\circ\prime}}{T_2 \delta_R \Delta S^{\circ\prime} - \delta_R \Delta H^{\circ\prime}} \tag{47}$$

Since $\delta_R \Delta H^\circ$ and $\delta_R \Delta H^{\circ\prime}$ are *independent* variables, they may assume any values whatsoever. Equation 47 can therefore be generally true only if the relationship of $\delta_R \Delta S^\circ$ to $\delta_R \Delta H^\circ$ and of $\delta_R \Delta S^{\circ\prime}$ to $\delta_R \Delta H^{\circ\prime}$ is such that the enthalpy quantities cancel out on both sides of the equation. The reader will convince himself that this happens if, and only if, the proportionality relationships (48) and (49) apply.

$$\delta_R \Delta H^\circ = \beta \delta_R \Delta S^\circ \tag{48}$$

$$\delta_R \Delta H^{\circ\prime} = \beta' \delta_R \Delta S^{\circ\prime} \tag{49}$$

An example of the validity of equation 48 has been shown in Figure 6-2.

There are two special cases, corresponding to extreme values of β. When $\beta = 0$, the free energy changes are entirely entropic; when $1/\beta = 0$, the free energy changes are entirely enthalpic.

Solving equations 44, 46, 48 and 49 for λ_1, we obtain (50).

$$\lambda_1 = \frac{[(T_1/\beta) - 1]}{[(T_1/\beta') - 1]} \cdot \frac{\delta_R \Delta H^\circ}{\delta_R \Delta H^{\circ\prime}} \tag{50}$$

It is immediately obvious from (50) that $\delta_R \Delta H^\circ$ is proportional to $\delta_R \Delta H^{\circ\prime}$. In other words, the existence of a precise linear free energy relationship at two temperatures implies the existence of a linear enthalpy relationship. By virtue of (48) and (49), a linear entropy relationship can also be shown to exist.

Analogous considerations apply in the case of rate constants.[27] If equation 51 gives a precise fit over a range of temperatures, and if each rate constant can be fitted to an Arrhenius equation, the constraints on the parameters are given in equations 52, where b and b' are proportionality constants. The expression for λ is given in equation 53.

$$\delta_R \log k = \lambda \, \delta_R \log k' \tag{51}$$

$$\log k = \log A - \frac{E}{2.303 \, RT}$$

$$\left.\begin{array}{l} \delta_R E = b \, \delta_R \log A \\ \delta_R E' = b' \, \delta_R \log A' \end{array}\right\} \tag{52}$$

$$\lambda = \frac{(2.303 \, RT/b) - 1}{(2.303 \, RT/b') - 1} \cdot \frac{\delta_R E}{\delta_R E'} \tag{53}$$

If the rate constants are treated by transition-state theory according to equation 18 of Chapter 4, and if $\Delta H^\ddagger$, $\Delta S^\ddagger$, and κ are constant, the constraints on the parameters are given in equations 54 and the expression for λ in equation 55.

$$\left.\begin{array}{l} \delta_R \Delta H^\ddagger = \beta \delta_R \Delta S^\ddagger \\ \delta_R \Delta H^{\ddagger\prime} = \beta' \delta_R \Delta S^{\ddagger\prime} \end{array}\right\} \tag{54}$$

$$\lambda = \frac{(T/\beta) - 1}{(T/\beta') - 1} \cdot \frac{\delta_R \Delta H^\ddagger}{\delta_R \Delta H^{\ddagger\prime}} \tag{55}$$

RELATIONSHIPS BETWEEN RATE AND EQUILIBRIUM CONSTANTS

One of the oldest extrathermodynamic ideas in chemistry is that reaction rates should tend to parallel equilibrium constants. It is well known that

[27] (a) J. E. Leffler, *J. Org. Chem.*, **20**, 1202 (1955); *J. Chem. Phys.*, **23**, 2199 (1955). (b) R. J. Cvetanović, *J. Chem. Phys.*, **30**, 19 (1959).

this idea is incorrect as a general proposition; yet it would be a mistake to reject it altogether. Often when a series of reactions involves only a single reaction mechanism, there is an unmistakable trend for rate constants to increase monotonically with equilibrium constants, and in some cases this trend even takes the form of an accurate linear relationship between the quantities $\Delta F^{\ddagger}$ and ΔF°.

In previous sections we have seen that free energy quantities are often conveniently described as linear combinations of the free energies associated with two model processes. The model processes considered so far have been chemical reactions, but there is no reason why they could not be changes of the reaction medium or of a substituent. Since a transition state has considerable resemblance to the reagents and products both in composition and structure, it is reasonable to suppose that any *changes* in its free energy can often be represented as a linear combination of the corresponding changes in the free energies of the reagents and products.

$$\delta F^{\ddagger} = a\, \delta \bar{F}_P{}^{\circ} + b\, \delta \bar{F}_R{}^{\circ} \tag{56}$$

The operator δ in equation 56 is either δ_M or δ_R.

We can impose a further constraint on $\delta F^{\ddagger}$ if we are willing to go beyond the formal approach characteristic of extrathermodynamics. Since the reaction coordinate of the transition state lies between that of the reagents and the products, we may safely assume that at least some other properties will be intermediate as well. Although the standard free energy of the transition state is of course at a maximum, it is not unlikely that any *changes* in its value due to medium or substituent effects will be *intermediate* to the corresponding changes for the reagents and products. This proposition may be stated mathematically as in equation 57, where the value of the parameter α is constrained to lie between zero and unity.[28]

$$\delta F^{\ddagger} = \alpha\, \delta \bar{F}_P{}^{\circ} + (1 - \alpha)\, \delta \bar{F}_R{}^{\circ} \tag{57}$$

Assuming that equation 57 is an adequate approximation, it can readily be shown that $\delta_R\, \Delta F^{\ddagger}$ or $\delta_M\, \Delta F^{\ddagger}$ is given by the *rate-equilibrium relationship*, equation 58.

$$\delta\, \Delta F^{\ddagger} = \alpha\, \delta\, \Delta F^{\circ} \tag{58}$$

It is clear from equation 57 that α is a parameter measuring the extent to which the transition state resembles the products with respect to its sensitivity to medium or structural changes. A value near unity indicates close resemblance to the products; a value near zero indicates close resemblance to the reagents. Moreover, the kind of resemblance measured by α should be a useful index of the kind of resemblance measured by the position of the transition state along the reaction coordinate. That is, α

[28] J. E. Leffler, *Science*, **117**, 340 (1953).

should be at least approximately equal to the fractional displacement of the transition state along the reaction coordinate from reagents to products. This conclusion will have two useful consequences. First of all, when the rate-equilibrium relationship is found to fit a set of experimental data, the value obtained for α can help to characterize the reaction mechanism. Secondly, we can predict approximately the value of α whenever we can predict the fractional displacement of the transition state along the reaction coordinate. A helpful guiding principle in this connection is that the transition state bears the greater resemblance to the less stable of the species in the chemical equilibrium. Thus α will be greater than one-half if $\Delta F°$ is positive and will approach unity as $\Delta F°$ becomes large; α will be less than one-half if $\Delta F°$ is negative and will approach zero as $\Delta F°$ becomes highly negative. It is particularly easy to appreciate these relationships for a reaction in which two bases struggle for possession of a proton.

$$B_1H^+ + B_2 \rightleftharpoons B_1 + B_2H^+$$

Since the transition state is by definition the position along the reaction coordinate for which the free energy is at a *maximum*, it follows that the bond to the weaker base is largely complete in the transition state, while the bond to the stronger base is largely broken.

It follows from the correlation of α with the extent to which the reaction is uphill in free energy, that α will be a constant only for small changes in structure or solvent.

A more precise insight into the meaning of α can be obtained by applying the formal theory of interaction mechanisms to the problem. For definiteness, we shall consider a substituent effect, although the treatment of a medium effect is closely similar.

$$RX_0 \rightleftharpoons RX^‡ \rightleftharpoons RX_P$$
$$R_0X_0 \rightleftharpoons R_0X^‡ \rightleftharpoons R_0X_P$$

If we express the free energy quantities as in equation 6, equation 57 is converted into an expression involving only the interaction terms.

$$(I_{R,X^‡} - I_{R_0,X^‡}) = \alpha(I_{R,X_P} - I_{R_0,X_P}) + (1 - \alpha)(I_{R,X_0} - I_{R_0,X_0}) \quad (59)$$

Equation 59 can be simplified further. The parameter α can be independent of the change in R only if each of the interaction terms is factorable in such a way that the factors characteristic of R cancel out on the two sides of the equation. Equation 60 shows how this might be done.

$$(I_R - I_{R_0})I_{X^‡} = [\alpha I_{X_P} + (1 - \alpha)I_{X_0}](I_R - I_{R_0}) \quad (60)$$

In deriving equation 60, each interaction term has been separated into two independent factors: $I_{R,X} = I_R \cdot I_X$. As pointed out earlier in the chapter, such a separation is permissible only if the interaction between R and X proceeds by a single mechanism. Let the interaction variable characteristic of the reaction zone be denoted by X. In first approximation, any change in the factor I_X is proportional to the change in this variable. (See, for example, equation 21.) Equation 60 therefore leads to equation 61, which expresses α in first approximation as a function of the interaction variable.[29]

$$\alpha = \frac{X^{\ddagger} - X_0}{X_P - X_0} \tag{61}$$

Having identified α as a ratio of changes in an interaction variable, we may conclude that in principle α can assume any value whatsoever, positive or negative. However, the authors are not aware of any examples of α outside of the range zero to one.

The most numerous examples of the rate-equilibrium relationship are proton transfer reactions. Since a proton transfer is necessarily attended by a change in charge number, it is probable that the variable X in this case is an electrical parameter, such as a charge or a dipole moment. An α-value of unity then implies that the electrical state of the reaction zone in the transition state closely resembles that in the product.

An example of the rate-equilibrium relationship outside the field of proton transfer is provided by the isomerization of variously substituted 5-aminotriazoles.[3] Figure 6-7 shows the relationship between log k and

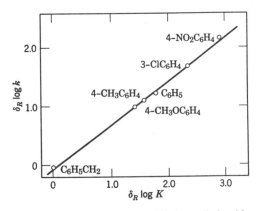

Figure 6-7. The rate-equilibrium relationship in the 5-aminotriazole rearrangement.

[29] The derivation of equation 61 has been less general than it need be. The same result is obtained if the interaction between R and X involves several independent variables characteristic of R, provided that there is only one independent variable characteristic of the reaction zone.

$\log K$ of reaction (62) at 423°K. The slope, α, is 0.77, indicating that insofar

$$
\begin{array}{ccc}
\text{Ar} & & \text{H} \\
| & & | \\
\text{N} & & \text{N} \quad \text{H} \\
\diagdown & & \diagdown \quad | \\
\text{N} \quad \text{C}-\text{NH}_2 & \underset{\substack{\text{ethylene} \\ \text{glycol}}}{\overset{k}{\rightleftharpoons}} & \text{N} \quad \text{C}-\text{N}-\text{Ar} \\
\| \quad \| & & \| \quad \| \\
\text{N} \text{---} \text{C}-\phi & & \text{N} \text{---} \text{C}-\phi
\end{array} \qquad (62)
$$

as the response of the free energy to substituents is a criterion, the transition state resembles the product much more than it does the reagent. Electron-withdrawing substituents both accelerate the reaction and make it more complete; both effects can be correlated with the Hammett substituent constants (Chapter 7).

If $\delta_R \Delta F^{\ddagger}$ remains proportional to $\delta_R \Delta F^{\circ}$ over a range of temperatures, a relationship similar to (50) can be derived between the activation *energy* and the heat of a reaction.

$$
\delta_R E = \alpha' \delta_R \Delta \bar{H}^{\circ} \qquad (63)
$$

Reaction (62) also illustrates this relationship, as shown in Figure 6-8. The slope α' is 0.9 as compared to 0.77 for α. The heats and activation energies are less precisely known than are the rate and equilibrium constants.

The rates and equilibrium constants for the initial addition step in the

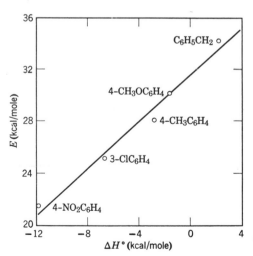

Figure 6-8. Enthalpy quantities of the rate-equilibrium relationship in the 5-aminotriazole rearrangement.

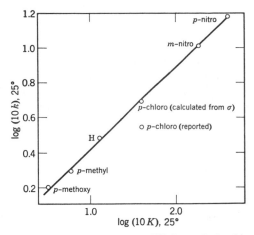

Figure 6-9. The rate-equilibrium relationship in the addition of semicarbazide to substituted benzaldehydes.

formation of semicarbazones are related as shown in Figure 6-9.[30] The slope α is about 0.47.

$$\underset{\substack{\text{O}\\\|}}{XC_6H_4\overset{\text{O}}{\overset{\|}{C}}H} + H_2NCNHNH_2 \underset{\substack{25\%\\ \text{ethanol}}}{\overset{k}{\rightleftarrows}} XC_6H_4\overset{\text{H}}{\underset{\underset{\text{H}}{\overset{|}{N}}\underset{\text{H}}{\overset{|}{-}}N\underset{\text{O}}{\overset{\|}{-}}C-NH_2}{\overset{|}{\underset{|}{C}}}-OH \tag{64}$$

Incidentally, we suspect that the rate constant reported for p-chloro is in error. If we estimate the rate constant extrathermodynamically by means of the Hammett $\rho\sigma$ relationship (Chapter 7), we obtain a point that falls on the line.

It should be noted that the rate-equilibrium parallelism is expected only within series of reactions proceeding by the same mechanism. If a given substrate is capable of giving two different products *by two different mechanisms*, it is by no means a foregone conclusion that the more stable product is formed at the faster rate.[31] In fact, there are numerous organic reactions whose product depends on whether or not the experimenter takes a coffee break before working up his reaction mixture.

[30] B. M. Anderson and W. P. Jencks, *J. Am. Chem. Soc.*, **82**, 1773 (1960).
[31] G. S. Hammond, *J. Am. Chem. Soc.*, **77**, 334 (1955); J. E. Leffler, *The Reactive Intermediates of Organic Chemistry*, Interscience Publisners, 1956.

SELECTIVITY VERSUS REACTIVITY

When a molecule of a reagent in a solution encounters a molecule of a substrate, there is a specific probability P that the encounter will result in the formation of a product molecule. When there are two different substrates in the solution, the reagent will have encounters with both of them, and a mixture of products results. The *selectivity* of the reagent for one substrate rather than the other is then measured by the ratio of probabilities, P_2/P_1, and may be defined as in equation 65.

Selectivity of the reagent for substrate S_2 rather than

$$\text{substrate } S_1 \equiv RT \ln (P_2/P_1) \qquad (65)$$

If the reaction mechanisms are similar and the substrate molecules have similar mobilities, the ratio P_2/P_1 is closely approximated by the ratio of rate constants, k_2/k_1, and the selectivity is then given by equation 66.

Selectivity of the reagent for substrate S_2 rather than

$$\text{substrate } S_1 = RT \ln k_2/k_1 \qquad (66)$$

$$= -(\Delta F_2^{\ddagger} - \Delta F_1^{\ddagger})$$

For activation-controlled reactions, P_2 is in general not equal to P_1.

Now let us imagine that the structure of the reagent is modified continuously so as to increase its reactivity. Eventually the reactivity will become high enough so that reaction occurs at each encounter. When this limit is reached, P_2 and P_1 are both unity, and the reagent is completely unselective. The question now arises whether the selectivity approaches its limiting value of zero as a continuous function. Or, to put the question in another way, we ask whether the magnitude of the selectivity decreases continuously towards zero with increasing reactivity when the reaction is activation-controlled.

Before trying to answer this question, we should like to define the concept of *reactivity*. Strictly speaking, there is no such thing as reactivity in an absolute sense, because reactivity exists only in the presence of a substrate and is specific for that substrate. Yet any chemist knows that the relative reactivities of various reagents do tend to be in a definite sequence that is independent of the nature of the substrate, provided that the change in the structure of the substrate is suitably limited. Thus when we say that the allyl radical is less reactive than the methyl radical, we mean that those substrates which react with the allyl radical will tend also to react with the methyl radical, and at a faster rate. We may therefore define the *reactivity* of a reagent quite arbitrarily in terms of its rate constant or activation free energy with a suitable standard substrate, S_0. We prefer the definition

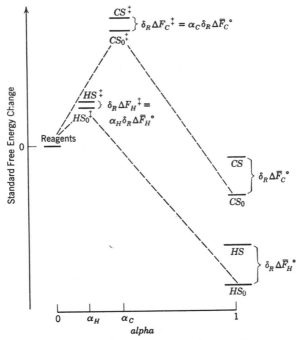

Figure 6-10. Graphical representation of selectivity and reactivity.

given in equation 67, where k_0 is the rate constant for reaction with the standard substrate.

$$\text{Reactivity of a reagent} \equiv k_0 \qquad (67)$$

In order to predict a monotonic relationship (other than an accidental one) between a decreasing magnitude of the selectivity and an increasing reactivity, we assume that the reactions in question obey the rate-equilibrium relationship and make use of the property of the parameter α to decrease with increasing resemblance of the transition state to the reagents. The problem involves two types of structural change, that of the reagent and that of the substrate. For example, in the alkylation of aromatic compounds, the reagent might be changed from $C_6H_5CH_2^+$ to $(C_6H_5)_3C^+$ and the substrate from benzene to anisole.[32]

The standard free energy changes involved in a selectivity-reactivity relationship might be similar to those shown in Figure 6-10. H (hot) represents the more reactive reagent, such as $C_6H_5CH_2^+$, and C (cold)

[32] Although we find $C_6H_5CH_2^+$ convenient as an example, we do not mean to express any opinion as to whether its reactions would be activation-controlled.

represents the less reactive reagent, such as $(C_6H_5)_3C^+$. S represents the less reactive substrate, for example, benzene, and S_0 represents the more reactive substrate, for example, anisole. We shall use the operator δ_R to denote the change in the substrate. Applying equation 66 and the rate-equilibrium relationship (58) to the reaction of H and C, we obtain equations 68.

$$\frac{\text{Selectivity of } H}{\text{Selectivity of } C} = \frac{\delta_R \Delta \bar{F}_H^{\ddagger}}{\delta_R \Delta \bar{F}_C^{\ddagger}} = \frac{\alpha_H \delta_R \Delta \bar{F}_H^{\circ}}{\alpha_C \delta_R \Delta \bar{F}_C^{\circ}} \qquad (68)$$

Since H is more reactive than C, the transition state will more nearly resemble the reagents in the reactions of H than in those of C. Hence α_H is less than α_C. Figure 6-10 makes use of this inequality but exaggerates it for emphasis.

With $\alpha_H < \alpha_C$, it is obvious from (68) that the selectivity of H is less than that of C provided that $\delta_R \Delta \bar{F}_H^{\circ}$ is less than or equal to $\delta_R \Delta \bar{F}_C^{\circ}$. In Figure 6-10 the latter two quantities are shown as equal. However, the selectivity of H can be less than that of C even if $\delta_R \Delta \bar{F}_H^{\circ}$ is greater than $\delta_R \Delta \bar{F}_C^{\circ}$ provided that α_H/α_C is small enough.

Returning to our example, equation 69, of benzyl or trityl cations adding either to benzene or to the *para*-position of anisole, we expect that the quantities $\delta_R \Delta \bar{F}_{\text{trityl}}^{\circ}$ and $\delta_R \Delta \bar{F}_{\text{benzyl}}^{\circ}$ will be about equal. The reason for this is that the effect of the p-methoxy substituent on the resonance energy depends largely on the presence of the positive charge and only to a lesser extent on the nature of R. The difference in free energy between the benzene and anisole systems will be less in the transition state than in the products

$$\text{R}^+ + \langle \bigcirc \rangle\text{—OCH}_3 \rightarrow \overset{\overset{\text{H}}{|}}{\underset{\overset{|}{\text{R}}}{\langle \bigcirc \rangle}} = \overset{+}{\underset{\text{CH}_3}{\text{O}}} \qquad (69)$$

because the positive charge in the transition state has been only partially transferred to the substrate.

An analogous example for which actual data exist is the selectivity of various cationic intermediates in $s_N 1$ solvolysis.[33] The rate constant for azide ion divided by the rate constant for water is 2.8×10^5 for trityl cation, 2.4×10^2 for p,p'-dimethylbenzhydryl cation, 1.7×10^2 for benzhydryl cation, and only 3.9 for t-butyl cation. Longer series of this sort are rare because of a tendency for the reaction mechanism to change. Methyl cation is essentially unavailable because its formation and reaction are simultaneous; in any case its reactions would be diffusion-controlled.

[33] C. G. Swain, C. B. Scott, and K. H. Lohmann, *J. Am. Chem. Soc.*, **75**, 136 (1953).

The variation of selectivity with reactivity when there is also a rate-equilibrium relationship is shown most beautifully by some data for the base-catalyzed halogenation of ketones and keto-esters. In this reaction the rate-determining step is the transfer of a proton from the carbonyl compound (which is the *reagent* in this example) to a catalyzing base (the *substrate*).

$$H-\overset{\overset{\displaystyle O}{\|}}{\underset{|}{C}}-\overset{\overset{\displaystyle O}{\|}}{C}- + RCO^{\ominus} \overset{K}{\rightleftharpoons} R\overset{\overset{\displaystyle O}{\|}}{C}OH + \;\overset{\diagdown}{\diagup}C=C\overset{O^{\ominus}}{\underset{\diagdown}{\diagup}} \tag{70}$$

For any given carbonyl compound reacting with a series of carboxylate ions, the rate constants are related to the base dissociation constants of the carboxylate ions by the empirical relationship (71).[34]

$$\delta_R \log k = \mathsf{B}\, \delta_R \log K_B \tag{71}$$

Because the equilibrium constant K of reaction (70) is proportional to K_B we may write equation 72, and equation 71 thus implies the rate-equilibrium relationship (73).

$$\delta_R \log K_B = \delta_R \log K \tag{72}$$

$$\delta_R \log k = \mathsf{B}\, \delta_R \log K \tag{73}$$

The empirical coefficient B is therefore the same as the coefficient α previously used in the rate-equilibrium relationship. For the iodination of acetone catalyzed by carboxylate ions in water at 25°, B, or α, is 0.88.[35] This value of α is plausible, for acetone enolate ion is a much stronger base than any of the carboxylate ions. The value for the bromination of acetoacetic ester is lower ($\alpha = 0.59$), consistent with the reduction in the base strength of the enolate ion.

Figure 6-11 shows the variation of B with the reactivity of the substrate, the reactivity (k_0) being defined as the rate constant for the halogenation of the ketone or ester with a hypothetical catalyzing base of $K_B = 10^{-10}$. These rate constants are obtained by interpolation from the rate constants for actual bases, using equation 71 for the interpolation. The results are listed in Table 6-6.

An inspection of the table or of Figure 6-11 shows that as the reactivity of the carbonyl compound increases, the coefficient B decreases. The

[34] Equation 71 is an example of the Brønsted relationship. Further discussion of the Brønsted relationship will be found in Chapter 7.
[35] R. P. Bell, E. Gelles, and E. Möller, *Proc. Roy. Soc.*, **A198,** 308 (1949).

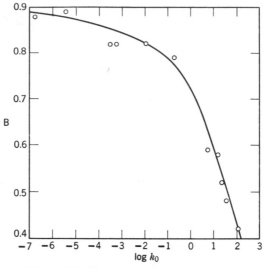

Figure 6-11. The decrease in the parameter B with increasing reactivity of the carbonyl compound.

Table 6-6. Change in B with Increasing Reactivity of the Carbonyl Compound[a]

SUBSTRATE	$\log k_0$[b]	B
Acetone	−6.78	0.88
2,5-Hexandione	−5.46	0.89
Chloroacetone	−3.51	0.82
Bromoacetone	−3.25	0.82
1,1-Dichloroacetone	−2.00	0.82
Diethyl malonate	−0.76	0.79
Ethyl acetoacetate	0.72	0.59
Ethyl 2-cyclopentanonecarboxylate	1.18	0.58
1-Phenyl-1,3-butandione	1.33	0.52
1-Phenyl-2-bromo-1,3-butandione	1.33	0.52
Acetylacetone	1.54	0.48
3-Bromo-2,4-pentandione	2.04	0.42

[a] R. P. Bell, E. Gelles, and E. Möller, *Proc. Roy. Soc. (London)*, **A198**, 310 (1949); R. P. Bell and H. L. Goldsmith, *ibid.*, **A210**, 322 (1952).
[b] k_0 in l/mole min., statistically corrected.

relationship is a selectivity-reactivity relationship because the coefficient B is proportional to the selectivity. To show this, let us consider that two bases R_1COO^- and R_2COO^- compete for reaction with a single carbonyl compound. The selectivity, or $RT \ln (k_2/k_1)$, is then given by equation 74. If now the carbonyl compound is changed while the two bases are kept

$$RT \ln \frac{k_2}{k_1} = B \times RT \ln \frac{K_{B2}}{K_{B1}} \tag{74}$$

the same, the selectivity of the carbonyl compound will vary in proportion to B, the proportionality constant being $RT \ln (K_{B2}/K_{B1})$.

The shape of the curve in Figure 6-11 suggests that as the reactivity becomes very small or very large, the selectivity parameter B approaches its limiting values unity and zero asymptotically.

Lest the reader be misled into attributing too great a generality to the correlation between increasing reactivity and decreasing magnitude of the selectivity, it should be noted that there are many exceptions. For example, atomic oxygen is highly reactive, yet it is also fairly selective. The rate constant for its reaction with 2,3-dimethyl-2-butene is reported to be about one hundred times that for its reaction with ethylene.[36]

Another example, typical in showing the rather limited success of most selectivity-reactivity correlations, comes from deuterium isotope effects. When a reagent $R\cdot$ attacks a hydrogen or deuterium atom in the substrate H—S or D—S, the strength of the $R \cdots H$ bond in the transition state $[R \cdots H \cdots S]^{\ddagger}$ should decrease as the reactivity of $R\cdot$ increases, while the strength of the $H \cdots S$ bond should approach that of the H—S bond in the substrate. In other words, for a highly reactive reagent equation 75 should be a good model for predicting the magnitude of the deuterium isotope effect.

$$\begin{aligned}
R\cdot + H—S &\rightleftharpoons [R \cdots H—S]^{\ddagger} \rightleftharpoons R—H + \cdot S \\
R\cdot + D—S &\rightleftharpoons [R \cdots D—S]^{\ddagger} \rightleftharpoons R—D + \cdot S
\end{aligned} \tag{75}$$

Since the bond to R is weak in the transition state, the deuterium isotope effect should be predictable largely from the zero-point energies associated with the H—S and D—S bonds in the transition state and in the ground state.[37] As the reactivity increases, these bonds should become more nearly alike, and hence the selectivity (k_H/k_D) should decrease. As a

[36] R. J. Cvetanović, *J. Chem. Phys.*, **30**, 19 (1959).

[37] In the transition state the asymmetric stretching vibration $\overrightarrow{R} \cdots \overleftarrow{H} \cdots \overrightarrow{S}$ is unstable and is replaced by translation along the reaction coordinate. But the symmetrical stretching vibration $\overleftarrow{R} \cdots H \cdots \overrightarrow{S}$ is stable and is treated in the usual way.

matter of fact, when hydrogen atoms are removed from the side chain of toluene in carbon tetrachloride solution, the isotope effect (k_H/k_D) is 1.30 when the reagent is the chlorine atom but 4.59 or 4.86 when the reagent is the less reactive bromine atom or N-succinimidyl radical.[38] The reactivity of the chlorine atom can be reduced by solvation in an aromatic solvent. Thus when toluene is used as the solvent instead of carbon tetrachloride, the isotope effect (k_H/k_D) in the chlorination reaction increases to 2.0.[39] On the other hand, qualitative predictions of (k_H/k_D) based on the selectivity-reactivity relationship are unreliable. For example, the methyl radical (highly reactive) and the allyl radical (much less reactive) discriminate between hydrogen and deuterium atoms in hydrocarbons to about the same extent.[40]

EXTRATHERMODYNAMIC RELATIONSHIPS INVOLVING PHYSICAL PROPERTIES

We have seen that such quantities as $\delta_R \Delta F°$ and $\delta_M \Delta F°$ consist entirely of interaction terms. In simple cases the relevant interactions involve only a single mechanism, and each interaction term can be separated into independent factors. Quantities such as $\delta_R \Delta F°$ are then proportional to the change, $R - R_0$, in the interaction variable characteristic of the R-group, and quantities such as $\delta_M \Delta F°$ are proportional to the change, $M - M_0$, in the interaction variable characteristic of the medium.

In discussing the conditions under which simple extrathermodynamic relationships might be expected to exist, it was not necessary to identify the variable R or M explicitly. On the other hand, when a simple relationship is actually observed, our understanding of it is not complete until the interaction variable has been identified. If R or M is a scalar variable, the criterion for deciding whether the variable has been correctly identified is simply that the plot of $\delta_R \Delta F°$ versus R or of $\delta_M \Delta F°$ versus M must be linear over the entire range of substituents or solvents to which the extrathermodynamic relationship applies. For example, if M is thought to be the reciprocal of the dielectric constant, the plot of $\delta_M \Delta F°$ versus $1/D$ must be a straight line.

The identification of R is more difficult than that of M because R, being characteristic of only a portion of the molecule, is not a macroscopic physical property. However, if the corresponding molar property happens

---

[38] K. B. Wiberg and L. H. Slaugh, *J. Am. Chem. Soc.*, **80**, 3033 (1958).

[39] H. C. Brown and G. A. Russell, *J. Am. Chem. Soc.*, **74**, 3995 (1952).

[40] A. S. Gordon, S. R. Smith, and J. R. McNesby, *J. Am. Chem. Soc.*, **81**, 5059 (1959).

to be an additive function of independent contributions assignable to the R- and X-portions of the molecule, the plot of $\delta_R \Delta F°$ against the appropriate *molar* property will be linear. To be specific, let the interaction variable be a partial volume characteristic of the substituent. The *molar* volume, V_{RX}, may be written in the form of equation 76, which is analogous to equation 6.

$$V_{RX} = V_R + V_X + V_{I(R,X)} \tag{76}$$

V_R and V_X are independent additive terms, and $V_{I(R,X)}$ is an interaction term. In first approximation the interaction term can be neglected, so that $\delta_R V_R \approx \delta_R V_{RX}$. A linear relationship with V_{RX} therefore implies a linear relationship with V_R.

Because of the considerable chance that a linear fit might be accidental, the success of a correlation with a physical property is less decisive than is the failure of such a correlation. Indeed, even when the failure is only partial, the physical property can be safely ruled out from the list of possible interaction variables. Judged by this harsh but fair criterion, very few of the correlations of free energy quantities with physical properties are really significant. Because of this difficulty it is often more fruitful to express the changes in the interaction variables by means of empirical parameters, such as σ-values for substituents (Chapter 7) or Y-values for solvents (Chapter 8).

Our skepticism of linear correlations with physical properties does not extend to spectral transitions. The excitation energy of a molecule is a Δ-quantity of virtually the same type as an energy or free energy of reaction or activation, except that the "reaction zone" is now that portion of the molecule in which the excitation takes place. A substituent or medium effect on a spectral transition may proceed by the same interaction mechanism as on an analogous chemical reaction, and linear relationships are then to be expected. For example, in the reaction of atomic oxygen with alkenes in the gas phase, the free energy of activation is roughly a linear function of the ionization potential as well as of the transition energy for the $\pi \rightarrow \pi^*$ electronic transition of the alkene.[36]

A Viscosity-Rate Relationship

An apparently significant correlation with a physical property has been observed in the photo-dissociation of dimesityl disulfide in a series of solvents.[41] The probable mechanism for this process is shown in equation 77.

[41] Y. Schaafsma, A. F. Bickel, and E. C. Kooyman, *Tetrahedron*, **10**, 76 (1960).

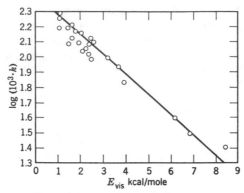

Figure 6-12. Logarithm of the rate constant of photochemical dissociation of dimesityl disulfide versus activation energy for kinematic viscous flow of the solvent, at 25°. Solvents include pure aliphatic and aromatic hydrocarbons, carbonyl compounds, halogen compounds, and mixtures of these compounds.

In the presence of the radiation, the disulfide is in steady-state photoequilibrium with the caged radical pair, the dissociation of which determines the observed rate. Since the factors tending to prevent escape of a molecule from a solvent cage are much the same as those causing resistance to flow of the solvent, it might be expected that the *net* decomposition rate at constant light intensity will parallel some function of the viscosity. As a matter of fact, the logarithms of the rate constants for the photodissociation of dimesityl disulfide in a series of solvents at constant light intensity are linearly related to the activation energy, $E_{vis.}$, for viscous flow of the solvent, as shown in Figure 6-12.[41]

7

Extrathermodynamic Free Energy Relationships. 1. Substituent Effects

He was pinched perspiringly in the epistemological dilemma of the skeptic, unable to accept solutions to problems he was unable to dismiss as unsolvable. He was never without misery and never without hope.

Joseph Heller
Catch 22

In Chapter 6 we constructed a formal system for generating extrathermodynamic relationships, based on the concept of physical interaction mechanisms with scalar independent variables. We shall now apply this system to the analysis of substituent effects. We shall find that simple linear free energy relationships need not be limited to closely related reactions but in some cases encompass *entire classes* of reactions. Thus our formal interaction mechanisms have exact counterparts in *real* interactions of broad scope.

When a linear free energy relationship applies to a group of reactions, we may choose the *model process* to be any convenient reaction in the group under a suitable single set of conditions. Normally we would choose the model process on the basis that accurate data are already available for a large number of substituents. Yet even so, we usually find that the existing data do not cover a sufficiently long list of structural changes. We therefore supplement the data for the model process with structural parameters whose operational definition varies somewhat from one structural change to another. For example, the temperature, the solvent, or even the nature of the reaction may be different, although an attempt is made to hold these variations to a minimum. In addition, an attempt is made to show that the parameters obtained from these variants of the standard or reference process are not very different from those that would have been obtained under fixed conditions.

THE HAMMETT $\rho\sigma$ RELATIONSHIP

The best known of the extrathermodynamic equations relating reaction rate or equilibrium to the structure of the reagent is the Hammett $\rho\sigma$ equation.[1] It describes the effect of a *meta-* or *para*-substituent on the rate or equilibrium constant of an aromatic side-chain reaction. It is based on the fact that, as the substituent is varied, the logarithms of the rate or equilibrium constants for a large number of aromatic side-chain reactions are linearly related to one another. Although any one of these reactions could have been used to define a set of substituent parameters in terms of which the free energy changes of the others might be described, the parameter σ has been defined on the basis of the acid dissociation equilibrium of benzoic acid in water at 25°.

$$C_6H_5COOH \overset{K_A{}^\circ}{\rightleftharpoons} C_6H_5COO^- + H^+$$

$$m\text{- or } p\text{-}XC_6H_4COOH \overset{K_A{}^x}{\rightleftharpoons} m\text{- or } p\text{-}XC_6H_4COO^- + H^+$$

$$\sigma_x \equiv \log \frac{K_A{}^x}{K_A{}^\circ}, \quad \text{in water at } 25° \tag{1}$$

The reason for this choice was the accessibility of the data and the belief, current at that time, that the mechanisms by which substituents exert their effects were known for this reaction. The substituents, X, must be in the *meta-* or *para*-position, since the linear relationships fail for *ortho*-substituents.

The effect of substitution on the rate or equilibrium constants for other benzene side-chain reactions can now be expressed as a function of σ. The resulting equation (2) is the well-known Hammett equation.

$$\log \frac{k_x}{k_0} \left(\text{or, } \log \frac{K_x}{K_0} \right) = \rho\sigma_x \tag{2}$$

The constants k_x and k_0 (or K_x and K_0) must be for the same experimental conditions, though not necessarily the conditions used in the definition of σ. The constant ρ is a function of the reaction and will also vary with the reaction conditions.

Values of the parameter σ for use in the Hammett equation are listed in Table 7-1.[2,3] Whenever possible, the values have been based on data for the standard model process, as defined in equation 1. Values obtained in this way will be referred to as *primary* σ-values. When the required data for the standard process are not available, the σ-value for that substituent is

[1] L. P. Hammett, *Physical Organic Chemistry*, McGraw-Hill, 1940.
[2] D. H. McDaniel and H. C. Brown, *J. Org. Chem.* **23**, 420 (1958).
[3] H. H. Jaffé, *Chemical Reviews*, **53**, 191 (1953).

Table 7-1[a]. Hammett Substituent Constants, σ, Based on the Ionization of Benzoic Acids[b]

SUBSTITUENT	σ_{meta}	*Meta*-Position ESTIMATED LIMIT OF UNCERTAINTY	σ_{para}	*Para*-Position ESTIMATED LIMIT OF UNCERTAINTY
CH_3	**−0.069**	0.02	**−0.170**	0.02
CH_2CH_3	−0.07	0.1	−0.151	0.02
$CH(CH_3)_2$	—	—	−0.151	0.02
$C(CH_3)_3$	−0.10	0.03	−0.197	0.02
$3,4-(CH_2)_4$	—	—	0.042	0.02
C_6H_5	0.06	0.05	−0.01	0.05
CF_3	0.43	0.1	0.54	0.1
CN	**0.56**	0.05	**0.660**	0.02
$COCH_3$	**0.376**	0.02	**0.502**	0.02
$CO_2C_2H_5$	0.37	0.1	0.45	0.1
CO_2H	(0.37)	0.1	(0.45)	0.1
$CO_2{}^-$	−0.1	0.1	0.0	0.1
$CH_2Si(CH_3)_3$	−0.16	>0.1	−0.21	>0.1
$Si(CH_3)_3$	−0.04	0.1	−0.07	0.1
$Si(C_2H_5)_3$	—	—	0.0	0.1
$Ge(CH_3)_3$	—	—	0.0	0.1
$Ge(C_2H_5)_3$	—	—	0.0	0.1
$Sn(CH_3)_3$	—	—	0.0	0.1
$Sn(C_2H_5)_3$	—	—	0.0	0.1
$N_2{}^{+ c}$	1.76	0.2	1.91	0.2
NH_2	−0.16	0.1	−0.66	0.1
$NHCH_3$	—	—	−0.84	0.1
$N(CH_3)_2$	—	—	−0.83	0.1
$NHCOCH_3$	0.21	0.1	0.00	0.1
$N(CH_3)_3{}^+$	0.88	>0.2	0.82	>0.2
NO_2	**0.710**	0.02	**0.778**	0.02
PO_3H^-	0.2	>0.1	0.26	>0.1
AsO_3H^-	—	—	−0.02	>0.1
OCH_3	**0.115**	0.02	**−0.268**	0.02
OC_2H_5	0.1	0.1	−0.24	0.1
$O(CH_2)_2CH_3$	0.1	0.1	−0.25	0.1
$OCH(CH_3)_2$	0.1	0.1	−0.45	0.1
$O(CH_2)_3CH_3$	0.1	0.1	−0.32	0.1
$O(CH_2)_4CH_3$	0.1	0.1	−0.34	0.1
OC_6H_5	**0.252**	0.02	**−0.320**	0.02(?)
OH	**0.121**	0.02	**−0.37**	0.04
$OCOCH_3$	0.39	0.1	0.31	0.1
SCH_3	0.15	0.1	0.00	0.1
SC_2H_5	—	—	0.03	0.1
$SCH(CH_3)_2$	—	—	0.07	0.1
SH	0.25	0.1	0.15	0.1
$SCOCH_3$	0.39	0.1	0.44	0.1
SCN	—	—	0.52	0.1
$SOCH_3$	0.52	0.1	0.49	0.1
SO_2CH_3	0.60	0.1	0.72	0.1
SO_2NH_2	0.46	0.1	0.57	0.1
$S(CH_3)_2{}^+$	1.00	>0.1	0.90	>0.1
$SO_3{}^-$	0.05	>0.1	0.09	>0.1
$SeCH_3$	0.1	0.1	0.0	0.1
F	**0.337**	0.02	**0.062**	0.02
Cl	**0.373**	0.02	**0.227**	0.02
Br	**0.391**	0.02	**0.232**	0.02
I	**0.352**	0.02	0.18	0.1
IO_2	0.70	0.1	0.76	0.1
$CH{=}CHNO_2{}^d$	**0.34**	0.03	0.26	0.03

[a] From the compilation of D. H. McDaniel and H. C. Brown, *J. Org. Chem.*, **23**, 420 (1958).
[b] Values in bold face are sigma constants based on *thermodynamic* ionization constants in *water* at 25°. It is recommended that reaction constants, ρ, be based on these values.
[c] From E. S. Lewis and M. D. Johnson, *J. Am. Chem. Soc.*, **81**, 2070 (1959).
[d] From R. Stewart and L. G. Walker, *Can. J. Chem.*, **35**, 1561 (1957).

obtained from some secondary reference process, chosen to resemble the ionization of benzoic acid in water at 25° as closely as possible. The procedure for obtaining such a *secondary* σ-value is as follows. First, a number of known primary σ-values are used to evaluate ρ for the secondary reference process by fitting a line according to equation 2. If the fit is satisfactory, that value of ρ and the known value of log (k_x/k_0) for the new substituent are then used to calculate the desired σ-value. Such secondary σ-values will depend somewhat on the choice of the secondary reference process, since the free energy relationships are only approximately linear.

Table 7-2. Solvent Dependence of the Secondary σ-Values for the OH-Group in Ethanol-Water Mixtures

% ETHANOL	SECONDARY VALUE OF	
	σ_{para}	σ_{meta}
0	(-0.37)[a]	$(+0.121)$[a]
40	-0.285	-0.014
50	-0.335	-0.055
70	-0.350	-0.102
80	-0.384	-0.111
90	-0.414	-0.126
95	-0.429	-0.129
100	-0.442	-0.134

[a] Primary value.

The secondary σ-values in Table 7-1 are all based on the ionization of substituted benzoic acids, although the solvent may have been an aqueous-organic mixture or the temperature different from 25°. A plot of log K_A for the ionization of benzoic acids in an aqueous-organic solvent against log K_A in water at 25° will usually give an excellent fit to a straight line, and a secondary σ-value derived from such a plot is probably a good approximation to the value that would have been obtained from the standard process. An important limitation should be noted, however. The accuracy may be poor for certain substituents that interact strongly with the solvent. A particularly good example of this is provided by the ionization of substituted benzoic acids in water-ethanol mixtures.[4] If we assume that the other substituents are well behaved, the $\rho\sigma$ plots for these mixed solvents can be used to evaluate secondary σ-values for the *para*- and *meta*-hydroxy substituents. Table 7-2 shows the systematic variation of these values as the solvent composition, and hence the solvation of the substituent, is changed.[3]

[4] W. L. Bright and H. T. Briscoe, *J. Phys. Chem.*, **37**, 787 (1933).

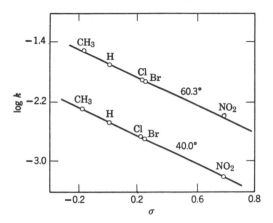

Figure 7-1. Decomposition rates of p-substituted α-diazoacetophenones in acetic acid. J. F. Lane and R. L. Feller, *J. Am. Chem. Soc.*, **73**, 4230 (1951).

Table 7-3 contains selected values of ρ, mostly from the compilation of Jaffé.[3] Unless otherwise noted, only examples for which the correlation is rated "excellent" (correlation coefficient .99 or better) have been included. An example of such an excellent fit is shown in Figure 7-1. A less excellent but somewhat more typical relationship with a correlation coefficient of .95 is shown in Figure 7-2. Jaffé does not regard a relationship as "satisfactory" unless the correlation coefficient is at least .95.

Although it should be possible, in principle, to predict the quality of a fit *a priori* from the resemblance of the given reaction to the ionization of benzoic acid in water, no single simple criterion, such as the nature of the functional group, the charge type, or the solvent, seems to work well. For example, the fit shown in Figure 7-2 is only barely satisfactory even though the reaction is a saponification in an aqueous-organic medium, whereas that shown in Figure 7-1 is excellent even though the process, a decomposition reaction in glacial acetic acid, is quite different from the standard process. In order to explain the relatively poor fit shown in Figure 7-2, one must assume that the substitution of methyl for hydrogen in the 2,6-positions greatly reduces the resemblance to the standard process, perhaps because it forces the carbonyl group out of the plane of the benzene ring.

Scope of the $\rho\sigma$ Relationship

Table 7-3 is only a small sample consisting of especially precise examples of the literally hundreds of applications of the Hammett $\rho\sigma$ relationship to

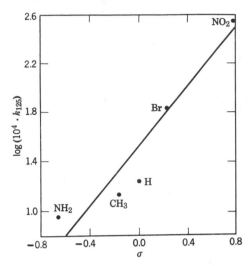

Figure 7-2. Saponification of methyl esters of 4-substituted-2,6-dimethyl benzoic acids in 60% dioxane at 125°. In this example, the correlation coefficient is .95 and the standard deviation is 0.24. (H. L. Goering, T. Rubin, and M. S. Newman, *J. Am. Chem. Soc.*, **76**, 787 (1954).)

be found in the literature. It is obvious from even a cursory examination of the reactions correlated that the important interaction mechanism (or mechanisms) is primarily polar. However, this does not mean that the correlation does not extend to free radical reactions, since these may often have a pronounced polar character. Thus the reaction of substituted aromatic aldehydes with *p*-chlorobenzoylperoxy radicals in acetic anhydride at 30° is reported to have a ρ-value of -1.668.[3,5] That is, this reaction responds more to a change in the polar character of a substituent than does the ionization of benzoic acid.

$$Cl-\left\langle\bigcirc\right\rangle-\overset{\overset{O}{\|}}{C}-OO\cdot + H-\overset{\overset{O}{\|}}{C}-\left\langle\bigcirc\right\rangle_X \longrightarrow$$

[5] C. Walling and E. A. McElhill, *J. Am. Chem. Soc.*, **73**, 2927 (1951).

$$
\left[Cl-\!\!\!\bigcirc\!\!\!-\!\!\underset{\underset{O}{\parallel}}{C}\!-\!OO^- \cdots \overset{\cdot}{H} \cdots \overset{+}{C}\!\!=\!\!O \right] \rightarrow
$$

with the X-substituted phenyl group below.

$$
Cl-\!\!\!\bigcirc\!\!\!-\!\!\underset{\underset{O}{\parallel}}{C}\!-\!OOH + \cdot\underset{\underset{O}{\parallel}}{C}\!-\!\!\!\bigcirc\!\!\!-X \tag{3}
$$

Since the ionization of benzoic acids in water, for which ρ is equal to $+1$, is facilitated by electron-withdrawing substituents, it would appear that a positive value of ρ indicates a reaction in which a negative charge (or partial negative charge) is developed on the side chain. Similarly, a negative value of ρ should correspond to the development of a partial positive charge on the side chain. The magnitude of ρ should therefore be a measure of the magnitude of the developing charge and of the extent to which it is able to interact with the substituents. This interpretation is at least qualitatively correct when ρ is far from zero. But it is doubtful when ρ is close to zero because the sign of ρ can then be inverted by a moderate change of temperature.

The Temperature Dependence of ρ

In Chapter 6 it was shown that a linear free energy relationship maintaining its linear form at more than one temperature implies a constraint on the enthalpies and entropies. Thus for a reaction which is correlated with σ, and which shows this correlation at more than one temperature, the enthalpies and entropies would be expected to obey equation 4.

$$
\delta_R \Delta H^\ddagger = \beta \, \delta_R \Delta S^\ddagger \tag{4}
$$

As a consequence of equation 4 we may write equation 5 as a first step in deriving the effect of temperature on ρ.

$$
\frac{\delta_R \Delta F_{T_1}^\ddagger}{\delta_R \Delta F_{T_2}^\ddagger} = \frac{(\beta - T_1) \, \delta_R \Delta S^\ddagger}{(\beta - T_2) \, \delta_R \Delta S^\ddagger} = \frac{-RT_1 \rho_{T_1} \sigma_R}{-RT_2 \rho_{T_2} \sigma_R} \tag{5}
$$

Since $\delta_R \Delta S^\ddagger$ and σ_R are independent of the temperature, they cancel out of equation 5.

$$
\frac{\rho_{T_1}}{\rho_{T_2}} = \frac{1 - \beta/T_1}{1 - \beta/T_2}
$$

Table 7-3. Values of ρ from Selected Examples of the Hammett Relationship[1]

REACTION	SOLVENT	TEMPERATURE °C	ρ	$-\log k_0$ (see note 2)	REF.
Ionization Equilibria					
ArCOOH	H_2O	25°	1.000	4.203	1
2-Methylbenzoic acids	50% EtOH	25°	1.673	5.772	4
2,6-Dimethylbenzoic acids	20% Dioxane	26°	1.116	3.974	5
Benzeneboronic acids	25% EtOH	25°	2.146	9.700	6
$ArPO(OH)_2$	H_2O	25°	0.755	1.836	7
$ArPO_2(OH)^-$	H_2O	25°	0.949	6.965	7
2-Chloro or 2-bromobenzenephosphonic acids	50% EtOH	25°	0.995	2.942	8
2-Chloro or 2-bromobenzenephosphonic acids, second proton	50% EtOH	25°	1.191	8.249	8
α-ArCH=N—OH	H_2O	25°	0.857	10.695	9
ArOH	H_2O	25°	2.113	9.847	10
$ArNH_3^+$	H_2O	25°	2.767	4.557	11
	20% Dioxane	25°	3.256	4.397	12
	45% Dioxane	25°	3.558	3.950	12
	70% Dioxane	25°	3.567	3.547	12
	82% Dioxane	25°	3.430	3.425	12
Other Equilibria					
Anilines with formic acid to give formanilides	67% C_5H_5N	100°	−1.429	−0.630	13
Conversion of 5-aryl-1-aminotetrazoles to 1-aryl-5-aminotetrazoles	$(CH_2OH)_2$	197°	0.978	−0.050	14
Dissociation of 1,1,4,4-tetraaryl-2,3-dibenzoyltetrazanes to form 1,1-diaryl-2-benzoylhydrazyl radicals	Acetone	−30°	−1.52[3]	3.409	15

Reaction Rates

Benzoic acids with diphenyldiazomethane to give benzhydryl benzoates	EtOH	30°	0.940	1.751	16
2-Methylbenzoic acids with diphenyl-diazomethane	EtOH	30°	0.946	1.812	17
Saponification of methyl benzoates	60% Acetone	0°	2.460	3.083	18
		15°	2.299	2.449	
		25°	2.229	2.075	
		40°	2.128	1.560	
		50°	1.920	1.247	
Saponification of ethyl benzoates	75% MeOH	25°	2.193	3.239	19
		40°	2.130	2.606	
		50°	2.116	2.227	
	85% EtOH	25°	2.537	3.217	20
		35°	2.464	2.748	
		50°	2.322	2.210	
	87.83% EtOH	30°	2.431	3.051	21
Saponification of *l*-menthyl benzoates	MeOH	30°	2.628	4.257	22
		40°	2.552	3.859	
		50°	2.476	3.473	
Saponification of ethyl arylacetates ($r = .974$)[1]	87.83% EtOH	30°	0.824	1.813	23
Saponification of ethyl 3-arylpropanoates ($r = .960$)[1]	87.83% EtOH	30°	0.489	2.198	23, 24
Saponification of ethyl cinnamates	87.83% EtOH	30°	1.329	2.752	23, 25
Hydrolysis of methyl benzenesulfonates	50% EtOH	50°	1.248	4.159	26
Ethanolysis of methyl benzenesulfonates	EtOH	70°	1.324	4.134	26
Hydrolysis of benzoyl chlorides	95% Acetone	25°	1.782	4.200	27
Ethanolysis of benzoyl chlorides	60% EtOH, 40% Et$_2$O	0°	1.922	3.997	28
		25°	1.499	2.969	

Table 7-3 (continued)

REACTION	SOLVENT	TEMPERATURE °C	ρ	$-\log k_0$ (see note 2)	REF.
2-Propanolysis of benzoyl chlorides	2-Propanol	25°	1.280	4.271	29
Benzoyl chlorides with aniline	Benzene	25°	1.219	2.980	30
		40°	1.114	2.723	
Hydrolysis of benzamides, base-catalyzed	60% EtOH	52.8°	1.364	5.190	31
		64.5°	1.273	4.661	
		80.3°	1.146	4.136	
		100.1°	1.100	3.523	
acid-catalyzed ($r = .998, .985, .847, .883$)[1]	60% EtOH	52.4°	−0.483	5.606	31
		65°	−0.310	5.072	
		79.5°	−0.298	4.428	
		99.6°	−0.222	3.806	
ArCN + H$_2$S + HO$^-$ → ArC(SH)=NH	EtOH	60.6°	2.147	3.698	32
Hydrolysis of benzoic anhydrides	75% Dioxane	58.25	1.568	5.113	33
		79.65	1.246	4.470	
ArO$^-$ + (CH$_2$)$_2$O → ArOCH$_2$CH$_2$OH	98% EtOH	70.4°	−0.947	4.254	34
ArO$^-$ + EtI → ArOEt + I$^-$	EtOH	42.5°	−0.994	3.955	35
Hydrolysis of aryl acetates, acid-catalyzed ($r = .992, .963, .986, .977$)[1]	60% Acetone	25°	−0.198	4.744	36
		40°	−0.203	3.588	
		60°	−0.232	3.244	
		80°	−0.277	2.588	
2,4-Dinitrophenyl aryl ethers + CH$_3$OK → ArOCH$_3$	MeOH	20°	1.450	2.319	37
p-Substituted-2,6-dinitroanisole + p-dimethylaminoanisole → p-methoxyphenyltrimethylammonium p-substituted-2,6-dinitrophenoxide	Acetone	35°	2.220	6.519	38
		45°	2.203	6.104	

Reaction	Solvent	Temp.			No.
Anilines plus benzoyl chloride to give benzanilides	Benzene	25° 40°	−2.781 −2.491	2.888 2.614	39
Anilines plus 2,4-dinitrophenyl fluoride to give secondary amine	99.8% EtOH	20° 30° 40°	−4.243 −4.106 −4.006	2.280 2.109 1.949	40
Anilines plus 2,4-dinitrophenyl chloride to give secondary amine (r = .967, .992, .993, .979)[1]	EtOH	25° 35° 45° 100°	−3.976 −3.204 −3.055 −2.415	3.952 3.930 3.670 2.324•	41
o-Toluidines plus benzoyl chloride to give amides	Benzene	25°	−2.727	3.435	42
N,N-Dimethylanilines plus 2,4,6-trinitro-anisole to give trimethylanilinium picrates	Acetone	15° 35°	−2.857 −2.378	5.352 4.505	43
N,N-Dimethylanilines plus methyl iodide	90% Acetone	35° 24.8°	−3.303 −3.069	3.366 6.802	44
	Nitrobenzene	40.1° 60°	−2.961 −2.710	6.365 5.812	45
N-benzylacetamides plus chlorine to give N-benzyl-N-chloroacetamides	CH₃COOH	18°	−1.051	−0.466	46
Phenyl isocyanates plus MeOH	(C₄H₉)₂O	20°	2.460	2.259	47
Benzaldehydes plus HCN	95% EtOH	20°	2.329	7.703	48
ArCOCH₃ + Br₂ → ArCOCH₂Br (r = .936, .942, .916)[1]	.5M HCl in 75% CH₃COOH	25° 35° 45°	−0.458 −0.441 −0.379	2.409 1.917 1.474	49

Table 7-3 (continued)

REACTION	SOLVENT	TEMPERATURE °C	ρ	$-\log k_0$ (see note 2)	REF.
2 ArCHO → ArCH$_2$OH + ArCOOH	50% MeOH	40° 60° 80.2° 100°	3.633 3.719 3.860 3.756	5.058 4.412 3.874 3.537	50
Diazoacetophenones plus acetic acid to give phenacyl acetates	99.85% CH$_3$COOH	40.05° 60.3°	−0.914 −0.911	4.275 3.480	51
ArCH=NCl + HO⁻ → ArCN	92.5% EtOH	0°	2.240	1.760	52
Substituted-benzyl p-chlorophenyl ethers plus chlorine to give the 2,4-dichloro-phenyl benzyl ethers	99% CH$_3$COOH	20°	−0.884	1.868	53
Chlorination of 4-methoxybenzophenones in the 3-position	99% CH$_3$COOH	20°	−0.696	1.655	54
Decomposition of t-butyl perbenzoates	(C$_6$H$_5$)$_2$O	100° 110.1° 120.2° 130.9°	−0.903 −0.678 −0.568 −0.525	5.192 4.597 4.056 3.532	55
ArCHO + p-ClC$_6$H$_4$COO· → ArCO· (r = .986)	(CH$_3$CO)$_2$O	30°	−1.668	−1.071	56
Oxidation of iso-hydrobenzoin by aryl iodosoacetates	CH$_3$COOH	34.7°	1.438	2.609	57
4-Substituted 2-nitrophenyl fluorides plus ethoxide ion to give the ethyl ethers	EtOH	49.6°	4.071	1.780	58

(1) Largely from the extensive table in the review by H. H. Jaffé, *Chemical Reviews*, **53**, 191 (1953). The symbol r denotes the correlation coefficient. (2) The intercept of the Hammett equation, i.e., value of log k for $\sigma = 0$. (3) Based on log $k/k_0 = \rho \times 2(\sigma_1 + \sigma_2)$. (4) J. Shorter and F. J. Stubbs, *J. Chem. Soc.*, 1180, (1949). (5) H. L. Goering, T. Rubin, and M. S. Newman, *J. Am. Chem. Soc.*, **76**, 787 (1954). (6) B. Bettman, G. E. K. Branch, and D. L. Yabroff, *J. Am. Chem. Soc.*, **56**, 1865 (1934); G. E. K. Branch, D. L. Yabroff, and B. Bettman, *ibid.*, **56**, 937 (1934); C. G. Clear and G. E. K. Branch, *J. Org. Chem.*, **2**, 522 (1938); D. E. Yabroff, G. E. K. Branch, and B. Bettman, *J. Am. Chem. Soc.*, **56**, 1850 (1934). (7) H. H. Jaffé, L. D. Freedman, and G. O. Doak, *J. Am. Chem. Soc.*, **75**, 2209 (1953). (8) H. H. Jaffé, L. D. Freedman, and G. O. Doak, *J. Am. Chem. Soc.*, **76**, 1548 (1954). (9) O. L. Brady and N. M. Chokshi, *J. Chem. Soc.*, 946 (1929); O. L. Brady and R. F. Goldstein, *J. Chem. Soc.*, 1918 (1926). (10) C. M. Judson and M. Kilpatrick, *J. Am. Chem. Soc.*, **71**, 3115 (1949); H. Kloosterziel and H. J. Backer, *Rec. trav. chim.*, **71**, 295 (1952); E. E. Sager, M. R. Schooley, A. S. Carr, and S. F. Acree, *J. Research Nat. Bur. Standards*, **35**, 521 (1945); R. Näsänen, P. Lumme, and A. L. Mukula, *Acta Chem. Scand.*, **5**, 1199 (1951). (11) F. G. Bordwell and G. D. Cooper, *J. Am. Chem. Soc.*, **74**, 1058 (1952); C. G. Clear and G. E. K. Branch, *J. Org. Chem.*, **2**, 522 (1938); R. C. Farmer and F. J. Warth, *J. Chem. Soc.*, **85**, 1713 (1904); N. F. Hall and M. R. Sprinkle, *J. Am. Chem. Soc.*, **54**, 3469 (1932); L. P. Hammett and M. A. Paul, *J. Am. Chem. Soc.*, **56**, 827 (1934); H. Kloosterziel and H. J. Backer, *Rec. trav. chim.*, **71**, 295 (1952); D. Pressman and D. H. Brown, *J. Am. Chem. Soc.*, **65**, 540 (1943); J. D. Roberts, R. A. Clement, and J. J. Drysdale, *J. Am. Chem. Soc.*, **73**, 2181 (1951); J. D. Roberts, R. L. Webb, and E. A. McElhill, *J. Am. Chem. Soc.*, **72**, 408 (1950). (12) J. C. James and J. G. Knox, *Trans. Faraday Soc.*, **46**, 254 (1950). (13) O. C. M. Davis, *Z. physik. Chem.*, **78**, 353 (1912). (14) R. A. Henry, W. G. Finnegan, and E. Lieber, *J. Am. Chem. Soc.*, **76**, 88 (1954). (15) W. K. Wilmarth and N. Schwartz, *J. Am. Chem. Soc.*, **77**, 4543 (1955). (16) J. D. Roberts, R. A. Clement, and J. J. Drysdale, *J. Am. Chem. Soc.*, **73**, 2181 (1951); J. D. Roberts and E. A. McElhill, *J. Am. Chem. Soc.*, **72**, 628 (1950); J. D. Roberts, E. A. McElhill, and R. Armstrong, *J. Am. Chem. Soc.*, **71**, 2923 (1949); J. D. Roberts, R. L. Webb, and E. A. McElhill, *J. Am. Chem. Soc.*, **72**, 408 (1950). (17) J. D. Roberts and J. A. Yancey, *J. Am. Chem. Soc.*, **73**, 1011 (1951). (18) E. Tommila, *Ann. Acad. Sci. Fennicae*, **A57**, No. 13 (1941), **A59**, No. 3 (1942); E. Tommila, L. Brehmer, and H. Elo, *ibid.*, **A59**, No 9 (1942), **AII**, No. 16 (1945); E. Tommila and C. N. Hinshelwood, *J. Chem. Soc.*, 1801 (1938). (19) E. Tommila and L. Ketonen, *Suomen Kemistilehti*, **18B**, 24 (1945). (20) D. P. Evans, J. J. Gordon, and H. B. Watson, *J. Chem. Soc.*, 1430 (1937); C. K. Ingold and W. S. Nathan, *J. Chem. Soc.*, 222 (1936). (21) R. L. Herbst, Jr. and M. E. Jacox, *J. Am. Chem. Soc*, **74**, 3004 (1952); K. Kindler, *Ann.*, **450**, 1 (1926); **452**, 90 (1927); **464**, 278 (1928); K. Kindler, *Ber.*, **69B**, 2792 (1936); H. McCombie and H. A. Scarborough, *J. Chem. Soc*, **107**, 156 (1915); F. H. Westheimer and R. P. Metcalf, *J. Am. Chem. Soc.*, **63**, 1339 (1941). (22) R. W. Taft, Jr., M. S. Newman, and F. H. Verhoek, *J. Am. Chem. Soc.*,

Table 7-3 (continued)

72, 4511 (1950). (23) K. Kindler, *Ann.*, **452**, 90 (1927). (24) K. Kindler, *Ann.*, **464**, 278 (1928). (25) K. Kindler, *Ber.*, **69B**, 2792 (1936). (26) R. E. Robertson, *Can. J. Chem.*, **31**, 589 (1953). (27) D. A. Brown and R. F. Hudson, *Nature*, **167**, 819 (1951); *J. Chem. Soc.*, 883 (1953). (28) G. E. K. Branch and A. C. Nixon, *J. Am. Chem. Soc.*, **58**, 2499 (1936). (29) J. F. Norris and D. V. Gregory, *J. Am. Chem. Soc.*, **50**, 1813 (1928). (30) E. G. Williams and C. N. Hinshelwood, *J. Chem. Soc.*, 1079 (1934). (31) I. Meloche and K. J. Laidler, *J. Am. Chem. Soc.*, **73**, 1712 (1951). (32) K. Kindler, *Ann.*, **450**, 1 (1926). (33) E. Berliner and L. H. Altschul, *J. Am. Chem. Soc.*, **74**, 4110 (1952). (34) D. R. Boyd and E. R. Marle, *J. Chem. Soc.*, **105**, 2117 (1914). (35) L. J. Goldsworthy, *J. Chem. Soc.*, 1254 (1926). (36) E. Tommila and C. N. Hinshelwood, *J. Chem. Soc.*, 1801 (1938). (37) Y. Ogata and M. Okano, *J. Am. Chem. Soc.*, **71**, 3212 (1949). (38) E. Hertel and H. Lührman, *Z. Elektrochem.*, **45**, 405 (1939). (39) F. J. Stubbs and C. N. Hinshelwood, *J. Chem. Soc.*, supplement p. 71 (1949); E. G. Williams and C. N. Hinshelwood, *J. Chem. Soc.*, 1079 (1934). (40) A. W. Chapman and R. E. Parker, *J. Chem. Soc.*, 3301 (1951). (41) A. L. Sklar, *J. Chem. Phys.*, **7**, 984 (1939). (42) F. J. Stubbs and C. N. Hinshelwood, *J. Chem. Soc.*, supplement p. 71 (1949). (43) E. Hertel and J. Dressel, *Z. physik. Chem.*, **B29**, 178 (1935); E. Hertel and H. Lührman, *Z. Elektrochem.*, **45**, 405 (1939). (44) W. C. Davies and W. P. G. Lewis, *J. Chem. Soc.*, 1599 (1934). (45) K. J. Laidler, *J. Chem. Soc.*, 1786 (1938). (46) G. Williams, *J. Chem. Soc.*, 37 (1930). (47) J. W. Baker and J. B. Holdsworth, *J. Chem. Soc.*, 743 (1947). (48) J. W. Baker and H. B. Hopkins, *J. Chem. Soc.*, 1089 (1949); W. L. Bright and H. T. Briscoe, *J. Phys. Chem.*, **37**, 787 (1933). (49) E. Tommila, *Ann. Acad. Sci. Fennicae*, **A59**, No. 8 (1942). (50) D. P. Evans, V. G. Morgan, and H. B. Watson, *J. Chem. Soc.*, 1167 (1935); W. S. Nathan and H. B. Watson, *J. Chem. Soc.*, 217 (1933). (51) J. F. Lane and R. L. Feller, *J. Am. Chem. Soc.*, **73**, 4230 (1951). (52) C. R. Hauser, J. W. Le Maistre, and A. E. Rainsford, *J. Am. Chem. Soc.*, **57**, 1056 (1935). (53) A. E. Bradfield and B. Jones, *J. Chem. Soc.*, 2903 (1931); B. Jones, *J. Chem. Soc.*, 1835 (1935); 1414 (1938). (54) B. Jones, *J. Chem. Soc.*, 1854 (1936). (55) A. T. Blomquist and I. A. Berstein, *J. Am. Chem. Soc.*, **73**, 5546 (1951). (56) C. Walling and E. A. McElhill, *J. Am. Chem. Soc.*, **73**, 2927 (1951). (57) K. H. Pausacker, *J. Chem. Soc.*, 107 (1953). (58) C. W. L. Bevan, *J. Chem. Soc.*, 655 (1953).

The dependence of ρ on the temperature is therefore expressed by equation 6.

$$\rho_T = \rho_\infty \left(1 - \frac{\beta}{T}\right) \tag{6}$$

Equation 6 appears to represent the temperature dependence of ρ for reactions whose fit to the $\rho\sigma$ relationship does not deteriorate as the temperature is changed.[3] However, we do not really have very extensive data for reactions at temperatures much different from room temperature.[3] Of the three hundred-odd reactions tabulated by Jaffé, only eight had been studied outside of the 0–100° range.

As will be seen in Chapter 9, the coefficient β typically has values in the range 400 to 800°K. When the experimental temperature is equal to β, equation 6 predicts that ρ will change sign.

If the sign of ρ can be changed by a moderate change in temperature, we must conclude either that the partial charge developed in the transition state changes its sign or, more plausibly, that the simple interpretation of the sign of ρ in terms of the sign of the charge developed in the transition state is not always sound. This matter will be treated more fully in Chapter 9.

Application of the $\rho\sigma$ Relationship to Free Radical Reactions

Although substituent effects in some free radical reactions are correlated directly by means of the $\rho\sigma$ relationship, in others polar and nonpolar interactions contribute in about equal degree, and simple extrathermodynamic relationships are not observed.[6,7] When the nonpolar interactions are essentially constant, $\rho\sigma$ relationships are observed as in the example cited on page 176. When the *polar* interactions are essentially constant in two reactions, proportionality between the *nonpolar factors* may also lead to a linear free energy relationship. An example of the latter sort comes from the rates of chain transfer of the radical intermediates produced in the polymerization of monomers, such as styrene and acrylonitrile. In reaction (7) the chain transfer agent is cyclohexane; in reaction (8) it is toluene.

$$R\cdot + C_6H_{12} \xrightarrow{k_7} RH + C_6H_{11}\cdot \tag{7}$$

$$R\cdot + C_6H_5CH_3 \xrightarrow{k_8} RH + C_6H_5CH_2\cdot \tag{8}$$

Figure 7-3 shows the relationship obtained when log k_7 is plotted against log k_8. Not only is the relationship nicely linear, but the slope is unity,

[6] C. H. Bamford, A. D. Jenkins, and R. Johnston, *Trans. Faraday Soc.*, **55**, 418 (1959)

[7] T. Alfrey, Jr. and C. C. Price, *J. Polymer Sci.*, **2**, 101 (1947).

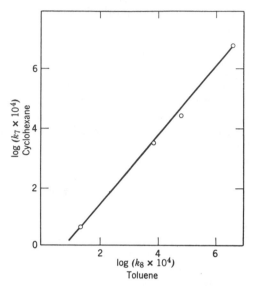

Figure 7-3. Reaction of a series of radicals with the substrates, cyclohexane and toluene, illustrating the linear relationships obtained when polar factors are nearly constant.

showing that the nonpolar substituent effects are equal rather than merely proportional in these reactions.

Use will be made of this fact in the analysis of substituent effects in reaction (9), a chain transfer reaction in which polar and nonpolar interactions are varying simultaneously.

$$R\cdot + CBr_4 \xrightarrow{k_9} R-Br + Br_3C\cdot \tag{9}$$

In reaction (9) polar effects are important because the ionic valence-bond structures (10b) and (10c) can make a significant contribution in the transition state.

$$\underset{(a)}{\dot{R}\ Br\text{-}CBr_3} \leftrightarrow \underset{(b)}{\overset{+}{R}\ \dot{Br}\ \overset{-}{C}Br_3} \leftrightarrow \underset{(c)}{\overset{+}{R}\ \overset{-}{Br}\ \dot{C}Br_3} \tag{10}$$

The presence of the nonpolar and polar substituent effects in reaction (9) makes the plot of log k_9 versus log k_8 (Figure 7-4) look like a scatter diagram. However, the pattern is not really a random one. The apparently complex substituent effects can be analyzed by the method of linear combination of two model processes, described in Chapter 6. In the present case the applicable equation is 11.

$$\delta_R \log k_9 = \delta_R \log k_8 + \rho \, \delta_R \sigma \tag{11}$$

In equation 11 the coefficient of the first term has been set equal to unity because of the observation of unit slope in the purely nonpolar extra-thermodynamic relationship of Figure 7-3.

The radicals R· of reactions (7) through (9) have at the growing end structures derived from their respective monomers, styrene, methyl methacrylate, vinyl acetate, methyl acrylate, methacrylonitrile, and acrylonitrile.[6] The σ-constants used in equation 11 are *para*-sigma constants corresponding to the groups attached to the double bond of the monomer, that is, σ_{phenyl} for styrene, σ_{CN} for acrylonitrile, etc. In Figure 7-5 the substrate is carbon tetrabromide, but ferric chloride and *n*-butyl mercaptan give similar results.

The Effect of Changes in Reaction Mechanism on the Hammett Relationship

When the mechanism of a reaction changes because of the presence of certain substituents or when the measured rate constant is actually a composite quantity depending on the rate and equilibrium

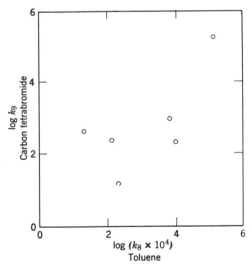

Figure 7-4. Reaction of a series of radicals with the substrates, carbon tetrabromide and toluene, illustrating the lack of correlation when polar factors vary.

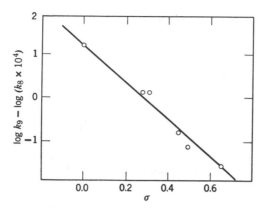

Figure 7-5. Substituent effects in a radical reaction expressed as a linear combination of polar and nonpolar models. k_9 refers to the reaction of R· with CBr_4, k_8 to the reaction of R· with toluene.

constants of several reaction steps, curvature in the $\rho\sigma$ relationship can occur. A striking example of an abrupt change in mechanism accompanied by an abrupt change in slope of the $\rho\sigma$ plot is the hydrolysis of benzoate esters in 99.9% sulfuric acid at 45°.[8] Hydrolysis of the *methyl* esters obeys a $\rho\sigma$ relationship with good precision, the rate decreasing consistently as the substituents are made more electron-attracting. This is to be expected for the acyl-oxygen fission mechanism:

$$\left[\begin{array}{c} \overset{\oplus}{O}H \\ \parallel \\ RC_6H_4\overset{}{C}OCH_3 \end{array} \rightleftharpoons \begin{array}{c} O \\ \parallel \\ RC_6H_4C-\overset{\oplus}{\underset{H}{O}}CH_3 \end{array} \right] \rightarrow \begin{array}{c} O \\ \parallel \oplus \\ RC_6H_4C \end{array} + CH_3OH \quad (12)$$

But for the *ethyl* esters it appears that strongly electron-attracting groups cause a sudden shift to alkyl-oxygen fission:

$$\begin{array}{c} O \\ \parallel \\ RC_6H_4C-\overset{\oplus}{\underset{H}{O}}-CH_2CH_3 \end{array} \rightarrow \begin{array}{c} O \\ \parallel \\ RC_6H_4C-OH \end{array} + CH_3CH_2{}^{\oplus} \quad (13)$$

The effect of the change in mechanism on the $\rho\sigma$ plot is shown in Figure 7-6.

[8] D. N. Kershaw and J. A. Leisten, *Proc. Chem. Soc.*, 84 (1960).

An example of a break in a $\rho\sigma$ plot even with constant mechanism is the behavior of the rate of formation of semicarbazones from substituted benzaldehydes.[9] This reaction has a two-step mechanism, and either step can be rate-determining:

$$\underset{\substack{\|\\ O}}{Ar-\overset{O}{\overset{\|}{C}}-H} + R-NH_2 \underset{k_{-1}}{\overset{k_1}{\rightleftharpoons}} R-\underset{\substack{|\\H}}{\overset{|}{\underset{|}{C}}}-NHR \xrightarrow{H^+,k_2} R-\underset{\substack{|\\H}}{\overset{|}{C}}=N-R + H_2O \tag{14}$$

In fairly acidic solutions (pH 1.75) step 2 is fast, and the observed rate constant is k_1; this gives a fairly precise linear $\rho\sigma$ relationship. In neutral solutions the observed rate is that of step 2, and the observed rate constant is a composite, $k_{obs} = K_1 k_2$. Since substituents have opposite effects on K_1 and k_2, they have almost no effect on k_{obs}, and ρ is only 0.07. At pH 3.9 a change in the nature of the substituents causes a change from k_1 rate-limiting to k_2 rate-limiting, and the $\rho\sigma$ plot has a sharp break at about $\sigma = 0$ (Figure 7-7). It should be noted that a change in mechanism always causes the curve to be concave up (Figure 7-6), while a change in rate-limiting step with otherwise constant mechanism can cause the curve to be

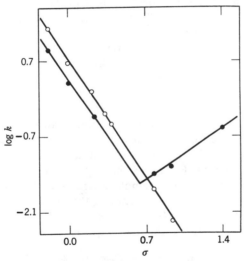

Figure 7-6. Hydrolysis of benzoate esters in 99.9% sulphuric acid. The open circles are for methyl esters, the filled circles for ethyl esters; k is in units of hr^{-1}.

[9] B. M. Anderson and W. P. Jencks, *J. Am. Chem. Soc.*, **82**, 1773 (1960).

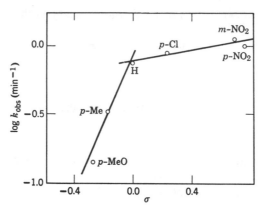

Figure 7-7. The formation of semicarbazones from substituted benzaldehydes at pH 3.9.

concave down, as in Figure 7-7. The condensation of substituted benz-aldehydes with methylethyl ketone[10] also gives a curved $\rho\sigma$ plot if the formal, mechanistically composite, rate constant is used.

The preceding examples dealt with the effect of abrupt changes in reaction mechanism. For some reactions it is believed that the reaction mechanism varies continuously. For example, in the nucleophilic displacement reaction the theoretical models, s_N1 and s_N2, are probably merely the endpoints of a continuous range of mechanism rather than discrete alternatives. A substituent can therefore alter the degree of resemblance of the actual mechanism to the s_N1 and s_N2 models. The effect will be small if the actual reaction mechanism is already close to either the s_N1 or the s_N2 limit. But for reactions near the middle of the range of mechanism, the effect can be quite appreciable.

An example of this effect, reflected in the shape of the $\rho\sigma$ plot, is the reaction of substituted benzyl halides with amines in nonhydroxylic solvents.[11,12] As is shown in Figure 7-8, the $\rho\sigma$ plot is markedly concave up. A *para*-methoxy substituent, for example, interacts strongly with positive charges and can reduce $\Delta F^{\ddagger}$ more by shifting the mechanism towards the s_N1 extreme than it can merely by interacting with the lesser amount of positive charge characteristic of the unchanged reaction mechanism. The

[10] D. S. Noyce and L. R. Snyder, *J. Am. Chem. Soc.*, **81**, 620 (1959).

[11] C. G. Swain and W. P. Langsdorf, Jr., *J. Am. Chem. Soc.*, **73**, 2813 (1951).

[12] G. S. Hammond, J. Peloquin, F. T. Fang, and J. G. Kochi, *J. Am. Chem. Soc.*, **82,** 443 (1960) and R. F. Hudson and G. Klopman, *J. Chem. Soc.*, 1062 (1962) have discussed related effects in the solvolysis of benzyl tosylates and in the reactions of benzyl bromides with substituted thiophenols.

dispersion into separate curves for the *meta-* and *para-*substituents may be due to the fact that the *meta-* and *para-*substituents differ in their ability to shift the reaction mechanism. The substituent effect on the reaction mechanism is analogous to a second interaction mechanism superimposed on the principal one. In the present example the two kinds of interaction are not simply correlated, and dispersion results. The $\rho\sigma$ plot for the solvolysis of substituted α-phenylethyl chlorides in a dioxanewater-formic acid mixture also seems to require separate lines for *meta-* and *para-*substituents, although the lines are not markedly curved.[13]

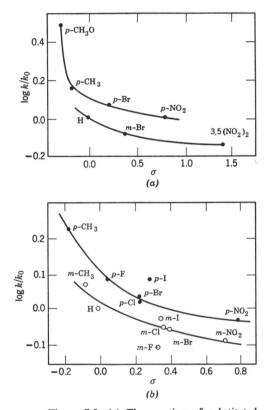

Figure 7-8. (*a*) The reaction of substituted benzyl chlorides with trimethylamine in benzene at 100°. (*b*) The reaction of substituted benzyl bromides with pyridine in acetone at 20°.

[13] C. Mechelynck-David and P. J. C. Fierens, *Tetrahedron*, **6,** 232 (1959).

The Effect of Multiple Substitution

When two or more substituents are introduced simultaneously into *meta-* or *para*-positions of an aromatic compound, it is usually found that their combined effect can be represented by the sum of their individual sigmas. Data are available for many reactions to confirm this simple additive relationship, including the ionization of benzoic acids, the saponification of ethyl benzoates, and the reaction of phenyl isothiocyanates with ethanol.[3] From a theoretical point of view, simple additivity is to be expected for those reactions for which the $\rho\sigma$ correlation is precisely linear. The constant slope, ρ, implies a constant value for the derivative of the free energy with respect to the interaction variable associated with the substituents. A constant value of the derivative implies that the effects of increments in the interaction variable are linear.

Deviations from additivity of substituent effects can be expected either if the $\rho\sigma$ plot is curved or if the interaction of the two substituents *with each other* is changed by the reaction. Circumstances in which the latter effect might occur can readily be imagined. For example, the two substituents might be attached to two different aromatic rings that are not conjugated in the reagent but become so as a result of the reaction. If the interaction variable involved in the interaction of the two substituents with each other is proportional to σ, the deviation from strict linearity would be given by a term proportional to the product of the two σ-values.

$$\delta_{(R_1+R_2)} \log k = \rho(\sigma_1 + \sigma_2) + q\sigma_1\sigma_2 \tag{15}$$

If the mechanism of the interaction between the two substituents differs from that involved in the $\rho\sigma$ term, then there should be no correlation between the correction term and the sigmas.

The Number of Interaction Mechanisms in the $\rho\sigma$ Relationships

Although a substituent on a benzene ring is expected to interact by at least two mechanisms (the well-known resonance and inductive effects), the simple linearity of many $\rho\sigma$ relationships might seem to imply the existence of only a single interaction mechanism. However, it can easily be shown that a linear relationship can be obtained even for two or more interaction mechanisms provided that their relative importance in the series of reactions remains constant.

For two interaction mechanisms the Hammett relationship will take the form of equation 16.

$$\delta_R \log k = \rho'\sigma' + \rho''\sigma'' \tag{16}$$

Since the two interaction mechanisms are independent, σ' and σ'' change independently. However, if we rearrange the equation as shown in equation 17, we see that the substituent effect can still be described by a *single* parameter as long as the *ratio* ρ'/ρ'' is constant. This is because, for constant ρ'/ρ'', the quantity $(\sigma' + \rho''\sigma''/\rho')$ is characteristic only of the substituent.

$$\delta_R \log k = \rho'\left(\sigma' + \frac{\rho''}{\rho'}\sigma''\right) \tag{17}$$

We shall now prove that the usual $\rho\sigma$ relationship is indeed the result of at least two independent interaction mechanisms whose relative importance, as measured by ρ'/ρ'', is constant. The proof will be indirect. By *assuming* that only a *single* interaction mechanism is involved in the $\rho\sigma$ equation, we shall derive a prediction that can be compared with experiment and shown to be contrary to fact.

The proof makes use of the equivalent logical status of operations involving the substituent and those involving the functional group. Although we have found it convenient to distinguish between these two types of operation by using different symbols for the operators, the distinction is arbitrary, for it is always possible to imagine a reaction that would bring about the desired change of substituent. Consequently, it is possible to relate the ρ-value for a reaction to the difference in σ between the final and the initial state of the functional group, treating the functional group as if it were a substituent. This method for probing the $\rho\sigma$ relationship has been developed and explored by Hine.[14] We shall now use it to derive a relationship for the special case that the interaction involves only a single mechanism.

It is helpful to make the problem symmetrical by selecting as substituents a pair of structures identical to the initial and final states of the functional group. An example of such a pair of structures might be COOH and COO^-. The relevant equilibria are then given by equations 18.[15]

$$X_1\phi X_1 \underset{\phantom{K_{X1}}}{\overset{K_{X1}}{\rightleftharpoons}} X_1\phi X_2 \tag{18a}$$

$$X_2\phi X_1 \underset{\phantom{K_{X2}}}{\overset{K_{X2}}{\rightleftharpoons}} X_2\phi X_2 \tag{18b}$$

In (18a) the substituent is X_1; in (18b) it is X_2.

The substituent effect on the standard free energy change may now be

[14] J. Hine, *J. Am. Chem. Soc.*, **81**, 1126 (1959).

[15] The constants K_{X_1} and K_{X_2} used in equations 18 and 19 are formal equilibrium constants corrected for statistical factors.

written as equation 19 in the form of a Hammett relationship or as equation 20 in terms of interaction factors.

$$\Delta F^\circ_{X_2} - \Delta F^\circ_{X_1} = -RT \ln \frac{K_{X_2}}{K_{X_1}} = -RT \rho (\sigma_{X_2} - \sigma_{X_1}) \qquad (19)$$

$$\Delta F^\circ_{X_2} - \Delta F^\circ_{X_1} = (I_{X_2} - I_{X_1})(I_{X_2} - I_{X_1}) \qquad (20)$$

Equation 20 of course contains the assumption that there is only a single interaction mechanism. The equation is obtained by making appropriate substitutions in equation 21, which was derived in Chapter 6, pages 140–141.

$$\delta_R \Delta F^\circ_{X_1 \to X_2} = (I_R - I_{R_0})(I_{X_2} - I_{X_1}) \qquad (21)$$

The symmetry of the right-hand side of equation 20 demands an equivalent symmetry of the right-hand side of equation 19. Since the factor $(\sigma_{X_2} - \sigma_{X_1})$ of equation 19 is proportional to $(I_{X_2} - I_{X_1})$, the other factor $-RT\rho$ must also be proportional to $(I_{X_2} - I_{X_1})$ and hence to $(\sigma_{X_2} - \sigma_{X_1})$.

$$\rho = \tau(\sigma_{X_2} - \sigma_{X_1}) \qquad (22)$$

The parameter τ is known as the *transmission coefficient*.[14]

Since the same value of ρ applies to *meta-* and *para*-substituents, equation 22 leads to 23.

$$(\sigma_{X_2} - \sigma_{X_1})_{meta} = (\sigma_{X_2} - \sigma_{X_1})_{para} \qquad (23)$$

According to equation 23 the difference between the σ-values for a given substituent in the *meta-* and *para*-positions must be the same for all substituents. That this is contrary to fact is readily seen from Table 7-4. *We therefore conclude that the $\rho\sigma$ relationship involves more than one interaction mechanism.*[16]

GENERAL TREATMENT OF SUBSTITUENT EFFECTS IN AROMATIC REACTIONS

On examining Table 7-4 we note that the sequence of decreasing $(\sigma_{meta} - \sigma_{para})$ for the various substituents is the same as the sequence of decreasing electron-releasing resonance. This revealing piece of evidence is part of the large body of fact on which the structural theory of substituent effects in aromatic reactions is based. According to this theory, the principal interaction mechanisms in the reactions of *meta-* and *para*-substituted aromatic compounds are the inductive and resonance mechanisms.

---

[16] "We make our own world; when we have made it awry, we can remake it, approximately truer, though it cannot be absolutely true, to the facts."—Havelock Ellis.

Table 7-4. Inequality of $(\sigma_{meta} - \sigma_{para})$
for Various Substituents[a]

SUBSTITUENT	$\sigma_{meta} - \sigma_{para}$
OC_6H_5	$0.57 \pm .03$
OH	$0.49 \pm .05$
OCH_3	$0.38 \pm .03$
F	$0.28 \pm .03$
Cl	$0.15 \pm .03$
Br	$0.16 \pm .03$
CH_3	$0.10 \pm .03$
$C(CH_3)_3$	$0.10 \pm .03$
C_6H_5	$0.07 \pm .07$
NO_2	$-0.07 \pm .03$
CN	$-0.10 \pm .05$
$COCH_3$	$-0.13 \pm .03$

[a] Calculated from primary σ-values of Table 7-1.

As we have seen, the success of the Hammett relationship depends on the coincidence that in a great many aromatic side-chain reactions, the relative importance of the inductive and resonance mechanisms is nearly, if not exactly, constant. However, there is no reason to suppose that the particular ratio of ρ'/ρ'' that gives rise to the σ-scale is universal. Indeed, we would expect that resonance is relatively less important in the ionization of benzoic acid than in reactions in which a formal positive charge or a nonbonding pair of electrons is produced on the ring or on an atom immediately adjacent to the ring. The reason for this may be seen from a consideration of resonance structures such as (24) to (28).

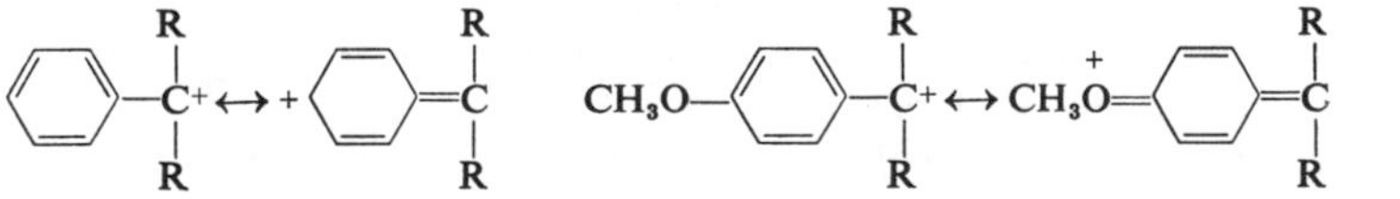

(24)

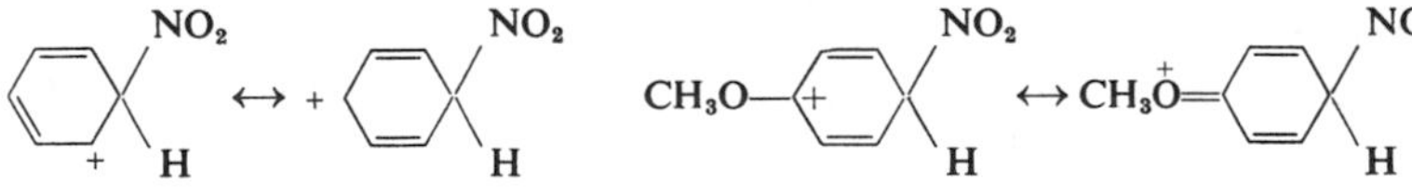

(25)

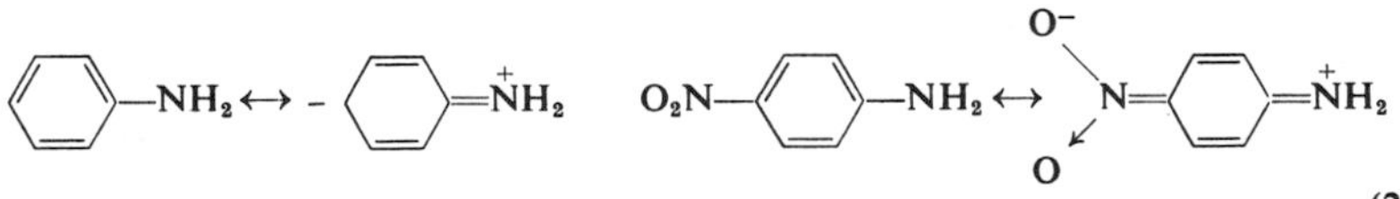

(26)

$$(27)$$

$$(28)$$

Because the relative importance of the resonance effect is variable, a general linear free energy relationship for substituent effects in aromatic reactions is at least as complicated as equation 16, with ρ' and ρ'' being *independent* functions of the reaction. It is quite possible, however, that the number of principal interaction mechanisms is greater than two. For example, the resonance effect itself could involve two mechanisms, depending on the sign of the charge to be stabilized, as shown by equations 24 and 25 versus 26 and 27.

Aromatic Substitution. The Brown Relationship

Although it may seem surprising that the ratio ρ'/ρ'' should ever be constant for a large class of reactions, we have not only the many side-chain reactions that obey the Hammett relationship but also another large class, the aromatic substitution reactions, which obey a similar relationship but with a different constant value of ρ'/ρ''. This relationship has been studied extensively by Brown and co-workers.[17] In the discussion of the Hammett relationship it was convenient to compare the effect of long series of substituents on pairs of reactions. In the discussion of aromatic substitution it is more convenient to compare the effect of pairs of substituents in a long series of reactions. The experimental problem of obtaining relative rates is further simplified if the two substituents are chosen to be the *same* substituent in the *meta-* and *para-*position. In this way it becomes practical to measure relative rates simply from the relative yields of *meta-* and *para-*substitution product.

The most thoroughly studied aromatic substitution relationship concerns the reactions of toluene and benzene with a series of reagents. The reaction rates are corrected for statistical factors and placed on a per-reactive-site basis by using *partial rate factors*, p_f, m_f, and o_f, defined by equations 29 through 31.

$$p_f = \frac{k_{\text{toluene}} \times (\text{fraction of } para\text{-substitution})}{k_{\text{benzene}} \times \frac{1}{6}} \tag{29}$$

[17] C. W. McGary, Jr., Y. Okamoto, and H. C. Brown, *J. Am. Chem. Soc.*, **77**, 3037 (1955).

$$m_f = \frac{k_{\text{toluene}} \times (\tfrac{1}{2})(\text{fraction of } meta\text{-substitution})}{k_{\text{benzene}} \times \tfrac{1}{6}} \tag{30}$$

$$o_f = \frac{k_{\text{toluene}} \times (\tfrac{1}{2})(\text{fraction of } ortho\text{-substitution})}{k_{\text{benzene}} \times \tfrac{1}{6}} \tag{31}$$

The partial rate factors need not be computed from measured rate constants but can be deduced from the yields of toluene and benzene substitution products obtained in competitive experiments in which both substrates are present simultaneously. When the latter method is used, the concentration of substrates should be kept low to avoid medium effects.

As can be seen from its definition, p_f is a measure of the ability of a reagent to discriminate between the *para*-position in a toluene molecule and a single position in a benzene molecule. It is therefore an *intermolecular* selectivity factor and is analogous to the quantity $k_{p\text{-methyl}}/k_0$ in the Hammett equation 2. The ratio p_f/m_f, in which the benzene rates cancel out, is an *intramolecular* selectivity factor, measuring the ability of the reagent to discriminate between the *para-* and *meta*-positions of toluene. Moreover, the free energy quantity $[-RT \ln p_f/m_f]$ associated with this ratio has a particularly simple physical significance, being simply equal to the difference between the standard free energies of the *para-* and *meta*-substituted transition states.

Brown and his co-workers[17,18] have found that the values of $\log p_f$ and $\log p_f/m_f$ vary in simple proportion as the aromatic substitution reaction (i.e., nitration, halogenation, Friedel-Crafts, etc.) is changed.

$$\log p_f = C_{\text{methyl}} \cdot \log \frac{p_f}{m_f} \tag{32}$$

Although we are now only concerned with the reactions of toluene, the Brown relationship will be seen to apply also to other aromatic substrates. We have therefore denoted the proportionality constant in (32) by C_{methyl} to emphasize its dependence on the nature of the directing substituent.

Table 7-5[18] gives some of the observed values of the partial rate factors for electrophilic substitution reactions of toluene and for a few closely related reactions in which the displaced group may be something other than a proton. The excellent fit to the $\log p_f$ versus $\log (p_f/m_f)$ relationship is shown in Figure 7-9. The slope is 1.325, based on 32 sets of data selected for reliability; if no data are excluded, the slope is 1.316.[18]

The fact that $\log p_f$ is proportional to $\log p_f/m_f$ implies that the relative importance of the inductive and resonance effects is constant throughout the series but does not by itself indicate any qualitative difference between

[18] L. M. Stock and H. C. Brown, *J. Am. Chem. Soc.*, **81**, 3323 (1959).

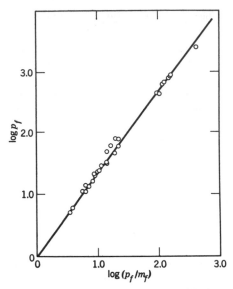

Figure 7-9. The Brown relationship for
the reactions of toluene.

these reactions and those correlated by the Hammett relationship. If the
Hammett relationship were applicable, it would predict equations 33.

$$\log p_f = \rho\sigma_{p\text{-methyl}} \tag{33a}$$

$$\log \frac{p_f}{m_f} = \rho(\sigma_{p\text{-methyl}} - \sigma_{m\text{-methyl}}) \tag{33b}$$

Hence the slope in equation 32, C_{methyl}, should be given by equation 34.

$$C_{methyl} = \frac{\sigma_{p\text{-methyl}}}{\sigma_{p\text{-methyl}} - \sigma_{m\text{-methyl}}} \tag{34}$$

Using σ-values from Table 7-1, we predict that $C_{methyl} = 1.68$, in contrast
to the observed value of 1.325. Hence we see that this series of reactions
can be described by an equation of the $\rho\sigma$ form, but that the substituent
constants are different from those used in the Hammett equation. The
reason for this is that the characteristic value of the ratio ρ'/ρ'' for the two
interaction mechanisms (equation 17) is different.

There are some indications that the selectivities measured by p_f and by
p_f/m_f decrease with increasing reactivity of the electrophilic reagent.
Unfortunately we do not have a precise experimental measure of the
reactivities of all the attacking reagents. However, in many cases we can

Table 7-5. Some Electrophilic Substitution Reactions of Toluene[18]

REACTION	o_f	m_f	p_f	$\dfrac{\log p_f}{\log (p_f/m_f)}$
Bromination, Br_2, 85% HOAc, 25°	600	5.5	2420	1.28
Chlorination, Cl_2, HOAc, 25°	617	4.95	820	1.31
Benzoylation, $\phi COCl$, $AlCl_3$, ϕNO_2, 25°	32.6	5.0	831	1.31
Benzoylation, $\phi COCl$, $AlCl_3$, 25°	30.7	4.8	589	1.33
Benzoylation, $\phi COCl$, $AlCl_3$, $C_2H_4Cl_2$, 25°	32.6	4.9	626	1.33
Acetylation, CH_3COCl, $AlCl_3$, $C_2H_4Cl_2$, 25°	4.5	4.8	749	1.31
Deuteration, D_2O, CF_3COOH, 70°	253	3.8	421	1.28
Chloromethylation, CH_2O, HCl, $ZnCl_2$, HOAc, 60°	117	4.37	430	1.32
Bromination, HOBr, $HClO_4$, 50% dioxane, 25°	76	2.5	59	1.29
Chlorination, HOCl, $HClO_4$, H_2O, 25°	134	4.0	82	1.46
Brominolysis of $ArB(OH)_2$, Br_2, 20% HOAc, 25°	—	3.3	78.5	1.38
Nitration, HNO_3, 90% HOAc, 45°	42	2.5	58	1.29
Nitration, $AcONO_2$, Ac_2O, 0°	47.3	3.56	60.3	1.45
Nitration, $AcONO_2$, Ac_2O, 30°	40.4	3.04	51.2	1.39
Nitration, HNO_3, CH_3NO_2, 30°	36.6	2.33	46.1	1.28
Sulfonylation, ϕSO_2Cl, $AlCl_3$, 25°	6.8	2.09	30.2	1.28
Detrimethylsilylation, $ArSiMe_3$, Br_2, 98.5% HOAc, 25°	—	3.2	49	1.43
Mercuration, $Hg(OAc)_2$, $HClO_4$, HOAc, 25°	4.98	2.25	32.9	1.30
Mercuration, $Hg(OAc)_2$, $HClO_4$, HOAc, 50°	4.20	2.41	28.8	1.36
Mercuration, $Hg(OAc)_2$, $HClO_4$, HOAc, 70°	3.24	2.23	24.5	1.34
Detrimethylsilylation, $ArSiMe_3$, $HClO_4$, 50% MeOH, 51.2°	—	2.3	21.2	1.38
Detrimethylsilylation, $ArSiMe_3$, HCl, HOAc, 25°	—	2.14	20.1	1.34
Mercuration, $Hg(OAc)_2$, HOAc, 25°	5.71	2.23	23.0	1.34
Mercuration, $Hg(OAc)_2$, HOAc, 50°	4.60	1.98	16.8	1.32
Mercuration, $Hg(OAc)_2$, HOAc, 70°	4.03	1.83	13.5	1.30
Mercuration, $Hg(OAc)_2$, HOAc, 90°	3.51	1.70	11.2	1.28
Detrimethylsilylation, $ArSiMe_3$, $p\text{-}C_7H_7SO_3H$, HOAc, 25°	17.5	2.0	16.5	1.33
Detrimethylsilylation, $ArSiMe_3$, $p\text{-}C_7H_7SO_3H$, H_2O, HOAc, 25°	15.7	2.19	14.3	1.42
Detrimethylsilylation, $ArSiMe_3$, $0.02M$ $Hg(OAc)_2$, HOAc, 25°	10.8	1.99	11.5	1.39
Methylation, CH_3Br, $GaBr_3$, ArH, 25°	9.51	1.7	11.8	1.27
Ethylation, C_2H_5Br, $GaBr_3$, ArH, 25°	2.84	1.56	6.02	1.33
Isopropylation, $i\text{-}C_3H_7Br$, $GaBr_3$, ArH, 25°	1.52	1.41	5.05	1.27

predict whether a given change in the nature of the attacking reagent will increase or decrease its reactivity. In all cases the selectivity appears to decrease with increasing reactivity, although the analysis is considerably complicated by uncertainty as to the exact nature of the attacking reagent in some cases.

For example, in the mercuration reaction the use of an acid catalyst reduces the selectivities (Table 7-5). In the Friedel-Crafts reaction catalyzed by gallium bromide, the kinetics and the dependence of the rate on the nature of the halide ion make it very likely that the reactions of the methyl and ethyl halides are concerted displacement reactions in which the aromatic substrate bonds to carbon at the same time that halide ion bonds to the catalyst. The *t*-butylation reaction and the *i*-propylation reaction, on the other hand, probably involve transition states resembling the corresponding carbonium ions.

$$
\underset{\substack{\text{H} \qquad \text{H}}}{\overset{\substack{\text{H} \quad \text{H}}}{\text{C}}} \cdots \text{Br} \cdots \text{GaBr}_3 \qquad\qquad \underset{\substack{\text{H} \quad \text{CH}_3}}{\overset{\substack{\text{CH}_3 \; \text{CH}_3}}{\text{C}^+}} \qquad (35)
$$

The expected order of increasing reactivity (and presumably of decreasing selectivity) is *t*-butyl < *i*-propyl < ethyl < methyl for carbonium ions. On the other hand, the reverse order is expected for the concerted reaction with the alkyl halide-catalyst complex. The mixed observed order of *selectivity* is methyl > ethyl > *i*-propyl < *t*-butyl and supports the suggested change in mechanism.[19]

If we postulate that the selectivities p_f and p_f/m_f always decrease with increasing reactivity, we can use the selectivity data to learn something about the nature of the attacking reagent. For example, in the benzoylation reaction the electrophilic intermediate produced in nitrobenzene is more selective and therefore apparently more solvent-stabilized than the intermediate using benzoyl chloride itself as a solvent.[20]

The selectivity of a reagent in discriminating between *ortho*- and *para*-positions depends not only on polar but also on steric interactions. The relationship between selectivity and reactivity is therefore more complicated. However, for reagents of similar steric bulk we still expect that the magnitude of the selectivity will decrease with increasing reactivity. Our reasons, briefly, are as follows. In Chapter 6 we showed that in the special case of a *single* interaction mechanism, a selectivity-reactivity relationship will exist only if certain fairly stringent conditions are satisfied. If there are several interaction mechanisms, a selectivity-reactivity relationship can still exist provided that these stringent conditions are satisfied independently by each mechanism. In the present case, the data for *para*- and *meta*-positions already suggest that polar effects acting alone will give a selectivity-reactivity relationship. Hence we now need to show only that steric effects acting alone can give one also. This can be done very easily as

[19] H. C. Brown and C. R. Smoot, *J. Am. Chem. Soc.*, **78**, 6255 (1956).
[20] H. C. Brown and F. R. Jensen, *J. Am. Chem. Soc.*, **80**, 2297 (1958).

follows. For reagents of fixed bulk, the steric retardation will depend on the degree of covalency and length of the bond established in the transition state. The longer the bond, the smaller will be the steric retardation. But the bond length will increase with the reactivity (other interactions being constant) because a more reactive species will generate a transition state that resembles the reagent state more.

An experimental result in agreement with the predicted effect of increasing reactivity on the product distribution is that benzenesulfonylation gives three times as much *ortho*-substitution as does benzoylation.[21] The actual reagents are believed to be $C_6H_5CO^+$ and $C_6H_5SO_2^+$, which are nearly equal in bulk, but the importance of bulk is less for the reactions of $C_6H_5SO_2^+$.

The relationship of Figure 7-9 involves p_f and m_f only. It might be expected that the use of o_f instead of p_f would produce scatter, at least whenever what might happen to be the critical degree of steric hindrance is exceeded. Figure 7-10 shows this effect.[19] Many of the points fall on a line through the origin, but those for ethylation, *i*-propylation, and *t*-butylation fall progressively farther below the line. The deviation of the point for chloromethylation is about the same as that for ethylation. There is a double reason for the difference between *i*-propylation and *t*-butylation: the *t*-butyl group is not only more bulky but probably forms a more covalently bonded transition state.

The Brown relationship has been extended to the reactions of *t*-butylbenzene[22] and anisole.[23] This extension is of theoretical interest because of the varying importance of resonance in the stabilizing effect of methyl, *t*-butyl, and methoxyl substituents.

Since resonance is particularly important in the effect of a *p*-methoxyl substituent (equations 24 and 25), the use of this substituent severely tests the idea that ρ'/ρ'' remains constant in a given series of reactions. The probable result of a drift in ρ'/ρ'' would be curvature in the plot of $\log p_f$ versus $\log p_f/m_f$. No such curvature is noticeable in the plots for toluene and *t*-butylbenzene. The plot for anisole (Figure 7-11) shows a slight curvature.[23]

Just as the Brown relationship can be used over a considerable range of directing substituents, it can also be used for a considerable range of attacking reagents. It can even be extended to the rearrangement of β-aryl carbonium ions and the formation of α-aryl carbonium ions. The resemblance between these processes and the processes conventionally regarded as aromatic substitution reactions is shown on page 203.

[21] F. R. Jensen and H. C. Brown, *J. Am. Chem. Soc.*, **80**, 4046 (1958).
[22] L. M. Stock and H. C. Brown, *J. Am. Chem. Soc.*, **81**, 5621 (1959).
[23] *Ibid.*, **82**, 1942 (1960).

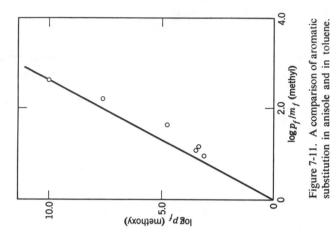

Figure 7-11. A comparison of aromatic substitution in anisole and in toluene.

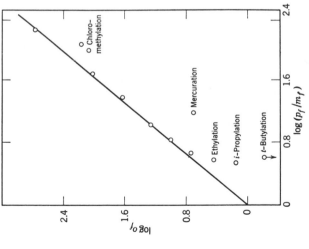

Figure 7-10. Perturbation by steric effects of the Brown relationship for the reactions of toluene.

substitution by X^+ rearrangement of Ar—$\overset{|}{C}$—$\overset{|}{C}$—X   ionization of Ar—$\overset{|}{C}$—X

$$(36)$$

The rates of such reactions can be correlated by a treatment formally similar to that used before. The relevant data are given in Table 7-6, in

Table 7-6. The Effect of Methyl-Substitution in Various Reactions of Aromatic Compounds[18]

REACTION	o_f	m_f	p_f	$\dfrac{\log p_f}{\log (p_f/m_f)}$
Ionization, Ar_3COH, H_2SO_4, 25°	—	1.24	10.5	1.10
Ionization, Ar_3CCl, SO_2, 0°	20.2	2.32	22.4	1.37
Migration aptitude, $(Ar_2COH)_2$	—	1.94	15.7	1.32
Rearrangement, $ArCHOHCH{=}CHCH_3$	2.00	1.21	8.99	1.10
Solvolysis reactions:				
$ArCMe_2Cl$, 90% acetone, 25°	3.63	2.00	26.0	1.27
Ar_2CHCl, 66.7% acetone, 0°	—	1.72	32.1	1.18
Ar_2CHCl, 70% acetone, 0°	—	1.76	32.5	1.19
Ar_2CHCl, 80% acetone, 0°	—	1.79	32.9	1.20
Ar_2CHCl, 90% C_2H_5OH, 0°	—	1.93	30.9	1.24
Ar_2CHCl, C_2H_5OH, 25°	2.9	2.1	16.2	1.36
$ArCH_2OTs$,[a] HOAc, 40°	—	2.6	56.5	1.31
$ArCH_2OTs$,[a] 76.6% acetone, 25.3°	—	1.79	29.8	1.21
$ArCH_2Cl$, 48% C_2H_5OH, 30°	4.95	1.30	9.37	1.13
$ArCH_2Cl$, 48% C_2H_5OH, 83°	4.78	1.39	10.6	1.16
$ArCH_2Cl$, 50% acetone, 30°	4.36	1.24	7.7	1.12
$ArCH_2Cl$, 50% acetone, 60°	4.11	1.20	8.7	1.09
$ArC(CH_3)_2CH_2OBs$,[b] HOAc, 75°	—	1.93	7.27	1.49

[a] Ts = $p\text{-}CH_3C_6H_4SO_2^-$.
[b] Bs = $p\text{-}BrC_6H_4SO_2^-$.

which p_f now represents $k_{p\text{-tolyl}}/k_{phenyl}$ and m_f represents $k_{m\text{-tolyl}}/k_{phenyl}$.[18] The fit, at least for the data available at present, is less satisfactory than the excellent fit for the electrophilic substitution reaction itself. The average value of $\log p_f/\log (p_f/m_f)$ is 1.23, in fair agreement with the slope, C_{methyl}, of Figure 7-9.

The $\rho\sigma^+$ Relationship

The same relative importance of resonance and inductive effects that is found in the aromatic substitution reaction is also found in a variety of aromatic side-chain reactions in which a positive charge capable of direct resonance interaction with the ring is generated. This makes it worthwhile to adopt a parameter σ^+ such that rates and equilibrium constants of these reactions are given by equation 37.

$$\log \frac{k}{k_0} = \rho\sigma^+ \tag{37}$$

Table 7-7. *Electrophilic Substituent Constants* (σ^+) *Compared with* σ [a,b]

SUBSTITUENT	σ_{meta}	σ^+_{meta}	σ_{para}	σ^+_{para}
Dimethylamino	—	—	−0.83	(−1.7)
Anilino	—	—	—	(−1.4)
Amino	−0.16	(−0.16)	−0.66	(−1.3)
Hydroxy	0.121	—	−0.37	(−0.92)
Acetylamino	0.21	—	−0.01	(−0.6)
Benzoylamino	—	—	—	(−0.6)
Methoxy	0.115	0.047	−0.268	−0.778
Phenoxy	0.252	—	−0.320	(−0.5)
Methylthio	0.15	0.158	0.00	−0.604
Methyl	−0.069	−0.066	−0.170	−0.311
Ethyl	−0.07	−0.064	−0.150	−0.295
Isopropyl	—	−0.060	−0.151	−0.280
t-Butyl	−0.10	−0.059	−0.197	−0.256
Phenyl	0.06	0.109	−0.01	−0.179
3,4-C_4H_4 (β-naphthyl)	—	—	0.042	−0.135
$CH_2COOC_2H_5$	—	(−0.01)	—	(−0.164)
Chloromethyl	—	(0.14)	—	(−0.01)
Hydrogen	0	0	0	0
Trimethylsilyl	−0.04	0.011	−0.07	0.021
Fluoro	0.337	0.352	0.062	−0.073
Chloro	0.373	0.399	0.227	0.114
Bromo	0.391	0.405	0.232	0.150
Iodo	0.352	0.359	0.18	0.135
Carboxy	—	0.322	—	0.421
Carbomethoxy	—	0.368	—	0.489
Carboethoxy	0.37	0.366	0.45	0.482
Trifluoromethyl	0.42	0.520	0.54	0.612
Cyano	0.56	0.562	0.660	0.659
Nitro	0.710	0.674	0.778	0.790
Carboxylate (K salt)	−0.1	−0.028	0.0	−0.023
$[(CH_3)_3N—]^+$ (chloride)	1.01	0.359[c]	0.88	0.408[c]

[a] From H. C. Brown and Y. Okamoto, *J. Am. Chem. Soc.*, **80**, 4979 (1958).
[b] Values in parenthesis are estimated from reactions other than the solvolysis of substituted 2-phenyl-2-propyl chlorides.
[c] The values of σ^+ for $[(CH_3)_3N—]^+$ are surprisingly small and deserve further study.

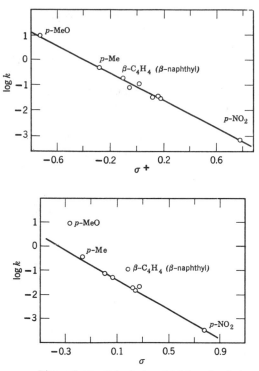

Figure 7-12. Solvolysis of triphenylmethyl chlorides.

The standard process for defining σ^+ is the solvolysis of substituted 2-phenyl-2-propyl chlorides in 90% acetone at 25°.[24] In order to make the σ^+ and σ scales as nearly as possible alike, a normalizing procedure was used. Since resonance effects are much less important for certain *meta*-substituents, these substituents were used to establish the value of ρ_0 for the σ^+ standard reaction. The plot of $\log k/k_0$ versus the Hammett σ-value for these substituents is nearly linear, and its average slope was adopted as the value of ρ_0. The σ^+-values are then obtained from equation 38.

$$\sigma_X^+ = \frac{\log (k_X/k_0)_{\text{standard reaction}}}{\rho_0} \qquad (38)$$

Values of σ^+ are given in Table 7-7.

As a result of the normalizing procedure, σ^+-values for most *meta*-substituents are very close to the corresponding σ-values. Substituents

[24] Y. Okamoto and H. C. Brown, *J. Org. Chem.*, **22**, 485 (1957); *J. Am. Chem. Soc.*, **80**, 4979 (1958).

such as nitro- or methoxy- for which resonance is particularly important, however, have substantially different σ- and σ^+-values even in the *meta*-position.

Figure 7-12 shows how the use of σ^+ rather than σ reduces scatter in the correlation of the triphenylmethyl chloride solvolysis rates in 40% ethanol-60% ethyl ether at 0°.[25] Similar or even better improvements are observed for the solvolysis of benzhydryl chlorides in ethyl or isopropyl alcohol.

A collection of ρ-values for the $\rho\sigma^+$ correlation is given in Table 7-8.[24] The values are always negative and their magnitudes are often large, in agreement with our expectation that these reactions are facilitated by electron-supplying substituents. The resemblance of the various reactions to the σ^+ standard reaction can be seen from the following sample formulas, which represent either transition states or products of the rate-determining steps.

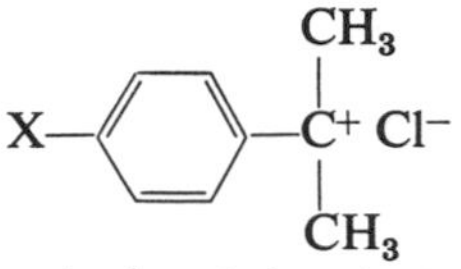

$$(39)$$

the σ^+ standard reaction

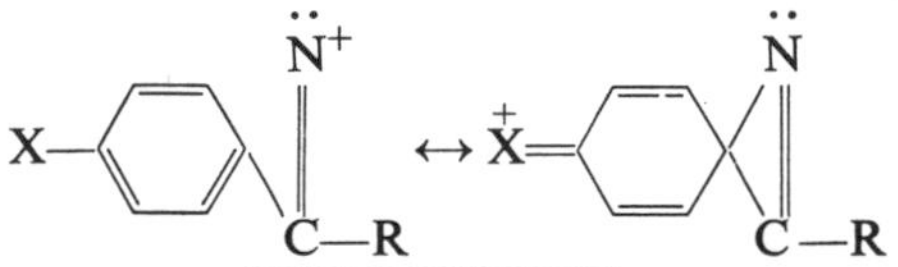

$$(40)$$

Beckmann rearrangement

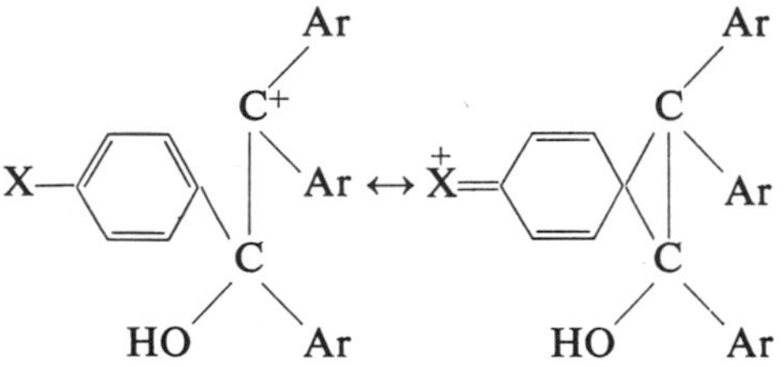

$$(41)$$

pinacol rearrangement

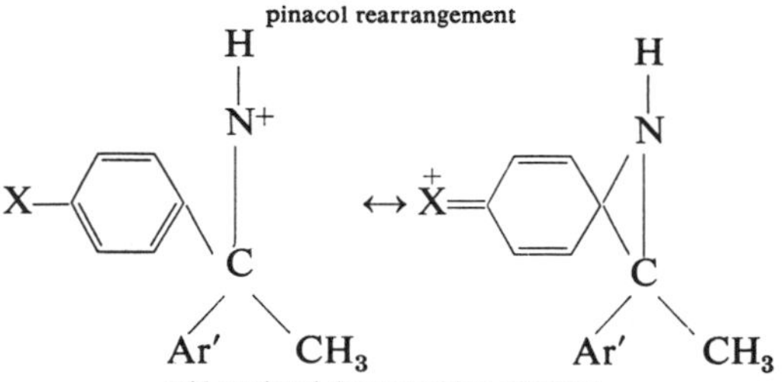

$$(42)$$

acid-catalyzed decomposition of azides

[25] A. C. Nixon and G. E. K. Branch, *J. Am. Chem. Soc.*, **58**, 492 (1936).

aromatic substitution

$$(43)$$

addition of bromine to a double bond

$$(44)$$

decomposition of α-phenylethyl chlorocarbonates

$$(45)$$

Although reaction (45) takes place in nonpolar solvents and with retention of configuration, the fact that σ^+ gives a much better fit than σ has been taken as supporting a mechanism involving intimate ion pairs.[26]

The resonance and inductive interaction mechanisms undoubtedly are the principal mechanisms involved in the $\rho\sigma^+$ relationship, but solvation also seems to be important for some substituents. Consider, for example, the effect of a *meta*-cyano substituent in the solvolysis of 2-phenyl-2-propyl chloride. The *meta*-cyano compound has the same $\Delta H^{\ddagger}$ as the *meta*-nitro compound but differs in $\Delta S^{\ddagger}$ by eight entropy units. Such large entropy effects in rigid molecules are almost certainly solvation effects. The possibility that the solvation might be susceptible to steric effects makes the behavior of the cyano-group flanked by two methyl groups particularly interesting. In the bromination of substituted mesitylenes the point for

[26] K. B. Wiberg and T. M. Shryne, *J. Am. Chem. Soc.*, **77**, 2774 (1955).

Table 7-8. Reaction Constants for the $\rho\sigma^+$ Correlation[1]

REACTION	ρ	CORRELATION COEFFICIENT	n^2	REF.
Solvolysis of 2-phenyl-2-propyl chlorides in 90% acetone at 25°[3]	−4.54	.992	8	4
Bromination of monosubstituted benzenes in acetic acid at 25°	−12.14	.987	8	5
Chlorination of monosubstituted benzenes in acetic acid at 25°	−8.06	.987	9	5
Equilibrium constants for carbonium ions from benzhydrols in H_2SO_4 at 25°	−6.67	.990	7	6
Nitration of monosubstituted benzenes by nitric acid in nitromethane or acetic anhydride at 0° or 25°	−6.22	.980	13	7
Solvolysis of p-alkylbenzhydryl chlorides in ethanol at 25°	−4.63	.994	5	8
Protonolysis of phenyltrimethylsilanes by perchloric acid in 72% methanol at 50°	−4.32	.972	12	9
Brominolysis of benzeneboronic acids in 20% acetic acid and 0.40M NaBr at 25°	−4.44	.992	16	10
Solvolysis of p-halobenzhydryl chlorides in 70% acetone at 25°	−4.11	.999	4	11
Solvolysis of benzhydryl chlorides in 2-propanol at 25°	−4.06	.992	4	12
Solvolysis of benzhydryl chlorides in ethanol at 25°	−4.05	.998	8	13
Addition of chlorine to cinnamic acids in acetic acid at 24°	−4.01	.999	4	14
Equilibrium constants for carbonium ions from triphenylcarbinols in aqueous H_2SO_4 at 25°	−3.64	.995	12	6
Addition of Cl_2 to benzalacetophenones in acetic acid at 24°	−3.66	.999	4	14
Thermal rearrangement of α-phenylethyl chlorocarbonates in dioxane at 80°	−3.01	.999	14	15
Acid-catalyzed rearrangement of phenyl-propenyl carbinols, $XC_6H_4CH(OH)CH\!=\!CHCH_3$, in 60% dioxane at 30°	−2.97	.978	7	16
Migration aptitudes in the rearrangement of sym. tetraarylpinacols by acetyl chloride in acetic acid or by iodine in acetic acid and benzene	−2.99	.979	9	17
Thermal rearrangement of α-phenylethyl chlorocarbonates in toluene at 80°	−2.87	.993	5	16
Migration aptitudes in the acid-catalyzed rearrangement of diphenylmethylcarbinyl azides	−2.69	.997	5	31
Solvolysis of triphenylcarbinyl chlorides in 40% ethanol-60% ethyl ether, at 60°	−2.57	.994	8	18
at 25°	−2.34	.992	7	
Reactivity ratio in the stannic chloride-catalyzed copolymerization of styrenes	−2.34	.991	5	19
Rearrangement of oximes in 94.5% H_2SO_4 at 50.9°	−1.76	.995	8	20
Reactivity ratio in the stannic chloride-catalyzed polymerization of styrenes with α-methylstyrene	−1.70	.969	5	19

Table 7-8 (continued)

REACTION	ρ	CORRELATION COEFFICIENT	n^2	REF.
Reaction of 1-aryl-1,3-butadienes with maleic anhydride in dioxane at 45°	−0.62	.997	4	21
Bromination of monosubstituted benzenes by hypobromous acid and perchloric acid in 50% dioxane at 25°	−5.78	.998	7	22
Protonolysis of aryltrimethylsilanes by sulfuric acid in acetic acid	−4.60	.980	14	23
Brominolysis of aryltrimethylsilanes in acetic acid at 25°	−6.04	.990	14	24
Bromination of polymethylbenzenes by bromine in nitromethane at 25°	−8.10	.995	11	25
Solvolysis of t-cumyl chlorides in ethanol at 25°	−4.67	.998	16	26
Equil. constants for formation of ion-pairs from triphenylcarbinyl chlorides in SO_2 at 0°	−3.73	.981	15	27
Frequency shifts in the infrared stretching frequency of the carbonyl group in substituted acetophenones	−12.30	.980	10	28
Acid-catalyzed decomposition of cumyl hydroperoxides in 50% ethanol at 50°	−4.57	.997	7	29
Relative reactivities of substituted toluenes towards $Cl_3C\cdot$	−1.46	.99	8	30

(1) Y. Okamoto and H. C. Brown, *J. Org. Chem.*, **22**, 485 (1957); *J. Am. Chem. Soc.*, **80**, 4979 (1958). (2) The number of compounds used in calculating ρ. (3) The value of ρ for this (standard) reaction was determined using the σ-values for *m*-Me, *m*-Et, *m*-F, *m*-Cl, *m*-I, *m*-Br, *m*-NO$_2$, and H. (4) H. C. Brown and Y. Okamoto, *J. Am. Chem. Soc.*, **79**, 1913 (1957). (5) P. B. D. de la Mare, *J. Chem. Soc.*, 4450 (1954); H. C. Brown and L. M. Stock, *J. Am. Chem. Soc.*, **79**, 1421, 5175 (1957). (6) N. C. Deno, J. J. Jaruzelski, and A. Schriesheim, *J. Am. Chem. Soc.*, **77**, 3044 (1955); N. C. Deno and W. L. Evans, *ibid.*, **79**, 5804 (1957). (7) J. D. Roberts, J. K. Sanford, F. L. J. Sixma, H. Cerfontain, and R. Zagt, *J. Am. Chem. Soc.*, **76**, 4525 (1954). (8) E. D. Hughes, C. K. Ingold, and N. A. Taher, *J. Chem. Soc.*, 949 (1940). (9) C. Eaborn, *J. Chem. Soc.*, 4858 (1956). (10) H. G. Kuivila and A. R. Hendrickson, *J. Am. Chem. Soc.*, **74**, 5068 (1952); H. G. Kuivila and C. E. Benjamin, *ibid.*, **77**, 4834 (1955). (11) E. D. Hughes and G. Kohnstam, as quoted by C. K. Ingold in *Structure and Mechanism in Organic Chemistry*, Cornell University Press, 1953, p. 332. (12) S. Altscher, R. Baltzly, and S. W. Blackman, *J. Am. Chem. Soc.*, **74**, 3649 (1952). (13) J. F. Norris and C. Banta, *J. Am. Chem. Soc.*, **50**, 1804 (1928); J. F. Norris and J. T. Blake, *ibid.*, **50**, 1808 (1928). (14) H. P. Rothbaum, I. Ting, and P. W. Robertson, *J. Chem. Soc.*, 980 (1948). (15) K. B. Wiberg and T. M. Shryne, *J. Am. Chem. Soc.*, **77**, 2774 (1955). (16) E. A. Braude and E. S. Stern, *J. Chem. Soc.*, 1096 (1947). (17) W. E. Bachmann and F. H. Mosher, *J. Am. Chem. Soc.*, **54**, 1124 (1932); W. E. Bachmann and J. W. Ferguson, *ibid.*, **56**, 2081 (1934). (18) A. C. Nixon and G. E. K. Branch, *J. Am. Chem. Soc.*, **58**, 492 (1936). (19) C. G. Overberger, L. H. Arond, D. Tanner, J. J. Taylor, and T. Alfrey, Jr., *J. Am. Chem. Soc.*, **74**, 4848 (1952). (20) D. E. Pearson, J. F. Baxter, and J. C. Martin, *J. Org. Chem.*, **17**, 1511 (1952); D. E. Pearson and J. D. Bruton, *ibid.*, **19**, 957 (1954). (21) E. J. DeWitt, C. T. Lester, and G. A. Ropp, *J. Am. Chem. Soc.*, **78**, 2101 (1956). (22) P. B. D. de la Mare and J. T. Harvey, *J. Chem. Soc.*, 36 (1956); *ibid.*, 131 (1957); P. B. D. de la Mare and M. Hassan, *ibid.*, 3004 (1957). (23) F. B. Deans and C. Eaborn, *J. Chem. Soc.*, 2299 (1959). (24) C. Eaborn and D. E. Webster, *J. Chem. Soc.*, 4449 (1957). (25) G. Illuminati and G. Marino, *J. Am. Chem. Soc.*, **78**, 4975 (1956); G. Illuminati, *Ricerca Sci.*, **26**, 2752 (1956). (26) Y. Okamoto, T. Inukai, and H. C. Brown, *J. Am. Chem. Soc.*, **80**, 4972 (1958). (27) N. N. Lichtin and M. J. Vignale, *J. Am. Chem. Soc.*, **79**, 579 (1957). (28) R. N. Jones, W. F. Forbes, and W. A. Mueller, *Can. J. Chem.*, **35**, 504 (1957); C. N. R. Rao and G. B. Silverman, *Curr. Sci. (India)*, **26**, 375 (1957). (29) A. W. de R. Van Steveninck and E. C. Kooyman, *Rec. trav. chim.*, **79**, 413 (1960). (30) E. S. Huyser, *J. Am. Chem. Soc.*, **82**, 394 (1960). (31) S. N. Ege and K. W. Sherk, *J. Am. Chem. Soc.* **75**, 354 (1953).

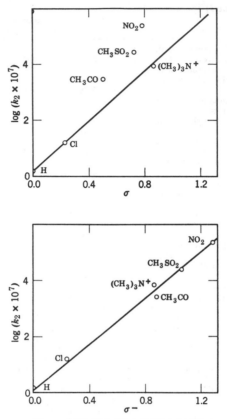

Figure 7-13. Substituent effects in the reactions of *para*-substituted 2-nitrophenyl
chlorides with methoxide ion.[29]

meta-cyano deviates from the best $\rho\sigma^+$ line to such an extent that a value of
0.686 rather than the usual 0.562 for σ^+_{m-CN} would seem to be needed.[27]

$$\text{(structure)} + Br_2 \xrightarrow[30°]{\text{nitromethane}} HBr + \text{(structure)} \quad (46)$$

It is probable that in sterically unhindered compounds solvation partly
counteracts electron withdrawal by *meta*-cyano.

[27] G. Illuminati, *J. Am. Chem. Soc.*, **80**, 4941 (1958).

The σ^- Parameter

Reactions involving unshared electron pairs on an atom next to the benzene ring are especially responsive to *para*-substituents capable of stabilizing a negative charge by resonance.[1,28] An example would be the marked effect of the *p*-nitro group on the basicity of phenoxide ion or of aniline in contrast to its lesser effect on the basicity of benzoate ion. Such reactions will give more nearly linear correlations with another reaction of the same type than they will with the ionization of benzoic acid. The special substituent constants for these reactions are designated by σ^-. As in the case of σ^+, the σ^-_{meta}-values are very close to the corresponding σ_{meta}. For groups not capable of strong electron-accepting resonance, σ_{para} is used for σ^-_{para}. Table 7-9 gives σ^- values taken from the compilation of Jaffé.[3]

Table 7-9. Substituent Constants, σ^-

Para-SUBSTITUENT	σ^-	*Para*-SUBSTITUENT	σ^-
COOH	0.728	$COCH_3$	0.874
$COOCH_3$	0.636	CN	1.000
$COOC_2H_5$	0.678	N_2^+	3.2[a]
$COOC_4H_9$	0.674	NO_2	1.270
$COOCH_2C_6H_5$	0.667	SO_2CH_3	1.049
$CONH_2$	0.627	$C_6H_4N{=}NC_6H_5$	1.088
CHO	1.126	$CH{=}CHC_6H_5$	0.619
		$CH{=}CHNO_2$	0.88[b]

[a] E. S. Lewis and M. D. Johnson, *J. Am. Chem. Soc.*, **81**, 2070 (1959).

[b] R. Stewart and L. G. Walker, *Can. J. Chem.*, **35**, 1561 (1957).

The effect of using σ^- rather than σ for aromatic nucleophilic substitution is shown in Figure 7-13.[29] The σ^- parameters give better linearity than σ for many reactions of phenols, phenolic esters, anilines, dimethylanilines, and anilides.

The Yukawa and Tsuno Equation

In previous sections we discussed two special cases of equation 17 corresponding to approximately constant values for the ratio ρ'/ρ''. These enabled us to establish scales of σ and σ^+. Reactions involving resonance stabilization of a positive charge in which the relative importance of the

[28] F. G. Bordwell and G. D. Cooper, *J. Am. Chem. Soc.*, **74**, 1058 (1952).
[29] J. F. Bunnett, F. Draper, Jr., P. R. Ryason, Paul Noble, Jr., R. G. Tonkyn, and R. E. Zahler, *J. Am. Chem. Soc.*, **75**, 642 (1953).

resonance interaction mechanism differs from that in the σ and σ^+ relationships can be handled by the method of linear combination of two models. The models will be the defining reactions for σ and σ^+.

$$\log \frac{k}{k_0} = a\sigma + b\sigma^+ \qquad (47)$$

Rewriting the relationship in the form used by Yukawa and Tsuno[30] gives equation 48.

$$\log \frac{k}{k_0} = \rho(\sigma + R[\sigma^+ - \sigma]) \qquad (48)$$

The coefficient a in equation 47 is equal to $\rho(1 - R)$ and b is equal to ρR. Since $\sigma^+ - \sigma$ for well-behaved *meta*-substituents is zero, these may be used to evaluate ρ. The remaining parameter R is the slope of the plot of

$$\left(\frac{1}{\rho} \log \frac{k}{k_0} - \sigma\right) \text{ versus } (\sigma^+ - \sigma).$$

Some representative values of R are given in Table 7-10, and an example of the improvement in fit over either of the single-model treatments is shown in Figure 7-14. Although the fit to an empirical equation can

Table 7-10. Some Values of the Parameter R in the Yukawa and Tsuno Equation

REACTION	R	REFERENCE
		(Standard
Solvolysis of 2-phenyl-2-propyl chlorides	1.000	σ^+ reaction)
Aromatic halogenation in acetic acid	1.74	[a]
Acid-catalyzed decomposition of		
diazoacetophenones	0.542	[b]
Triphenylcarbinol ionization equilibrium in		
aqueous acid	0.742	[c]
Rearrangement of phenyl propenyl carbinols	0.396	[d]
Brominolysis of benzeneboronic acids	2.29	[e]
Epoxidation of *cis*-dimethylstilbenes	0.323	[f]

[a] J. Miller, *Australian J. Chem.*, **9**, 61 (1956).
[b] Y. Tsuno, T. Ibata, and Y. Yukawa, *Bull. Chem. Soc., Japan*, **32**, 960 (1959).
[c] N. C. Deno and W. L. Evans, *J. Am. Chem. Soc.*, **79**, 5804 (1957).
[d] E. A. Braude and E. S. Stern, *J. Chem. Soc.*, 1096 (1947).
[e] H. G. Kuivila and A. R. Hendrickson, *J. Am. Chem. Soc.*, **74**, 5068 (1952); H. G. Kuivila and C. E. Benjamin, *ibid.*, **77**, 4834 (1955).
[f] O. Simamura (private communication).

[30] Y. Yukawa and Y. Tsuno, *Bull. Chem. Soc. Japan*, **32**, 965 (1959); **32**, 971 (1959).

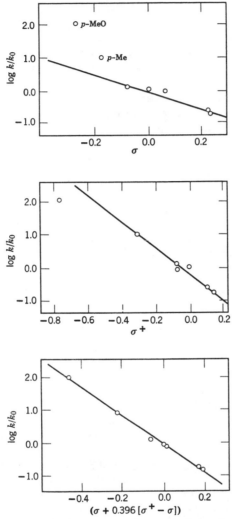

Figure 7-14. Rearrangement of phenyl-
propenylcarbinols.

always be improved by using more parameters, the improvement in this
instance is significant because of the theoretical soundness of the treatment.

It will be noted that several of the reactions included in Table 7-10 are
also included in Table 7-8 and give fairly good correlations with the single
parameter σ^+. Since the corresponding values of R are quite different from
unity, it is obviously possible to have a considerable variation in the relative

importance of the resonance and inductive mechanisms before the perturbation of a simple linear free energy relationship becomes noticeable.

In reactions involving negative charges and nonbonding electron pairs, the fit of the data could similarly be improved by using a linear combination of σ and σ^-. At present, however, the data are too scanty to warrant this.

Normal Substituent Constants, σ^0

From time to time we have mentioned that certain *meta*-substituents are well behaved. These substituents, which give better $\rho\sigma$ correlations than do substituents in general, are fluoro, chloro, bromo, iodo, methyl, acetyl, and nitro.[31,32] With other substituents, considerably more scatter is found, not only in reactions of the type generating charges directly adjacent to the ring but even for the ionization of benzoic acid in solvents other than water. The well-behaved substituents are thought to exert an effect in which variations in solvation or resonance are relatively unimportant. Resonance effects are able to persist even in the *meta*-position, perhaps because of an inductive relay of the charges placed on the ring.

It has been proposed to designate the well-behaved substituents as "normal" and their substituent constants as *normal* substituent constants, σ^0.[31]

For certain reactions the reaction site is insulated from the ring to such an extent that there should be only a minimal and constant amount of resonance involving the ring and its substituents. In such reactions almost any substituent can be regarded as exerting only a "normal" effect. Table 7-11 lists both the σ^0-values for the well-behaved substituents and the additional σ^0-values obtained from substituent effects in the following reactions:

Ionization of $ArCH_2COOH$ in water at 25° ($\rho = 0.46$).
Saponification rates of $ArCH_2COOEt$ in 88 % ethanol at 30° ($\rho = 1.00$).
Ionization of $ArCH_2CH_2COOH$ in water at 25° ($\rho = 0.24$).
Saponification rates of CH_3COOCH_2Ar in 60% acetone at 25° ($\rho = 0.73$).

[31] R. W. Taft, Jr., and I. C. Lewis, *J. Am. Chem. Soc.*, **81**, 5343 (1959).
[32] H. van Bekkum, P. E. Verkade, and B. M. Wepster, *Rec. trav. chim.*, **78**, 815 (1959); R. O. C. Norman, G. K. Radda, D. A. Brimacombe, P. D. Ralph, and E. M. Smith, *J. Chem. Soc.*, 3247 (1961).

It is noteworthy that weakly conjugating substituents like *para*-chloro, bromo, and methyl have σ^0-values very close to σ. Substituents in which resonance *removes* electrons from the ring also have nearly normal σ values: In the ionization of benzoic acid, the conjugation of such groups with the carboxyl and carboxylate groups is not important enough to produce much effect.

Table 7-11. Normal Substituent Constants,[a] σ^0

SUBSTITUENT	σ^0_{meta}	σ^0_{para}
$N(CH_3)_2$	−0.15	−0.44
OCH_3	0.13[b] (0.06)[c]	−0.12[b] (−0.16)[c]
NH_2	−0.14	−0.38
OH	(—)[d] (0.04)[e]	(—)[d] (−0.13)[e]
F	0.35	0.17
Cl	0.37	0.27
Br	0.38	0.26
I	0.35	0.27
CH_3	−0.07	−0.15
H	0	0
CN	0.62	0.69[f] (0.63)[e]
COOR	0.36	0.46[f]
$COCH_3$	0.34	0.46[f] (0.40)[e]
NO_2	0.70	0.82[f] (0.73)[e]

[a] From R. W. Taft, Jr., *J. Phys. Chem.*, **64,** 1805 (1960).
[b] Value for pure aqueous solutions only.
[c] Value for nonhydroxylic media and most mixed aqueous-organic solvents.
[d] The value is strongly dependent on the nature of hydroxylic solvents.
[e] Value for nonhydroxylic solvents only.
[f] Value for pure aqueous and most aqueous-organic solvents.

In addition to the reactions used to define σ^0, several other reactions are much better correlated by means of σ^0 than by means of σ.[33] Examples are the oxidation of substituted triarylsilanes to silanols in wet piperidine,[34] the ionization of $ArPO_2OH^-$ in water,[35] and the formation of ion-pairs from benzoic acids and 1,3-diphenylguanidine in benzene,[36] Figure 7-15.

[33] R. W. Taft, Jr., S. Ehrenson, I. C. Lewis, and R. E. Glick, *J. Am. Chem. Soc.*, **81,** 5352 (1959).
[34] H. Gilman and G. E. Dunn, *J. Am. Chem. Soc.*, **73,** 3404 (1951).
[35] H. H. Jaffé, L. D. Freedman, and G. O. Doak, *J. Am. Chem. Soc.*, **75,** 2209 (1953).
[36] M. M. Davis and H. B. Hetzer, *J. Res. Natl. Bur. Stand.*, **60,** 569 (1958).

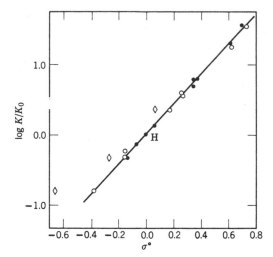

Figure 7-15. Ion-pair formation from ArCOOH and 1,3-diphenylguanidine in benzene at 25°. The slope, ρ, is 2.24. Open circles, *para*-substituents; solid circles, *meta*-substituents; diamonds, abscissa is σ rather than $\sigma°$.

The unimportance of "abnormal" resonance interaction in the latter reaction in contrast to its importance in the ionization of benzoic acids in water means that dipolar resonance structures such as (50) are stabilized by the more polar solvent.

$$CH_3O-\langle\ \rangle-C\overset{O}{\underset{OH}{\diagup}} \leftrightarrow CH_3\overset{+}{O}=\langle\ \rangle=C\overset{O^-}{\underset{OH}{\diagup}} \tag{50}$$

SUBSTITUENT EFFECTS IN ALIPHATIC REACTIONS

An indication that a modification of the approach used in the aromatic series should also work with aliphatic compounds is the correlation occasionally obtained between aliphatic reactivities and σ. For example, in the reaction of *trans*-3-substituted acrylic acids with diphenyldiazomethane, $\log k$ is a good linear function of the *para*-sigma constant for the substituent.[37] The benzoic acid and *trans*-3-substituted acrylic acid

[37] J. Hine and W. C. Bailey, Jr., *J. Am. Chem. Soc.*, **81**, 2075 (1959).

systems resemble each other in that the molecules are fairly rigid, steric effects are nearly constant, and the substituent is conjugated with the carboxyl group through a π-electron system.

The $\rho'\sigma'$ Relationship

An aliphatic system geometrically not so very different from the benzoic acids, but in which no π-electrons are available for resonance effects, is the series of 4-substituted bicyclo[2.2.2]octane-1-carboxylic acids.[38] Substituent effects in this series should be largely inductive. The fact that log K_A for the ionization of these acids is linearly related to log k for the ester hydrolyses, and to log k for the reaction of the acids with diphenyldiazomethane, proves that it is not necessary for a substituent effect to have a resonance component in order to give a linear free energy relationship.

$$\text{(51)}$$

Substituent effects in the bicyclo[2.2.2]octane system are correlated with the parameter σ', which is given a definition analogous to that of σ. The standard process is the ionization of the acid in 50% ethanol at 25°.

$$\sigma_X' \equiv \left(\frac{1}{1.464}\right) \log \left(\frac{K_A^x}{K_A^\circ}\right) \tag{52}$$

The divisor 1.464 is the value of ρ for the ionization of benzoic acids in 50% ethanol and is incorporated in the definition of σ' in order to facilitate the comparison of σ'-values with the corresponding σ-values. For other reactions in this system, the substituent effect is given by equation 53.

$$\log \left(\frac{k_x}{k_0}\right) = \rho'\sigma_X' \tag{53}$$

The values of σ' are compared with Hammett σ-values in Table 7-12 and Figure 7-16.

It can readily be seen from the figure that the differences between σ_{meta} and σ' are relatively small. It can also be seen that the sign of $(\sigma_{para} - \sigma')$ corresponds to the direction of the resonance effect of the substituent and that, with one insignificant exception, it is the same as the sign of $(\sigma_{meta} - \sigma')$.

---

[38] J. D. Roberts and W. T. Moreland, Jr., J. Am. Chem. Soc., 75, 2167 (1953).

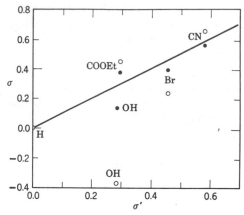

Figure 7-16. A comparison of substituent effects in the benzoic acid and bicyclo[2.2.2]-octane-l-carboxylic acid systems. The open circles are for *para*-substituents, the solid circles for *meta*-substituents.

It is clear that σ' is correlated in some way with the inductive part of the *meta*- and *para*-substituent effects in the benzoic acid system. Indeed, in view of the similarities of the two systems in size, rigidity, and colinearity of the bonds to the substituent and functional group (at least in the *para*-position), one might be tempted to go even further and equate σ' with the

Table 7-12. Comparison of σ'- and σ-Values

SUBSTITUENT	σ'	σ_{para}	σ_{meta}	$(\sigma_{para}-\sigma')$	$(\sigma_{meta}-\sigma')$
H	0.000	0.000	0.000	...	...
OH	0.283	−0.37	0.121	−0.65	−0.162
COOEt	0.297	0.45	0.37	0.15	0.07
Br	0.454	0.232	0.391	−0.222	−0.063
CN	0.579	0.660	0.56	0.081	−0.02

inductive component of sigma.[39] This assumption is justified to some extent by a comparison of the inductive component of sigma obtained in this way with that obtained by experiments in which the resonance effect is damped by steric hindrance.[40,41]

The σ'-values, although not very many of them are available, are particularly valuable because of their theoretical simplicity. They will serve in

[39] R. W. Taft, Jr., *J. Am. Chem. Soc.*, **79**, 1045 (1957).
[40] B. M. Wepster, *Rec. trav. chim.*, **76**, 335, 337 (1957).
[41] R. W. Taft, Jr. and H. D. Evans, *J. Chem. Phys.*, **27**, 1427 (1957).

the next section as a primary standard with which to compare inductive substituent constants obtained in a more devious and complicated way.

The Polar Substituent Constant $\sigma*$

Although systems consisting of rigid molecules give simple correlations between the logarithms of the rate and acid ionization constants, similar correlations do not work well for most aliphatic systems. For example, Figure 7-17[42] shows a nearly complete lack of correlation between substituent effects on the rate constants of the alkaline hydrolysis of a series of ethyl esters and on the acid ionization constants.

The reason for the scatter in Figure 7-17 is the presence of additional interaction mechanisms whose relative importance varies in an uncorrelated way in the two reactions. Since scatter is not observed in rigid systems in which the structural alterations take place in a part of the molecule that remains at all times remote from the reaction site, it is obviously due to some sort of a proximity effect. The effect could be a direct interaction of the sort usually envisaged when steric hindrance is mentioned or an indirect effect on the solvation at the reaction site.

In order to find a parameter expressing the purely polar effects of a substituent in aliphatic reactions (or in the reactions of *ortho*-substituted benzene derivatives), it is necessary to combine data from several reactions in such a way that the steric factors will cancel out. The suitability of the acid and base-catalyzed hydrolysis of esters for this purpose was first

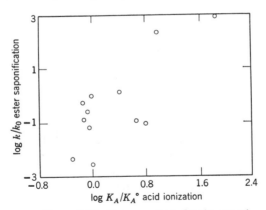

Figure 7-17. Rate constants for the saponification of ethyl esters in 85% ethanol at 25° and equilibrium constants for the ionization of the acids in water at 25°.

[42] R. W. Taft, Jr., *J. Am. Chem. Soc.*, **75**, 4231 (1953).

suggested by Ingold[43] and further applied by Taft.[44] Ingold suggested that the mechanisms for the acid and base-catalyzed hydrolyses are so very similar that the steric effects should cancel out in the ratio of the rate constants. That is, the rate constant k_B for the base-catalyzed hydrolysis divided by the rate constant k_A for the acid-catalyzed hydrolysis should be a measure of polar effects alone.

We can see that the foregoing proposition is at least plausible by inspecting the formulas for the transition states for the two processes.[45] It is

$$RC\overset{O}{\underset{OR'}{<}} + H_2O \rightarrow RC\overset{O}{\underset{OH}{<}} + R'OH \qquad (54)$$

$$\left[RC\overset{OH}{\underset{OH\;H}{<}} \cdots O{-}R' \right]^+ \qquad \left[R{-}C\overset{OH}{\underset{O}{<}} \cdots OR' \right]^-$$

transition state for acid-catalyzed hydrolysis   transition state for base-catalyzed hydrolysis

true that the two structures are very similar, but some serious questions remain. First of all, one wonders whether the free energies of solvation will be equal for the two transition states, since the charge types, numbers of protons, and numbers of nonbonding electron pairs are different. In order for the canceling hypothesis to be valid it is necessary that solvation effects cancel out exactly. Another requirement is that the steric and polar effects do not interact. If this requirement were not satisfied, one would not obtain a linear measure of the polar effect, and the nonlinear measure obtained would probably not be very useful for correlating substituent effects in other reactions. If any part of the polar effect were dependent on the *conformation* of R, that part at least should vary from one reaction to another. Moreover, the susceptibility of a reaction to polar effects might well depend on the amount of steric interaction with the reaction zone, since this steric repulsion could change the bond angles and bond lengths. Despite these serious theoretical reservations, however, the canceling hypothesis has had surprisingly fruitful results.

The polar substituent constants σ^* are defined by equation 55.[46] In

[43] C. K. Ingold, *J. Chem. Soc.*, 1032 (1930).
[44] (a) R. W. Taft, Jr., *J. Am. Chem. Soc.*, **74**, 2729 (1952). (b) R. W. Taft, in *Steric Effects in Organic Chemistry*, M. S. Newman (Ed.), Wiley, 1956, ch. 13.
[45] M. L. Bender, *J. Am. Chem. Soc.*, **73**, 1626 (1951).
[46] R. W. Taft, Jr., *J. Am. Chem. Soc.*, **74**, 3120 (1952); **75**, 4231 (1953).

equation 55, k_0 represents the rate constant for hydrolysis of the acetate (R = CH_3).

$$\sigma^* \equiv \left(\frac{1}{2.48}\right)\left[\log\left(\frac{k}{k_0}\right)_B - \log\left(\frac{k}{k_0}\right)_A\right]$$

$$\equiv \left(\frac{1}{2.48}\right)\left[\log\frac{k_B}{k_A} - \log\frac{k_{0B}}{k_{0A}}\right] \tag{55}$$

Although σ^* is quite different from σ in its operational definition, the arbitrary factor 1/2.48 has been included to make the σ^* scale as much like the σ scale as possible. The factor 2.48 is an average difference in the Hammett ρ-values for base and acid-catalyzed ester hydrolyses.

To complete the operational definition of σ^* it is necessary to specify the temperature, the solvent, and the structure of the alcohol part of the ester. This has not been done for the time being because data are not available for a sufficiently extensive list of substituents for any constant set of conditions. Instead, the various σ^*-values obtained by completing the definition with various specifications of temperature, solvent, and alcohol have been averaged. The justification for this procedure (which we adopt only with reluctance and as a temporary expedient) is that the dispersion in the values of σ^* so obtained for a given substituent is small. At least the value of σ^* is reasonably constant in a series of aqueous organic solvents, although its value may be disconcertingly different in pure water.[47] The presently available σ^*-values have about the same status as "secondary" σ-values.

The reactions used by Taft[44b] for the evaluation of σ^* were the rates of hydrolysis, alcoholysis, or formation of ethyl, methyl, cyclohexyl, benzyl, or 1-menthyl esters in 60% aqueous acetone, methanol, ethanol, 85% ethanol, or cyclohexanol at 25 or 30°.[48] Table 7-13 lists these and other

[47] R. W. Taft, Jr., *J. Am. Chem. Soc.*, **74**, 2729 (1952); S. A. Bernhard and L. P. Hammett, *J. Am. Chem. Soc.*, **75**, 1798, 5834 (1953). For example, σ^* for the ethyl group in 70% acetone is −0.097, while its value in water is −0.028, both calculated from data on the hydrolysis of $RCOOC_2H_5$ at 25°. Since the ethyl group and the methyl group (the standard R for $k = k_0$) are relatively well behaved with respect to solvent effects in the Hammett equation, we would expect still greater dispersion of σ^* for other groups, particularly those capable of participating in hydrogen-bonding. The cancellation of effects due to steric hindrance to solvation may also be incomplete and hence contribute additional dispersion as the solvent is changed.

[48] R. W. Taft, Jr., M. S. Newman, and F. H. Verhoek, *J. Am. Chem. Soc.*, **72**, 4511 (1950); H. A. Smith and R. R. Myers, *J. Am. Chem. Soc.*, **64**, 2362 (1942); G. Davis and D. P. Evans, *J. Chem. Soc.*, 339 (1940); V. C. Haskell and L. P. Hammett, *J. Am. Chem. Soc.*, **71**, 1284 (1949); H. A. Smith and J. Burn, *J. Am. Chem. Soc.*, **66**, 1491 (1944); H. A. Smith, *ibid.*, **61**, 254 (1939); H. A. Smith and R. B. Hurley, *ibid.*, **72**, 112 (1950); R. J. Hartman and A. G. Gassman, *ibid.*, **62**, 1559 (1940); R. J. Hartman and

Table 7-13. Polar Substituent Constants,[44] $\sigma*$

SUBSTITUENT[a]	$\sigma*$	SUBSTITUENT[a]	$\sigma*$
Cl_3C	2.65	H	0.490
F_2CH	2.05	$C_6H_5CH{=}CH$	0.410
CH_3OOC	2.00	$(C_6H_5)_2CH$	0.405
Cl_2CH_+	1.940	$ClCH_2CH_2$	0.385
$(CH_3)_3NCH_2$	1.90	$CH_3CH{=}CH$	0.360
$trans\text{-}O_2NCH{=}CH$	1.70[b]	$CF_3CH_2CH_2$	0.32
CH_3CO	1.65	$C_6H_5CH_2$	0.215
$C_6H_5C{\equiv}C$	1.35	$CH_3CH{=}CHCH_2$	0.13
$CH_3SO_2CH_2$	1.32	$CF_3CH_2CH_2CH_2$	0.12
$N{\equiv}CCH_2$	1.300	$C_6H_5(CH_3)CH$	0.11
$trans\text{-}Cl_3CCH{=}CH$	1.188[b]	$C_6H_5CH_2CH_2$	0.080
FCH_2	1.10	$C_6H_5(C_2H_5)CH$	0.04
$HOOCCH_2$	1.05	$C_6H_5CH_2CH_2CH_2$	0.02
$ClCH_2$	1.050	CH_3	(0.000)[c]
$trans\text{-}HOOCCH{=}CH$	1.012[b]	$cyclo\text{-}C_6H_{11}CH_2$	-0.06
$BrCH_2$	1.000	C_2H_5	-0.100
CF_3CH_2	0.92	$n\text{-}C_3H_7$	-0.115
$trans\text{-}ClCH{=}CH$	0.900[b]	$i\text{-}C_4H_9$	-0.125
$trans\text{-}Cl_2CHCH{=}CH$	0.882[b]	$n\text{-}C_4H_9$	-0.130
$C_6H_5OCH_2$	0.850	$cyclo\text{-}C_6H_{11}$	-0.15
ICH_2	0.85	$t\text{-}C_4H_9CH_2$	-0.165
$C_6H_5(OH)CH$	0.765	$i\text{-}C_3H_7$	-0.190
$CH_2{=}CH$	0.653[b]	$cyclo\text{-}C_5H_9$	-0.20
CH_3COCH_2	0.60	$sec\text{-}C_4H_9$	-0.210
C_6H_5	0.600	$(C_2H_5)_2CH$	-0.225
$HOCH_2$	0.555	$(CH_3)_3SiCH_2$	-0.26
CH_3OCH_2	0.520	$(t\text{-}C_4H_9)(CH_3)CH$	-0.28
$O_2NCH_2CH_2$	0.50	$t\text{-}C_4H_9$	-0.300

[a] Defined so as to be the R-group of structure (54).
[b] J. Hine and W. C. Bailey, Jr., *J. Am. Chem. Soc.*, **81**, 2075 (1959).
[c] By definition.

A. M. Borders, *ibid.*, **59**, 2107 (1937); B. V. Bhide and J. J. Sudborough, *J. Indian Inst. Sci.*, **8**, 89 (1925) (*Zentr.*, **97**, I, 80 (1926)); J. J. Sudborough and L. L. Lloyd, *J. Chem. Soc.*, **75**, 467 (1899); J. J. Sudborough and M. K. Turner, *ibid.*, **101**, 237 (1912); W. B. S. Newling and C. N. Hinshelwood, *J. Chem. Soc.*, 1357 (1936); R. J. Hartman, H. M. Hoogsteen, and J. A. Moede, *J. Am. Chem. Soc.*, **66**, 1714 (1944); D. P. Evans, J. J. Gordon, and H. B. Watson, *J. Chem. Soc.*, 1439 (1938); *ibid.*, 1430 (1937); H. S. Levenson and H. A. Smith, *J. Am. Chem. Soc.*, **62**, 2324 (1940); C. K. Ingold and W. S. Nathan, *J. Chem. Soc.*, 222 (1936); K. Kindler, *Ann.*, **464**, 278 (1928); E. Tommila, *Ann. Acad. Sci. Fennicae, Ser. A.* **57**, No. 13, 3 (1941), *A.* **59**, No. 3, 3 (1942), *A. 59*, No. 4, 3 (1942).

values of σ^* for various substituents.[44] The substituents listed in this table are the R groups of structure (54). It is seen that values of σ^*, in contrast to σ', are available for a large number of substituents.

Let us now consider what kind of polar interaction mechanism might be measured by σ^*. We begin with the defining equation, (55), in which σ^* is seen to be proportional to the difference, $\delta_R \Delta F_B^{\ddagger} - \delta_R \Delta F_A^{\ddagger}$. On taking this difference we find that the free energy of the reagents cancels out and that σ^* is therefore also proportional to the quantity, $\delta_R F_B^{\ddagger} - \delta_R F_A^{\ddagger}$, which is characteristic only of the transition states. Since the reaction zones in the transition states are much less unsaturated than the carbonyl groups of the esters, resonance interactions can make only a small contribution to σ^*. The resonance effect is reduced even further if we consider substituents of the type XCH_2, because the X-group is now insulated against resonance interaction with the reaction zone by an intervening CH_2-group. Hence $(\sigma^*_{XCH_2} - \sigma^*_{CH_3})$ should be a measure of the inductive effect of X relative to hydrogen. It is therefore expected that a plot of σ_X' versus $\sigma^*_{XCH_2}$ should be a straight line if σ^* is indeed a measure of polar effects alone. As shown in Figure 7-18, the σ^* parameters pass this test with flying colors, although only four points are available. The slope, 0.45, of this plot is a measure of the relative attenuation of the inductive effect on passing through the cage structure of the bicyclic system (51) as compared to the attenuation produced by a single CH_2 group.

The attenuation of the inductive effect by one methylene group can

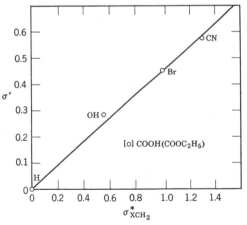

Figure 7-18. The correlation between σ_X' and $\sigma^*_{XCH_2}$. The point in brackets is σ'_{COOEt} and $\sigma^*_{CH_2COOH}$. The line is drawn with a slope of 0.45.

be calculated from the σ^*-values for the pairs, $ClCH_2CH_2 \div ClCH_2$, $CF_3CH_2CH_2CH_2 \div CF_3CH_2CH_2$, and $C_6H_5CH_2CH_2 \div C_6H_5CH_2$. The results are 0.37, 0.27, and 0.37, or an average of 0.34 ± 0.05. Using the σ^*-values for $C_6H_5CH_2CH_2CH_2$ and $C_6H_5CH_2$, and assuming that the attenuation by n-methylene groups is the nth power of the attenuation by a single methylene group, we obtain an attenuation of 0.33. We have no assurance, however, that the nth power model is correct, for it fails rather badly in predicting the slope 0.45 of the σ_X' versus $\sigma^*_{XCH_2}$ relationship.[49]

Since inductive effects are additive in first approximation, it should be possible to calculate σ^* for a bulky group from the σ^*-values of its part-structures. Of course this procedure will work only if the inductive effect

Table 7-14. *Additivity of* σ^* *Values for Bulky Substituents*

GROUP —CXYZ	OBSERVED σ^*	PREDICTED σ^*
—C(CH$_3$)$_3$	−0.300	$3 \times (-0.100) = -0.300$
—CH(C$_6$H$_5$)(CH$_3$)	+0.11	$0.215 - 0.100 = +0.115$
—CH(C$_6$H$_5$)(C$_2$H$_5$)	+0.04	$0.215 - 0.115 = +0.100$
—CH(C$_6$H$_5$)(OH)	+0.765	$0.215 + 0.555 = +0.770$
—CHCl$_2$	+1.940	$2 \times (1.050) = +2.10$

is not modified by steric factors. The success of such a procedure is therefore a test of the assumption that σ^* is a linear measure of the polar effect alone.

In a group XYZC containing substituents X, Y, and Z in place of hydrogen, the additivity of the inductive effect implies equation 56.

$$\sigma^*_{CXYZ} - \sigma^*_{CH_3} = (\sigma^*_{CH_2X} - \sigma^*_{CH_3})$$
$$+ (\sigma^*_{CH_2Y} - \sigma^*_{CH_3}) + (\sigma^*_{CH_2Z} - \sigma^*_{CH_3}) \quad (56)$$

Since $\sigma^*_{CH_3}$ is zero by definition, we obtain

$$\sigma^*_{CXYZ} = \sigma^*_{CH_2X} + \sigma^*_{CH_2Y} + \sigma^*_{CH_2Z} \quad (57)$$

The only comparisons relevant to a test of the independence of the polar and steric effects are those in which the triple substitution involves considerable crowding. We have therefore chosen combinations such as $X = Y = Z = CH_3$ and $X = C_6H_5$, $Y = CH_3$, $Z = H$ for the test. The results are shown in Table 7-14. The fit of equation 57 is seen to be fairly good.

[49] J. C. McGowan, *J. Appl. Chem. (London)*, **10**, 312 (1960) has discussed various methods of estimating the attenuation of the inductive effect per CH$_2$-group and has concluded that a value near 0.50 is in better agreement with most facts. This article also contains a useful survey of the previous literature.

As a further test of the independence of polar and steric effects we can calculate values of σ^* for *ortho*-substituted phenyl groups. From the difference in log k_B/k_A for the various *ortho*-substituted esters and the *ortho*-toluate ester we obtain $(\sigma^*_{o-X} - \sigma^*_{o-CH_3})$. Because of the normalizing factor of 2.48 used in defining σ^*, the quantity $(\sigma^*_{m-X} - \sigma^*_{m-CH_3})$ for *meta*-substituents is equal to $(\sigma_{m-X} - \sigma_{m-CH_3})$. Similarly, the quantity $(\sigma^*_{o-X} - \sigma^*_{o-CH_3})$ should represent the experimentally unavailable purely polar part of σ_{ortho}. Since the *ortho*- and *para*-substituents have similar resonance mechanisms, there should be a marked similarity between the quantities $(\sigma^*_{o-X} - \sigma^*_{o-CH_3})$ for *ortho*-substituents and the quantities $(\sigma_{p-X} - \sigma_{p-CH_3})$ for *para*-substituents. These quantities are compared in Table 7-15.[44b] The

Table 7-15. A Comparison of σ^* and σ

SUBSTITUENT	$\sigma^*_{o-X} - \sigma^*_{o-CH_3}$	$\sigma_{p-X} - \sigma_{p-CH_3}$
CH_3O	−0.22	−0.10
C_2H_5O	−0.18	−0.07
F	+0.41	+0.23
Cl	+0.37	+0.40
Br	+0.38	+0.40
I	+0.38	+0.35
NO_2[a]	+0.97	+0.95
NO_2[b]	+1.39	+1.46

[a] For benzene derivatives other than anilines or phenols.
[b] For anilines and phenols.

general trend is toward equality between the two quantities, except for the first three entries. For methoxy and ethoxy, no real independence of steric and polar effects is expected, since the resonance interaction of these substituents requires a coplanar conformation. No such explanation is possible for the discrepancy in the case of the fluoro-substituent, but this substituent might form an intramolecular hydrogen bond in the transition state for the acid-catalyzed process. On the whole, the predominant result is similarity rather than difference between the two quantities, a significant result, for the total *ortho*-effects on the rates are large. (See Table 7-17.)

The Taft $\rho^*\sigma^*$ Equation

A large number of aliphatic reaction rates can be correlated by means of equation 57, in which ρ^* is an empirical parameter dependent on the nature of the reaction and on the reaction conditions.[46]

$$\log \frac{k}{k_0} = \rho^*\sigma^* \tag{57}$$

Table 7-16. Reactions Correlated by the $\rho^*\sigma^*$ Equation[a]

REACTION[b]	ρ^*
Aliphatic Compounds	
Ionization equilibrium of carboxylic acids in water at $25°$[c]	$1.721 \pm .025$
Hydrolysis of $RCH(OEt)_2$, acid-catalyzed, in 50% dioxane at $25°$[d]	$-3.652 \pm .085$
Hydrolysis of $RC(CH_3)(OEt)_2$ (mono-substituted acetonals), acid-catalyzed, in 50% dioxane at $25°$[d]	$-3.541 \pm .176$
Reaction of diphenyldiazomethane with carboxylic acids in ethanol at $25°$[e]	$1.175 \pm .043$
Alkaline hydrolysis of acetatopentamine cobalt (III) ions $[Co(NH_3)_5X]^{++}$, where X is $RCOO^-$, in water at $25°$[f]	$0.786 \pm .023$
Acidities of alcohols RCH_2OH in isopropyl alcohol at $27°$[g]	$1.364 \pm .027$
Alkaline hydrolysis of ethyl 4-substituted bicyclo-[2.2.2]-octane-1-carboxylates in 87.83% ethanol at $30°$[h]	$0.975 \pm .042$
Sulfation of alcohols with equimolar H_2SO_4, initial H_2O/H_2SO_4 1.290, at $25°$[i]	$4.600 \pm .149$
Catalysis of dehydration of acetaldehyde hydrate by $RCOOH$ in acetone at $25°$[j]	$0.801 \pm .015$
Catalysis of iodination of acetone by $RCOOH$ in water at $25°$[k]	$1.143 \pm .022$
Catalysis of decomposition of nitramide by $RCOO^-$ in water at $15°$[l]	$-1.426 \pm .035$
Catalysis of depolymerization of dimeric dihydroxy-acetone by $RCOO^-$ in water at $25°$[m]	$-1.362 \pm .069$
Decomposition of cholesteryl xanthates, $RSCSOC_{27}H_{45}$, at $176°$[n]	$1.438 \pm .092$
Hydrolysis of formals, $H_2C(OR)_2$, acid-catalyzed, in water at $25°$[o]	$-4.173 \pm .150$
Bromination of ketones, $C_6H_5COCHR_1R_2$, base-catalyzed, in water at $25°$[p]	$1.590 \pm .079$
Vapor phase reaction of RCl with Na[q]	$-2.480 \pm .174$
$\delta_R\Delta H$ (kcal) for the dissociation of $(CH_3)_3B$ addition complexes with straight-chain amines, RNH_2. (Equilibrium)[r]	$-7.26 \pm .21$
Acid-catalyzed hydrolysis of monosubstituted ethylene oxides in $0.7M$ aqueous $HClO_4$ at $0°$[s]	$-1.827 \pm .079$
Solvolysis of secondary R_1R_2-carbinyl p-bromobenzenesulfonates in acetic acid at 70–$75°$[t]	-3.49
Solvolysis of t-alkyl halides $R_1R_2R_3CX$ (Cl or Br) in 80% ethanol at $25°$[t]	-3.29
Solvolysis of primary alkyl p-toluenesulfonates, RCH_2OTs, in ethanol at $100°$[t]	-0.742
Displacement of primary alkyl bromides, RCH_2Br, by thiophenoxide ion in methanol at $20°$[t]	-0.606
Acidities of primary alcohols, RCH_2OH, in water at $25°$[u]	1.42
ortho-Substituted Benzene Derivatives	
Ionization equilibrium of benzoic acids in water at $25°$[c]	$1.787 \pm .13$
Catalysis of dehydration of acetaldehyde hydrate by benzoic acids in acetone at $25°$[j]	$0.771 \pm .019$
Ionization equilibrium of anilinium ions in water at $25°$[v]	$2.898 \pm .15$
Benzanilide formation from anilines with benzoyl chloride in benzene at $25°$[w]	$2.660 \pm .22$

(a) Mostly from reference 44b. (b) Rates unless otherwise noted. (c) Landolt-Börnstein, *Physikalisch-Chemische Tabellen*, Springer, 1936; J. F. J. Dippy, *Chem. Revs.*, **25**, 151 (1939); G. E. K. Branch and M. Calvin, *The Theory of Organic Chemistry*, Prentice-Hall, 1941;

The same equation can be applied to the reactions of certain *ortho*-substituted benzene derivatives, using the σ^*-values of Table 7-15. Table 7-16 lists some of the reactions that have been successfully correlated by means of equation 57.

Formally the $\rho^*\sigma^*$ equation is a correlation of free energies by the method of linear combination. By using the definition of σ^* and expanding the equation, it can be written as (58).

$$\delta_R \Delta F^{\ddag} = \frac{\rho^*}{2.48} \delta_R \Delta F_B{}^{\ddag} - \frac{\rho^*}{2.48} \delta_R \Delta F_A{}^{\ddag} \qquad (58)$$

In deriving equation 58 we have ignored the complications due to the looseness of specification of temperature and solvent in the definition of σ^*.

Just as the particular blend of interaction mechanisms represented by σ is quite common in the reactions of aromatic compounds, the particular blend of interaction mechanisms represented by σ^* is quite common in aliphatic reactions. In the preceding section we have justified the identification of σ^* as a polar substituent parameter. However, the theoretical soundness of using equation 57 in no way depends on the correctness of the identification of the interaction mechanisms associated with σ^*. The theoretical soundness of (57) depends merely on the correctness of the assumption that there are two principal interaction mechanisms and that these two mechanisms differ in their relative importance in the two model processes. Whether or not the difference in $\delta_R \Delta F^{\ddag}$ for the two models isolates the effect of one of the mechanisms in pure form is irrelevant for the adequacy of the representation by linear combination. The fact that the validity of equations 57 and 58 is independent of the *identification* of the interaction mechanisms reflects the generality of the extrathermodynamic approach, which in this respect resembles thermodynamics.

A. L. Henne and C. J. Fox, *J. Am. Chem. Soc.*, **76**, 479 (1954); M. Kilpatrick and J. G. Morse, *ibid.*, **75**, 1854 (1953); E. Gelles, *ibid.*, **75**, 6201 (1953); L. Sommer et al., *ibid.*, **71**, 1509 (1949). (d) M. M. Kreevoy and R. W. Taft, Jr., *J. Am. Chem. Soc.*, **77**, 5590 (1955). (e) R. W. Taft, Jr. and D. J. Smith, *J. Am. Chem. Soc.*, **76**, 305 (1954). (f) F. Basolo, J. G. Bergmann, and R. G. Pearson, *J. Phys. Chem.*, **56**, 22 (1952). (g) J. Hine and M. Hine, *J. Am. Chem. Soc.*, **74**, 5266 (1952). (h) J. D. Roberts and W. T. Moreland, Jr., *J. Am. Chem. Soc.*, **75**, 2167 (1953) (using σ^* for XCH$_2$). (i) N. C. Deno and M. S. Newman, *J. Am. Chem. Soc.*, **72**, 3852 (1950). (j) R. P. Bell and W. C. E. Higginson, *Proc. Roy. Soc. (London)*, **A197**, 141 (1949). (k) H. M. Dawson, G. V. Hall and A. Key, *J. Chem. Soc.*, 2849 (1928). (l) A. A. Frost and R. G. Pearson, *Kinetics and Mechanism*, Wiley, 1953, p. 222. (m) R. P. Bell and E. C. Baughan, *J. Chem. Soc.*, 1947 (1937). (n) G. L. O'Connor and H. R. Nace, *J. Am. Chem. Soc.*, **75**, 2118 (1953). (o) A. Skrabal and H. H. Eger, *Z. physik. Chem.*, **122**, 349 (1926). (p) D. P. Evans and J. J. Gordon, *J. Chem. Soc.*, 1434 (1938). (q) H. V. Hartel, N. Meer, and M. Polanyi, *Z. physik. Chem.*, **B19**, 139 (1932). (r) H. C. Brown, M. D. Taylor, and S. Sujishi, *J. Am. Chem. Soc.*, **73**, 2464 (1951). (s) J. G. Pritchard and F. A. Long, *J. Am. Chem. Soc.*, **78**, 2667 (1956). (t) A. Streitweiser, Jr., *J. Am. Chem. Soc.*, **78**, 4935 (1956) (u) P. Ballinger and F. A. Long, *J. Am. Chem. Soc.*, **82**, 795 (1960). (v) M. Kilpatrick and C. A. Arenberg, *J. Am. Chem. Soc.*, **75**, 3812 (1953). (w) F. J. Stubbs and C. N. Hinshelwood, *J. Chem. Soc.*, 1949 supplement p. 71.

Steric Effects in Aliphatic Reactions

For some reaction series it is possible to correlate deviations from equation 57, apparently due to steric effects, in one reaction series with deviations due to the same effects in another reaction series. The acid-catalyzed hydrolysis of esters has been proposed as a suitable standard of steric effects, for the polar effects in this reaction are comparatively small.[50] This insensitivity to polar effects can be seen from the low ρ-values for the acid-catalyzed hydrolysis of *meta*- and *para*-substituted benzoates and from the fact that the rates of the acid-catalyzed hydrolysis of ethyl

Table 7-17. *Steric Substituent Constants at* 25° [44b]

SUBSTITUENT	E_s	SUBSTITUENT	E_s	SUBSTITUENT	E_s
		Aliphatic Compounds, RCOOR' (substituent = R)			
H	1.24	ϕCH_2	−.38	$\phi(Et)CH$	−1.50
CH_3	(.00)	ϕCH_2CH_2	−.38	Cl_2CH	−1.54
C_2H_5	−.07	$\phi CH_2CH_2CH_2$	−.45	$(CH_3)_3CCH_2$	−1.74
Cyclobutyl	−.06	*i*-Propyl	−.47	ϕ_2CH	−1.76
CH_3OCH_2	−.19	C/clopentyl	−.51	Me(neopentyl)CH	−1.85
$ClCH_2$, FCH_2	−.24	F_2CH	−.67	Br_2CH	−1.86
$BrCH_2$	−.27	Cyclohexyl	−.79	Et_2CH	−1.98
CH_3SCH_2	−.34	$MeOCH_2CH_2$	−.77	Cl_3C	−2.06
ICH_2	−.37	$ClCH_2CH_2$	−.90	$(n\text{-Propyl})_2CH$	−2.11
n-Propyl	−.36	*i*-Butyl	−.93	$(i\text{-Butyl})_2CH$	−2.47
n-Butyl	−.39	*cyclo*-$C_6H_{11}CH_2$	−.98	Br_3C	−2.43
n-Amyl	−.40	Me(Et)CH	−1.13	$Me_2(\text{neopentyl})C$	−2.57
i-Amyl	−.35	F_3C	−1.16	$(\text{neopentyl})_2CH$	−3.18
n-Octyl	−.33	Cycloheptyl	−1.10	Me(*t*-Bu)CH	−3.33
$(CH_3)_3CCH_2CH_2$	−.34	$\phi(Me)CH$	−1.19	$Me_2(t\text{-Bu})C$	−3.9
ϕOCH_2	−.33	*t*-Butyl	−1.54	Et_3C	−3.8
				Me(*t*-Bu)(neopentyl)C	−4.0
		ortho-Substituted Benzoates			
CH_3O	.99	Cl	.18	I	−.20
C_2H_5O	.90	Br	.00	NO_2	−.75
F	.49	CH_3	(.00)	C_6H_5	−.90

chloroacetate and ethyl propionate differ by only about 20%. By assuming that substituent effects in these reactions are to a good approximation due solely to steric effects, it is possible to establish a set of steric substituent constants, E_s.[50a] Table 7-17 lists E_s values defined by equation 59, in which k_0 is again the rate constant for the methyl substituent.

$$E_s = \log \left(\frac{k}{k_0}\right)_A \qquad (59)$$

The identification of E_s as a steric parameter is supported by the fact that E_s values are not at all additive, in contrast to polar substituent parameters.

[50] R. W. Taft, Jr., *J. Am. Chem. Soc.*, **75**, 4538 (1953); *ibid.*, **74**, 3126 (1952).

[50a] It should be noted, however, that the function E_s as defined in equation 59 includes contributions from resonance interactions between the R-group and the carbonyl group of the ester.

As can be seen from Table 7-17, the values of E_s increase faster than additively as bulky groups are attached to the α-carbon atom of the acyl group.

For reactions in which the steric effects of substituents involve the same mechanism as in ester hydrolysis, equation 60 is applicable.

$$\log \frac{k}{k_0} = \rho^* \sigma^* + sE_s \tag{60}$$

Equation 60 is simply another correlation by linear combination in which the relative importance of the two interaction mechanisms differs from that

Table 7-18. Some Reactions Correlated by the sE_s Equation

REACTION	S
Acid-catalyzed hydrolysis of *ortho*-substituted benzamides in water at 100°[a]	0.812 ± .032
Acid-catalyzed methanolysis of RCOO-β-$C_{10}H_7$ at 25°[b]	1.376 ± .057
Acid-catalyzed 1-propanolysis of RCOO-β-$C_{10}H_7$ at 25°[b]	1.704 ± .093
Acid-catalyzed 2-propanolysis of RCOO-β-$C_{10}H_7$ at 25°[b]	1.882 ± .106
Reaction of methyl iodide with 2-alkylpyridines in nitrobenzene at 30°[c]	2.065 ± .096
$\Delta H°$ for reaction of 2-alkylpyridines with diborane[d]	3.322 ± .365
$\Delta H°$ for reaction of 2-alkylpyridines with BF_3 in nitrobenzene[e]	5.49 ± .36
$\Delta H°$ for reaction of 2-alkylpyridines with trimethylboron in nitrobenzene[f]	6.36 ± .34

[a] E. E. Reid, *Am. Chem. J.*, **24**, 397 (1900).
[b] M. Harfenist and R. Baltzly, *J. Am. Chem. Soc.*, **69**, 362 (1947).
[c] H. C. Brown and A. Cahn, *J. Am. Chem. Soc.*, **77**, 1715 (1955); if E_{act} is used, s = 2.235 ± .089.
[d] H. C. Brown and L. Domash, *ibid.*, **78**, 5384 (1956).
[e] H. C. Brown and R. H. Horowitz, *ibid.*, **77**, 1733 (1955).
[f] H. C. Brown, D. Gintis and L. Domash, *ibid.*, **78**, 5387 (1956).

in equation 58. If we substitute the definitions for σ^* and E_s, we obtain (60a).

$$\log \frac{k}{k_0} = \frac{\rho^*}{2.48} \log \left(\frac{k}{k_0}\right)_B + \left(s - \frac{\rho^*}{2.48}\right) \log \left(\frac{k}{k_0}\right)_A \tag{60a}$$

If $\rho^* \sigma^*$ is small compared to sE_s, equation 60 reduces to (61), which seems to correlate substituent effects fairly well for the reactions listed in Table 7-18.[44b]

$$\log \frac{k}{k_0} = sE_s \tag{61}$$

A reaction requiring the use of the more general equation (60) is the base-catalyzed methanolysis of 1-menthyl esters of aliphatic acids.[51]

Steric interaction mechanisms occur in greater variety than do polar interaction mechanisms. Thus equations 60 and 60a, which represent just one of the possibly infinite set of steric interaction mechanisms, fail to correlate substituent effects in many reactions. Examples are the alkaline hydrolysis of the N-methylanilides of various acids,[52] and the acid-catalyzed hydrolysis of a series of esters of thiolacetic acid.[53]

The assignment of deviations from equation 60 to changes in the nature of the steric interaction mechanism can be tested in at least one example. In the hydrolysis[54] of glycerates, the substituent effects do not obey equation 60 no matter what value of the mixing parameter s is used.

$$CH_2-CH-\overset{\overset{\displaystyle O}{\displaystyle \|}}{C}-O-R'$$
$$\underset{\displaystyle OH}{|}\ \ \underset{\displaystyle OH}{|}$$

The lack of correlation is due to a change in the steric interaction mechanism when the substituent is in the alkyl rather than the acyl part of the ester. If the steric effects are eliminated by assuming again that they are identical in the acid and base-catalyzed reactions, the difference

$$\left[\log \left(\frac{k}{k_0} \right)_B - \log \left(\frac{k}{k_0} \right)_A \right]$$

is proportional to σ^*, equation 62.

$$\log \frac{(k/k_0)_B}{(k/k_0)_A} = (3.33 \pm 0.11)\sigma^* \tag{62}$$

In (62) a term analogous to sE_s, but depending in a *different* way on the nature of the substituent, has canceled out, leaving only the polar effect. The *polar effect*, however, involves the same polar interaction mechanism as before, as shown by its proportionality to σ^*.

We have just seen that steric effects in the alkyl and acyl parts of an ester involve different steric interaction mechanisms. A useful device for predicting whether or not two systems will depend on the same steric interaction mechanisms is the principle of isosterism: the steric mechanisms should be

[51] W. S. Puvelich and R. W. Taft, Jr., *J. Am. Chem. Soc.*, **79**, 4935 (1957).
[52] S. S. Biechler and R. W. Taft, Jr., *J. Am. Chem. Soc.*, **79**, 4927 (1957).
[53] P. N. Rylander and D. S. Tarbell, *J. Am. Chem. Soc.*, **72**, 3021 (1950); J. R. Schaefgen, *ibid.*, **70**, 1308 (1948).
[54] C. M. Groocock, C. K. Ingold, and A. Jackson, *J. Chem. Soc.*, 1057 (1930).

identical in two structures of the same size and shape, regardless of the particular elements from which they are formed.[55] For example, structure (63) should be a good model, sterically speaking, for structure (64). Two such structures are said to be *isosteres*.

$$(63)$$

$$(64)$$

Structure (63) shows a complex between a 2-alkylpyridine and BF_3, whereas structure (64) shows an important resonance structure of the transition state for the reaction of the 2-alkylpyridine with methyl iodide. Because polar effects in the formation of the complex and in the formation of the isosteric transition state are small, the linear relationship found between $\Delta F\ddagger$ for the reaction with CH_3I and $\Delta \bar{H}°$ for the reaction with BF_3 indicates that the steric interaction mechanisms are identical.[55]

Dispersion from the $\rho^*\sigma^*$ Equation due to Resonance and α-Hydrogen Effects

In the acid-catalyzed hydrolysis of diethyl acetals, R—$CH(OEt)_2$, a plot of log k against σ_R^* gives a rather rough linear correlation for saturated substituents, but the points for the α,β-unsaturated substituents 2-methyl-vinyl, 2-phenylvinyl, and phenyl are nowhere near the line, the rates being much higher than predicted.[56] This dispersion is almost certainly due to resonance enhancement of the substituent effect, as might be expected from the transition state formula (65).

$$(65)$$

[55] H. C. Brown and R. H. Horowitz, *J. Am. Chem. Soc.*, **77**, 1733 (1955).
[56] M. M. Kreevoy and R. W. Taft, Jr., *J. Am. Chem. Soc.*, **77**, 5590 (1955).

Failure of the $\rho^*\sigma^*$ correlation has also been noted for the ionization of α,β-unsaturated acids.[44b]

In some reactions there is a distinctive kind of dispersion from the $\rho^*\sigma^*$ equation that depends on the number of hydrogen atoms bonded to the α-carbon atom of the substituent. For example, in the acid-catalyzed hydrolysis of the diethyl acetals and ketals $R_1R_2C(OEt)_2$, where R_1 is a saturated alkyl group and R_2 is methyl or hydrogen, rate constants varying by over seven orders of magnitude are fitted to better than 0.2 log unit by equation 66, where n is the number of α-hydrogen atoms.

$$\log \frac{k}{k_0} = -3.60(\sigma_{R_1}^* + \sigma_{R_2}^*) + (0.54)(n - 6) \qquad (66)$$

The rate constant k_0 in equation 66 refers to acetone ketal. It will be noted that in this compound there are six α-hydrogen atoms; since $\sigma_{CH_3}^*$ is zero by definition, the right side of the equation correctly reduces to zero. To show how equation 66 can also be applied to acetals, let us consider acetal itself. There are three α-hydrogen atoms in $CH_3CH(OEt)_2$; hence the value of $\log k/k_0$ is given by

$$[-3.60(\sigma_{CH_3}^* + \sigma_H^*) - 0.54 \cdot 3] = -3.38.$$

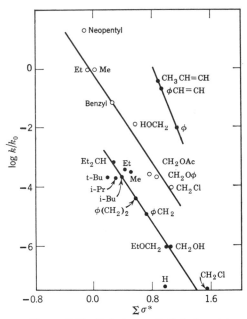

Figure 7-19. Hydrolysis of diethyl acetals (solid circles) and diethyl ketals (open circles) in 49.6% dioxane at 25°.

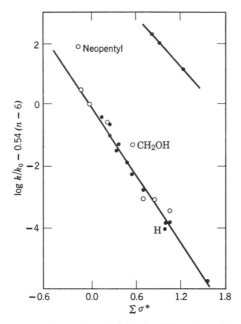

Figure 7-20. Hydrolysis of diethyl acetals (solid circles) and diethyl ketals (open circles) in 49.6% dioxane at 25°, with correction for α-hydrogen atoms. The compounds are the same as in Figure 7-19.

For comparison, the experimental value of log k/k_0 for acetal is -3.48.[56] Figures 7-19 and 7-20 show plots of the data before and after addition of the $0.54(n - 6)$ term.

The distinctive feature of the α-hydrogen effect, as illustrated by these data, is that the plot of log k/k_0 versus $\Sigma\,\sigma^*$ consists of a series of *parallel* lines, each line being characterized by the number of α-hydrogen atoms in the reagent. In equation 66 the vertical displacements of the lines amount to a constant increment per α-hydrogen atom. However, later in this section we shall have an example in which the vertical displacements are not equal.

We now come to the difficult, vexing, and moot question of interpreting the rate effects associated with changes in the number of α-hydrogen atoms. It has been pointed out that the parameter σ^* measures an almost purely inductive effect, since the transition states for acid and base-catalyzed ester hydrolysis have very little unsaturation in the reaction zone.

The transition state for the hydrolysis of an acetal or a ketal, on the other hand, has considerable unsaturation (equation 65), and it is tempting to ascribe the dispersion noted in equation 66 to a resonance effect. For alkyl substituents such resonance would involve hyperconjugation, as shown in structures (67) and (68).[57]

$$
\begin{array}{ccc}
& H^+ & \\
CH_3 & & CH_2 \\
\diagdown & & \diagdown\diagup \\
\overset{+}{C}-OEt & \leftrightarrow & C-OEt \\
\diagup & & \diagup \\
CH_3 & & CH_3
\end{array} \qquad (67)
$$

$$
\begin{array}{ccc}
CH_3 & & CH_3{}^+ \\
\diagdown & & \diagdown\diagup \\
CH_2-\overset{+}{\underset{|}{C}}-OEt & \leftrightarrow & CH_2=C-OEt \\
| & & | \\
CH_3 & & CH_3
\end{array} \qquad (68)
$$

We shall use h_{CH} to denote the stabilization of the transition state by hyperconjugation *per* carbon-hydrogen bond, and h_{CC} to denote the analogous stabilization *per* carbon-carbon bond. We can then interpret equation 66 on the basis that the coefficient 0.54 is equal to $h_{CH}/2.303\,RT$ and that h_{CC} is negligibly small.[58] We are reluctant to make this interpretation, however, because the value obtained for h_{CC} seems rather small in comparison with h_{CH}, although both values are still within the rather large uncertainty limits of their quantum mechanical calculation. More likely, the effects of hyperconjugation are perturbed or even overwhelmed by additional interaction mechanisms. Prominent among the possible perturbing mechanisms that have been mentioned is steric hindrance to solvation.[58a]

An equilibrium showing a closely analogous α-hydrogen effect is the hydrogenation of aldehydes and ketones in toluene at 60°.[59] The equilibrium constants are correlated by equation 69, in which K_0 is the

[57] Although the effects of hyperconjugation are small, the evidence that they are real has recently been reinforced somewhat by means of deuterium isotope effects in bridgehead compounds. V. J. Shiner, Jr., *J. Am. Chem. Soc.*, **82**, 2655 (1960).

[58] R. W. Taft, Jr., and I. C. Lewis, *Tetrahedron*, **5**, 228 (1959) have fitted the data of reference (56) to the equation, $\log k/k_0 = -3.60(\sigma_{R_1}^* + \sigma_{R_2}^*) + 0.62(n-6) + 0.24n'$, where n' is the number of *alpha* carbon-carbon bonds. They suggest that the ratio of the parameters, 0.62/0.24, is a better estimate of h_{CH}/h_{CC}.

[58a] W. M. Schubert, J. M. Craven, R. G. Minton, and R. B. Murphy, *Tetrahedron*, **5**, 194 (1959); R. Clement and J. N. Naghizadeh, *J. Am. Chem. Soc.*, **81**, 3154 (1959); V. J. Shiner, Jr. and C. J. Verbanic, *ibid.*, **79**, 369 (1957).

[59] H. Adkins, R. M. Elofson, A. G. Rossow, and C. C. Robinson, *J. Am. Chem. Soc.*, **71**, 3622 (1949).

equilibrium constant for the conversion of acetone to 2-propanol.

$$\log \frac{K}{K_0} = 4.681 \sum \sigma^* - 0.393(n - 6) \qquad (69)$$

A qualitatively different α-hydrogen effect is observed when pK_A for substituted ammonium ions in water at $25°$ is plotted against $\sum \sigma^*$ for the substituents. The points fall on or near three separate and virtually parallel lines, corresponding to the number of hydrogen atoms bonded to the nitrogen atom.[60]

Primary amines $pK_A = -3.14 \sum \sigma^* + 13.23$

Secondary amines $pK_A = -3.23 \sum \sigma^* + 12.13$

Tertiary amines $pK_A = -3.30 \sum \sigma^* + 9.61$

However, in contrast to the effects described above, the vertical displacements of the lines are not equal, as shown by the unequal differences in the intercepts. The correlation with $\sum \sigma^*$ is excellent for tertiary amines, but for primary or secondary amines the points for compounds with more than minimum steric requirements tend to scatter somewhat.

The identification of the mechanism responsible for the dispersion is again a moot question. A useful hint comes from the proton transfer reactions of amines or anilines with acids such as 2,6-dinitrophenol in the solvents, anisole, and chlorobenzene. The equilibrium constants for the formation of ion pairs, BH^+A^-, in these aprotic solvents are in a fairly good inductive order, with no evidence for appreciable dispersion due to an α-hydrogen effect.[61] One is therefore tempted to ascribe the dispersion observed for pK_A in water to a solvation mechanism dependent on the number of N—H bonds. The fact that the correlations with $\sum \sigma^*$ deteriorate somewhat for certain amines with bulky substituents could then be explained as steric hindrance to solvation. The additional fact that deviations are observed only for bulky primary and secondary amines but not for tertiary amines could suggest, furthermore, that the solvation mechanism susceptible to steric hindrance involves N—H $\cdots$ O hydrogen bonds.

THE BRØNSTED RELATIONSHIP

Expressions for the rate of acid or base-catalyzed reactions often contain not only terms dependent on the hydronium and hydroxide ion concentrations but also terms proportional to the concentrations of various weak

[60] H. K. Hall, Jr., *J. Am. Chem. Soc.*, **79**, 5441 (1957). This paper contains an extensive and useful survey of pK_A values of amines.

[61] R. P. Bell and J. W. Bayles, *J. Chem. Soc.*, 1518 (1952).

acids or bases that may also be present in the solution. The rate constants k_A and k_B obtained for the kinetic processes involving the molecular acid or base are found to depend on the strength of these catalysts. The dependence often takes the form of a linear free energy relationship, such as equations 70 or 71.

$$\log k_A = B_A \log K_A + \log g_A \tag{70}$$

$$\log k_B = B_B \log K_B + \log g_B \tag{71}$$

Equations 70 and 71 are known as the Brønsted catalysis relationships.[62] The quantities g and B are parameters characteristic of the particular series of reactions. B, the Brønsted coefficient, is always positive and, as far as the authors are aware, never greater than unity.[63]

A good example of a reaction obeying the Brønsted relationship is the acid-catalyzed dehydration of 1,1-ethanediol.[64]

$$CH_3\!-\!\underset{\underset{OH}{\displaystyle |}}{\overset{\overset{OH}{\displaystyle |}}{C}}\!-\!H \xrightarrow[\text{acid catalyst}]{92.5\% \text{ acetone}} CH_3\!-\!\overset{\overset{O}{\displaystyle \|}}{C}\!-\!H + H_2O \tag{72}$$

The rate expression for reaction (72) contains terms of the form k_A(HA)(1,1-diol), in which k_A is a rate constant characteristic of the acid catalyst, HA. Figure 7-21 shows the linear relationship between the logarithms of these rate constants, measured in aqueous acetone at 25°, and the logarithms of the dissociation constants in water at 25°, for a series of aliphatic acids. Although most of the points fall on a good straight line, a few deviate. The acids corresponding to the deviant points are also atypical in structure. For example, pivalic acid is conspicuously hindered to solvation. Formic acid is conspicuously unhindered to solvation and indeed owes its great acidity in comparison to acetic acid to a more favorable entropy of ionization, attributable to a difference in solvation.[65] Other deviant points are those for crotonic, cinnamic, and phenylpropiolic acids, all of which are unsaturated and have possibilities for resonance not available to the other acids. Restrictions on the range of permissible variation in structure are of course characteristic of linear free energy relationships in general, but the restrictions seem to be particularly stringent in the Brønsted relationship.

---

[62] J. N. Brønsted and K. J. Pedersen, *Z. physikal. Chem.* **108**, 185 (1924).

[63] Relationships formally analogous to the Brønsted relationship, *but not involving proton transfer*, do sometimes have slopes greater than unity.

[64] R. P. Bell and W. C. E. Higginson, *Proc. Roy. Soc.*, **A197**, 141 (1949).

[65] See Table 3-2.

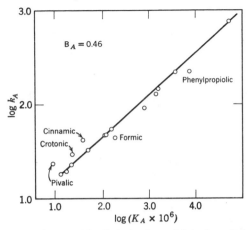

Figure 7-21. Dehydration of 1,1-ethanediol catalyzed by aliphatic acids.

Figure 7-22 shows a plot for the same reaction catalyzed by substituted benzoic acids. There are two lines, the upper one for *meta-* and *para-* substituents and the lower for *ortho-*substituents. The line of Figure 7-21 would fall between the two lines shown in Figure 7-22, if they were plotted on the same graph.

Reaction (72) is also catalyzed by phenols. Although the correlation is rough, it extends over four orders of magnitude in the rate constant and six

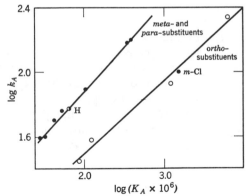

Figure 7-22. Dehydration of 1,1-ethanediol catalyzed by aromatic acids. Solid circles, *meta-* and *para-*substituents; open circles, *ortho-*substituents.

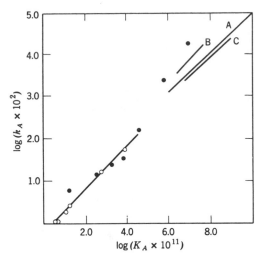

Figure 7-23. Dehydration of 1,1-ethanediol catalyzed by phenols. Solid circles, *ortho*-substituted phenols: open circles, *meta*- and *para*-substituted phenols. The lines of Figure 7-21 (line A) and 7-22 (lines B and C) are included.

in the dissociation constant. Figure 7-23 shows the experimental points for the phenols, including those with *ortho*-substituents; the lines for the three classes of carboxylic acids are included for comparison.

The Interpretation of the Brønsted Slopes, B

Many, if not most, acid- or base-catalyzed reactions have reaction mechanisms consisting of several steps. Assuming a particular reaction mechanism, it is often possible to show that the existence of the Brønsted relationship implies the existence of a rate-equilibrium relationship for one of the individual steps in the reaction. To illustrate this point, let us consider the following possible reaction mechanism for the acid-catalyzed dehydration of 1,1-ethanediol.

(a) $\underset{\overset{|}{OH}}{\overset{OH}{|}}CH_3-CH + H^+ \underset{\substack{rapid,\\ reversible}}{\overset{1/K'_a}{\rightleftharpoons}} \underset{\overset{|}{O-H}}{\overset{\overset{\oplus}{H-O-H}}{|}}CH_3-CH$

(b) $\underset{\underset{\displaystyle OH}{|}}{\overset{\overset{\displaystyle H-\overset{\oplus}{O}-H}{|}}{CH_3-CH}} + RCO_2^- \xrightarrow[\text{(rate determining)}]{k'_b} CH_3CHO + HOH + RCOOH$

The equilibrium constant for step (a) is written as $1/K'_a$, in order to let K'_a be the acid dissociation constant of $CH_3CH(OH)OH_2^+$.

The reaction rate by this two-step mechanism is given by equation 73.

$$\text{Rate} = k'_b[RCO_2^-][CH_3CH(OH)OH_2^+] \tag{73}$$

Making use of the expression for $1/K'_a$, equation 73 can be transformed into (74).

$$\text{Rate} = \frac{k'_b}{K'_a}[RCO_2^-][H^+][CH_3CH(OH)_2] \tag{74}$$

Furthermore, the quantity $[RCO_2^-][H^+]/[RCO_2H]$ is equal to the acid dissociation constant K_A. Hence we obtain equation 75, which may be compared with the empirical rate law (76). The result is (77).

$$\text{Rate} = \frac{k'_b K_A}{K'_a}[RCO_2H][CH_3CH(OH)_2] \tag{75}$$

$$\text{Rate} = k_A[RCO_2H][CH_3CH(OH)_2] \tag{76}$$

$$k_A = \frac{k'_b K_A}{K'_a} \tag{77}$$

Next we wish to relate K'_b, the *equilibrium* constant for step (b), to K_A. The final expression is (78),

$$K'_b = \frac{[CH_3CHO][RCOOH]}{[CH_3CH(OH)OH_2^+][RCO_2^-]}$$

$$K'_b = \frac{[CH_3CHO][H^+]}{[CH_3CH(OH)OH_2^+]} \cdot \frac{[RCOOH]}{[RCO_2^-][H^+]}$$

$$K'_b = \frac{K'_c}{K_A} \tag{78}$$

where K'_c is the equilibrium constant for reaction (c).

(c) $\underset{\underset{\displaystyle OH}{|}}{\overset{\overset{\displaystyle OH_2^+}{|}}{CH_3-CH}} \overset{K'_c}{\rightleftharpoons} CH_3CHO + H_2O + H^+$

We are now in a position to show that the observed Brønsted relationship between k_A and K_A implies the existence of a rate-equilibrium relationship between k'_b and K'_b. This is done as follows, using equations 70, 77, and 78.

$$\log k_A = B_A \log K_A + \log g_A$$
$$\log k'_b = (B_A - 1) \log K_A + \log g_A + \log K'_a$$
$$\log k'_b = (1 - B_A) \log K'_b - (1 - B_A) \log K'_c + \log g_A + \log K'_a \quad (79)$$

Since the last three terms on the right in equation 79 involve only constants characteristic of the substrate or parameters characteristic of the Brønsted relationship, they are constant, and the rate-equilibrium relationship (80) results.

$$\log k'_b = (1 - B_A) \log K'_b + a\ constant \quad (80)$$

To be consistent with our previous notation[66] we shall use the symbol α to denote the slope of the rate-equilibrium relationship. Hence in this particular mechanism, $\alpha = 1 - B$.

The reader will undoubtedly be able to visualize alternative reaction mechanisms. In some of these, for example the one shown in equation 81, the Brønsted relationship implies a rate-equilibrium relationship with $B = \alpha$. In still others, such as that shown in equation 82, no rate-equilibrium relationship is implied.

$CH_3CH(OH)_2 + H_2O + RCOOH$

$$\rightleftharpoons CH_3CHO + H_3O^+ + H_2O + RCO_2^- \quad (81)$$

$CH_3CH(OH)_2 + RCOOH \rightleftharpoons$

$$CH_3CHO + HOH + RCOOH \quad (82)$$

[66] Chapter 6, pp. 156–159.

In general, a rate-equilibrium relationship is implied by a Brønsted relationship whenever the equilibrium constant for the rate-determining step of the reaction mechanism is proportional to K_A or to K_A^{-1}. When this is the case, the Brønsted slope B is related to the slope α of the corresponding rate-equilibrium relationship by one or the other of the following equations, depending on the details of the reaction mechanism.

$$B = \alpha \tag{83}$$

$$B = 1 - \alpha \tag{84}$$

The conversion of a Brønsted relationship into a rate-equilibrium relationship is sometimes of value in diagnosing the reaction mechanism. It will be recalled[66] that there exists a simple principle for predicting the *magnitude* of α. For proton transfer reactions this principle takes the following form: α measures the degree of resemblance of the transition state to the products; and the position of the transition state along the reaction coordinate is always such that the stronger base is the less neutralized of the two bases between which the proton is being transferred. If the magnitude of α anticipated for a particular mechanism is incompatible with the experimental value of B, that mechanism can be ruled out.

Since the rate-equilibrium relationship embodies one of the most important qualitative ideas of chemistry, the scope and limitations of the Brønsted relationship are of special interest. As is clear from the examples shown in Figures 7-21 to 7-23, there are severe limitations on the permissible changes in the structure of the acid or base catalyst. However, within these limitations the fit is very good, the deviations of the points from the line being no greater than the experimental error in many reaction series. The success of the Brønsted relationship is even more striking when we remember the innate complexity of the processes that are being compared. Particularly for acid and base dissociation in water, the thermodynamic quantities $\Delta H°$, $\Delta S°$, and $\Delta C_p°$ prove by their very complexity that interaction mechanisms depending on the detailed microscopic structure of the solvent play an important role.[67]

It follows from the connection between the Brønsted and rate-equilibrium relationships that the slope B should not remain precisely constant for very large changes in reactivity. No such curvature has in fact been found, but of course large changes in K_B or K_A are attainable only by going outside of a single structural class.

Solvent Effects in the Brønsted Relationship

Ordinarily the K_A or K_B values are those for aqueous solutions, but some data are available on the effect of solvent changes on the rate constants k_A

[67] F. S. Feates and D. J. G. Ives, *J. Chem. Soc.*, 2798 (1956).

or k_B. An example is the solvent effect in the base-catalyzed decomposition of nitramide.

$$H_2N—NO_2 \rightleftharpoons \overset{\displaystyle OH}{\overset{\displaystyle |}{HN}}=N\rightarrow O \qquad (85)$$

$$\overset{\displaystyle OH}{\overset{\displaystyle |}{HN}}=N\rightarrow O + B \longrightarrow BH^+ + OH^- + N_2O \qquad (86)$$

The Brønsted slopes B for the relationships of equation 87 vary with the solvent, as shown in Table 7-19.[68,69]

$$\log k_{B,solvent} = B \log K_{B,water} + g \qquad (87)$$

Table 7-19. Solvent Effects on the Brønsted Slope B for the Decomposition of Nitramide at 25°

SOLVENT	VALUE OF B[a]	
	ANIONS	AMINES
Water	0.80	0.75
m-Cresol	0.78	0.84
i-Amyl alcohol	0.83	0.92
Anisole	—	0.64
Benzene	—	0.7 ± 0.1

[a] Plotted against K_B for aqueous solutions.

The practice of plotting $\log k_{B,solvent}$ against $\log K_{B,water}$ has two undesirable effects. One is that the magnitude of B can no longer be interpreted on some simple basis, such as the rate-equilibrium relationship, for we do not know what fraction of the medium effect on B is due to a medium effect on $\delta_R \log K_B$. The other is that the fit tends to deteriorate as the solvents in which k_B and K_B are measured become less alike. For example, when the rate constants in anisole as solvent are compared with the base dissociation constants in m-cresol rather than in water, the precision of the fit to a straight line improves. The improved fit using the dissociation constants for the amines in m-cresol rather than in water has been attributed to peculiarities of water as a solvent.[69] For some reactions, aqueous dissociation constants give a set of parallel Brønsted lines for primary, secondary, and tertiary amines, whereas nonaqueous dissociation constants give a single line.[61,70]

[68] E. F. Caldin and J. Peacock, *Trans. Faraday Soc.*, **51**, 1217 (1955).
[69] G. C. Fettes, J. A. Kerr. A. McClure, J. S. Slater, C. Steel, and A. F. Trotman-Dickenson, *J. Chem. Soc.*, 2811 (1957).
[70] A. F. Trotman-Dickenson, *J. Chem. Soc.* 1293 (1949).

REACTIVITY IN NUCLEOPHILIC SUBSTITUTION REACTIONS

Correlation with Base Dissociation Constants

The rate at which a nucleophile attacks a carbon atom sometimes parallels the affinity of the nucleophile for protons. We shall therefore begin our extrathermodynamic discussion of nucleophilic activity by assuming that base dissociation is a suitable model process.

A reaction in which there is a parallelism between nucleophilic activity and basicity is the hydrolysis of acetic anhydride catalyzed by nitrogen heterocycles. The rate-determining step in this reaction is believed to be the formation of an acylium compound.[71]

$$CH_3CO\text{-}O\text{-}COCH_3 + RC_5H_4N \xrightarrow{k_c} CH_3CO\text{-}\overset{+}{N}C_5H_4R + CH_3COO^-$$

$$CH_3CO\text{-}\overset{+}{N}C_5H_4R + H_2O \xrightarrow{\text{fast}} CH_3COOH + H^+ + RC_5H_4N \qquad (88)$$

The rate constants for various substituted pyridines in water at $0°$ are correlated with the basicity constants in water at $25°$, as shown in Figure 7-24. Two lines are required, one for unhindered pyridines, the other for *ortho*-substituted pyridines. As is the case for most extrathermodynamic relationships in which proximity effects produce dispersion into two lines rather than scatter, the lines are very nearly parallel. The slopes are about 0.95.

There is a strong analogy between the relationship shown in Figure 7-24 and that of a typical Brønsted plot, such as Figure 7-22. In both figures the points fall on a single line only if the catalysts belong to a narrowly defined structural type. In both figures *ortho*-substitution displaces the relationship to a parallel line. However, while the Brønsted slopes B are generally between zero and unity, the slopes of the relationships between nucleophilic rate constants and K_B are not so limited. For example, the hydrolysis of *p*-nitrophenyl acetate in water is also catalyzed by nitrogen heterocycles and proceeds by a similar mechanism.[72]

$$B: + O_2NC_6H_4\text{-}O\text{-}\overset{\overset{O}{\|}}{C}CH_3 \xrightarrow{k_c} \overset{+}{B}\text{-}\overset{\overset{O}{\|}}{C}\text{-}CH_3 + O_2NC_6H_4O^-$$

$$\overset{+}{B}\text{-}\overset{\overset{O}{\|}}{C}\text{-}CH_3 + H_2O \xrightarrow{\text{fast}} B: + H^+ + CH_3\text{-}\overset{\overset{O}{\|}}{C}\text{-}OH \qquad (89)$$

[71] V. Gold and E. G. Jefferson, *J. Chem. Soc.*, 1409 (1953).
[72] M. L. Bender and B. W. Turnquest, *J. Am. Chem. Soc.*, **79**, 1656 (1957).

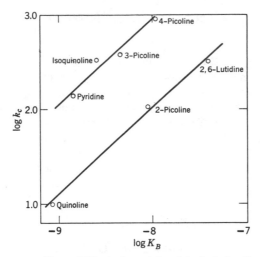

Figure 7-24. Amine-catalyzed hydrolysis of acetic anhydride.

Good linearity is observed for the relationship between $\log k_c$ and $\log K_B$ for the bases pyridine, 4-picoline, imidazole, N-methylimidazole, and 3-picoline, but the slope is now 1.62. Again there are structural limitations. Trimethylamine is much less effective as a catalyst than would be predicted from its base strength and is not on the line of the heterocycles.

The close parallelism between basicity and nucleophilicity within narrow structural classes is sometimes used to resolve questions of reaction mechanism. For example, the interaction of a proton donor with an aromatic hydrocarbon results either in the formation of a covalently bonded ion, as in equations 90 and 91, or in the formation of a π-complex, as in equation 92.[73,74,75]

$$\text{ArH} + \text{HF (liquid)} \overset{K_{\text{HF}}}{\rightleftharpoons} \text{ArH}_2{}^+ + \text{F}^- \qquad (90)$$

$$\text{ArH} + \text{HF} + \text{BF}_3 \overset{K_{\text{HBF4}}}{\rightleftharpoons} \text{ArH}_2{}^+ + \text{BF}_4{}^- \qquad (91)$$

$$\text{ArH} + \text{HCl} \overset{K_\pi}{\rightleftharpoons} \text{ArH·HCl} \qquad (92)$$

The two mechanisms of interaction with a proton donor can serve as alternative models for the mechanism of aromatic halogenation. As a matter of fact, a graph of $\log K_{\text{HBF}_4}$ against $\log k_{\text{halogenation}}$ is approximately linear; a graph of $\log K_\pi$ against $\log k_{\text{halogenation}}$ is rather scattered.[75]

[73] E. L. Mackor, A. Hofstra, and J. H. van der Waals, *Trans. Faraday Soc.*, **54**, 66 (1958).
[74] D. A. McCaulay and A. P. Lien, *J. Am. Chem. Soc.*, **73**, 2013 (1951).
[75] H. C. Brown and J. D. Brady, *ibid.*, **74**, 3570 (1952).

It appears that the π-complex between the hydrocarbon and HCl is less satisfactory as a model for the electrophilic substitution transition state than is the covalently bonded "σ-complex" obtained with HF and BF_3. The latter substance is simply a mixture of the carbonium ions formed by bonding a proton to definite carbon atoms of the aromatic ring rather than to the π-electron cloud generally. Similarly, the logarithms of the basicity constants K_{HF} for a series of alkyl-substituted benzenes vary linearly with the logarithms of the rate constants for the dedeuteration of the hydrocarbons in mixtures of trifluoroacetic and sulfuric acids, and for their reaction with trichloromethyl free radical.[76]

In this connection it is interesting to note that the quantities $\delta_R \log K_{HF}$ and $\delta_R \log K_{HBF_4}$ for a series of polynuclear aromatic hydrocarbons are proportional to the difference in π-electron energy between the hydrocarbon and its conjugate acid ion, as calculated by molecular orbital theory.[73] This might lead to the expectation that in the reactions correlated by these quantities, the activation energy is largely due to a change in potential energy. However, this expectation is contrary to fact, at least for the acid-catalyzed dedeuteration reaction. The activation energies for the dedeuteration reaction are almost independent of the nature of the hydrocarbon, and the differences in rate are largely entropy effects. Each methyl group in an *ortho* or *para* position relative to the departing deuteron increases $\Delta S\ddagger$ by about 10 cal/mole deg. The small changes in the enthalpy of activation show no correlation with the changes in the heat of formation of the protonated hydrocarbon.

When the structures of the nucleophile and the substrate are varied more widely, it becomes clear that the preceding approach of direct comparison with base dissociation constants is inadequate. A good example is furnished by the nucleophilic displacement of chloride ion from chloroacetate ion.[77]

$$B^- + ClCH_2CO_2^- \xrightarrow{k} BCH_2CO_2^- + Cl^- \tag{93}$$

If B^- is a carboxylate ion, a fairly good linear relationship is obtained between the logarithm of the rate constant at 45° and log K_B at 25°, as shown in Figure 7-25. With the exception of the point for formate ion, none of the values of log k deviate from the line drawn in the figure by more than 0.16 unit, even though the correlation includes a rather wide variety of structural types. Points falling on or near the line are for aliphatic carboxylate ions, *meta*- and *para*-substituted benzoate ions, several *ortho*-substituted benzoate ions, and (after suitable correction for

[76] E. L. Mackor, P. J. Smit, and J. H. van der Waals, *Trans. Faraday Soc.*, **53**, 1309 (1957).
[77] G. F. Smith, *J. Chem. Soc.*, 521 (1943).

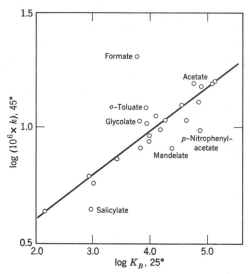

Figure 7-25. Reaction of carboxylate ions with chloroacetate ion. The straight line is drawn with a slope of 0.20.

symmetry) tartrate, malate, phthalate, malonate, and oxalate. Evidently here is a correlation that is only slightly perturbed by proximity effects or by changes in the electrical charge type of the nucleophilic ion. Yet the rate constants for reaction with sulfite ion, and particularly with thiosulfate ion, are greater by several orders of magnitude than would be expected from the basicity of these nucleophiles.[77]

Correlation with the Reactivity of Methyl Bromide

The extraordinarily high reactivity of sulfur bases as compared to oxygen bases of the same strength indicates that the process of base dissociation is not a sufficiently accurate or sufficiently complete model for nucleophilic activity. To improve on this situation, two approaches are now possible. One is to search for a better model process; the other, to use the process of base dissociation in linear combination with some other process. Both approaches have been explored.

Swain and Scott have given a treatment in which the standard process is the nucleophilic displacement on methyl bromide in water at 25°.[78] Let $k_N°$ be the rate constant for the reaction of methyl bromide with a given nucleophile N, and $k_0°$ be the rate constant for its reaction with the nucleophile water, both constants being measured in aqueous solution at

[78] C. G. Swain and C. B. Scott, *J. Am. Chem. Soc.*, **75,** 141 (1953).

25° and expressed as *second-order* rate constants. The nucleophilic activity of N may then be expressed by means of a nucleophilic parameter n, defined by equation 94.

$$n \equiv \log \frac{k_N{}^\circ}{k_0{}^\circ} \tag{94}$$

In this treatment the standard nucleophile, for which n is zero by definition, is water.

If the nucleophilic displacement reactions of methyl bromide are a good model for the nucleophilic displacement reactions of other substrates, we may hope to find linear free energy relationships, which we can write in the form of equation 95.

$$\log \frac{k_N}{k_0} = s \cdot n \tag{95}$$

In equation 95, k_N and k_0 are the rate constants for the reaction of the given substrate with the nucleophiles N and water, and s is a parameter characteristic of the substrate that measures the susceptibility of the reaction rate to changes in the nucleophilic activity.

Table 7-20. Nucleophilic Parameters n[78]

NUCLEOPHILE	n[a]	NUCLEOPHILE	n[a]
H_2O	$0 \pm .03$	HO^-	$4.20 \pm .15$
CH_3COO^-	$2.72 \pm .14$[b]	ϕNH_2	$4.49 \pm .27$
Cl^-	$3.04 \pm .09$[c]	SCN^-	$4.77 \pm .09$[c]
Br^-	$3.89 \pm .09$[c]	I^-	$5.04 \pm .21$
N_3^-	4.00	$S_2O_3^=$	$6.36 \pm .04$

[a] Calculated from CH_3Br reaction rates unless otherwise noted.
[b] Calculated from data for the reaction of various nucleophiles with epichlorohydrin.
[c] Calculated from data for the reaction of various nucleophiles with glycidol.

Values of the nucleophilic parameter n are listed in Table 7-20. Some of the values in the table are *secondary* values, that is, they are based on data for solvents that are only partly aqueous and for temperatures other than 25°, or even on data for substrates other than methyl bromide. For this reason the accuracy of the n-values is only fair. However, the range in rate constants to be fitted by the n-values is so large that the values should be

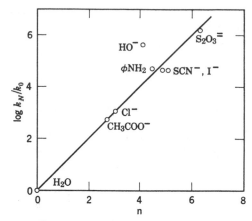

Figure 7-26. Displacement reactions of mustard ion with various nucleophiles. See eq. 96.

useful at least for semiquantitative purposes. It will be noted that the order of increasing n is quite different from that of increasing basicity.

Figure 7-26 shows a plot according to equation 95 for the reactions of mustard ion with various nucleophiles.[78,79]

$$N: + \quad \underset{\text{(mustard ion)}}{\begin{array}{c} CH_2 \\ | \quad \overset{\oplus}{\diagdown} \\ \diagup S\text{—}CH_2CH_2Cl \\ CH_2 \end{array}} \rightarrow \underset{\text{(charge } z+1)}{NCH_2CH_2SCH_2CH_2Cl} \qquad (96)$$

$$\underset{\text{(charge } z)}{}$$

The values of k_N/k_0 span six orders of magnitude. Compared to this range the fit is quite good, with the exception of the point for hydroxide ion. The point for hydroxide ion deviates even more in a similar plot for the reactions of β-propiolactone but is well behaved in the plots for other substrates. Values of the parameter s and of log k_0 for a number of substrates are listed in Table 7-21.

Dependence on Polarizability

Although the rate constants for nucleophilic displacement on methyl bromide provide a more realistic measure of nucleophilic activity than do the base dissociation constants, there are enough discrepancies to indicate that even this measure is not generally useful. An example that has already

[79] A. G. Ogston, E. R. Holiday, J. St. L. Philpot, and L. A. Stocken, *Trans. Faraday Soc.*, **44**, 45 (1948); P. D. Bartlett and C. G. Swain, *J. Am. Chem. Soc.*, **71**, 1406 (1949).

been mentioned is the "erratic" behavior of hydroxide ion for some substrates. In other cases the fit to equation 95 is poor for all nucleophiles.[80a] For example, triphenylmethyl cation as a substrate does not give a good fit, perhaps because of the unusual degree of resonance dispersion of the charge or the high polarizability of this ion.[78] Another striking example comes from nucleophilic substitution in aromatic systems. Thiophenoxide ion is much more reactive than methoxide ion with 2,4-dinitrochlorobenzene but less reactive with p-nitrofluorobenzene.[80b,c]

Table 7-21. *Substrate Constants for Nucleophilic Displacement Reactions*[78]

SUBSTRATE	NUMBER OF NUCLEOPHILES	s	$\log k_0$	MOLARITY OF WATER
Methyl bromide	6	(1.00)	−5.04[a]	55.5
Ethyl tosylate	5	0.66 ± .28	−5.22	21.6
Benzyl chloride	3	0.87 ± .10	−5.94	21.6
β-Propiolactone	7	0.77 ± .30	−4.25	55.5
Epichlorohydrin	6	0.93 ± .03	−6.01	55.5
Glycidol (2,3-epoxypropanol)	6	1.00 ± .01	−6.55	55.5
Mustard cation (equation 96)	7	0.95 ± .07	−2.54	52.7
Benzenesulfonyl chloride	3	1.25 ± .42	−4.57	27.8
Benzoyl chloride	4	1.43 ± .20	−3.28	27.8

[a] In 99% water–1% acetone at 49.8°.

In order to do justice to these and other facts, we must conclude that the interaction of a nucleophile with a substrate in a displacement process involves at least two independent mechanisms whose relative importance varies with the substrate. If the relative importance of the mechanisms happens to be like that in the reactions of methyl bromide, then equation 95 will give a good fit. However, to treat the general case of nucleophilic activity we need a linear combination of at least two model processes.

In the next higher approximation we shall use a linear combination of *two* model processes. It is convenient that one of these be base dissociation in water because values of K_B are available for a wide variety of nucleophiles. We have already seen that nucleophilic substitution rates can be correlated with base dissociation constants within narrow structural

[80] (a) R. F. Hudson and M. Green, *J. Chem. Soc.*, 1055 (1962), and references cited therein. (b) J. D. Reinheimer and J. F. Bunnett, *J. Am. Chem. Soc.*, **81**, 315 (1959); (c) C. W. Brown and J. Hirst, *J. Chem. Soc.*, 254 (1956).

classes. It is therefore likely that at least some of the interaction mechanisms involved in nucleophilic displacements are also present in base dissociation.

For the second process we may in principle choose any plausible model process for which enough data are available. However, we are more likely to fit a wide variety of reactions if we choose the second process so that the relative importance of the various interaction mechanisms is as different as possible from that in base dissociation. Let us therefore digress briefly and compare the stretched bond to B that is formed in the transition state structure (97a) with the much shorter bond in the conjugate acid.

$$B \cdots \overset{\diagdown \diagup}{\underset{|}{C}} \cdots X \qquad B\text{—}H^+ \qquad (97)$$

$$(a) \qquad\qquad\qquad (b)$$

Because of the extension of the B $\cdots$ C bond in the transition state, the overlap of bonding orbitals forming the covalent bond is small unless they are polarized so as to increase the overlap.[81] In BH^+, on the other hand, the bond is short and the overlap is considerable even without polarization. We therefore expect that a highly polarizable reagent B is a better nucleophile in relation to its base strength than is a less polarizable reagent. There are numerous facts to support such an analysis. For example, the nucleophilic parameters n increase with increasing polarizability within a given column of the periodic table even though the base strength decreases. Thus in Table 7-19 the n values increase in the sequence $Cl^- < Br^- < I^-$; the values for SCN^- and $S_2O_3^=$ (where the nucleophilic atom is sulfur) are greater than those for oxygen bases, including even the hydroxide ion.

The polarizability of the nucleophile becomes less important as the polarizability of the electrophilic atom in the substrate increases, because the driving force due to the polarization of a bonding orbital can now be furnished by the substrate. Thus, as the electrophilic atom becomes more polarizable, the reactivity sequence for a series of nucleophiles becomes more nearly like that of the base strengths. For example, nucleophilic

---

[81] J. O. Edwards and R. G. Pearson, *J. Am. Chem. Soc.*, **84**, 16 (1962) have given a particularly lucid quantum-mechanical description of this phenomenon. Under the influence of an electric field due to an adjacent atom, the ground state wave function of a polarizable atom can mix with excited state wave functions of that atom to give a new set of wave functions in which the electron distribution is shifted in a direction opposite to that of the electric field. The effect is the more pronounced the smaller the energy gap between the highest filled orbitals and the lowest empty orbitals of the atom. The result is to polarize not only the filled orbitals but also the empty ones. Thus a high polarizability can improve the efficiency of an atom not only as a nucleophile but also as an electrophile.

substitution in neopentyl p-toluenesulfonate can take place either on the carbon atom or on the more polarizable sulfur atom.

$$(CH_3)_3CCH_2OSO_2C_7H_7 \begin{cases} \xrightarrow[\text{(S—O fission)}]{CH_3O^-} (CH_3)_3CCH_2O^- + C_7H_7SO_3CH_3 \\ \\ \xrightarrow[\text{(C—O fission)}]{C_6H_5S^-} (CH_3)_3CCH_2SC_6H_5 + C_7H_7SO_3^- \end{cases}$$

It appears that polarizable reagents like thiophenoxide ion react preferentially at the carbon atom, whereas the less polarizable but more strongly basic methoxide ion attacks at the sulfur atom.[82] An even more striking correlation between the tendency to attack carbon rather than sulfur and the polarizability of the reagent has been found for the reactions of 2,4-dinitrophenyl p-toluenesulfonate.[83] However, the interpretation is more complicated in this case because the effectiveness of a polarizable reagent in aromatic nucleophilic substitution can be enhanced by interaction with the π-electrons of the aromatic ring.

A closely related effect is the interaction of a polarizable nucleophile with a polarizable substituent in the substrate. The nucleophile will interact with the substituent by London dispersion forces and thereby reduce the free energy of activation.[84] The effect is particularly important when the polarizable substituent is situated near the reaction zone. Since the introduction of a polarizable substituent is often equivalent to the introduction of a bulky substituent, the beneficial effect of polarizability should also be present in the complex phenomenon of steric hindrance. Thus it may be relevant that in the nucleophilic reactions of alkyl bromides, the use of neopentyl bromide instead of ethyl bromide reduces the rate constant by a factor of 10^{-5} in the reaction with ethoxide ion, but only by 10^{-4} in the reaction with the more polarizable iodide ion.[85]

The Edwards Equation

In the Edwards equation, nucleophilic reactivity is correlated with a linear combination of two model processes.[86]

$$\log \frac{k_N}{k_0} = a E_N + b H_N \tag{98}$$

[82] F. G. Bordwell, B. M. Pitt, and M. Knell, *J. Am. Chem. Soc.*, **73**, 5004 (1951).

[83] J. F. Bunnett and J. Y. Bassett, Jr., *J. Am. Chem. Soc.*, **81**, 2104 (1959).

[84] J. F. Bunnett, *ibid.*, **79**, 5969 (1957).

[85] I. Dostrovsky and E. D. Hughes, *J. Chem. Soc.*, 161 (1946).

[86] J. O. Edwards, *J. Am. Chem. Soc.*, **76**, 1540 (1954); **78**, 1819 (1956).

As before, the rate or equilibrium constant k_N refers to a process in which the substance N is the nucleophile, and k_0 refers to the analogous process in which water is the nucleophile. H_N is a measure of the basicity of the nucleophile, that is, its nucleophilic reactivity towards protons, and is defined by equation 99 in which K_A is the ionization constant of the conjugate acid in water at 25° and 1.74 is a correction term for the pK_A of H_3O^+.

$$H_N = pK_A + 1.74 \tag{99}$$

The parameter E_N, another measure of nucleophilic reactivity, is the standard electrode potential of the reagent X^- in the equilibrium reaction (100), relative to that of a similar equilibrium for water.[86]

$$2X^- \underset{}{\overset{E°}{\rightleftharpoons}} X_2 + 2e^- \tag{100}$$

$$2H_2O \rightleftharpoons H_4O_2^{++} + 2e^-; \quad E° = -2.60 \text{ v} \tag{100a}$$

E_N is defined by equation 101.

$$E_N \equiv E° + 2.60 \tag{101}$$

The parameters a and b are empirical and characteristic of the substrate undergoing reaction with the nucleophile.

The use of an oxidation half cell, equation 100, as a model for nucleophilic reactivity is plausible when it is remembered that a nucleophile is formally oxidized in a displacement process. The connection between E_N and polarizability is not immediately obvious but becomes so when the empirical relationship (102) is considered.[86,87]

$$E_N = 3.60 \, P_N + 0.0624 \, H_N \tag{102}$$

The quantity P_N in equation 102 is a measure of the polarizability of the nucleophile relative to that of water and is defined in equation 103, where (MR) denotes the molar refraction extrapolated to infinite wavelength.

$$P_N = \log (MR)_N - \log (MR)_{H_2O} \tag{103}$$

The molar refraction, (MR), is proportional to the polarizability of the molecule averaged over all directions. Presumably the component of polarizability along the direction of the newly forming chemical bond would be more valuable, but this quantity is not generally available. Equation 102 has been found to predict the value of E_N for six monatomic or diatomic molecules to within a few hundredths of a volt.

From equation 102, which relates E_N to P_N and H_N, we can see that the original equation 98 is not in a form in which the contribution of H_N has

[87] E_N appears also to be linearly related to the nucleophilic parameter n, as defined by Swain and Scott.[78]

Table 7-22. Nucleophilic Parameters for Use in the Edwards Equation[86]

NUCLEOPHILE	E_N[a]	H_N[b]	P_N	$(MR)_N$	NUCLEOPHILE	E_N	H_N	P_N	$(MR)_N$
F^-	-0.27	4.9	-0.150	2.6	SCN^-	1.83	(1.00)	—	—
H_2O	0.00	0.0	0.000	3.67	NH_3	1.84	11.22	—	—
NO_3^-	0.29[c]	(0.40)	—	—	$(MeO)_2POS^-$	2.04	(4.00)	—	—
$SO_4^=$	0.59	3.74	—	—	$EtSO_2S^-$	2.06	(-5.00)	—	—
$ClCH_2COO^-$	0.79[d]	4.54	—	—	I^-	2.06	(-9.00)	0.718	19.2
CH_3COO^-	0.95[e]	6.46	—	—	$(EtO)_2POS^-$	2.07	(4.00)	—	—
C_5H_5N	1.20[f]	7.04	—	—	$MeC_6H_4SO_2S^-$	2.11	(-6.00)	—	—
Cl^-	1.24	(-3.00)	0.389	9.0	$SC(NH_2)_2$	2.18	0.80	—	—
$C_6H_5O^-$	1.46[g]	11.74	—	—	$S_2O_3^=$	2.52	3.60	—	—
Br^-	1.51	(-6.00)	0.539	12.7	$SO_3^=$	2.57	9.00	—	—
N_3^-	1.58[h]	6.46	—	—	CN^-	2.79	10.88	—	—
HO^-	1.65	17.48	0.143	5.1	$S^=$	3.08	14.66	0.611	15.0
NO_2^-	1.73	5.09	—	—					
$C_6H_5NH_2$	1.78[h]	6.28	—	—					

[a] From the electrode potentials according to equation 101 unless otherwise noted.
[b] Values in parentheses are estimates.
[c] Calculated from the equilibrium constant for the $Hg(NO_3)_2$ complex.
[d] Calculated from the equilibrium constant for the silver *bis*-chloroacetate complex.
[e] Calculated from the epichlorohydrin rate constant.
[f] Calculated from the mustard cation rate constant.
[g] Calculated from the iodoacetate rate constant.
[h] Calculated from the methyl bromide rate constant.

been separated so as to appear in only one term. This can be achieved, however, by combining equations 98 and 102 to give equation 104, in which A equals 3.60a, and B equals (b + 0.0624a).

$$\log \frac{k_N}{k_0} = A\, P_N + B\, H_N \tag{104}$$

The earlier form (98) of the equation had caused some consternation because for some substrates b turns out to be negative. This Schönheitsfehler is avoided by the use of equation 104, for the parameter B is always positive within experimental error. However, it should be pointed out that there is no *extrathermodynamic* objection to a negative coefficient in a correlation by linear combination of two model processes. There is however an objection to the use of equation 104, since the quantity P_N does not refer to any physically real model process.

Table 7-22 gives values of E_N, H_N, P_N, and $(MR)_N$ for various nucleophiles. Table 7-23 gives the parameters a and b for various substrates. Equation 98 also serves to correlate equilibrium constants for the formation of complexes of the nucleophiles with various inorganic cations.[86]

Table 7-23. Substrate Parameters for Nucleophilic Displacement Reactions According to the Edwards Equation[86]

SUBSTRATE	a	b	SUBSTRATE	a	b
Ethyl tosylate[a]	1.68	0.014	Mustard cation[e]	2.45	0.074
Benzyl chloride[b]	3.53	−0.128	Methyl bromide[f]	2.50	0.006
β-Propiolactone[c]	2.00	0.069	Benzoyl chloride[g]	3.56	0.008
Epichlorohydrin[d]	2.46	0.036	Diazoacetone[h]	2.37	0.191
Glycidol[d]	2.52	0.000	Iodoacetate ion[i]	2.59	−0.052

[a] H. R. McCleary and L. P. Hammett, *J. Am. Chem. Soc.*, **63**, 2254 (1941).

[b] G. W. Beste and L. P. Hammett, *ibid.*, **62**, 2481 (1940).

[c] P. D. Bartlett and G. Small, *ibid.*, **72**, 4867 (1950).

[d] J. N. Brønsted, M. Kilpatrick, and M. Kilpatrick, *ibid.*, **51**, 428 (1929).

[e] C. G. Swain and C. B. Scott, *ibid.*, **75**, 141 (1953); A. G. Ogston, E. R. Holiday, J. St. L. Philpot, and L. A. Stocken, *Trans. Faraday Soc.*, **44**, 45 (1948).

[f] C. G. Swain and C. B. Scott, *loc. cit.*; E. A. Moelwyn-Hughes, *Trans. Faraday Soc.*, **45**, 167 (1949); A. Slator and D. F. Twiss, *J. Chem. Soc.*, **95**, 93 (1909).

[g] C. G. Swain and C. B. Scott, *loc. cit.*

[h] C. E. McCauley and C. V. King, *J. Am. Chem. Soc.*, **74**, 6221 (1952); in water at 25°.

[i] H. J. Backer and W. H. van Mels, *Rec. trav. chim.*, **49**, 177, 363, 457 (1930); C. Wagner, *Z. physik. Chem.*, **A115**, 121 (1925); in water at 25°.

RELATIONSHIPS INVOLVING SPECTRAL PROPERTIES

Infrared Spectra

Substituent effects on the frequency of an infrared transition can often be correlated with substituent effects on some other infrared transition or with substituent effects on an equilibrium or rate constant. In some cases the correlation is best described in terms of a parameter like σ or σ^*. The interpretation of the infrared frequency shifts is simplest, and the correlations are usually most precise, when the reduced mass associated with the transition is largely independent of the mass of the molecule, as is the case for the vibration of an atom or a small group on the periphery of a large molecule.

Simple linear relationships are often observed for the substituent effects on two different modes of vibration of the same molecule.[88] For example, in a series of benzoyl compounds, ϕCOR, the carbonyl frequency is in approximately linear relationship to the frequency of one of the vibrations of the aromatic hydrogens. Or again, the frequencies of the symmetric and antisymmetric vibrations of the sulfone group in sulfones, thiolsulfonates, sulfonates, sulfonic acids, and sulfonamides are linearly related. An example of an external relationship between the frequency of vibration of a structural element in one molecule and a similar structural element in another molecule is found in a comparison of the infrared spectra of $RCOCH_3$ and RNO.[88] A series of the $N{=}O$ and $C{=}O$ stretching frequencies have been found to be linearly related, with the exception of those for R equal to CF_3.

Relationships between the vibrational frequency of a bond and $\Delta \bar{H}°$ or $\Delta \bar{F}°$ for a reaction involving the bond are often observed, provided that the structural change is suitably limited. For example, the enthalpies of hydrogenation of nonconjugated, noncyclic olefins bear a simple though nonlinear relation to the $C{=}C$ stretching frequency. A similar curvilinear relationship is found for the enthalpies of chlorine addition to fluorinated olefins.[89]

The shift in the OH-stretching frequency caused by hydrogen bond formation to a carbonyl oxygen atom appears to increase approximately as a linear function of $\Delta \bar{F}°$ for the formation of the hydrogen-bonded complex, although relatively few data are available.[90] The points for complexes in which the acceptor is another hydroxyl group or an amine nitrogen fall on a separate line.

[88] L. J. Bellamy and R. L. Williams, *J. Chem. Soc.*, 863 (1957).
[89] L. J. Bellamy and R. L. Williams, *J. Chem. Soc.*, 2463 (1958).
[90] E. Grunwald and W. C. Coburn, Jr., *J. Am. Chem. Soc.*, **80**, 1322 (1958).

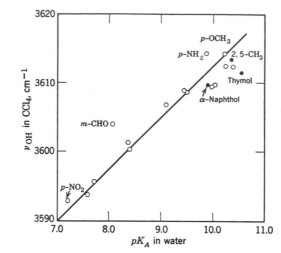

Figure 7-27. OH-stretching frequency of monomeric phenols in carbon tetrachloride versus pK_A in water. Data for *ortho*-substituted phenols are indicated by solid circles.

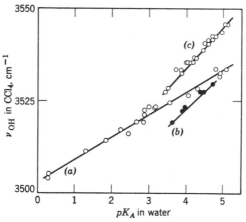

Figure 7-28. OH-stretching frequency of monomeric carboxylic acids in carbon tetrachloride plotted versus pK_A in water.
(*a*) Saturated aliphatic acids and acetylenic acids, except for (*b*). (*b*) Aliphatic acids with one or more aryl groups in the 2-position. (*c*) Aromatic acids and unsaturated acids having the COOH group attached to the carbon of a double bond.

The OH-stretching frequency of monomeric phenols and carboxylic acids in carbon tetrachloride has been compared with the pK_A values of these acids in water.[91] The relationship disperses into several lines, corresponding to more or less well-defined structural classes. Figure 7-27 shows the fairly good linear correlation obtained for a series of substituted phenols. Those with *ortho*-substituents are correlated about as well as those with *meta*- or *para*-substituents. Data for carboxylic acids are shown in Figure 7-28. Aliphatic carboxylic acids as diverse as trifluoroacetic, pivalic, 2-phenoxypropanoic, phenylpropiolic and methylpropiolic fall on one line, but two additional lines are needed for still other acids. One of these is for acids with one or more aryl groups in the 2-position. The other is for aromatic acids and unsaturated acids having the carboxyl group attached to the carbon of a double bond, although acetylenic acids apparently behave like ordinary saturated acids. The frequencies of the saturated acids can also be correlated with the parameter σ^*.

The relationships between the OH-stretching frequencies and the standard free energy changes in reactions of the OH-group are of theoretical interest. The square of the stretching frequency is almost precisely proportional to the force constant of the OH-bond, and hence it provides a measure of the potential energy of that bond. The existence of linear correlations between ν_{OH} and pK_A, for example, might justify the postulate that substituent effects on pK_A are also largely potential energy effects. However, the notable dispersion in Figure 7-28 shows that such a postulate, if valid at all, is severely limited in its scope to narrow structural classes.

The correlations between infrared spectral data and rate or equilibrium constants can be expressed as correlations between spectral data and substituent parameters. Thus in the aromatic series, linear correlations are often obtained when the substituent effect on a spectral property is plotted against σ. An example comes from studies of intramolecular hydrogen bonding in *ortho*-hydroxy aromatic Schiff bases of the type shown in equation 105.[92] The introduction of an electron-withdrawing substituent

$$(105)$$

in ring B will reduce the electron density on the nitrogen atom and hence weaken the O—H $\cdots$ N hydrogen bond. The weakening of the hydrogen

[91] J. D. S. Goulden, *Spectrochimica Acta*, **6**, 129 (1954).
[92] A. W. Baker and A. T. Shulgin, *J. Am. Chem. Soc.*, **81**, 1523 (1959).

bond in turn is expected to result in an increase in the OH-stretching frequency of the complex.[93] As a matter of fact, substituents in ring B cause shifts in ν_{OH} that are quite well correlated with σ. The introduction of a nitro-group in the 4-position of ring A increases the strength of the intramolecular hydrogen bond but does not change the sensitivity of ν_{OH} to substitution in ring B; the plot of ν_{OH} versus σ is parallel to that obtained before.

In the aliphatic series, infrared spectral data often give good correlations with σ^*. Thus infrared frequencies associated with carbonyl or nitrosyl groups in aliphatic compounds are represented by the following *simple linear equations*.[94]

$$\nu = 1690 + 52.56\sigma^* \ (\text{RCOCH}_3 \text{ vapor})$$

$$\nu = 1500 + 110.1\sigma^* \ (\text{RNO vapor})$$

$$\nu = 1655 + 53.71\sigma^* \ (\text{RCOCH}_3 \text{ in solution})$$

$$\nu = 1458 + 124.3\sigma^* \ (\text{RNO in solution})$$

For aliphatic alcohols of the formula $R_1R_2R_3COH$ in CCl_4 solution, the intensity of the infrared band due to OH is correlated with $\Sigma\sigma^*$.[95]

$$\text{A} = 1270\sum\sigma^* + 3200 \qquad (106)$$

The quantity A in (106) is the integrated band intensity in units of l mole^{-1} cm^{-2}. Since the value of $\Sigma\sigma^*$ will depend on the position of the OH-group in the molecule, the correlation may be useful in investigations of structure. Intramolecular hydrogen bonds have been suggested as the explanation for the deviation of 3-chloro-1-propanol and 2-nitro-2-methyl-1-propanol from relationship (106). The points for 2-chloroethanol and furfuryl alcohol fall on the line, but hydrogen bonds in these compounds would involve five-membered rather than six-membered rings.

Visible and Ultraviolet Spectra

Optical excitation of a molecule in the visible or ultraviolet range frequently leads to excited electronic states in which the distribution of electrons relative to the nuclei is sufficiently different from that of the ground state to make the energy of the excitation process responsive to the influence of polar substituents or polar solvents. The excitation process therefore resembles a polar reaction, except that the excited state is not an equilibrium state. The deviation of the excited state from a state having the same electronic energy, but consisting of vibrational, rotational, and solvational subspecies in an equilibrium distribution, can be predicted from the Franck-Condon principle. The basic idea is that the time required for

[93] R. M. Badger and S. H. Bauer, *J. Chem. Phys.*, **5**, 839 (1937).
[94] D. G. O'Sullivan and P. W. Sadler, *J. Chem. Soc.*, 4144 (1957).
[95] T. L. Brown, *J. Am. Chem. Soc.*, **80**, 6489 (1958).

the excitation of electrons is much shorter than that required for the displacement of atomic nuclei. Hence the electronically excited state is produced with the atomic nuclei in virtually the same relative positions that existed in the ground state just before the excitation. Two important corollaries can be derived. First, the transition probability for the excitation is at or near a maximum when the excited state has several quanta of vibrational energy in addition to the electronic energy. Hence the frequency, ν_{max}, of maximum absorption is not a measure purely of the difference in electronic energy but also reflects a contribution from molecular vibrations. The latter can be evaluated when the vibrational fine-structure of the electronic absorption band can be resolved.[96]

The second corollary applies when the absorbing molecule is a solute in solution. Since the surrounding solvent molecules will not have time to change their positions and orientations during the excitation, the excited state is produced in a solvent shell that is still arranged as if it were in equilibrium with a charge distribution like that of the ground state. This particular arrangement of solvent molecules is not likely to be the optimum one for the new charge distribution. Hence the change in solvation energy during optical excitation can be quite different from that associated with an otherwise analogous equilibrium process.

In spite of these differences, the optical and chemical processes have enough in common to give rise to extrathermodynamic relationships. A particularly instructive example has been discussed by Scheibe and Brück.[97] A series of methine dyestuffs is formed by the loss of a proton from compounds of the type shown in equation 107, in which the rings are

$$+ \; H^+ \qquad (107)$$

[96] See, for example, F. A. Matsen, W. W. Robertson, and R. L. Chuoke, *Chem. Revs.*, **41**, 273 (1947); W. F. Hamner and F. A. Matsen, *J. Am. Chem. Soc.*, **70**, 2482 (1948).
[97] G. Scheibe and D. Brück, *Zeitschrift für Elektrochemie*, **54**, 403 (1950).

closed by the groups X_1, $X_2 =$ —CH=CH—, CH_2, or S and the substituents R_1 and R_2 are hydrogen, methoxyl, methyl, or dimethylamino. The absorption maxima of the colored bases vary over a wide range, from 450 to 600 mμ. Similarly, the pK_A values of the parent acids vary widely, from -2 to $+10$. Yet in spite of these large variations, it is found that the absorption maximum, λ_{max}, of the methine bases increases smoothly with the pK_A of the parent acids.

The implied extrathermodynamic relationship may be obtained as follows. In order to establish a formal analogy between the optical excitation, B $\rightarrow$ B*, of the methine base and a chemical reaction, it is convenient to define an energy function, F_{B*}^*, as in equation 108, where N_0 is Avogadro's number and c_0 the velocity of light in a vacuum.

$$N_0 h\nu_{max} = \frac{N_0 h c_0}{\lambda_{max}} \equiv F_{B*}^* - \bar{F}_{B}^{\circ} \tag{108}$$

The function F_{B*}^* is neither a standard energy nor a standard free energy, for the excited state produced with the maximum probability is not a state of thermodynamic equilibrium.

Next we express K_A of the parent acid A in terms of standard free energies, as in equation 109.

$$-RT \ln K_A = \bar{F}_{B}^{\circ} + \bar{F}_{H^+}^{\circ} - \bar{F}_{A}^{\circ} \tag{109}$$

On adding equations 108 and 109 we obtain 110, which in turn implies 111.

$$N_0 h\nu_{max} - RT \ln K_A = F_{B*}^* + \bar{F}_{H^+}^{\circ} - \bar{F}_{A}^{\circ} \tag{110}$$

$$N_0 h \cdot \delta_R \nu_{max} - RT \cdot \delta_R \ln K_A = \delta_R F_{B*}^* - \delta_R \bar{F}_{A}^{\circ} \tag{111}$$

By means of equation 111, substituent effects on the excited state can be compared with those on the acid. As a matter of fact, Scheibe and Brück find that the quantity $(N_0 h\nu_{max} - RT \ln K_A)$ is perfectly constant and equal to 60 kcal, in spite of the wide variation both of ν_{max} and of $\ln K_A$. In other words, equation 112 fits the data.

$$\delta_R F_{B*}^* = \delta_R \bar{F}_{A}^{\circ} \tag{112}$$

Admittedly the thermodynamic significance of $\delta_R F_{B*}^*$ is somewhat in doubt, but the simple equality (112) strongly implies that the effects of substituents on B* and A are very similar.

The most conspicuous difference between the electronic structure of the methine base and its conjugate acid is the extent of resonance interaction across the central =CH— or —CH_2— group. In the dye, the interaction is strong. In the conjugate acid it is weak, since the —CH_2— group acts as an insulator. Indeed, the ultraviolet spectrum of the acid is very similar to that of a one-to-one mixture of the corresponding unjointed quaternary

ammonium compounds. It is reasonable to suppose therefore that resonance interaction across the central carbon atom is very weak also in the
excited state B*. A plausible valence bond structure is (113).

$$
\begin{array}{c}
\text{(113)}
\end{array}
$$

(B*)

   Electronic absorption spectra can sometimes be used to elucidate the
interaction mechanisms between different parts of the same molecule. For
example, in p-nitroaniline the nitro- and amino-group interact both by an
inductive and by a resonance mechanism. The two effects reinforce each
other, both acting so as to reduce the electron density on the amino-group
and hence to decrease the base strength. The resonance effect can be
damped by using steric hindrance to impose a nnnplanar conformation
unfavorable to resonance. Thus it is found that the introduction of
sterically hindering alkyl groups next to the nitro-group increases the
basicity of the amino-group. The introduction of steric hindrance also
affects the absorption spectrum, particularly the band at 3800 Å whose
intensity decreases markedly. It is found that the decrease in the extinction
coefficient $\epsilon_{\max}$ of that band as bulky groups are introduced next to the
nitro-group varies linearly with the increase in log K_B. By assuming that
complete damping of the resonance between the nitro- and the amino-group
would reduce $\epsilon_{\max}$ to zero, it is possible to estimate the value of log K_B for a
hypothetical p-nitroaniline species in which only the inductive effect is
active. It is found in this way that in ordinary p-nitroaniline, the resonance
and inductive effects of the nitro-group are about equally important.[98]
   There is of course no guarantee that substituents in electronically excited
molecules will exert their effect by the same interaction mechanisms that
are also important in analogous molecules belonging to electronic ground
states. When the significant interaction mechanisms are different, simple
extrathermodynamic relationships between optical and rate or equilibrium
data cannot be expected to exist. An example in which substituent effects
in electronically excited molecules appear to be qualitatively different from
those in the analogous unexcited molecules comes from the "principal
band" in the ultraviolet spectrum of *para*-substituted nitrobenzenes,

[98] B. M. Wepster, *Rec. trav. Chim.*, **76**, 335, 337 (1957).

acetophenones and anisoles.[99,100,101] The excitation process giving rise to this band is believed to be a charge transfer. Thus for nitrobenzene and acetophenone, the excitation process is believed to polarize the molecule in such a way as to move electrons away from the substituent; for anisole, the polarization is in the opposite direction. The excited states can be described at least approximately by assuming that the resonance structures (114) have

(114)

an enhanced importance. In spite of the reversal of the direction of polarization, *para*-halogen substituents are able to stabilize the excited state not only in nitrobenzene and acetophenone, but also in anisole. Moreover, they do so in about the same order of effectiveness. The actual sequences of increasing ν_{max} in the gas phase are $I > Br > Cl > F > H$ for acetophenone and nitrobenzene, and $I > Br \approx Cl > H \approx F$ for anisole. The interaction appears to involve a principal mechanism in which charges on the benzene ring can be stabilized regardless of their sign. The values of ν_{max} can be correlated approximately with the polarizability of the *para*-substituents. That is, plots of ν_{max} versus the molar refraction of the substrate are approximately linear.[101] On the other hand, the values of ν_{max} are not correlated with Hammett's σ.

For *para*-alkyl substituted acetophenones and nitrobenzenes, the values of ν_{max} are in the sequence, hydrogen > methyl > ethyl > *i*-propyl > *t*-butyl,[99] which is also the sequence of inductive effects. However, the neopentyl group in both nitrobenzene and pyridinium perchlorate reduces the value of ν_{max} by a greater amount than does either methyl or *t*-butyl,[100] whereas the inductive effect of the neopentyl group is similar to that of the methyl group. The particular effectiveness of the neopentyl group in stabilizing the electronically excited state may again be attributed to a high polarizability.[100]

[99] W. M. Schubert, J. Robins, and J. L. Haun, *J. Am. Chem. Soc.*, **79**, 910 (1957).
[100] W. M. Schubert and J. Robins, *ibid.*, **80**, 559 (1958).
[101] W. M. Schubert, J. M. Craven and H. Steadly, *ibid.*, **81**, 2695 (1959).

8

Extrathermodynamic
Free Energy Relationships.
II. Medium Effects

*—Ah! Voilà quatre-vingts volumes de recueils
d'une académie des sciences, s'écria Martin; il
se peut qu'il y ait là du bon.*

Voltaire
Candide

In the preceding chapter we discussed a number of theoretically sound and useful extrathermodynamic relationships involving substituent effects. We found it possible, in many cases, to interpret the formal extrathermodynamic interaction mechanisms in terms of specific theoretical models, such as the inductive effect or the resonance effect. In this chapter we shall discuss the extrathermodynamic correlation of medium effects. However, the discussion will be more formal than it has been for substituent effects, and we shall hardly ever try to describe a particular interaction mechanism in terms of a specific theoretical model. The reason for the more cautious approach in the case of medium effects is that it is often difficult to choose between alternative theoretical models because the effects that the models predict are so very similar.

RELATIONSHIPS DERIVED FROM THEORETICAL SOLUTION MODELS

Although we shall not make extensive use of theoretical solution models, it is worthwhile to show that a number of such models predict linear free energy relationships. It will be recalled (Chapter 6) that simple linear relationships can be derived only if the relevant interaction terms in the expressions for the standard free energy can be separated into independent factors. We shall therefore examine the theoretical models from this point of view, inquiring whether or not the terms representing the medium effect on the solvent-solute interaction are factorable.

The Regular Solution

A regular solution[1] is one for which the entropy of mixing is a simple statistical quantity and for which the heat of mixing is a simple function of the molar heats of vaporization of the pure components. The resulting expressions for the partial molal entropy and standard partial molal free energy of the *solute* (with the mole fraction as the unit of concentration) are given in equations 1 and 2,

$$\bar{S}_2 = S^\circ_{2,\,\text{pure liquid}} - R \ln N_2 \tag{1}$$

$$\bar{F}_2{}^\circ = F^\circ_{2,\,\text{pure liquid}} + V_2 \left[\left(\frac{\Delta H^\circ_{1,\,\text{vap}}}{V_1} \right)^{1/2} - \left(\frac{\Delta H^\circ_{2,\,\text{vap}}}{V_2} \right)^{1/2} \right]^2 \tag{2}$$

where N_2 is the mole fraction of solute, and $\Delta H^\circ_{\text{vap}}$ is the molar heat of vaporization of the respective component as a pure liquid. The quantities $\Delta H^\circ_{\text{vap}}/V$ are the average cohesive energies per unit volume of the pure liquid components and have the dimensions of pressure. They are closely related to the internal pressures of the liquid components. From equation 2 it is possible to derive the quantity $\delta_M(\bar{F}_2{}^\circ/V_2)$ for a given change of solvent. The difference between two such quantities for two solutes, $\delta_R\delta_M(\bar{F}^\circ_{\text{solute}}/V_{\text{solute}})$, is given by equation 3.

$$\delta_R\delta_M\left(\frac{\bar{F}^\circ_{\text{solute}}}{V_{\text{solute}}} \right) = -2\delta_R \left(\frac{\Delta H^\circ_{\text{vap, solute}}}{V_{\text{solute}}} \right)^{1/2} \cdot \delta_M \left(\frac{\Delta H^\circ_{\text{vap, solvent}}}{V_{\text{solvent}}} \right)^{1/2} \tag{3}$$

Equation 3 factors the quantity $\delta_R\delta_M(\bar{F}^\circ_{\text{solute}}/V_{\text{solute}})$ into a term dependent solely on the two solvents, times a term dependent solely on the two solutes. Unlike the parameters in most extrathermodynamic relationships, the two factors in (3) are not adjustable; they can be calculated from properties of the pure components. Equations 1 through 3 and the regular solution model on which they are based fit reasonably well only for mixtures of nearly spherical molecules having small or zero electric dipole moment. Relationships analogous to equation 3 in which the factors are separable but not calculable from the properties of the pure liquid components are observed more frequently.[2] Equations of this form containing adjustable parameters will be encountered in a later section.

Values of $\Delta H^\circ_{\text{vap}}/V$ for various substances are given in Table 8-1.

---

[1] J. H. Hildebrand and R. L. Scott, *Solubility of Nonelectrolytes*, Reinhold, 1950.

[2] For example, see the empirical formula developed by F. Körösy, *Trans. Faraday Soc.*, 33, 416 (1937), which relates the solubilities of a large number of gases in a wide range of organic solvents to solvent parameters and the critical temperature of the gas.

Table 8-1. Square Roots of the Average Cohesive Energies per Cubic Centimeter,[a] at 25°

SUBSTANCE	$\left(\dfrac{\Delta H_{vap}^{\circ}}{V}\right)^{1/2}$	SUBSTANCE	$\left(\dfrac{\Delta H_{vap}^{\circ}}{V}\right)^{1/2}$
n-Butane	6.7	Methyl chloride	8.6
1-Butene	6.7	Methylene chloride	9.7
n-Pentane	7.05	Chloroform	9.3
Isopentane	6.75	Carbon tetrachloride	8.6
Neopentane	6.25	Methyl bromide	9.4
Cyclopentane	8.10	1,2-Dibromoethane	10.4
n-Octane	7.55	Nitromethane	12.6
Benzene	9.15	Nitrobenzene	10.0
Napthalene	9.9	Dimethyl sulfide	9.0
Phenanthrene	9.8	Diethyl ether	7.4
Perfluoro-*n*-butane	5.2	Pyridine	10.7

[a] In $(cal/cc)^{1/2}$. For a more comprehensive list, see J. H. Hildebrand and R. L. Scott, *Solubility of Nonelectrolytes*, Reinhold, 1950, p. 435.

Even though the regular solution model may fail for a polar molecule RX, it is quite possible that it will still apply successfully to the contribution F_R in the expression,

$$\bar{F}_{RX}^{\circ} = F_R + F_X + I_{R,X} \tag{4}$$

Dispersion Forces

An alternative way of estimating the cohesive free energy of a mixture of nonpolar molecules is calculation of the London dispersion interaction on the basis of a continuum model.[3] Although the static dipole moment of a nonpolar molecule may be zero, its instantaneous dipole moment fluctuates rapidly because of the orbital motion of the electrons. The fluctuating dipole interacts with the polarizable solvent continuum in which it is embedded in such a way as to minimize the free energy of the system. The resulting expression for the cohesive part of the partial molar free energy of the solute, that is, $\bar{F}_2^{\circ} - F_{2,gas}^{\circ}$, is equation 5.

$$\bar{F}_2^{\circ} - F_{2,\,gas}^{\circ} \approx -\frac{16}{12}\pi^2 N_0^{\,2} \cdot \left(\frac{\mathscr{I}_1 \mathscr{I}_2}{\mathscr{I}_1 + \mathscr{I}_2}\right) \cdot \left(\frac{\alpha_1}{V_1}\right) \cdot \left(\frac{\alpha_2}{V_2}\right) \tag{5}$$

$\mathscr{I}_1$ and $\mathscr{I}_2$ are the ionization potentials of the solvent and solute molecules, and α_1 and α_2 are their polarizabilities. For most of the commonly used organic solvents $\mathscr{I}_1$ is nearly constant, and the quantity $\mathscr{I}_1 \mathscr{I}_2/(\mathscr{I}_1 + \mathscr{I}_2)$ is therefore characteristic of the solute, to a good approximation. Hence the

[3] B. Linder, *J. Chem. Phys.*, **33**, 668 (1960); *ibid.*, **35**, 371 (1961); *ibid.*, **37**, 963 (1962).

dispersion interaction is factorable into a solute-contribution and a solvent-contribution.

The Born Charging Model

The static part of the electrical interaction between an ionic or dipolar solute and the medium can also be estimated from a continuum model. The procedure is to calculate the work required to generate the actual charge (or dipole) of the initially uncharged solute molecules, treating the surrounding solvent medium as a continuum whose dielectric constant is everywhere equal to the macroscopic value. The resulting expression for spherical ions is equation 6.[4] The charge is represented by z times the

$$\bar{F}^{\circ}_{\text{electrostatic}} = \frac{N_0(ze)^2}{2Db} \tag{6}$$

charge e of an electron, the dielectric constant by D, and the radius of the ion by b. The expression for dipolar molecules in a dielectric continuum is equation 7.[5]

$$\bar{F}^{\circ}_{\text{electrostatic}} = \frac{N_0\bar{\mu}^2(D - D_i)}{(2D + D_i)D_i b^3} \tag{7}$$

Equation 7 applies to spherical molecules of radius b and internal dielectric constant D_i, immersed in a continuum of dielectric constant D. Each molecule bears an electric doublet, of dipole moment $\bar{\mu}$, and located at its center.

Although equations 6 and 7 only very rarely succeed in reproducing the entire medium effect, their qualitative properties are enlightening. Differentiation of (6) with respect to D generates a function of $1/D^2$; hence a given change in dielectric constant, δD, produces a large effect at low dielectric constant and a much smaller effect at high dielectric constant. Since even the least polar of solvents has a dielectric constant of about 2 as compared to unity for the gas phase, the change from the gas phase to any solvent whatsoever is an extremely large one electrostatically. Transferring an ion from the gas phase to a medium of $D = 2$ produces a change in electrostatic free energy proportional to $(1 - \frac{1}{2}) = \frac{1}{2}$. Transferring an ion from a medium of $D = 2$ to one of $D = \infty$ similarly produces a change in electrostatic free energy proportional to $(\frac{1}{2} - 1/\infty) = \frac{1}{2}$. Hence the change from the gas phase to even the least polar of solvents is fully as drastic a medium change, electrostatically, as any change of solvent that can be made.[6]

[4] M. Born, Z. Physik, **1**, 45 (1920).
[5] J. G. Kirkwood, J. Chem. Phys., **2**, 351 (1934).
[6] R. G. Pearson and D. C. Vogelsong, J. Am. Chem. Soc., **80**, 1038 (1958).

Most organic ions have their charge localized on an exposed functional group rather than at the center. According to the formal electrostatic theory, an eccentrically located charge on a spherical ion is in first approximation equivalent to a centrally located charge superposed on a centrally located dipole.[5] The electrostatic work required to produce an eccentric charge is therefore a sum consisting of separate terms for a central charge and a central dipole and is evidently greater than the work required to produce a central charge distribution.

The same qualitative result may be obtained by a less formal but more intuitive approach. In this approach the ionic charge is treated simply as a point charge, and the burden of accounting for the eccentricity of the charge is placed on the ionic radius. Since equation 6 involves $1/b$, the effective average required is an average over $1/b$, that is, b is a sort of harmonic mean. Now the harmonic mean of two unequal numbers is always less than the algebraic mean. Hence the effective radius of an ion decreases as the charge (treated as a point charge) moves towards the surface, and $F^\circ_{electrostatic}$ correspondingly increases. Conversely, any mechanism that, by dispersion of charge, produces a more central charge distribution (for example, a resonance mechanism or even one involving a very rapid subsidiary equilibrium that changes the position of the charge) will reduce $F^\circ_{electrostatic}$. Such a dispersion of charge results in the greatest reduction in $F^\circ_{electrostatic}$, and therefore proceeds most readily, in solvents of low dielectric constant.

It should be recognized that equations 6 and 7 refer only to the electrostatic part of the medium effect and that other contributions, such as the London dispersion energies, can be of the same order of magnitude. The nonelectrostatic free energies can sometimes be estimated empirically by means of solvent effects on the free energies of uncharged molecules whose structures are closely similar to those of the ions. For example, the nonelectrostatic parts of the $\delta_M \bar{F}^\circ$ for tetraphenylboride ion and for tetraphenylphosphonium ion are probably very close to the total $\delta_M \bar{F}^\circ$ for tetraphenylmethane.[7]

If the ion or dipolar molecule is small, the fields acting on adjacent solvent dipoles may be so strong as to orient these. If the energy required to destroy this orientation is greater than kT, the ion plus the oriented solvent molecules tend to behave as a single kinetic unit. In such a case it is more fruitful to apply the electrostatic continuum model to the solvated unit rather than to the bare ion. A more cumbersome way of dealing with this kind of solvation is to treat the dielectric constant as a function of the distance from the bare ion rather than to use the constant macroscopic value.

[7] E. Grunwald, G. Baughman, and G. Kohnstam, *J. Am. Chem. Soc.*, **82**, 5801 (1960).

SOME $\delta_M \bar{F}^\circ$ RELATIONSHIPS EMPLOYING A SINGLE PARAMETER

In the first approximation discussed in Chapter 6 the free energy of a solute was expressed by equation 8,

$$\bar{F}^\circ_{RX} = F_R + F_X + I_{R,X} + I_{R,M} + I_{X,M} \tag{8}$$

in which each of the terms I is factorable. Where this approximation is adequate, it was found that $\delta_M \Delta \bar{F}^\circ$ for any reaction involving RX should be independent of the nature of R. The medium effect $\delta_M \Delta \bar{F}^\circ$ depends only on the nature of the changes taking place at the reaction site.

Table 8-2. Substituent Effects on Acid Dissociation Constants in Various Solvents, 25°C

	BENZOIC ACIDS			ANILINIUM IONS		
SUBSTITUENT	WATER	METHANOL	ETHANOL	WATER	METHANOL	ETHANOL
	pK_A Values of Reference Acid[a]					
H	4.20	9.41	10.25	4.61	5.80	5.70
	$\delta_R pK_A$ Values[b]					
o-NO$_2$	−2.03	−1.83	−1.77	−4.85	−5.61	−6.41
p-NO$_2$	−0.78	−1.02	−1.17	−3.60	−4.52	−5.21
m-NO$_2$	−0.71	−1.05	−1.17	−2.13	−2.91	−3.24
o-Cl	−1.26	−1.21	−1.12	−1.96	−2.40	−2.43
m-Cl	−0.37	−0.59	−0.63	−1.26	−1.53	−1.55
p-Cl	−0.23	−0.34	−0.42	−0.77	−1.00	−1.07
m-F	−0.34	−0.51	−0.53	−1.20	−1.45	−1.47
p-F	−0.06	−0.18	−0.23	−0.06	−0.32	−0.35
o-CH$_3$	−0.29	−0.09	−0.02	−0.20	−0.10	−0.10
m-CH$_3$	0.07	0.09	0.06	0.09	0.19	0.29
p-CH$_3$	0.17	0.18	0.18	0.50	0.54	0.57
o-OCH$_3$	−0.11	−0.17	−0.24	−0.09	—	—
p-OCH$_3$	0.27	0.36	0.32	0.73	—	—

[a] Based on data from M. Kilpatrick and C. A. Arenberg, *J. Am. Chem. Soc.*, **75**, 3812 (1953); E. Grunwald and B. J. Berkowitz, *ibid.*, **73**, 4939 (1951); B. Gutbezahl and E. Grunwald, *ibid.*, **75**, 559 (1953); A. L. Bacarella, E. Grunwald, H. P. Marshall, and E. L. Purlee, *J. Org. Chem.*, **20**, 747 (1955).
[b] Taken from the compilation of M. M. Davis and H. B. Hetzer, *J. Res. Natl. Bur. Standards (U.S.)*, **60**, 569 (1958); M. Kilpatrick and C. A. Arenberg, *J. Am. Chem. Soc.*, **75**, 3812 (1953); and A. I. Biggs, *J. Chem. Soc.*, 2572 (1961).

One process for which the first approximation is good enough for some purposes is the ionization equilibrium of an acid.[8] For example, the $\delta_M pK_A$ for the ionization of a series of carboxylic acids in two solvents of such widely divergent properties as water and *meta*-cresol is constant with an average deviation from the mean of 0.2 units. The corresponding quantity for the ionization of a series of *ortho*-, *meta*-, and *para*-substituted anilinium ions is constant with an average deviation from the mean of 0.05 units. A deviation of 0.05 units in pK corresponds to about 12% in the equilibrium constant and a deviation of 0.2 corresponds to about 60%.

Table 8-2 contains the ionization constants for a number of substituted aromatic acids in water, methanol, and ethanol. For the majority of acids, the $\delta_R pK_A$ values are seen to differ only slightly in these solvents, but differences amounting to as much as 1 pK unit are not unusual.[9]

The Acidity Function H_0

Mixtures of mineral acids or other strong acids with a less acidic solvent component, such as water, or ethanol, or glacial acetic acid, are frequently used as reaction media. The extent to which a solute is protonated by such a medium can be predicted (to the first approximation) by means of a parameter characteristic of the medium. The best known of these parameters is the acidity function H_0, which applies to the protonation of uncharged bases.

We shall use the symbol RB to denote the base, where B is the basic functional group.

$$RBH^+ \overset{K_c}{\rightleftharpoons} RB + H^+$$

$$K_c = \frac{c_{RB}c_{H^+}}{c_{RBH^+}}$$

The ionization of RBH^+ in a strongly acidic medium, such as 50% sulfuric acid, differs from that in an ordinary aqueous solution because it is not possible to bring about any important change in the hydrogen ion activity without changing the medium. That is, a solvent like 50% sulfuric acid is a strongly acidic buffer. The quantity $\delta_M \log K_c$ is therefore not independent of $\delta_M \log c_{H^+}$, and it is not particularly convenient to treat these quantities separately. A more useful quantity than $\delta_M \log K_c$ is

[8] J. N. Brønsted, A. Delbanco, and A. Tovborg-Jensen, *Z. physik. Chem.*, **A169**, 361 (1934).

[9] A more complete summary of medium effects on $\delta_R pK_A$ for substituted benzoic acids has been given by M. M. Davis and H. B. Hetzer, *J. Res. Natl. Bur. Stand. (U.S.)*, **60**, 569 (1958).

$(\delta_M \log K_c - \delta_M \log c_{H^+})$, which is equal to the experimentally accessible $\delta_M \log (c_{RB}/c_{RBH^+})$.

$$\delta_M \log \frac{c_{RB}}{c_{RBH^+}} = \delta_M \log K_c - \delta_M \log c_{H^+} \qquad (9)$$

In a typical experiment, c_{RB}/c_{RBH^+} is measured spectrophotometrically, and the concentrations of RB and RBH^+ are very low. Hence $\delta_M \log c_{H^+}$ is, to a very good approximation, independent of the presence of these

Table 8-3. Protonation of Uncharged Bases in Sulfuric Acid–Water Mixtures at 25° [a]

	$\delta_M \log c_{RB}/c_{RBH^+}$			
$c_{H_2SO_4}$	p-Nitro-aniline	o-Nitro-aniline	2-Nitro-4-chloroaniline	N,N-Dimethyl-2,4-dinitro-aniline
0.9 and 1.5	0.406	0.388	—	—
1.5 and 2.1	0.309	0.331	0.347	—
2.1 and 2.7	—	0.299	0.297	0.361
2.7 and 3.3	—	0.297	0.285	0.356
3.3 and 3.9	—	—	0.286	0.291
3.9 and 4.5	—	—	0.316	0.256

[a] Obtained by interpolation from data of K. N. Bascombe and R. P. Bell, *J. Chem. Soc.*, 1096 (1959).

solutes. Moreover, if the basic group B is kept constant, $\delta_M \log K_c$ is, in the first approximation, independent of the nature of R. The quantity $\delta_M \log (c_{RB}/c_{RBH^+})$ is therefore characteristic only of B and of the change in the medium.

As a matter of fact, for uncharged bases the analysis can be simplified even further. For a limited but important group of solvents (including all mixtures of water with a mineral acid), the experimental values of $\delta_M \log c_{RB}/c_{RBH^+}$ appear to be independent of the nature of B with about the same accuracy that they are independent of the nature of R. It is therefore appropriate to define a parameter H_0, which is now characteristic only of the medium, so that for uncharged bases

$$\delta_M H_0 = \delta_M \log \frac{c_{RB}}{c_{RBH^+}} \qquad (10)$$

The extent to which actual values of $\delta_M \log c_{RB}/c_{RBH^+}$ are independent of the nature of the base is illustrated in Table 8-3 for a series of increments of 0.6 molar in the concentration of a sulfuric acid-water mixture. The values for *o*-nitroaniline and 2-nitro-4-chloroaniline agree well within their experimental errors. The values for N,N-dimethyl-2,4-dinitroaniline deviate significantly from those for the primary anilines, but the deviations are relatively small.

Because of experimental difficulties, the measurement of c_{RB}/c_{RBH^+} for any given base is limited to a relatively narrow range of acidities. As a result, the function $\delta_M H_0$ cannot be given a simple operational definition in terms of a single substance. However, the ranges covered by the different bases overlap, and the definition of $\delta_M H_0$ can be based on an average result. In practice the substances used to establish the average are mostly, but not solely, ring-substituted *o*-nitroanilines.[10,11,12,13] Since the *o*-nitroanilines cover a wide range of base strength, present a minimum of experimental problems, and have been shown to behave in good accordance with equation 10,[12,13] we recommend that the definition of H_0 be based exclusively on data for these bases.

Absolute values of H_0 are obtained by supplementing equation 10 with the statement that the standard medium is the dilute aqueous solution and that H_0 equals $-\log c_{H^+}$ in that medium. Since in dilute aqueous solution, $\log c_{RB}/c_{RBH^+} + \log c_{H^+} = \log K_{c(H_2O)}$, we may calculate absolute values of H_0 from equation 11.

$$H_0 = \log \frac{c_{RB}}{c_{RBH^+}} - \log K_{c(H_2O)} \tag{11}$$

By virtue of equation 10, equation 11 applies not only to dilute aqueous solutions but also to any medium.

Values of H_0 for a number of mixtures of water with a strong acid, based largely on data for *o*-nitroanilines, are listed in Tables 8-4 and 8-5. Values of H_0 satisfying our definitions have also been reported for mixtures of water with HCOOH,[14] Cl_3CCOOH,[12] $Cl_2CHCOOH$,[12] $ClCH_2COOH$,[12] and H_3PO_3;[12] for mixtures of strong acids with trifluoracetic acid,[15,16] acetic acid,[11] formic acid,[11] methanol,[17] methyl isobutyl ketone,[18] and nitromethane;[19] and for ternary systems consisting of a strong acid in a

[10] L. P. Hammett and A. J. Deyrup, *J. Am. Chem. Soc.*, **54**, 2721 (1932).
[11] M. A. Paul and F. A. Long, *Chem. Revs.*, **57**, 1 (1957).
[12] K. N. Bascombe and R. P. Bell, *J. Chem. Soc.*, 1096 (1959).
[13] E. Högfeldt and J. Bigeleisen, *J. Am. Chem. Soc.*, **82**, 15 (1960).
[14] R. Stewart and T. Mathews, *Can. J. Chem.*, **38**, 602 (1960).
[15] E. L. Mackor, P. J. Smit, and J. H. van der Waals, *Trans. Faraday Soc.*, **53**, 1309 (1957).
[16] H. H. Hyman and R. A. Garber, *J. Am. Chem. Soc.*, **81**, 1847 (1959).
[17] C. A. Bunton, J. B. Ley, A. J. Rhind-Tutt, and C. A. Vernon, *J. Chem. Soc.*, 2327 (1957).
[18] A. Mörikofer, W. Simon, and E. Heilbronner, *Helv. Chim. Acta*, **42**, 1737 (1959).
[19] H. Van Looy and L. P. Hammett, *J. Am. Chem. Soc.*, **81**, 3872 (1959).

Table 8-4. *Values of* H_0 *for Aqueous Solutions of Acids at Rounded Molar Concentrations,* 25° [a]

ACID (moles/liter)	HNO_3	HCl	HBr	$HClO_4$	H_2SO_4	H_3PO_4	HF
0.1	+0.98	+0.98	+0.98	—	+0.83	+1.45	—
0.25	+0.55	+0.55	+0.55	—	+0.44	+1.15	—
0.5	+0.21	+0.20	+0.20	—	+0.13	+0.97	—
0.75	−0.02	−0.03	−0.03	−0.04	−0.07	+0.78	—
1.0	−0.18	−0.20	−0.20	−0.22	−0.26	+0.63	+1.20
1.5	−0.45	−0.47	−0.48	−0.53	−0.56	+0.41	+1.04
2.0	−0.67	−0.69	−0.71	−0.78	−0.84	+0.24	+0.91
2.5	−0.85	−0.87	−0.91	−1.01	−1.12	+0.07	+0.74
3.0	−1.02	−1.05	−1.11	−1.23	−1.38	−0.08	+0.60
3.5	−1.17	−1.23	−1.31	−1.47	−1.62	−0.22	+0.49
4.0	−1.32	−1.40	−1.50	−1.72	−1.85	−0.37	+0.40
4.5	−1.46	−1.58	−1.71	−1.97	−2.06	−0.53	+0.34
5.0	−1.57	−1.76	−1.93	−2.23	−2.28	−0.69	+0.28
5.5	−1.69	−1.93	−2.14	−2.52	−2.51	−0.84	+0.21
6.0	−1.79	−2.12	−2.38	−2.84	−2.76	−1.04	+0.15
6.5	−1.89	−2.33	—	−3.22	−3.03	−1.24	+0.08
7.0	−1.99	−2.53	−2.85	−3.61	−3.32	−1.45	+0.02
7.5	—	−2.73	—	−3.98	−3.60	−1.66	−0.04
8.0	—	−2.93	−3.34	−4.33	−3.87	−1.85	−0.11
8.5	—	−3.13	—	−4.69	−4.14	−2.04	−0.17
9.0	—	−3.31	−3.89	−5.05	−4.40	−2.22	−0.24
9.5	—	−3.46	—	−5.42	−4.65	−2.40	−0.30
10.0	—	−3.60	−4.44	−5.79	−4.89	−2.59	−0.36

[a] From M. A. Paul and F. A. Long, *Chem. Revs.*, **57**, 1 (1957).

dioxane-water mixture,[17] alcohol-water mixture,[20,21] or acetic acid-water mixture.[22] The analogous acidity function D_0 for deuterium acids has also been established for $DCl-D_2O$ and for $D_2SO_4-D_2O$.[13] The numerical values are very similar to those for H_0.

The effect of temperature on H_0 has been investigated for sulfuric acid-water mixtures in the range 20–80°C. To a good approximation, the variation of H_0 with temperature is linear. Values of dH_0/dT are listed in Table 8-6. The values are negative at the lower sulfuric acid concentrations but become positive at about 45 wt % H_2SO_4.

[20] D. P. N. Satchell, *J. Chem. Soc.*, 3524 (1957).
[21] S.-J. Yeh and H. H. Jaffé, *J. Am. Chem. Soc.*, **81**, 3274 (1959). This scale is based largely on data for substituted azobenzenes.
[22] K. B. Wiberg and R. J. Evans, *ibid.*, **80**, 3019 (1958).

Table 8-5. Values of H_0 for Mixtures of Water with a Strong Acid at Rounded Weight Percentages

ACID (weight %)	H_2SO_4[a] (25°)	HNO_3[b] (20°)	CH_3SO_3H[c] (25°)
20	−1.01	−1.28	−0.61
25	−1.37	−1.57	−0.77
30	−1.72	−1.85	−0.96
35	−2.06	−2.10	−1.17
40	−2.41	−2.36	−1.41
45	−2.85	−2.62	−1.69
50	−3.38[d]	−2.88	−2.00
55	−3.91[d]	−3.13	−2.34
60	−4.46[d]	−3.42	−2.71
65	−5.04	−3.72	−3.12
70	−5.65	−3.99	−3.60
75	−6.30	−4.30	−4.05
80	−6.97	−4.62	−4.53
85	−7.66	−4.96	−5.35
90	−8.27	−5.31	−6.33
95	−8.86	−5.75	−7.30
97	−9.14	—	—
99	−9.74	—	—
99.8	−10.36	—	—
100.0	−11.10	−6.3	−7.86

[a] From M. A. Paul and F. A. Long, *Chem. Revs.*, **57**, 1 (1957). These values are based largely on the data of Hammett and Deyrup, obtained in 1932.[10] Values obtained by more modern methods[12,13] agree quite well.
[b] From J. G. Dawber and P. A. H. Wyatt, *J. Chem. Soc.*, 3589 (1960).
[c] From K. N. Bascombe and R. P. Bell, *ibid.*, 1096 (1959).
[d] H_0 obtained largely from data for bases other than primary *o*-nitroanilines.

It has been pointed out that according to the formal theory beginning with equation 8, $\delta_M \log c_{RB}/c_{RBH^+}$ need not be independent of the nature of B even in first approximation. Since the H_0 function is based largely on data for *o*-nitroanilines, it is of interest to examine substances with other basic groups. A convenient test of the assumed independence is the degree of constancy of the quantity [$\log c_{RB}/c_{RBH^+} - H_0$], that is, of

$$\log \frac{c_{RB}}{c_{RBH^+}} - \left[\log \frac{c_{RB}}{c_{RBH^+}} - \log K_{c(H_2O)}\right]_{o\text{-nitroanilines}}$$

Table 8-6. Temperature Dependence of Acidity Functions in Sulfuric Acid–Water Mixtures

H_2SO_4 (wt %)	dH_0/dT (degrees^{-1})[a]	dH_R/dT (degrees^{-1})[b]
20	-0.0015	$+0.01_0$
30	-0.0005	$+0.00_5$
40	0.0000	$+0.00_5$
50	$+0.0002$	$+0.00_5$
55	$+0.0015$	0.00_0
60	$+0.0025$	0.00_0
65	$+0.003_8$	-0.00_5
70	$+0.005_0$	-0.01_5
75	$+0.005_0$	-0.02_0
80	$+0.006_2$	-0.03_5
85	$+0.007_0$	-0.04_5
90	$+0.007_7$	-0.05_0
95	$+0.009$	-0.05_5
98	$+0.010$	-0.05_5
100	$+0.012$	—

[a] A. I. Gelbshtein, G. G. Sheglova, and A. I. Temkin, *Doklady Akad Nauk. S.S.S.R.*, 107, 108 (1956).

[b] R. D. Bushick, doctoral dissertation, University of Pittsburgh, 1961.

This quantity is equal to log $K_{c(H_2O)}$, the ionization constant of RBH^+ in dilute aqueous solution, if the nature of the basic group truly makes no difference. For the protonation of phenyl ethers, aryl carboxylic acids, benzaldehydes, benzophenones, benzamides, and azobenzenes in sulfuric acid-water mixtures, variations in [log c_{RB}/c_{RBH^+} − H_0] as the per cent of sulfuric acid is changed appear to be random and within the experimental error over a tenfold range in c_{RB}/c_{RBH^+}.[11,23,24,25] For the protonation of secondary and tertiary anilines, for propionamide, and for some *ortho*-substituted benzoic acids, the variations appear to be nonrandom and significant.[26] But it is difficult to be sure of this, because it is difficult to

[23] E. M. Arnett and C. Y. Wu, *J. Am. Chem. Soc.*, 82, 5660 (1960).

[24] L. A. Flexser, L. P. Hammett, and A. Dingwall, *ibid.*, 57, 2103 (1935); 60, 3097 (1938). L. A. Flexser and L. P. Hammett, *ibid.*, 60, 885 (1938). J. T. Edward, H. S. Chang, K. Yates, and R. Stewart, *Can. J. Chem.*, 38, 1518 (1960).

[25] W. M. Schubert and R. E. Zahler, *J. Am. Chem. Soc.*, 76, 1 (1954). W. M. Schubert, J. Donohue and J. D. Gardner, *ibid.*, 76, 9 (1954).

[26] For example, see data for N,N-dimethyl-2,4-dinitroaniline in Table 8-3 and standard deviations of $pK_{c(H_2O)}$ values reported by E. Högfeldt and J. Bigeleisen, *J. Am. Chem. Soc.*, 82, 17 (1960), Table I. Also, R. Stewart and M. R. Granger, *Can. J. Chem.*, 39, 2508 (1961); J. T. Edward and I. C. Wang, *ibid.*, 40, 966 (1962).

distinguish between nonrandom deviations from the H_0 function and systematic medium effects on the absorption spectra.

On the other hand, for the protonation of a series of diaryl olefins in sulfuric acid-water mixtures, the variations are so large that the H_0 function cannot be said to describe the acid-base behavior even in first approximation.[27] Evidently the substitution of a hydrophobic group such as $>C{=}C<$ for $-NH_2$ is too large a structural change.

Acidity Functions in Nonaqueous Media. So far we have emphasized the protonation of bases in *aqueous* acids. However, it is often desirable to substitute an organic liquid for the water component. The new component may be a pure substance, such as trifluoroacetic acid, or it may be a mixture of constant composition, such as a liquid consisting always of 24.4 mole % ethanol and 75.6 mole % water. There are now two equally reasonable ways of defining an acidity scale. One is to retain the previous definition that the acidity scale approaches $-\log c_{H^+}$ in dilute aqueous solution; the other is to change the reference state so that the acidity scale approaches $-\log c_{H^+}$ in dilute solution in the new solvent component. For example, the reference state for an acidity scale in CF_3COOH-H_2SO_4 mixtures could be either the dilute aqueous solution or the dilute solution in CF_3COOH. We shall continue to use the symbol H_0 for the scale based on the dilute aqueous reference state and shall use H_0' to denote the scale based on the dilute solution in the new solvent component.

The advantage of using the H_0 scale even in nonaqueous solvents is that the proton-donating power of any medium is then always expressed on the same scale. For example, an $0.01 M$ solution of sulfuric acid in trifluoroacetic acid has the same H_0 value as a $9.5 M$ solution in water, indicating that the two media are equally efficient at transferring a proton to a neutral base. The disadvantage of retaining the H_0 scale in a nonaqueous solvent is that this entails considerable loss of accuracy in the prediction of acid-base equilibria.

In analyzing this problem, let us use the symbol M to denote the organic liquid to which the strong acid is added. The quantity $\log c_{RB}/c_{RBH^+}$ may be expressed on the H_0 scale by equation 12 and on the H_0' scale by equation 13.

$$\log \frac{c_{RB}}{c_{RBH^+}} = H_0 + \log K_{c(H_2O)} \tag{12}$$

$$\log \frac{c_{RB}}{c_{RBH^+}} = H_0' + \log K_{c(M)} \tag{13}$$

$K_{c(M)}$ is the acid dissociation constant of RBH^+ in dilute solution in M. The experimental data on solutions of strong acids in organic liquids[17,20,21]

[27] N. C. Deno, P. T. Groves, and G. Saines, *J. Am. Chem. Soc.*, **81**, 5790 (1959).

indicate that equation 13 is an excellent first approximation if the dielectric constant is at least moderately high, say about 20 or greater. In other words, the quantity $[\log c_{RB}/c_{RBH^+} - \log K_{c(M)}]$ is satisfactorily independent of the nature of the base.

If equation 13 fits a set of experimental data, equation 12 can fit the same set of data with equal accuracy if and only if the quantity $[\log K_{c(M)} - \log K_{c(H_2O)}]$ is a constant, characteristic of M and independent of the nature of the base. Unfortunately this is not usually the case. For example, the values of $[\log K_{c(M)} - \log K_{c(H_2O)}]$ have been found to be rather sensitive functions of the nature of RBH^+ in methanol,[28] ethanol,[20,29] and dioxane-water mixtures.[30]

Salt Effects and Polar Solute Effects on the Acidity Function. The acidity of a dilute solution of a strong acid in water can be greatly increased by the substitution of a concentrated aqueous salt solution for the water component. For example, the substitution of $4N$ lithium chloride reduces the pH of $0.1N$ hydrochloric acid by nearly one unit.[31] The increase in acidity caused by different salts can be roughly correlated with the molar heat of solution of the salt in water.[31]

Salt effects on K_c for the ionization of an acid are correspondingly large and specific for each salt. For acids of the type RBH^+, the quantities $[\log K_{c(\text{salt solution})} - \log K_{c(H_2O)}]$ for a given salt solution are also sensitive to the nature of RBH^+.[32] Hence an H_0 scale cannot be established for mixtures of a strong acid with a concentrated aqueous salt solution, except as a rough approximation.

In solvents of low dielectric constant, proton transfer is accompanied by ionic association. For example, in glacial acetic acid containing perchloric acid, the proton transfer to aniline is best treated as an equilibrium involving ion pairs.

$$H^+ClO_4^- + C_6H_5NH_2 \rightleftharpoons C_6H_5NH_3^+ClO_4^- \tag{14}$$

A spectrophotometric measurement of the acid/base ratio in such a solution gives a result that is much closer to the value of $c_{RBH^+ClO_4^-}/c_{RB}$ than to that of c_{RBH^+}/c_{RB}. Since the equilibrium constant for the formation of the ion pair $RBH^+ClO_4^-$ from the free ions depends specifically on the constitution and structure of RBH^+, it can be demonstrated that the quantity $[\log K_{c(H_2O)} + \log (\text{acid/base ratio})]$ is a specific function of the nature of

[28] A. L. Bacarella, E. Grunwald, H. P. Marshall, and E. L. Purlee, *J. Org. Chem.*, **20**, 747 (1955).

[29] B. Gutbezahl and E. Grunwald, *J. Am. Chem. Soc.*, **75**, 559 (1953).

[30] H. P. Marshall and E. Grunwald, *ibid.*, **76**, 2000 (1954).

[31] F. E. Critchfield and J. B. Johnson, *Anal. Chem.*, **31**, 570 (1959).

[32] M. A. Paul, *J. Am. Chem. Soc.*, **76**, 3236 (1954).

the base.[33] The experimental results for glacial acetic acid[11] confirm this analysis and suggest that it is not very fruitful to establish an H_0 function in solvents of low dielectric constant on the basis of spectrophotometric measurements of acid/base ratios.

In aprotic solvents, such as carbon tetrachloride or nitromethane, proton transfer from an acid to a base is accompanied by hydrogen-bond formation involving additional molecules of the acid. For example, the addition of triethylamine to an excess of acetic acid in carbon tetrachloride leads to the formation of the solvated ion pair, $Et_3NH^+OAc^-\cdot HOAc$, the structural formula for which is probably that shown in equation 15.[34]

$$CH_3-C\underset{O \cdots H \cdots O}{\overset{O \cdots H----O}{\underset{\oplus N(C_2H_5)_3}{\overset{(?)}{|}}}}C-CH_3 \qquad (15)$$

Addition of pyridine or of 2,4-dinitroaniline to an excess of sulfuric acid in nitromethane leads to a reaction that is best represented by equation 16.[35]

$$RB + 3H_2SO_4 \rightleftharpoons RBH^+OSO_3H^-\cdot 2H_2SO_4 \qquad (16)$$

The major product of this reaction is an ion pair solvated by sulfuric acid rather than a pair of free ions, even though nitromethane is a highly polar solvent, having a dielectric constant greater than that of methanol. The example proves that a high dielectric constant alone is not enough to promote electrolytic dissociation. The solvent molecules must also be able to stabilize the free ions by molecular complex formation.

When the acid performs a dual function as hydrogen-bond donor as well as proton donor, the hydrogen bonding provides an entire set of additional mechanisms in which the quantities $\delta_M \log c_{RB}/c_{RBH^+}$ may depend specifically on the nature of the base. A priori one would therefore guess that it is not very fruitful to extend the concept of an acidity function to solutions of a strong acid in an aprotic solvent.

Other Acidity Functions

Equation 8, which expresses the free energy of a solute in first approximation, predicts that the quantities $\delta_M \log c_{RB}/c_{RBH^+}$ are independent

[33] S. Bruckenstein, ibid., **82**, 307 (1960).
[34] G. M. Barrow and E. A. Yerger, ibid., **76**, 5211 (1954).
[35] H. Van Looy and L. P. Hammett, ibid., **81**, 3872 (1959).

Table 8-7. Tentative Values of the Acidity Function H_

1. Hydrazine-Water Mixtures, 25° [37]

Wt % HYDRAZINE	H_	Wt % HYDRAZINE	H_
5	11.18	35	13.56
10	11.55	40	14.03
15	11.93	45	14.52
20	12.29	50	14.99
25	12.72	55	15.43
30	13.15	60	15.93

2. Ethylenediamine-Water Mixtures, 25° [38]

Wt % $(CH_2NH_2)_2$	H_	Wt % $(CH_2NH_2)_2$	H_
20	13.3	40	14.8
30	14.0	50	15.8

3. Ethanolamine-Water Mixtures, 25° [39]

Wt % $HOCH_2CH_2NH_2$	H_	Wt % $HOCH_2CH_2NH_2$	H_
20	12.60	80	14.89
40	13.45	100	15.35
60	14.25		

4. Alkali Hydroxide–Water Mixtures, 20° [36]

Wt % ALKALI HYDROXIDE	H_	
	NaOH	KOH
20	15.2	15.4
30	16.0	16.2
40	17.3	17.5
45	18.1	18.4

5. Sodium Methoxide in Methanol, 25° [40]

$c_{NaOMe}(M)$	H_	$c_{NaOMe}(M)$	H_
0.5	14.1	2.0	15.5
1.0	14.7	2.5	15.9
1.5	15.1	3.0	16.3

6. $0.01N$ $C_6H_5N(CH_3)_3OH$ in 5 Mole % Water-95 Mole % Tetramethylene Sulfone (sulfolane) at 25°: H_ = 18.7 [41]

7. Sulfuric Acid–Water Mixtures,[a] 25°

Wt % H_2SO_4	H_	Wt % H_2SO_4	H_
30	−0.58	46	−1.43
34	−0.85	50	−1.66
38	−1.03	54	−1.88
42	−1.22	58	−2.15

[a] Based on data for nitrate ion[42] and log K_A = 1.34 for nitric acid [G. C. Hood, O. Redlich, and C. A. Reilly, J. Chem. Phys., **22**, 2067 (1954)].

of the nature of R but not of B. In deriving the acidity function H_0 we made the additional assumption that these quantities are also independent of the nature of B. We cannot of course expect this assumption to hold if the change in the nature of B is too drastic, for example, if two bases of different charge type are compared. However, it is reasonable to assume that there exists a set of acidity functions, each of which corresponds to the constancy of $\delta_M \log (c_{\text{base}}/c_{\text{conjugate acid}})$ for a given charge type.

For acids and bases of the charge type $RBH\text{-}RB^-$, this function will be denoted by H_-. Although the constancy of $\delta_M \log c_{RB^-}/c_{RBH}$ has been less thoroughly tested, we have compiled in Table 8-7 some tentative values of H_- for various solvents. These values are based on a scale according to which H_- approaches $-\log c_{H^+}$ in dilute aqueous solution. Most of the values have been obtained in rather basic solvents[36-41] and are reasonably independent of the nature of RB^-. Some data are also available for sulfuric acid-water mixtures, but here the values do seem to depend on the nature of the base. Thus in 30 wt % H_2SO_4, an H_- scale based on nitrate ion differs by about 0.5 unit from one based on picrate ion,[42] and an H_- scale based on cyanocarbon bases seems to differ by even more.[43] Resonance of the negative charge is undoubtedly responsible for part of these differences. The values listed in Table 8-7 are based on nitrate ion.

For other bases, such as the triarylcarbinols, ionization is accompanied by the loss of a water molecule.

$$Ar_3COH + H^+ \rightleftharpoons Ar_3C^+ + H_2O \tag{17}$$

Values of $\log c_{ROH}/c_{R^+}$ for substances ionizing in this way cannot be expected to be correlated by the H_0 acidity function because the process is too dissimilar from the ionization of o-nitroanilines. A new acidity function, H_R,[44] has therefore been established from data for the ionization

[36] G. Schwarzenbach and R. Sulzberger, *Helv. Chim. Acta*, **27**, 348 (1944).

[37] N. C. Deno, *J. Am. Chem. Soc.*, **74**, 2039 (1952).

[38] R. Schaal, *Compt. rend.*, **238**, 2156 (1954).

[39] F. Masure and R. Schaal, *Bull. Soc. Chim. France*, 1138 (1956).

[40] J. H. Ridd, *Chem. and Ind.*, 1268 (1957).

[41] C. H. Langford and R. L. Burwell, Jr., *J. Am. Chem. Soc.*, **82**, 1503 (1960).

[42] N. C. Deno, H. J. Peterson, and E. Sacher, *J. Phys. Chem.*, **65**, 199 (1961).

[43] R. H. Boyd, *J. Am. Chem. Soc.*, **83**, 4288 (1961).

[44] H_R is the same function sometimes symbolized by C_0[45] and is closely related to the function J_0.[48] The symbol H_R was introduced by A. M. Lowen, M. A. Murray, and G. Williams, *J. Chem. Soc.*, 3321 (1950); the symbol C_0 was used in the early publications of N. C. Deno and co-workers,[45] and J_0 was defined by Gold and Hawes.[48] The discontinuance of the symbol C_0 in favor of H_R has recently been advocated.[46]

of triarylcarbinols and diarylcarbinols.[45–47,14] This function is defined by equations 18.

$$\delta_M H_R = \delta_M \log \frac{c_{ROH}}{c_{R^+}} \tag{18a}$$

$$H_R = -\log c_{H^+} \text{ in dilute aqueous solution} \tag{18b}$$

The constancy of $\delta_M \log c_{ROH}/c_{R^+}$ for different bases ionizing by this mechanism is as satisfactory as that of $\delta_M \log c_{RB}/c_{RBH^+}$. Values of H_R

Table 8-8. Values of the Acidity Function H_R for Aqueous Acids, 25° [45,46]

ACID,	H_R		
Wt %	H_2SO_4-H_2O	$HClO_4$-H_2O	HNO_3-H_2O
5	−0.07	−0.24	−0.49
10	−0.72	−1.04	−1.30
15	−1.32	−1.74	−2.00
20	−1.92	−2.43	−2.63
25	−2.55	−3.08	−3.23
30	−3.22	−3.79	−3.83
35	−4.00	−4.61	−4.45
40	−4.80	−5.54	−5.07
45	−5.65	−6.60	−5.70
50	−6.60	−7.86	−6.40
55	−7.67	−9.36	−7.17
60	−8.92	−11.14	−8.05
70	−11.52	—	—
80	−14.12	—	—
90	−16.72	—	—
95	−18.08	—	—
98	−19.64	—	—

for mixtures of water with a strong acid will be found in Table 8-8. Values have also been reported for formic acid-water mixtures[14] and for perchloric acid-water-dioxane mixtures.[47] The temperature dependence of H_R has been studied for the system H_2SO_4-H_2O. Average values of dH_R/dT for the range 0–45°C are listed in Table 8-6.

Because of the involvement of water in the reaction (equation 17) used to define the H_R function, it is logical to seek relationships between H_R and

[45] N. C. Deno, J. J. Jaruzelski, and A. Schriesheim, *J. Am. Chem. Soc.*, **77**, 3044 (1955).
[46] N. C. Deno, H. E. Berkheimer, W. L. Evans, and H. J. Peterson, *ibid.*, **81**, 2344 (1959).
[47] H. Dahn, L. Loewe, and G. Rotzler, *Chem. Ber.*, **93**, 1572 (1960).

the activity of water in the medium. For example, Gold and Hawes[48] in their early work assumed the relationship (19).

$$H_R = H_0 + \log a_{H_2O} \qquad (19)$$

Even though that equation is not general, it has been found experimentally that the closely related equation (20) applies to sulfuric acid-water mixtures in the range 83 to 100 wt % H_2SO_4.[45] This is the concentration range in which there is at least one sulfuric acid molecule for each water molecule, and the water molecules are largely in the form of hydronium bisulfate.

$$H_R = H_0 + \log a_{H_2O} - 4.92 \qquad (83\text{--}100\% \ H_2SO_4) \qquad (20)$$

Acid Catalysis

Acidity functions have also played an important role in the investigation of the mechanisms of acid-catalyzed reactions. For reactions in which a weakly basic substrate, S, is in rapid equilibrium with its conjugate acid, the question arises as to whether the latter reacts unimolecularly to give product or whether it reacts with a water molecule. In principle, if certain additional assumptions can be made, the medium effect on the reaction rate should be able to decide the question. One of the necessary assumptions is that in the unimolecular mechanism for the reaction of SH^+, the transition state closely resembles the SH^+ molecule. That is, $\delta_M \bar{F}^\circ_{SH^+}$ is equal to $\delta_M \bar{F}^\ddagger$. The other assumption is that the substrate behaves as an H_0-base. Given these assumptions, $\delta_M \log k_\psi$ for the unimolar mechanism should be equal to $-\delta_M H_0$. The symbol k_ψ denotes the pseudo first-order rate constant for the disappearance of substrate.

On the other hand, if the protonated substrate reacts with a water molecule, it can be argued that $\delta_M \log k_\psi$ should be more nearly equal to $\delta_M \log c_{H^+}$ than to $-\delta_M H_0$.[49,50] This mechanistic criterion, which was first proposed by Zucker and Hammett in 1939,[49] appeared to agree with the mechanisms of many reactions as deduced in other ways[50] but has now been abandoned because of the numerous exceptions that have since been discovered.[51]

[48] V. Gold and B. W. V. Hawes, *J. Chem. Soc.*, 2102 (1951).

[49] L. Zucker and L. P. Hammett, *J. Am. Chem. Soc.*, **61**, 2791 (1939).

[50] F. A. Long and M. A. Paul, *Chem. Revs.*, **57**, 935 (1957).

[51] E. Grunwald, A. Heller, and F. S. Klein, *J. Chem. Soc.*, 2604 (1957); J. Koskikallio and E. Whalley, *Trans. Faraday Soc.*, **55**, 815 (1959); G. Archer and R. P. Bell, *J. Chem. Soc.*, 3228 (1959); R. H. Boyd, R. W. Taft, Jr., A. P. Wolf, and D. R. Christman, *J. Am. Chem. Soc.*, **82**, 4729 (1960); J. F. Bunnett, *ibid.*, **82**, 499 (1960); A. J. Kresge and Y. Chiang, *ibid.*, **83**, 2877 (1961).

SOME $\delta_M F^\circ$ RELATIONSHIPS EMPLOYING MORE THAN ONE PARAMETER

In the second approximation discussed in Chapter 6, $\bar{F}^\circ_{RX}$ is given by equation 21.

$$\bar{F}^\circ_{RX} = \mathsf{F}_R + \mathsf{F}_X + \mathsf{I}_{R,X} + \mathsf{I}_{R,M} + \mathsf{I}_{X,M} + \mathsf{II}_{R,X,M} \tag{21}$$

In this approximation the first-order interaction terms $\mathsf{I}_{R,X}$ etc. are no longer factorable, but the second-order interaction term $\mathsf{II}_{R,X,M}$ is separable into three factors.

In applying the second approximation to acid-base equilibria, let us first consider proton-transfer reactions in those solvents in which the H_0 acidity function was previously found to be useful. The nature of the R-group is now an important variable, and we therefore compare the ionization of the base RB with that of a reference base, R_0B.

$$RBH^+ \underset{}{\overset{K_e}{\rightleftharpoons}} RB + H^+ \tag{22}$$

$$R_0BH^+ \underset{}{\overset{K_e^\circ}{\rightleftharpoons}} R_0B + H^+ \tag{23}$$

$$\delta_R \Delta F^\circ = \Delta \bar{F}_R{}^\circ - \Delta \bar{F}^\circ_{R_0} = (\bar{F}^\circ_{RB} - \bar{F}^\circ_{RBH^+}) - (\bar{F}^\circ_{R_0B} - \bar{F}^\circ_{R_0BH^+}) \tag{24}$$

Expanding the free energies of equation 24 in terms of equation 21, we obtain

$$\delta_R \Delta \bar{F}^\circ = (\mathsf{I}_{R,B} - \mathsf{I}_{R,BH^+}) - (\mathsf{I}_{R_0,B} - \mathsf{I}_{R_0,BH^+})$$
$$+ (\mathsf{II}_{R,B,M} - \mathsf{II}_{R,BH^+,M}) - (\mathsf{II}_{R_0,B,M} - \mathsf{II}_{R_0,BH^+,M}) \tag{25}$$

On separating the second-order interaction terms as shown in equations 26, we obtain (27) for $\delta_R \Delta F^\circ$.

$$\mathsf{II}_{R,B,M} = \mathsf{II}_R \cdot \mathsf{II}_B \cdot \mathsf{II}_M$$
$$\vdots$$
$$\vdots \tag{26}$$

$$\mathsf{II}_{R_0,BH^+,M} = \mathsf{II}_{R_0} \cdot \mathsf{II}_{BH^+} \cdot \mathsf{II}_M$$

$$\delta_R \Delta F^\circ = (\mathsf{I}_{R,B} - \mathsf{I}_{R,BH^+}) - (\mathsf{I}_{R_0,B} - \mathsf{I}_{R_0,BH^+})$$
$$+ \mathsf{II}_M(\mathsf{II}_B - \mathsf{II}_{BH^+})(\mathsf{II}_R - \mathsf{II}_{R_0}) \tag{27}$$

The medium effect on $\delta_R \Delta F^\circ$ is given by an equation, (28), containing only the second-order interaction terms.

$$\delta_M \delta_R \Delta F^\circ = \delta_M \mathsf{II}_M[(\mathsf{II}_B - \mathsf{II}_{BH^+})(\mathsf{II}_R - \mathsf{II}_{R_0})] \tag{28}$$

It will be noted that in the second approximation, $\delta_M \delta_R \Delta F^\circ$ no longer vanishes but is equal to the product of a factor dependent on the medium change and one dependent on the nature of the solutes.

In order to show the relationship of the present treatment to that involving the H_0 function, we express the free energy quantity $\delta_M \delta_R \Delta F^\circ$ in terms of the acid/base ratios as follows:

$$\delta_M \, \delta_R \, \Delta F^\circ = \delta_M (\Delta \bar{F}_R{}^\circ - \Delta \bar{F}_{R_0}^\circ)$$

$$= -2.303 \, RT \delta_M \left(\log \frac{K_c}{K_c{}^\circ} \right)$$

$$= -2.303 \, RT \left(\delta_M \log \frac{c_{RB}}{c_{RBH^+}} - \delta_M \log \frac{c_{R_0B}}{c_{R_0BH^+}} \right) \quad (29)$$

Let us assume that the reference base, R_0B, is an H_0-base. That is, $\delta_M \log (c_{R_0B}/c_{R_0BH^+})$ is precisely equal to $\delta_M H_0$.[52] Equation 29 then reduces to 30.

$$\delta_M \log \frac{c_{RB}}{c_{RBH^+}} = \delta_M H_0 - \frac{1}{2.303RT} \delta_M \delta_R \Delta F^\circ$$

$$= \delta_M H_0 - \frac{(\|_B - \|_{BH^+})(\|_R - \|_{R_0})}{2.303RT} \cdot \delta_M \|_M$$

$$= \delta_M H_0 + f(\text{structure of the base}) \cdot \varphi(\text{solvent change}) \quad (30)$$

That is, $\delta_M \log c_{RB}/c_{RBH^+}$, instead of being simply equal to $\delta_M H_0$, is equal to that quantity plus a term consisting of a structural parameter multiplied by a solvent parameter.

It would of course be desirable to express the parameters as explicit functions of appropriate macroscopic properties of the base and of the solvent. Whether or not this can be done depends on the nature of the solvent-solute interaction, which, owing to the use of equations 26, is assumed to proceed by a single mechanism. In a two-component solvent system such as H_2O-H_2SO_4, the interaction variable characteristic of the solvent might very well be a monotonic function of the solvent composition. If so, the solvent parameter can usually be represented by means of a suitable macroscopic property that also varies monotonically with the solvent composition. The success of such a representation depends only on a similarity of the functional relationships to the solvent composition, not necessarily on any fundamental connection in terms of solvation mechanism. We shall find, for example, that for proton transfer to a neutral base

[52] Because the operational definition of $\delta_M H_0$ depends on the average behavior of a number of substances rather than on the actual behavior of a single substance, an actual substance that is precisely an H_0-base probably does not exist. It is therefore not quite proper, from a logical point of view, to build the second approximation on a foundation consisting of the H_0 function as a first approximation. In the present treatment we proceed to do so because the result obtained is useful. It would be desirable, however, to redefine $\delta_M H_0$ in terms of the actual behavior of a single reference substance.

in strong acid-water mixtures, the solvent parameter is represented adequately by the logarithm of the water activity.

Returning to the formal treatment, we wish next to obtain a relationship for the rate constants of acid-catalyzed reactions. Let us consider first a reaction in which a weakly basic substrate RB is reversibly protonated to a species RBH^+, the unimolecular decomposition of which determines the rate.

$$H^+ + RB \rightleftharpoons RBH^+ \rightarrow R^{\ddagger}B^{\ddagger}H^+ \tag{31}$$

We distinguish formally between R and $R^{\ddagger}$ as well as between B and $B^{\ddagger}$, for decomposition may occur in either part of the molecule. The pseudo-first-order rate constant, k_ψ, is defined by equation 32, where c_{RB} is the actual molar concentration of unprotonated substrate.[53]

$$k_\psi \equiv \frac{\text{rate}}{c_{RB}} = \frac{\kappa kT}{h} \frac{c_{R^{\ddagger}B^{\ddagger}H^+}}{c_{RB}} \tag{32}$$

$$-2.303RT\,\delta_M \log k_\psi = \delta_M(\bar{F}^{\circ}_{R^{\ddagger}B^{\ddagger}H^+} - \bar{F}^{\circ}_{RB}) \tag{33}$$

$$= \delta_M(\bar{F}^{\circ}_{R^{\ddagger}B^{\ddagger}H^+} - \bar{F}^{\circ}_{RBH^+}) + \delta_M(\bar{F}^{\circ}_{RBH^+} - \bar{F}^{\circ}_{RB}) \tag{34}$$

The medium effect represented by the last term in equation 34 can be treated by equation 30. There remains then the problem of expressing the medium effect on the difference in free energy between transition state and protonated substrate. Applying equation 21 to the problem, we obtain

$$\delta_M(\bar{F}^{\circ}_{R^{\ddagger}B^{\ddagger}H^+} - \bar{F}^{\circ}_{RBH^+}) = \delta_M(I_{B^{\ddagger}H^+,\,M} - I_{BH^+,\,M})$$

$$+ \delta_M(I_{R^{\ddagger},\,M} - I_{R,\,M}) + \delta_M \mathbb{I}_M[(\mathbb{I}_{B^{\ddagger}H^+}\mathbb{I}_{R^{\ddagger}} - \mathbb{I}_{BH^+}\mathbb{I}_R)] \tag{35}$$

In (35) the last term is seen to be factored already, and the solvent-sensitive factor $\delta_M \mathbb{I}_M$ is identical to the one that will be used later in connection with RBH^+ and RB. On the other hand, according to the second approximation which we are now using, the first-order interaction terms would not in general be factorable. However, we believe that we may nevertheless factor them in this special instance, assuming that the change from ground

---

[53] Usually c_{RB} is equal to the *formal* concentration of the substrate because the steady-state concentration of RBH^+ is small compared to that of RB. If this condition is not met and if pK_A of RBH^+ is known, a correction can be made easily by means of the H_0 function. However, in making such a correction, it is important to keep in mind that the H_0 function represents the data only in the first approximation, whereas we are trying to develop a second approximation. The use of the H_0 function is therefore justified only if the correction to be made is small.

to transition state is a very small one as structural changes go. We will therefore replace equation 35 with the following.

$$\delta_M(F^{\circ}_{R\ddagger B\ddagger H^+} - F^{\circ}_{RBH^+}) = \delta_M I_M[(I_{B\ddagger H^+} - I_{BH^+}) + (I_{R\ddagger} - I_R)]$$
$$+ \delta_M II_M[(II_{B\ddagger H^+} \cdot II_{R\ddagger} - II_{BH^+} II_R)] \quad (36)$$

Since we are assuming only a single solvent-solute interaction mechanism, there is only one important independent variable characteristic of the solvent. For small changes in this variable, the factors $\delta_M I_M$ and $\delta_M II_M$ are both proportional to the change in the variable and hence to each other. We will therefore replace equation 36 in its brief turn with the following:

$$\delta_M(F^{\circ}_{R\ddagger B\ddagger H^+} - F^{\circ}_{RBH^+}) = \delta_M II_M[\lambda(I_{B\ddagger H^+} - I_{BH^+})$$
$$+ \lambda(I_{R\ddagger} - I_R) + (II_{B\ddagger H^+} \cdot II_{R\ddagger} - II_{BH^+} II_R)] \quad (37)$$

The proportionality constant relating $\delta_M I_M$ to $\delta_M II_M$ is λ. Combining equations 30, 33, 34, and 37, we obtain

$$\delta_M \log k_{\psi} = -\delta_M H_0 - \frac{\delta_M II_M}{2.303 RT}[\lambda(I_{B\ddagger H^+} - I_{BH^+}) + \lambda(I_{R\ddagger} - I_R)$$
$$+ (II_{B\ddagger H^+} \cdot II_{R\ddagger} - II_{BH^+} II_R) - (II_B - II_{BH^+}) \cdot (II_R - II_{R_0})] \quad (38)$$

The essential features of equation 38 are shown more clearly by (39).

$$\delta_M \log k_{\psi} = -\delta_M H_0 + f(\text{substrate and transition state structure})$$
$$\cdot \varphi(\text{solvent change}) \quad (39)$$

The function φ(solvent change) is the same as in equation 30.

The Bunnett Equation

A relationship of this form has been discovered empirically by Bunnett.[54] The function of the reaction medium is the logarithm of the activity of the water in the medium. The slope, or parameter characteristic of the structure of the reagents and the transition state, is represented by the symbol w.

$$\delta_M \log k_{\psi} = -\delta_M H_0 + w \, \delta_M \log a_{H_2O} \quad (40)$$

Figure 8-1, for the hydrolysis of methylal in a series of perchloric acid-water mixtures, is typical of the relationships obtained for reactions proceeding by the unimolecular mechanism assumed in our derivation.

$$\underset{\displaystyle \text{CH}_2 \big\langle \genfrac{}{}{0pt}{}{\text{OCH}_3}{\text{OCH}_3}}{} + H^+ \rightleftharpoons \underset{\displaystyle \text{CH}_2 \big\langle \genfrac{}{}{0pt}{}{\overset{\text{H}}{\overset{|}{+}\text{OCH}_3}}{\text{OCH}_3}}{} \rightarrow CH_3OCH_2^+ + CH_3OH \quad (41)$$

[54] J. F. Bunnett, *J. Am. Chem. Soc.*, **82**, 499 (1960); **83**, 4956 (1961).

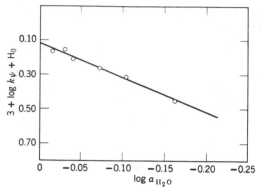

Figure 8-1. Rate constants for the hydrolysis
of methylal in perchloric acid-water mixtures
at 25°, plotted according to equation 40.

**Table 8-9. Values of log a_{H_2O} for Aqueous Solutions of
Acids at 25°** [54]

ACID	$\log a_{H_2O}$		
MOLARITY	HCl	H_2SO_4	$HClO_4$
0.5	−0.008	−0.008	−0.008
1.0	−0.017	−0.018	−0.018
1.5	−0.027	−0.030	−0.030
2.0	−0.039	−0.043	−0.043
2.5	−0.053	−0.063	−0.060
3.0	−0.070	−0.085	−0.081
3.5	−0.087	−0.111	−0.106
4.0	−0.107	−0.142	−0.135
4.5	−0.130	−0.176	−0.172
5.0	−0.155	−0.219	−0.215
5.5	−0.181	−0.267	−0.271
6.0	−0.211	−0.320	−0.330
6.5	−0.244	−0.377	−0.411
7.0	−0.279	−0.439	−0.496
7.5	−0.318	−0.510	−0.602
8.0	−0.358	−0.587	−0.714
8.5	−0.399	−0.670	−0.842
9.0	−0.444	−0.761	−0.983
9.5	−0.490	−0.859	−1.150
10.0	−0.539	−0.968	—
10.5	−0.591	−1.082	—

In this example the term $w \log a_{H_2O}$, which measures the improvement of the second approximation over the first approximation, amounts to as much as 0.3 unit. An improvement of this magnitude is equivalent to an improvement by a factor of two in the prediction of k_ψ. Values of $\log a_{H_2O}$ as a function of acid molarity are given in Table 8-9.

It is found empirically that the values of the parameter w associated with reactions that are believed to go by the unimolecular reaction of the protonated oxygen or nitrogen base fall in the range -2.5 to 0. A number of examples are included in Table 8-10. Although, for this mechanism, the

Table 8-10. *Correlation of the Rate Constants of Acid-Catalyzed Reactions According to the Bunnett Equation*[a]

REACTION	MINERAL ACID	w	r[b]	PROBABLE REACTION MECHANISM[c]
Hydrolysis of methylal, 25.0°	HCl, 0.4–3.9M	$-5.26 \pm .68$	0.06	A
	H_2SO_4, 1.2–2.0M	$-0.09 \pm .59$	0.01	
	$HClO_4$, 0.9–4.3M	$-2.04 \pm .19$	0.02	
Hydrolysis of methoxy-methyl acetate, 25.0°	HCl, 0.6–5.0M	$-2.39 \pm .23$	0.03	A
	H_2SO_4, 0.7–3.3M	$-0.67 \pm .54$	0.04	
Depolymerization of trioxane, 25.0°	HCl, 0.6–5.9M	$-2.10 \pm .20$	0.05	A
	H_2SO_4, 0.5–4.6M	$-1.02 \pm .09$	0.02	
	$HClO_4$, 0.8–5.1M	$-1.96 \pm .14$	0.04	
Hydrolysis of sucrose, 25.0°	HCl, 0.5–2.1M	$-0.43 \pm .26$	0.01	A
sec-Butyl alcohol, oxygen exchange, 99.8°	$HClO_4$, 1.1–3.5M	$+0.82 \pm .43$	0.03	B
t-Butyl alcohol, oxygen exchange, 25.0°	H_2SO_4, 1.0–4.0M	$-2.32 \pm .86$	0.08	A
Hydrolysis of benz-amide,[d] 25.0°	HCl, 1.0–8.6M	$+1.18 \pm .05$	0.02	B
	H_2SO_4, 1.0–8.0M	$+2.02 \pm .12$	0.06	
	$HClO_4$, 5.9–8.5M	$+1.75 \pm .13$	0.05	
Hydrolysis of diethyl ether, 50 atm, 120.1°	$HClO_4$, 1.1–5.6M	$+2.72 \pm .18$	0.04	B
Hydrolysis of ethyl acetate, 25.0°	HCl, 0.7–10.2M	$+4.15 \pm .09$	0.07	
	H_2SO_4, 1.4–6.8M	$+4.50 \pm .21$	0.07	
Hydrolysis of t-butyl acetate, 25.0°	HCl, 0.06–6.9M	$-1.17 \pm .31$	0.07	A
Bromination (enolization) of acetone, 25.0°	HCl, 1.0–8.0M	$+3.83 \pm .09$	0.03	C
Benzoic acid, oxygen exchange, 73.0°	H_2SO_4, 1.1–3.0M	$+8.8 \pm 1.4$	0.07	

[a] J. F. Bunnett, *J. Am. Chem. Soc.*, **83**, 4956 (1961).
[b] r = standard error of fit of $\log k_\psi$ to equation 40.
[c] A: Reversible protonation, followed by unimolecular reaction of the protonated substrate.
  B: Reversible protonation, followed by nucleophilic attack by a water molecule on the protonated substrate.
  C: Reversible protonation, followed by nucleophilic attack of a water molecule on a *different* proton.
[d] The fact that proton transfer to benzamide is considerable at the higher acidities was taken into account in these calculations.

plots are satisfactorily linear for mixtures of water with a given acid, a change from one acid to another changes w. For example, in reaction (41), in perchloric acid-water mixtures w is -2.04 ± 0.19. while in sulfuric acid-water mixtures w is -0.09 ± 0.59. As will be discussed later, such a dependence of w on the nature of the acid is expected if there is a contribution from more than one solvation mechanism. Such a dependence might also be observed in the case of a single solvation mechanism if the linear relationship to log a_{H_2O} is accidental rather than fundamental.

Reactions in which the protonated oxygen or nitrogen base is believed to undergo nucleophilic attack by a water molecule, rather than to decompose unimolecularly, seem to give values of w in the range $+1.2$ to $+3.3$ and occasionally give curvilinear plots.[55]

Reactions of unsaturated compounds in which the rate-determining step is believed to be the protonation of a multiple bond seem to give w values near zero.[55]

Reactions of oxygen or nitrogen bases for which the mechanism is believed to consist of a reversible protonation of the substrate followed by nucleophilic attack by a water molecule on a *different* proton seem to give curvilinear plots with w $> +3.3$.[55]

Plots of $\delta_M \log (c_{RB}/c_{RBH^+}) - \delta_M H_0$ should, according to equations 30 and 38, also be linear in log a_{H_2O}, for the solvent-sensitive parameter is the same in both equations.

$$\delta_M \log \frac{c_{RB}}{c_{RBH^+}} = \delta_M H_0 + w' \, \delta_M \log a_{H_2O} \qquad (42)$$

Equation 42 has not yet been tested over a wide range of structure. For substituted *o*-nitroanilines, the available data[10,12,13] indicate that w' is close to zero. For N-substituted anilines w' appears to be about -1. It would be interesting to test equation 42 by means of data for aliphatic bases, such as the dialkyl ethers or ketones, where the w' term might be more substantial.[56]

The preceding treatment assumes of course that only a single solvation mechanism is operative, which is almost certainly not true. In a rigorous treatment the right-hand sides of equations like (21) or (30) should be summations, with individual terms for each interaction mechanism. Neglecting the existence of multiple interaction mechanisms will have two effects on the free energy relationships with which we are dealing. One effect is that for any two-component solvent system the relationship will

[55] J. F. Bunnett, *J. Am. Chem. Soc.*, **83**, 4968 (1961).
[56] E. M. Arnett and C. Y. Wu, *J. Am. Chem. Soc.*, **82**, 4999 (1960); H. J. Campbell and J. T. Edward, *Can. J. Chem.*, **38**, 2109 (1960); J. T. Edward and I. C. Wang, *ibid.*, **40**, 966 (1962); R. W. Taft, Jr., *J. Am. Chem. Soc.*, **82**, 2965 (1960).

actually be curved, although the curvature may be too slight to be apparent from the data. We would expect the curvature to be slight and perhaps insignificant particularly when the relative importance of the various solvation mechanisms changes only monotonically with the composition. On the other hand, if instead of merely changing the composition of the two-component solvent we actually substitute a different compound, that point will be found to deviate significantly from the line. The second effect of multiple solvation mechanisms is therefore that plots for different two-component solvent systems will have different slopes.

In the Bunnett equation, the parameter characteristic of the medium is equated to the logarithm of the water activity. The success of the equation suggests that interaction with some physical quantity associated with water is the major solvation mechanism. The fact that w is different for different two-component solvent systems means, however, that other interaction mechanisms also operate to a significant extent. It would probably be an oversimplification to interpret the parameter w as a measure of the average number of water molecules involved in the reaction, although the number of such water molecules is undoubtedly one of the possible determinants of the magnitude of w.

The mY Equation for Acids in Aqueous-Organic Solvent Systems[57,58,59]

The ionization of organic acids in aqueous-organic solvent systems is analogous to the ionization of organic acids in aqueous-inorganic acid systems except that the hydrogen ion concentration is no longer characteristic of the medium. The quantity useful in this problem is $K_c = c_{H^+} c_{RB} / c_{RBH^+}$ rather than the concentration ratio c_{RB}/c_{RBH^+}. The medium effect on K_c is given by equation 43.

$$- \frac{1}{2.303RT} \delta_M(\Delta F_R^\circ - \Delta F_{R_0}^\circ) = \delta_M(\log K_{c,R} - \log K_{c,R_0})$$

$$= \delta_R \delta_M \log K_c \qquad (43)$$

According to equation 28, the quantity $\delta_M(\Delta F_R^\circ - \Delta F_{R_0}^\circ)$ is equal to the product of a factor characteristic only of the medium change and a factor characteristic only of RB and R_0B. The symbol m is used to represent the parameter characteristic of RB, and the symbol Y is used to represent the parameter characteristic of the medium. Because it was expected that the latter would be a function of several independent physical quantities, no

[57] E. Grunwald and B. J. Berkowitz, *J. Am. Chem. Soc.*, **73**, 4939 (1951).
[58] B. Gutbezahl and E. Grunwald, *ibid.*, **75**, 559 (1953).
[59] H. P. Marshall and E. Grunwald, *ibid.*, **76**, 2000 (1954).

attempt was made in the beginning to equate Y with any single physical property of the solvent.

$$-\delta_M \, \delta_R \log K_c = \delta_R m \cdot \delta_M Y \tag{44}$$

A scale of $\delta_M Y$ values can be constructed by assigning an arbitrary numerical value to $\delta_M Y$ for a certain medium change chosen as a standard. We shall use the operator δ_M° to denote the standard medium change. Equation 45 constitutes a definition of $\delta_M^\circ Y$ for all applications to the ionization of acids in aqueous-organic solvent systems.

$$\delta_M^\circ Y = Y_{\text{(pure organic solvent component)}} - Y_{\text{(pure water)}}$$
$$= 1, \text{ by definition} \tag{45}$$

A scale of Y values can be constructed by assigning, in addition, the value Y = 0 to the reference solvent, which in this case is water. Taking the ratio of two equations of the form (44) we find that $\delta_{M_1} \, \delta_R \log K_c$ is proportional to $\delta_{M_2} \, \delta_R \log K_c$.

$$\delta_{M_1} \, \delta_R \log K_c = \frac{\delta_{M_1} Y}{\delta_{M_2} Y} \cdot \delta_{M_2} \, \delta_R \log K_c \tag{46}$$

In order to test the proportionality we do a series of experiments with acids containing various groups R. A typical result is shown in Figure 8-2 for a number of substituted anilinium and ammonium ions in the system dioxane-water. It will be noted that the effects treated by equation 46 are

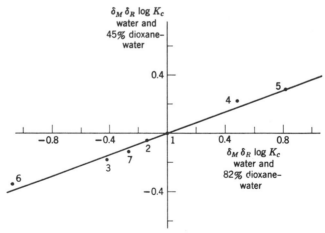

Figure 8-2. Representative plot of $\delta_M \delta_R \log K_c$ according to eq. 46. 1. aniline (the reference substance), 2. p-toluidine, 3. p-anisidine, 4. N-methylaniline, 5. N-dimethylaniline, 6. ammonia, 7. trimethyl-amine. Data from Reference 59. The slope is equal to $\delta_{M_1} Y_0 / \delta_{M_2} Y_0$.

quite large in this system, ranging over two units of log K_c, in spite of the rather limited variation of structure.

The Y function for use with neutral bases RB is denoted by Y_0, and those for use with charged bases by Y_- and Y_+, and so on. Table 8-11 gives Y_0 values for the ionization of ammonium and anilinium ions in methanol-water, ethanol-water, and dioxane-water mixtures. It has been noted[59] that these Y_0 values fit an empirical equation 47, in which x_{H_2O} is the weight

Table 8-11. Values of Y_0 and Y_- for Partly Aqueous Organic Solvents, 25° [57,58,59]

| WT % ORGANIC COMPONENT | Y_0 | | | | Y_- |
	DIOXANE-WATER	ETHANOL-WATER	METHANOL-WATER	$-(1 - x_{H_2O})^2$	ETHANOL-WATER
0	0.000	0.000	0.000	0.000	0.000
20	−0.052	−0.057	−0.054	−0.040	0.349
45	−0.230	−0.220	−0.225	−0.202	0.750
70	−0.522	−0.445	−0.494	−0.490	0.940
82	−0.635	−0.605	−0.660	−0.672	0.969
95	—	−0.882	—	−0.902	—
100	—	−1.000	−1.000	−1.000	1.000

fraction of water.

$$Y_0 = -(1 - x_{H_2O})^2 \tag{47}$$

After the establishment of a scale of Y values, values of $\delta_R m$ are obtained from the slopes of plots of equation 44. Like the w values of the Bunnett equation, $\delta_R m$ values depend on the nature of the nonaqueous solvent component,[59] and probably for the same reasons. Again, the fact that the empirical values of Y_0 are some function of the amount of water present suggests that interaction with some physical quantity associated with water is the major solvation mechanism.

The Y_- scale for the ionization of carboxylic acids has been established only for ethanol-water mixtures.[57] These Y_- values are given in Table 8-11 and are represented with fair accuracy by equation 48.

$$Y_- = 1 - x_{H_2O}^2 \tag{48}$$

Just as the Bunnett equation can be used to learn something about the participation of water in acid-catalyzed reactions in strong acid-water mixtures, so the mY equation can be used for the same purpose in aqueous-organic solvent systems.[60,61] One method is to compare the medium effect

[60] B. Gutbezahl and E. Grunwald, *J. Am. Chem. Soc.*, **75**, 565 (1953).

[61] (a) H. Kwart and L. B. Weisfeld, *ibid.*, **80**, 4670 (1958). (b) H. Kwart and A. L. Goodman, *ibid.*, **82**, 1947 (1960).

on the pseudo-first-order rate constant k_ψ (defined in equation 32) with the medium effect on an indicator ratio, c_I/c_{IH^+}.[61] A typical solvent system is $0.1M$ $HClO_4$ in a series of methanol-water mixtures, and a typical indicator is p-nitroaniline. For reactions whose mechanism is the *unimolecular* reaction of a protonated substrate, the rate constant should be correlated by equation 49, the derivation of which involves the same assumptions as those used in the derivation of equations 38 and 39.

$$\delta_M \log k_\psi + \delta_M \log \frac{c_I}{c_{IH^+}} = \mathrm{m}_{S,I} \cdot \delta_M Y_0 \qquad (49)$$

The parameter $\mathrm{m}_{S,I}$ is characteristic of the nature of the substrate and of the indicator. Except for this dependence on the nature of the indicator, $\mathrm{m}_{S,I}$ is analogous to the parameter w of the Bunnett equation.

Reactions that are believed, on other grounds, to go by the postulated mechanism have been found to give straight-line plots according to equation 49. An example is shown in Figure 8-3. Other reactions whose mechanisms are believed to involve the reaction of a protonated substrate with a water molecule, for example, the acid-catalyzed solvolysis of phenyl acetate, give curved plots, as in Figure 8-4.[61] This procedure seems to give enlightening results for reactions whose plots are either accurately linear

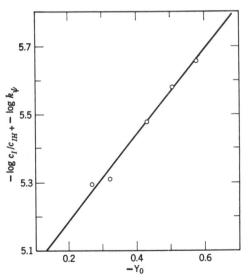

Figure 8-3. Kinetic plot for the hydration of 3-p-menthene in ethanol-water mixtures, 44.15° (Reference 61a).

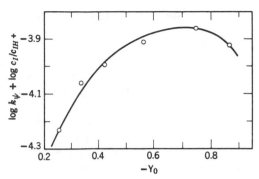

Figure 8-4. Kinetic plot for the solvolysis of phenyl acetate in ethanol-water mixtures, 25°. [Data taken from W. A. Waters, *J. Chem. Soc.*, 1014 (1936).]

or markedly curved, but it would be dangerous to give a mechanistic interpretation to plots with only a slight curvature.

An Extrathermodynamic Acidity Scale for the System Ethanol-Water

So far we have considered acidity functions such as H_0 that depend not only on the acidity of the medium, as measured by the proton activity, but also on the charge type and perhaps the nature of the proton acceptor. It would obviously be more satisfying to express the acidity solely as a function of the proton activity of the medium. Unfortunately the proton activity cannot be measured directly because the proton is a charged particle. If we imagine an experiment in which a quantity of protons is added to an isolated phase, the free energy of the phase will change not only because of the addition of the protons in their role as a chemical component but also because of the attendant addition of electrical charge. The total free energy change is the sum of an electrical and of a chemical free energy term, and of the two it is the chemical free energy term that we would like to know. Although the total free energy change can in principle be measured, there is no way in which the electrical and chemical terms can be obtained separately from such an experiment. The blame must be placed on the electrical term, which is equal to the integral

$$\int (\text{potential}) \cdot d\,(\text{charge}).$$

This integral cannot be evaluated absolutely because there is no way of knowing the absolute value of the electrical potential, and this is because there is no way of measuring absolutely the net electrical charge of an isolated phase.

Whereas the electrical part of the free energy change depends on the unknown initial value of the electrical charge of the phase, the chemical part is independent of this value for all systems that we are likely to encounter.[62] The reason is that the number of charged particles needed to produce even a very high electrical potential is so small as not to alter significantly the chemical composition of the system. The indeterminacy of the absolute value of the electrical potential therefore presents no problem if the chemical potential of the protons can be evaluated by a suitable extra-thermodynamic method. In this section we shall describe one such method, based on the mY equation for acid dissociation constants.

In any general discussion of acidity we must distinguish between the *formal* hydrogen ion species, which includes all protons whether solvated or not, and the unsolvated protons. Since we have been using the symbol H^+ to denote the formal species, we shall use a different symbol, p, to denote the unsolvated protons.

The acidity of a solution is measured by the proton activity, which in turn is related to the partial molar free energy of the protons by equation 50.

$$F_p = F_p{}^\circ + RT \ln a_p \tag{50}$$

In order that the proton activity of all solutions be expressed on the same scale, the standard term, $F_p{}^\circ$, must be independent of the solvent. It is convenient to choose the reference state for a_p in such a way that $F_p{}^\circ$ vanishes, as in equation 50a.

$$F_p = RT \ln a_p \qquad (F_p{}^\circ = 0 \text{ by appropriate choice of reference state}) \tag{50a}$$

It is obvious from (50a) that the acidity of a solution is known when F_p is known.

Our next task is to relate F_p to the partial molar free energy of the formal hydrogen ion species. The relationship is simple: the two quantities are equal.

$$F_p = F_{H^+} \tag{51}$$

Equation 51 follows immediately from the theorem that the partial molar free energy of a formal species is equal to that of a subspecies consisting of the unsolvated monomer, which in this case is simply the proton. (See Chapter 2, pp. 33–34.) Thus the relationship between F_p and the hydrogen ion concentration is given for dilute solutions by equation 52.

$$F_p = F_{H^+} = F^\circ_{H^+} + RT \ln c_{H^+} \tag{52}$$

[62] J. N. Brønsted, *Z. Physik. Chem.*, **A143**, 301 (1929).

The quantity c_{H^+} can be measured by a variety of methods,[63,64] so that the acidity is known if $F^\circ_{H^+}$ is known.

In any formal thermodynamic treatment, $F^\circ_{H^+}$ is simply the standard partial molar free energy of the hydrogen ion and is completely analogous to the formal quantity $\bar{F}^\circ_{C^+}$ for any other cation. However, when we try to interpret numerical values for $F^\circ_{H^+}$, and particularly for the solvent dependence of $F^\circ_{H^+}$, we must remember that the hydrogen ion differs from most other cations in that there is an extra large discrepancy between the species as represented by the formula and the species as it actually exists in solution. For a "normal" ion, such as potassium ion, the species existing in solution is best described as the solvate of the formal species, and $\delta_M \bar{F}^\circ_{C^+}$ depends partly on electrostatic interactions of long range whose magnitude is a function of the macroscopic dielectric constant and partly on the formation of relatively weak ion-solvent "bonds" at short range. For hydrogen ion, the species actually existing in solution is best described as the solvated lyonium ion, and $F^\circ_{H^+}$ therefore depends also on the strength of the covalent bond between the proton and the solvent molecule, that is, on the solvent basicity.[65]

To evaluate the medium effect on $F^\circ_{H^+}$, we begin with the medium effect on the standard free energy change for the ionization of a carboxylic acid, RCOOH.

$$\delta_M \Delta F^\circ_{RCOOH} = -2.303RT \delta_M \log K_{c,\,RCOOH}$$
$$= \delta_M F^\circ_{H^+} + \delta_M F^\circ_{RCOO^-} - \delta_M F^\circ_{RCOOH} \qquad (52)$$

By expressing $\bar{F}^\circ_{RCOOH}$ and $\bar{F}^\circ_{RCOO^-}$ in the terminology of equation 21, equation 52 becomes 53.

$$-2.303RT \delta_M \log K_{c,\,RCOOH}$$
$$= \delta_M F^\circ_{H^+} + \delta_M(l_{M,\,COO^-} - l_{M,\,COOH})$$
$$+ \delta_M(ll_{R,\,COO^-,M} - ll_{R,\,COOH,\,M})$$
$$+ \delta_M(l'_{M,\,COO^-} - l'_{M,\,COOH})$$
$$+ \delta_M(l''_{M,\,COO^-} - l''_{M,\,COOH}) + \cdots \qquad (53)$$

Equation 53 includes terms l', l'', ... to allow for additional interaction mechanisms. However, the success of the mY_- equation proves empirically that the second-order interaction terms are significant in only one interaction mechanism with the solvent, this of course being the mechanism in which the variable characteristic of the solvent is proportional to the parameter Y_-. Since all first-order interaction terms appearing in equation

[63] R. G. Bates, *Chem. Revs.*, **42**, 1 (1948).
[64] W. M. Clark, *The Determination of Hydrogen Ions*, Williams and Wilkins Company, Baltimore, 1920.
[65] J. N. Brønsted, *Chem. Revs.*, **16**, 287 (1935).

53 are independent of the nature of R, we may gather up all of the R-dependence in a single term proportional to Y_-, as in equation 54.

$$-2.303RT\,\delta_M \log K_{c,\,RCOOH} = \delta_M \bar{F}^\circ_{H^+} + 2.303RT\,m_{RCOOH}Y_-$$
$$+ \delta_M(\text{function of } COOH) \quad (54)$$

The physical significance of the term $2.303RT\,m_{RCOOH}Y_-$ is complex. The term accounts for the entire second-order interaction and for the part of the first-order interaction that is proportional to Y_-. The remainder of the first-order interaction is represented by the term $\delta_M(\text{function of } COOH)$.

Table 8-12. *Values of $\delta_M \bar{F}^\circ_{H^+}$ and Fit of Equations 56 and 57 in the System Ethanol-Water at 25° [60]*

(Reference solvent, water; $m_{CH_3COOH} = 1.035$; $m_{\varphi NH_3^+} = 3.614$)

			$\delta_M \log K_{c,CH_3COOH}$		$\delta_M \log K_{c,\varphi NH_3^+}$	
WT % ETHANOL	$\delta_M \bar{F}^\circ_{H^+}$ (kcal/mole)	$\delta_M \log f_H{}^a$	(obs)	(calc, eq. 56)	(obs)	(calc, eq. 57)
0.0	0.000	0.000	0.00	0.00	0.00	0.00
20.0	0.011	0.008	−0.37	−0.37	0.22	0.20
35.0	0.057	0.042	−0.67	−0.66	0.48	0.45
50.0	0.343	0.251	−1.08	−1.10	0.71	0.71
65.0	0.739	0.542	−1.53	−1.50	0.84	0.83
80.0	1.57	1.152	−2.11	−2.15	0.89	0.91
100.0	6.42	4.71	−5.6	−5.74	−1.06	−1.09

^{a} $\delta_M \log f_H = \delta_M \bar{F}^\circ_{H^+}/2.303RT$.

In a completely analogous way we may represent the medium effect on the dissociation of acids like anilinium ion, RBH^+, by equation 55.

$$-2.303RT\,\delta_M \log K_{c,\,RBH^+} = \delta_M \bar{F}^\circ_{H^+} + 2.303RT\,m_{RBH^+}Y_0$$
$$+ \delta_M(\text{function of } BH^+) \quad (55)$$

The quantity $\delta_M \bar{F}^\circ_{H^+}$ can be evaluated on the basis of equations 54 and 55 only in the special case that the functions of $COOH$ and BH^+ are negligibly small. This means that the first-order interaction terms must be closely proportional to the second-order interaction terms. Although there is no logical necessity that this should ever happen, there is also no reason why it could not happen, and the Bunnett equation and the mY equation for solvolysis rates (next section) provide actual examples where it does happen. As a matter of fact, the dissociation constants for acids of the types $RCOOH$ and RBH^+ in ethanol-water mixtures can be represented within their experimental errors if the functions of $COOH$ and BH^+ in equations 54 and 55 are ignored. The fit of the experimental data for acetic acid and for anilinium ion to the empirical equations 56 and 57 is shown in Table 8-12.[60]

$$\delta_M \log K_{c,CH_3COOH} = -\delta_M \log f_H - m_{CH_3COOH}Y_- \quad (56)$$
$$\delta_M \log K_{c,\varphi NH_3^+} = -\delta_M \log f_H - m_{\varphi NH_3^+} \cdot Y_0 \quad (57)$$

The quantity $\delta_M \log f_H$ is an empirical parameter, adjusted so as to obtain best fit to the data.[60] However, on comparing equations 56 and 57 with 54 and 55, we see that $\delta_M \log f_H$ is very probably equal to $\delta_M \bar{F}^\circ_{H^+}/2.303RT$, for it is most unlikely that the function of $COOH$ in (54) should be nearly equal to the function of BH^+ in (55) unless both functions are small. The resulting values of $\delta_M \bar{F}^\circ_{H^+}$ are listed in the second column of Table 8-12.

Values of $\delta_M \bar{F}^\circ_{H^+}$ obtained in this way have been compared with values obtained by a completely independent approach. Using two extrathermodynamic methods based ultimately on the Born equation, Izmailov[66]

Table 8-13. Values of $\delta_M \bar{F}^\circ_{H^+}$ Obtained by Extrathermodynamic Methods Based on the Born Equation, 25° [66]

(Reference solvent, water)

SOLVENT	$\delta_M \bar{F}^\circ_{H^+}$	SOLVENT	$\delta_M \bar{F}^\circ_{H^+}$
H_2O	0.0	$n\text{-}C_4H_9OH$	6.6
NH_3	−23.0	$i\text{-}C_4H_9OH$	6.3
CH_3OH	4.6	$i\text{-}C_5H_{11}OH$	6.2
C_2H_5OH	5.9	$C_6H_5CH_2OH$	4.3
$n\text{-}C_3H_7OH$	5.9	$HCOOH$	12.0

concludes that $\delta_M \bar{F}^\circ_{H^+}$ for the transfer of hydrogen ion from water to ethanol is about 5.9 kcal, in good agreement with the value of 6.4 kcal obtained by the preceding approach. Results obtained for other hydrogen-bonding solvents are listed in Table 8-13.

The mY Equation for Solvolysis Reaction Rates

The unimolecular, or s_N1, or limiting mechanism for solvolysis is equation 58.

$$R - X \xrightarrow[k_1]{\text{slow}} \overset{+\delta}{R_+} - \overset{-\delta}{X_+} \longrightarrow R^+ + X^-$$

(transition state)   (carbonium ion or ion pair intermediate)

$$R^+ + X^- + \text{solvent} \xrightarrow{\text{fast}} \text{solvolysis product} + HX$$

(58)

The medium effect on the rate constant k_1 can be shown, on the basis of the second approximation, to be factorable into separate functions of the medium and of the structure of reagent and transition state. The derivation

[66] N. A. Izmailov, *Proc. Acad. Sci. USSR, Phys. Chem. Sect.* (Eng. Transl.), **126**, 509 (1959); N. A. Izmailov, *Proc. Acad. Sci. USSR, Chem. Sect.* (Engl. Transl.), **127**, 501 (1959).

Table 8-14. Values of Y Used in the Correlation of Solvolysis Rates[68]

SOLVENT SYSTEM, A-B	VOL % COMPONENT A[a]	Y
Ethanol-water	100	−2.033
	95	−1.287
	90	−0.747
	80	(0.000)
	60	1.124
	40	2.196
	20	3.051
H_2O	—	3.493
Methanol-water	100	−1.090
	80	0.381
	60	1.492
	40	2.391
	20	3.025
CH_3COOH—$HCOOH$	100	−1.65
	90	−0.929
	75	−0.175
	50	0.757
	25	1.466
$HCOOH$-water	100	2.054
	50	2.644
	25	3.100
CH_3COOH-water	2.00M H_2O	−0.863
	8.00M H_2O	0.193
	16.00M H_2O	0.984
	50	1.938
	25	2.843
Dioxane-water	90	−2.030
	80	−0.833
	70	0.013
	60	0.715
	50	1.361
	40	1.945
	20	2.877
Acetone-water	95.2	−2.76
	90	−1.856
	80	−0.673
	70	0.130
	60	0.796
	50	1.398
	40	1.981
	20	2.913
Dioxane-HCOOH	80	−2.296
	60	−0.677
	40	0.402
	20	1.291
Formamide-water	100	0.604
	80	1.383
n-C_3F_7COOH	—	1.7
i-Propyl alcohol	—	−2.73
t-Butyl alcohol	—	−3.26

[a] x vol % means that x volumes of component A have been added to 100-x volumes of component B, both volumes at 25° before mixing.

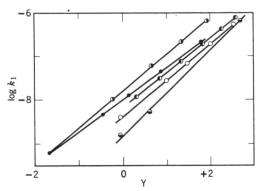

Figure 8-5. Plot of log k_1 versus Y for the solvolysis of neophyl chloride at 50.0° in: EtOH-H$_2$O, ○; MeOH-H$_2$O, ◑; dioxane-H$_2$O, ◒; AcOH-H$_2$O, ●; AcOH-HCOOH, ◐ (Reference 69c).

of this equation is completely analogous to that of equation 37 for the unimolecular decomposition of a protonated substrate.

$$\delta_M(F^\circ_{R\ddagger X\ddagger} - F^\circ_{RX}) = \delta_M \|_M[\lambda(I_{X\ddagger} - I_X) + \lambda(I_{R\ddagger} - I_R)$$
$$+ (\|_{X\ddagger}\|_{R\ddagger} - \|_X\|_R)] \quad (59)$$

Denoting the structure-dependent factor by $-2.303RT \, m_{RX}$ and the medium-dependent factor by $\delta_M Y$ converts equation 59 to 60.

$$-\frac{1}{2.303RT} \delta_M(F^\circ_{R\ddagger X\ddagger} - F^\circ_{RX}) = \delta_M \log k_1 = m_{RX} \cdot \delta_M Y \quad (60)$$

For solvolysis reactions, $\delta_M Y$ is evaluated by arbitrarily setting m_{RX} equal to unity for the solvolysis of t-butyl chloride. Values of $\delta_M Y$ obtained in this way are converted to a scale of Y values by choosing Y = 0 for an ethanol-water mixture prepared by adding 20 cc of water to 80 cc of ethanol at 25°.[67,68]

Table 8-14 gives values of Y for a wide variety of solvents. Figure 8-5 is a typical plot of log k_1 versus Y in several two-component solvent systems. The figure shows dispersion of log k_1 into several accurately straight lines corresponding to the several two-component solvent systems. Table 8-15 gives some values of m_{RX}.[67,69]

[67] E. Grunwald and S. Winstein, *J. Am. Chem. Soc.*, **70**, 846 (1948).

[68] A. H. Fainberg and S. Winstein, *ibid.*, **78**, 2770 (1956).

[69] (a) A. H. Fainberg and S. Winstein, *J. Am. Chem. Soc.*, **79**, 1597 (1957). (b) *ibid.*, **79**, 1602 (1957). (c) *ibid.*, **79**, 1608 (1957). (d) S. Winstein, A. H. Fainberg, and E. Grunwald, *ibid.*, **79**, 4146 (1957).

The dispersion of the data into separate lines for different solvent systems is analogous to the dispersion of w values and of $\delta_R m$ values in the relationships discussed previously and has the same significance with regard to the existence of multiple solvation mechanisms. To date the Y values for solvolysis have not been shown to bear a simple functional relationship to any single physical property of the solvent. However, the term "ionizing

Table 8-15. Some Values of the Parameter m_{RX} for Solvolysis

SUBSTRATE	SOLVENT SYSTEM	m_{RX}
α-Phenylethyl	Ethanol-water, 25°	0.966
chloride	50°	0.856
	Methanol-water, 25°	0.912
	Acetic acid-water, 25°	1.136
	Acetic acid-formic acid, 25°	1.194
	Dioxane-water, 25°	1.136
	Formic acid-water, 25°	1.073
t-Butyl bromide	Ethanol-water, 25°	0.941
	Methanol-water, 25°	0.947
	Acetic acid-formic acid, 25°	0.946
	Acetic acid-water, 25°	1.067
	Dioxane-water, 25°	0.931
	Acetone-water, 25°	0.913
Neophyl chloride	Ethanol-water, 50°	0.833
	Methanol-water, 50°	0.790
	Acetic acid-formic acid, 50°	0.837
	Acetic acid-water, 50°	0.733
	Dioxane-water, 50°	0.961

power of the medium" is qualitatively descriptive of Y. The appropriateness of the description can be seen from the fact that Y values are well correlated with the solvent parameter Z. The parameter Z is a measure of the ability of the medium to stabilize ion pairs relative to less polar electronically excited states produced by charge transfer.[70] Z will be discussed more fully in a later section.

PARAMETRIC EQUATIONS MAKING EXPLICIT ALLOWANCE FOR MULTIPLE SOLVATION MECHANISMS

If solvation of a solute RX is assumed to involve two solvation mechanisms, then application of the first approximation to $\Delta F^{\ddagger}$ for solvolytic reactions

[70] E. M. Kosower, *J. Am. Chem. Soc.*, **80**, 3253 (1958).

gives equation 61, in which the second mechanism of interaction with the solvent is indicated by primed subscripts.

$$F^\circ_{R\ddagger X\ddagger} - F^\circ_{RX} = F_{X\ddagger} - F_X + F_{R\ddagger} - F_R + I_{R\ddagger,X\ddagger} - I_{R,X} + I_{R\ddagger,M}$$
$$+ I_{X\ddagger,M} - I_{R,M} - I_{X,M} + I'_{R\ddagger,X'\ddagger} - I_{R',X'} + I_{R'\ddagger,M'}$$
$$+ I_{X'\ddagger,M'} - I_{R',M'} - I_{X',M'} \tag{61}$$

Assuming that the interaction terms are separable, we obtain equation 62.

$$\delta_M(F^\circ_{R\ddagger X\ddagger} - F^\circ_{RX}) = \delta_M I_M[(I_{R\ddagger} - I_R) + (I_{X\ddagger} - I_X)]$$
$$+ \delta_M I_{M'}[(I_{R'\ddagger} - I_{R'}) + (I_{X'\ddagger} - I_{X'})] \tag{62}$$

Replacing the free energy increments with the logarithms of rate constants, we see that there should exist a relationship having the form (63).

$$\delta_M \log k = cd + c'd' \tag{63}$$

In equation 63, c and c' are characteristic only of RX, and d and d' are characteristic only of the solvent. Swain, Mosely, and Bown,[71] working on the hypothesis that a solvent should exert both a nucleophilic push and an electrophilic pull, fitted certain existing data to an equation of this form. The fact that some data can be fitted successfully to equation 63 indicates that there are indeed two major mechanisms for interaction with the solvent. Of course this fact taken by itself is not sufficient to justify the description of the two interaction mechanisms as nucleophilic and electrophilic solvent participation. In order to construct scales of absolute values of the parameters it is necessary, just as in the case of the parameters m and Y, to assign arbitrary values to certain solvents and structures. Table 8-16 lists the results obtained by Swain, Mosely, and Bown. In a *very* rough way the parameters d and d' parallel our expectations for the relative nucleophilic and electrophilic properties of the solvents.

In equation 62 it is quite possible that the terms corresponding to one of the two mechanisms for interaction with the medium will cancel out in the comparison of the ground and transition states. For example, the quantity

$$\delta_M(I_{R\ddagger,M} + I_{X\ddagger,M} - I_{R,M} - I_{X,M})$$

might vanish even though the terms $\delta_M(I_{R,M} + I_{X,M})$ and $\delta_M(I_{R\ddagger,M} + I_{X\ddagger,M})$ for the reagent and transition state do not vanish separately. In such a situation it is fruitful to discuss medium effects on reaction rates in terms of a single solvation interaction mechanism even though such a procedure would not do for the solvent stabilization of the reagent alone. For example, medium effects on the solubilities of gaseous alkyl halides are

[71] C. G. Swain, R. B. Mosely, and D. E. Bown, *J. Am. Chem., Soc.*, **77**, 3731 (1955).

known to be very large, too large to be accounted for solely by the interaction of their permanent electric dipole moments with the solvent continuum.[72] Nevertheless, as is well known,[73] there is a rough parallelism between their solvolysis rates and the dielectric properties of the solvent. The reason for this is the partial cancellation of the nonelectrostatic factors

Table 8-16. Parameters in the Correlation of Solvolysis Rates by Means of Equation 63 [71]

(The terms $c_1 d_1$ and $c_2 d_2$ are thought to measure nucleophilic push and electrophilic pull by solvent, respectively.)

COMPOUND	c_1	c_2	COMPOUND	c_1	c_2
C_6H_5COCl	0.81	0.52	trans-2-Methoxy-cyclohexyl OBs	0.57	0.57
C_6H_5COF	1.36	0.66			
$CH_3Br^{a,b}$	0.80	0.27	$C_6H_5CHClCH_3$	1.47	1.75
C_2H_5Br	0.80	0.36	$(C_6H_5)_2CHCl$	1.24	1.25
C_2H_5OTs	0.65	0.24	$t\text{-}C_4H_9Cl^b$	(1.00)	(1.00)
$i\text{-}C_3H_7Br$	0.90	0.58	$(C_6H_5)_3CSCN$	0.19	0.28

SOLVENT	d_1	d_2	AQUEOUS SOLVENT	WATER, VOL %	d_1	d_2
CH_3OH	-0.05	-0.73	CH_3OH	30.5	-0.06	1.32
C_2H_5OH	-0.53	-1.03	C_2H_5OH	20^b	(0.00)	(0.00)
				40	-0.22	1.34
H_2O	-0.44	4.01		60	-0.26	2.13
CH_3COOH	-4.82	3.12	$(CH_3)_2CO$	10	-0.53	-1.52
$HCOOH$	-4.40	6.53		20	-0.45	-0.68
				30	-0.09	-0.75
				50	-0.25	0.97

[a] c_1 and c_2 are adjusted so that $c_1/c_2 = 3.00$ for CH_3Br.
[b] Reference compound or solvent.

in $\delta_M \Delta F^{\ddagger}$. The reason that the correlation of the rates with the dielectric properties of the solvent is only rough is that the cancellation of the nonelectrostatic factors is usually incomplete. As will be seen in Chapter 9, the complicating effect of a second interaction mechanism can sometimes also be reduced by the approximate cancellation of its effects on the enthalpy and entropy of solvation.

[72] (a) R. A. Clement and M. R. Rice, *J. Am. Chem. Soc.*, **81**, 326 (1959). (b) E. Grunwald and S. Winstein, *ibid.*, **70**, 846 (1948).
[73] C. K. Ingold, *Structure and Mechanism in Organic Chemistry*, Cornell University Press, Ithaca, New York, 1953, pp. 345–350.

If the analysis in terms of two separate interaction mechanisms for the solvent is pushed to the second approximation, second-order interaction terms must be added to equation 61. The full equation will contain interaction terms of three types. The first type consists of first-order interaction terms of the same form as used previously, but these are no longer separable into two factors. The second type consists of the second-order interaction terms, such as $\mathrm{II}_{R,X,M}$ and $\mathrm{II}_{R',X',M'}$, which are separable into three factors. The third type consists of cross terms that express the effect of one interaction mechanism on the other interaction mechanism. These have a form such as $\mathrm{II}_{R,X,M'}$ or $\mathrm{II}_{R,X',M}$ and are all separable into three factors. Because of the obvious complexity of the second approximation, no corresponding parametric equations have yet been tested.

For a reaction consisting of a pre-equilibrium with equilibrium constant K followed by a rate-determining step with rate constant k, the solvent variable which determines the medium effect on K may not be the same as that which determines the effect on k. If $k_{\mathrm{obs}} = Kk$, the observed medium effect will be a composite one, but the possibility of interaction between the two medium effects does not arise. This becomes obvious when the medium effect is written as in equation 64.

$$\delta_M \log k_{\mathrm{obs}} = \delta_M \log K + \delta_M \log k \qquad (64)$$

An example of a reaction for which the medium effect should be divided into two independent parts is the Baeyer-Villiger reaction.[74] The effect of

$$\underset{\text{ArCR}}{\overset{\text{O}}{\overset{\|}{\text{Ar}\!-\!\text{C}\!-\!\text{R}}}} + \underset{\text{CF}_3\text{COOH}}{\overset{\text{O}}{\overset{\|}{\text{CF}_3\text{COOH}}}} \underset{\rightleftharpoons}{\overset{K}{}} \underset{\text{OH}}{\overset{\overset{\text{O}}{\overset{\|}{\text{OOCCF}_3}}}{\overset{|}{\underset{|}{\text{Ar}\!-\!\text{C}\!-\!\text{R}}}}}$$

$$\underset{\text{OH}}{\overset{\overset{\text{O}}{\overset{\|}{\text{OOCCF}_3}}}{\overset{|}{\underset{|}{\text{Ar}\!-\!\text{C}\!-\!\text{R}}}}} \xrightarrow[\text{acid}]{k} \underset{}{\overset{\text{O}}{\overset{\|}{\text{Ar}\!-\!\text{O}\!-\!\text{C}\!-\!\text{R}}}} + \text{CF}_3\text{COOH}$$

the medium on the first step of the reaction should resemble that on a typical carbonyl addition equilibrium. The effect of the medium on the second step, however, might be quite different. This is probably the reason for the poor correlation of the rates of the Baeyer-Villiger reaction for a series of ketones in ethylene chloride and in acetonitrile.[74]

---

[74] M. F. Hawthorne and W. D. Emmons, *J. Am. Chem. Soc.*, **80**, 6398 (1958).

RELATIONSHIP OF THE MEDIUM EFFECT ON THE RATE OF A REACTION TO THAT ON THE EQUILIBRIUM CONSTANT

If the solvation of the reagent, transition state, and product of a reaction can be described by a single interaction mechanism, then the medium effect on the rate constant, $\delta_M \log k$, bears a simple relationship to the medium effect on the equilibrium constant, $\delta_M \log K$. For definiteness, consider the unimolecular rearrangement of a substrate, as represented by equation 65.

$$S \rightleftharpoons T^{\ddagger} \rightleftharpoons P \tag{65}$$

For the present purpose nothing is gained by subdividing the molecules of S, $T^{\ddagger}$, and P into smaller zones, and we shall therefore represent the standard partial molar free energies by equations of the form 66.

$$\bar{F}_S{}^{\circ} = \mathsf{F}_S + \mathsf{I}_{S,M} \tag{66}$$

In first approximation, the interaction term $\mathsf{I}_{S,M}$ is factorable. Hence we obtain equations 67 and 68.

$$\delta_M(\bar{F}_{T^{\ddagger}}^{\circ} - \bar{F}_S{}^{\circ}) = -RT\,\delta_M \log k = (\delta_M \mathsf{I}_M)[\mathsf{I}_{T^{\ddagger}} - \mathsf{I}_S] \tag{67}$$

$$\delta_M(\bar{F}_P^{\circ} - \bar{F}_S{}^{\circ}) = -RT\,\delta_M \log K = (\delta_M \mathsf{I}_M)[\mathsf{I}_P - \mathsf{I}_S] \tag{68}$$

Since $\delta_M \log k$ and $\delta_M \log K$ are both proportional to $\delta_M \mathsf{I}_M$, they are proportional to each other.

Actual examples in which rate and equilibrium constants for a reaction have been measured in a series of solvents are rare, so that we cannot say how probable it is that our assumptions will be valid. In the dissociation of hexaphenylethane, a roughly proportional relationship is indeed observed, with $\delta_M \log k \approx 0.2\,\delta_M \log K$.[75] Since the equilibrium constants are much more solvent-dependent than the rate constants, the transition state for dissociation seems to resemble the hexaphenylethane rather than the pair of radicals. This conclusion is also consistent with purely kinetic evidence. If the transition state resembled a pair of nearly planar radicals, it should be stabilizable by interaction with π-complexing agents. However, the acceleration of the rate of dissociation by added 1,3,5-trinitrobenzene is only very slight.[76]

The quantities $\delta_M \log k$ and $\delta_M \log K$ need not be restricted to the same reaction. Linear relationships among the medium effects are to be expected in first approximation for any two processes in which the interaction with

[75] K. Ziegler, A. Seib, F. Knoevenagel, P. Herte, and F. Andreas, *Ann.*, **551**, 150 (1942); K. Ziegler and L. Ewald, *ibid.*, **473**, 163 (1929).

[76] J. E. Leffler and R. A. Hubbard, II, *J. Org. Chem.*, **19**, 1089 (1954).

the solvent proceeds by the same single mechanism. It is not even necessary that these processes be simple one-step reactions. Under appropriate conditions linear relationships are observed even for radical chain reactions. For example, the photochemical monochlorination of 2,3-dimethylbutane has been studied in a series of solvents.[77] In *aliphatic* solvents the product composition is nearly independent of the solvent, about 60 % of the product

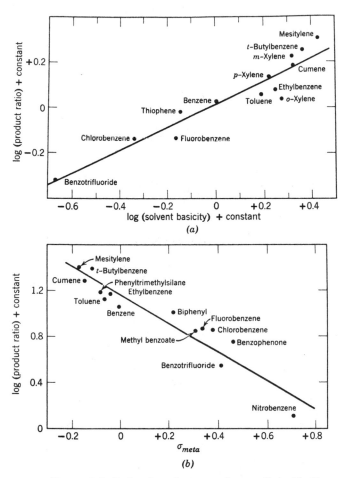

Figure 8-6. Ratio of tertiary to primary alkyl chloride produced in the photochlorination of 2,3-dimethylbutane at 55° compared with (a) solvent basicity, (b) σ_{meta} (from Reference 77).

[77] G. A. Russell, *J. Am. Chem. Soc.*, **80**, 4987 (1958).

being the primary monochloride. In *aromatic* solvents, on the other hand,

$$
\underset{CH_3}{\overset{CH_3}{H-C-C-H}} \xrightarrow[hv]{Cl_2} \underset{CH_3}{\overset{ClCH_2 \quad CH_3}{H-C-C-H}} + \underset{CH_3}{\overset{CH_3 \quad CH_3}{Cl-C-C-H}}
$$

(primary monochloride)   (tertiary monochloride)

the major product is the tertiary monochloride. Moreover, the ratio of tertiary to primary product varies greatly with the aromatic solvent, the ratio being greatest in just those solvents that are best able to form π-complexes with chain carriers such as the chlorine atom. In fact, as shown in Figure 8-6a, the logarithm of the product ratio is linearly related to the logarithm of the basicity of the solvent, the latter being measured by the equilibrium constant for reaction with hydrogen chloride. Another version of this relationship is the linear plot, Figure 8-6b, of the logarithm of the product ratio versus σ_{meta} for the substituent attached to the benzene ring of the aromatic solvent.[77]

The extrathermodynamic relationship between σ_{meta} and the solvent effect is one of a type that we have not previously discussed. It is possible to foresee the discovery of other relationships of the same type, in which $\delta_M \Delta F°$ is a function of $\delta_R \Delta F°$. The quantity $\delta_R \Delta F°$ might be the effect of various substituents Y on the reactivity of some compound such as m-$YC_6H_4N{=}NR$ in benzene solution. The quantity $\delta_M \Delta F°$ might be the effect of the same substituents on the reactivity of the unsubstituted reagent, $C_6H_5N{=}NR$, in a series of solvents, C_6H_5Y.

EXTRATHERMODYNAMIC RELATIONSHIPS FOR SALT SOLUTIONS

Medium effects brought about by the addition of a salt to a nominal solvent have been studied extensively.[78] For *nonpolar* nonelectrolytes in water the salt effects, $\delta_M \bar{F}°$, can be explained by a single interaction mechanism involving the internal pressure of the medium.[78,79] The addition of small ions such as Na^+Cl^- to pure water produces changes in the physical properties of the liquid that are in many ways analogous to those produced by applying a greater external pressure to pure water.[80] Thus there exists a value of the applied external pressure that causes pure water to have the same compressibility, specific heat capacity, viscosity, and temperature of

[78] F. A. Long and W. F. McDevit, *Chem. Reviews*, **51**, 119 (1952).
[79] W. F. McDevit and F. A. Long, *J. Am. Chem. Soc.*, **74**, 1773 (1952).
[80] G. Tammann, *Z. anorg. allgem. Chem.*, **158**, 25 (1926).

maximum density that the salt solution has at atmospheric pressure. The applied pressure (in excess of atmospheric pressure) needed to produce this effect is regarded as equal to the "internal pressure" of the salt solution minus that for pure water.

A plausible explanation of the ability of salts to have an effect like that of an applied pressure is that their ions exert a tension on the surrounding solvent molecules, drawing them closer together.

Any increase in internal pressure has an effect on solubility because it becomes more difficult to produce the cavity that is to be occupied by the

Table 8-17. Salts Arranged in Order of Decreasing Effectiveness at Salting-out Nonpolar Nonelectrolytes from Aqueous Solution

(Based on data in reference 79.)

Na$_2$SO$_4$, BaCl$_2$, NaOH, NaF, NaCl, KCl, NaBr, LiCl, RbCl, KBr, NaNO$_3$, NaClO$_4$, NH$_4$Cl, NaI, CsCl, HCl, NH$_4$Br, CsI, K benzoate, HClO$_4$, (CH$_3$)$_4$NBr

nonpolar solute molecule. The $\delta_M \bar{F}^\circ$ due to the presence of a given salt is proportional to the change in internal pressure, which is in turn proportional to the concentration of the salt.

$$\frac{1}{RT} \delta_M \bar{F}^\circ = k c_{salt} \qquad (69)$$

The proportionality constant k of equation 69 depends on the internal pressure at unit salt concentration, which is characteristic of the salt, and on the size of the cavity needed, which is characteristic of the solute. The constant k is therefore separable into independent factors corresponding to the nature of the salt and of the solute. This makes possible two extrathermodynamic relationships. If the nature of the salt is kept constant while its concentration is increased, $\delta_M \bar{F}^\circ$ is proportional to the change in internal pressure, which in turn is proportional to the salt concentration. Hence the $\delta_M \bar{F}^\circ$ quantities for any two nonpolar nonelectrolytes are proportional to each other. If the salt concentration is kept constant but the nature of the ions is changed, each salt will produce a characteristic change in internal pressure. Since the quantities $\delta_M \bar{F}^\circ$ are proportional to these changes in internal pressure, the order of increasing effects for a series of salts is independent of the nature of the nonpolar solute. Table 8-17 shows the "salting order" for a series of univalent salts as deduced from data for aromatic hydrocarbons.

Salt effects in solutions of nonpolar nonelectrolytes in aqueous-organic solvents seem to involve two major interaction mechanisms. One of these is again an internal pressure effect, the other seems to depend on the changes in the ratio of activities of the water and the organic solvent component. For example, NaCl added to a dioxane-water mixture lowers the activity of the water and raises that of the dioxane. The effect is thus analogous to that produced by adding more dioxane to the medium in the absence of salt. That is, this salt effect mechanism tends to increase the solubility of nonpolar solutes, whereas the effect on the internal pressure tends to reduce the solubility of nonpolar solutes. Since the increase in the activity of the dioxane solvent-component is analogous to the salting-out of dioxane from its aqueous solution, there is a moderately good correlation between the change in internal pressure in water and the change in the water/dioxane activity ratio in the water-dioxane solvent. As a result of this correlation, salts that are good agents for salting-out nonelectrolytes from aqueous solution are also good agents for increasing the dioxane activity in the mixed solvent. Hence the net effect of a salt that is a salting-out agent in water may be to salt-in in the dioxane-water solvent.[81]

For *polar* nonelectrolytes there are deviations from the salting order (Table 8-17) even for water as the solvent. This suggests that the interaction between salts and *polar* nonelectrolytes involves additional mechanisms other than that involving the internal pressure even when pure water is the nominal solvent. The pattern of the deviations suggests, further, that there may be two additional interaction mechanisms, one for substances having an acidic functional group and another for substances having a basic functional group.[78]

The salt effect on the standard partial molar free energy of *electrolytes* in solvents of high dielectric constant can be considered the sum of a colligative term, dependent on the charge type of the electrolyte and the ionic strength, and a specific term analogous to the salt effect on nonelectrolytes. However, the magnitudes of the specific noncolligative interactions of a given electrolyte with a series of salts do not fall even approximately in the "salting order" of Table 8-17.[82]

In summary, as the polarity of the solute is increased from nonpolar to polar to electrolytic, the relative importance of interaction mechanisms not involving the internal pressure increases. Hence salt effects on reaction rates or equilibria of the type that involve changes in polarity are difficult to predict, although the salt-effect sequence is sometimes the same as for nonpolar solutes.[83-85]

[81] E. Grunwald and A. F. Butler, *J. Am. Chem. Soc.*, **82**, 5647 (1960).

[82] See, for example, H. S. Harned and B. B. Owen, *The Physical Chemistry of Electrolytic Solutions*, Reinhold, 1943, ch. 14.

[83] F. A. Long, F. B. Dunkle, and W. F. McDevit, *J. Phys. Chem.*, **55**, 829 (1961).

[84] G. Akerlöf, *J. Am. Chem. Soc.*, **48**, 3046 (1926); R. A. Robinson, *Trans. Faraday Soc.*, **26**, 217 (1930).

[85] E. F. J. Duynstee, E. Grunwald, and M. L. Kaplan, *J. Am. Chem. Soc.*, **82**, 5654 (1960).

EXTRATHERMODYNAMIC RELATIONSHIPS INVOLVING MEDIUM EFFECTS ON PHYSICAL PROPERTIES

An easily obtainable experimental measure of the polarity of a solvent is the parameter Z, which is the transition energy, in kcal/mole, of the charge transfer absorption band of 1-ethyl-4-carbomethoxypyridinium iodide ion pair.[70] Solvent molecules are oriented about the ion pair in its ground

state in such a way as to minimize the electrostatic free energy of the dipole. Excitation of the ion pair produces a species the major component of whose dipole moment is in the plane of the aromatic ring rather than

Table 8-18. Z Values for Various Solvents[70]

SOLVENT	Z (kcal/mole)	SOLVENT	Z (kcal/mole)
HOH	94.6	$CHCl_3(0.13M\ C_2H_5OH)$	63.2
CH_3OH	83.6	CH_3COCH_3	65.7
CH_3CH_2OH	79.6	$HCO\cdot N(CH_3)_2$	68.5
$CH_3CH_2CH_2OH$	78.3	CH_3CN	71.3
$CH_3CH_2CH_2CH_2OH$	77.7	C_5H_5N	64.0
$(CH_3)_2CHOH$	76.3	$(CH_3)_3CCH_2CH(CH_3)_2$	60.1
$(CH_3)_3COH$	71.3	CH_3SOCH_3	71.1
$HOCH_2CH_2OH$	85.1	$HCO\cdot NH_2$	83.3
$HCF_2CF_2CH_2OH$	86.3	cyclo-$C_3H_5COCH_3$	65.4
$HCF_2(CF_2)_3CH_2OH$	84.8	CH_3COOH	79.2
		CH_2Cl_2	64.2

perpendicular to it. Since the solvent molecules do not have time to adjust to the new orientation of the dipole, the solvent stabilization of the excited state is much less than that of the ground state and may even be a destabilization. Hence solvent effects on the transition energy are a sensitive measure of the electrostatic part of the solvation of an ion pair or a dipole. Table 8-18 contains the Z values for a series of solvents. Some of these, for solvents in which 1-ethyl-4-carbomethoxypyridinium iodide is insufficiently soluble, have been measured by means of other ion pairs

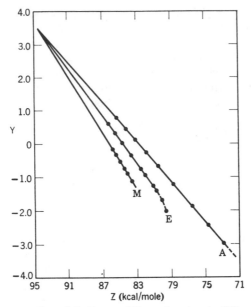

Figure 8-7. Z versus Y in methanol-water (M), ethanol-water (E), and acetone-water (A). The equations for the lines are

$$Y_M = 0.41632\, Z - 35.877,$$
$$Y_E = 0.35338\, Z - 29.946,$$
and
$$Y_A = 0.29887\, Z - 24.758$$

(Reference 70).

whose transition energies are linearly related to those of the standard ion pair. Z is also linearly related to the energies of the n to π^* transition of mesityl oxide and of methyl cyclopropyl ketone, although the point for water deviates.[86]

For two-component solvents in which both can be determined, Z is linearly related to Y, as shown in Figure 8-7.[70] That the correlation is not merely coincidental is demonstrated by the fact that the lines of Figure 8-7 for different aqueous organic solvent systems converge at a single point, corresponding to the Y value for pure water.

For nonhydroxylic solvents the rate of decomposition of p-methoxy-neophyl p-toluenesulfonate provides a measure of polarity analogous to the use of Y for hydroxylic solvents.[87] A plot of $\log k_1$ for this reaction versus Z gives a fairly good linear relationship.

[86] E. M. Kosower, *J. Am. Chem. Soc.*, **80**, 3261 (1958).
[87] S. G. Smith, A. H. Fainberg, and S. Winstein, *J. Am. Chem. Soc.*, **83**, 618 (1961).

$$(CH_3)_2C{=}CH{-}\langle\bigcirc\rangle{-}OCH_3 + HOTs$$

Z has also been observed to be a linear function of the activation energy for the isotopic exchange reaction $I^{*-} + CH_3I \rightarrow CH_3I^* + I^-$ in a series of hydroxylic solvents and acetone.[88] A similar plot of Z versus log k for this reaction is also linear, with the exception of the point for acetone.[88] Z is also linear in log k for the Menschutkin reaction of pyridine with ethyl iodide in a series of hydroxylic solvents, but the point for acetone is again found to deviate.[88] Apparently there is a contribution from the specific solvation of iodide ion by acetone to the entropy of activation of these reactions, which is not paralleled in its effect on Z. Corresponding to the dispersion of the Z versus Y plot for two-component solvents in Figure 8-7, a plot of Z versus Y for a series of single-component solvents is strongly curved but apparently monotonic.[88]

An important property of a solvent for hydroxylic compounds is its ability to act as a hydrogen-bond acceptor. This ability can be measured roughly by the shift, $\Delta\nu$, in the frequency of the OH or OD stretching band in the spectrum of a hydroxylic solute with respect to the value, ν_0, for the uncomplexed OH or OD.[89] Because of the resemblance between the partial proton transfer in a hydrogen bond and the complete proton transfer in an acid-base equilibrium, the shift $\Delta\nu/\nu_0$ in the OH (or OD) stretching frequency of a given hydroxyl compound in a series of solvents is linearly related to log K_B for the various solvents ionizing as bases in water. For CH_3OD the empirical relationship is

$$\log K_B + 14.0 = pK_A = -15.1 + 301\left(\frac{\Delta\nu}{\nu_0}\right) \qquad (70)$$

[88] E. M. Kosower, *J. Am. Chem. Soc.*, **80**, 3267 (1958).
[89] W. Gordy and S. C. Stanford, *J. Chem. Phys.*, **9**, 204 (1941); W. Gordy, *ibid.*, **9**, 215, 440 (1941).

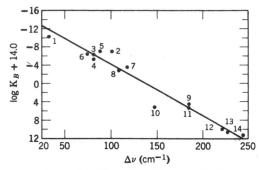

Figure 8-8. Plot of log K_B versus $\Delta\nu$ for the fundamental O-D stretching vibration of CH_3OD in a series of basic solvents. 1. nitrobenzene, 2. acetone, 3. acetophenone, 4. p-methylacetophenone, 5. benzaldehyde, 6. anisole, 7. diethyl ether, 8. p-dioxane, 9. aniline, 10. N-dimethylaniline, 11. pyridine, 12. tri-n-butylamine, 13. tri-n-propylamine, 14. piperidine.

The fit of the data for a series of ketones, nitro-compounds, and amines to equation 70 is shown in Figure 8-8.

The frequency shifts for HCl and D_2O as hydrogen-bond donors in the same series of solvents are also linearly related to each other and to log K_B.[89]

RELATIONSHIPS BETWEEN STRUCTURAL AND SOLVENT PARAMETERS

In the first approximation, structural and solvent effects are merely additive. In the second approximation there are interaction terms corresponding to the third derivatives, such as $\partial^3 F°/\partial M\ \partial R\ \partial X$, $\partial^3 F°/\partial M\ \partial R^2$, and so on. In order to apply the second approximation in a consistent manner, it is necessary to use a treatment in which *all* the third derivatives are retained. Hence when simultaneous structural and medium changes are made, an equation like the $\rho\sigma$ equation should contain a term in σ^2 as well as one in $\delta_{M}\rho$. However, it appears that the term in σ^2 is smaller than that in $\delta_{M}\rho$, and the Hammett equation customarily contains only a term in σ. Using this approximation, and applying the mY and $\rho\sigma$ equations to the ionization of aromatic acids, we can derive relationships between any two of the free energy quantities with the aid of the following diagram. In this diagram the solvents are characterized by their Y-values and the aromatic acids by the σ-values of their substituent.

Solvent Parameter	Structural Parameter	
	σ_0	σ
Y	$\log K_Y{}^\circ \xrightarrow{\delta_R \log K_Y} \log K_Y$	
	$\uparrow \delta_M \log K_{\sigma_0}$	$\uparrow \delta_M \log K_\sigma$
Y_0	$\log K_{Y_0}^\circ \xrightarrow{\delta_R \log K_{Y_0}} \log K_{Y_0}$	

As can be seen from the diagram, the free energy change corresponding to the change in parameters from σ_0, Y_0 to σ, Y can be expressed in two ways, corresponding to the two paths indicated by the arrows. Equating these two results gives us equation 71.[57,58,90,91]

$$\log \left(\frac{K_Y}{K_{Y_0}^\circ}\right) = \delta_M \log K_{\sigma_0} + \delta_R \log K_Y$$

$$= \delta_R \log K_{Y_0} + \delta_M \log K_\sigma \quad (71)$$

Hence

$$\delta_M \log K_{\sigma_0} + \rho_Y(\sigma - \sigma_0)$$

$$= \rho_{Y_0}(\sigma - \sigma_0) + \delta_M \log K_{\sigma_0} + (m - m_0)(Y - Y_0) \quad (72)$$

Combining terms and rearranging, we obtain equation 73.

$$\frac{\sigma - \sigma_0}{m - m_0} = \frac{Y - Y_0}{\rho_Y - \rho_{Y_0}} \quad (73)$$

The left side of (73) is a function solely of the given structural change, whereas the right side is a function solely of the given solvent change. Since the structural change and the solvent change are independent variables, equation 73 can be true only if both sides are independently equal to a constant as well as to each other.

$$m - m_0 = C \cdot (\sigma - \sigma_0) \quad (74)$$

$$\rho_Y - \rho_{Y_0} = C \cdot (Y - Y_0) \quad (75)$$

Equations 74 and 75 have been tested for the ionization of substituted benzoic acids[57] and anilines[58] in a series of ethanol-water mixtures. The results, as shown in Table 8-19, conform to the equations. The results for the solvolysis rates of *meta-* and *para-*substituted α-phenylethyl chlorides

[90] L. Wilputte-Steinert, P. J. C. Fierens, and H. Hannaert, *Bull. Soc. Chim. Belg.*, **64**, 628 (1955).

[91] C. Mechelynck-David and P. J. C. Fierens, *Tetrahedron*, **6**, 232 (1959).

in a series of dioxane-water and dioxane-water-formic acid mixtures also conform well to these equations.[91]

The analogous relationships between m and σ^* for aliphatic compounds has been tested for the acid dissociation of substituted ammonium ions in ethanol-water mixtures.[92]

Table 8-19. Relationship between ρ and Y Parameters in the System Ethanol-Water, $25°$[57,58]

SOLVENT	ρ	
	(Experimental)	(calc., eq. 75)
Benzoic acids, C = 0.638		
Water	1.000	(1.000)
50 vol % EtOH	1.464	1.463
100% EtOH	1.626	1.628
Anilinium ions, C = −0.573		
Water	3.03 ± .14	3.18
24.7 wt % EtOH	3.43 ± .14	3.23
93.3 wt % EtOH	3.45 ± .13	3.66
97.7 wt % EtOH	3.74 ± .16	3.72
100.0% EtOH	3.90 ± .13	3.75

The general problem of the relationship between any pair of linear free energy equations involving independent variables, such as temperature, ionic strength, purely structural parameters, or purely medium parameters, has been discussed by Miller[93] for the special assumption that the free energy equations are exactly linear. However, because actual free energy relationships are linear only in first approximation, the equations resulting from this treatment, and the quality of their fit to the data, are still within the scope of the first approximation.

[92] R. W. Taft, Jr., *J. Am. Chem. Soc.*, **75**, 4231 (1953). See Table III, entry number 5.
[93] S. I. Miller, *J. Am. Chem. Soc.*, **81**, 101 (1959).

9

Extrathermodynamic Analysis

of Enthalpy and Entropy Changes

> *Data have an ephemeralness, a rhapsodic spontaneity, a nakedness so utterly at variance with the orderly instincts that pervade our being and with the given unity of our own experience as to be unfit for use in the building of reality. The constructs, on the other hand, are foot-loose, subjective, and altogether too fertile with logical implications to serve in their indiscriminate totality as material for the real world. They do, however, contain the solid logical substance which a stable reality must contain.*
>
> Henry Margenau
> *The Nature of Physical Reality*

The extrathermodynamic approach consists in expressing the changes in one thermodynamic quantity as a function of those in another. In the preceding two chapters the approach was applied to free energies of activation or of reaction. Since only one thermodynamic quantity—the free energy—was considered, it was always necessary to compare two or more different processes. However, when data for two or more thermodynamic quantities are available, the extrathermodynamic approach can give useful information even for a single process.

In this chapter we shall analyze the behavior of enthalpies and entropies of activation or of reaction. We shall give examples of the relationship of these quantities in a single process and also of their behavior in those sets of processes in which linear free energy relationships had been noted previously. We shall begin with a brief outline of the underlying theory.

THE FIRST APPROXIMATION[1]

In the discussion of linear free energy relationships, it was useful to classify the relationships on the basis of the number of interaction mechanisms and

[1] In this section the symbols ΔH, ΔS, ΔF, and ΔC_p without superscripts will denote the *standard* changes in these functions in a chemical equilibrium or in the formation of a transition state complex. The operator δ without a subscript will denote either δ_R or δ_M.

the nature of the interaction variables. We would like to retain this method of classification for enthalpy and entropy relationships. However, due to the fact that ΔH and ΔS are related thermodynamically to the *temperature derivative* of the free energy, a problem now arises because the temperature derivatives of the interaction variables are not necessarily negligible. For example, it is conceivable that the purely inductive effect of a mobile substituent such as —OCH_3 is displaced relative to that of a rigid substituent such as —H or —Cl as the temperature changes. If the effect is large enough, it will make a significant contribution to the magnitude of $(\partial \delta_R \Delta F/\partial T) = -\delta_R \Delta S$ and must be allowed for in the extrathermodynamic analysis.

Fortunately it is possible to approach this problem from a phenomenological point of view. In principle, the quantities ΔH and ΔS for an equilibrium or rate process are functions of the temperature.

$$\left(\frac{\partial \Delta H}{\partial T}\right) = \Delta C_p$$

$$\left(\frac{\partial \Delta S}{\partial T}\right) = \frac{\Delta C_p}{T}$$

However, the temperature dependence of ΔH and ΔS is usually no greater than the error of measurement, except for data of uncommonly high accuracy or covering a very wide temperature range. (Chapter 3, pages 43–45.) It is therefore customary to calculate ΔH and ΔS quantities in first approximation on the assumption that $\Delta C_p = 0$, that is, ΔH and ΔS are constant. Higher approximations are generated by assuming, first that ΔC_p is constant, next that $d \Delta C_p/dT$ is constant, and so on. These approximations are summarized as follows:

First approximation: $\Delta H =$ constant; $\Delta S =$ constant; $\Delta C_p = 0$

Second approximation: $\Delta C_p =$ constant; $d \Delta C_p/dT = 0$

Third approximation: $d \Delta C_p/dT =$ constant; $d^2 \Delta C_p/dT^2 = 0$

Since reactions for which ΔC_p is known are rather scarce, it rarely pays at our present stage of knowledge to carry the analysis of $\delta \Delta H$ and $\delta \Delta S$ beyond the first approximation. Moreover, we shall see in the next section that the basic principles for deriving information about the microscopic mechanisms from the enthalpy-entropy data are the same in all approximations, the difference between the approximations consisting only in the number of disposable parameters. We shall therefore concern ourselves chiefly with the first approximation throughout this chapter.

Let us assume then that $\Delta C_p = 0$ and derive the logical consequences of that assumption. For definiteness, let us begin by considering a substituent effect due to a single interaction mechanism. We may then write equation 1 in first approximation.

$$\delta_R \Delta F = \left(\frac{\partial \Delta F}{\partial R}\right)_{R_0 X, T} (R - R_0) \tag{1}$$

The quantity ΔF might be a standard free energy of reaction or of activation. R and R_0 are the interaction variables in the presence and absence of the substituent, respectively; we shall assume that they are functions of the temperature. The derivative $(\partial \Delta F/\partial R)_{R_0 X, T}$ is always evaluated at the unsubstituted or reference reagent $R_0 X$ and hence is a function of the temperature only.

$\delta_R \Delta F$ is a function both of the temperature and of the substituent.

$$\delta_R \Delta F = \delta_R \Delta H - T\delta_R \Delta S \tag{2}$$

However, since $\delta_R \Delta H$ and $\delta_R \Delta S$ are by hypothesis independent of the temperature, the variables in equation 2 are neatly separated.

The functions R and R_0 cannot be written *a priori* in an analogous separated form. However, it is useful to define a temperature-independent variable r such that

$$R = r + \phi(T, RX)$$
$$R_0 = r_0 + \phi(T, R_0 X) \tag{3}$$

Thus r is the temperature-independent part of R, and the function ϕ represents the temperature-dependent part. The latter of course depends also on the substituent R in the reagent RX.

If we write equations 1, 2, and 3 at any two temperatures T_1 and T_2 in the range where $\Delta C_p = 0$, we obtain the ratios (4).

$$\frac{\delta_R \Delta H - T_1 \delta_R \Delta S}{\delta_R \Delta H - T_2 \delta_R \Delta S} = \frac{(\partial \Delta F/\partial R)_{R_0 X, T_1}}{(\partial \Delta F/\partial R)_{R_0 X, T_2}} \cdot \frac{(r - r_0) + \delta_R \phi(T_1, RX)}{(r - r_0) + \delta_R \phi(T_2, RX)} \tag{4}$$

In equation 4 two functions of the *independent* variables R and T are equated. These functions can remain equal for all arbitrary changes in R and T only if certain conditions are met. The various sets of conditions under which (4) can be valid are listed below.

1. The ratios in equation 4 are in fact independent of R. This will happen if and only if equations 5 and 6 apply.

$$\delta_R \Delta H = \beta \, \delta_R \Delta S \tag{5}$$

$$\delta_R \phi(T, RX) = (r - r_0) \cdot \psi(T) \tag{6}$$

The parameter β in equation 5 is a constant, for $\delta_R \Delta H$ and $\delta_R \Delta S$ are both independent of the temperature. The function $\psi(T)$ in (6) is a function solely of T. Equation 4 thus reduces to (7), in which the temperature is the only independent variable.

$$\frac{\beta - T_1}{\beta - T_2} = \frac{(\partial \Delta F/\partial R)_{R_0 X, T_1}}{(\partial \Delta F/\partial R)_{R_0 X, T_2}} \cdot \frac{1 + \psi(T_1)}{1 + \psi(T_2)} \tag{7}$$

From equations 3 and 6 it can be shown that R is given by equation 8.[2]

$$R = r[1 + \psi(T)] \tag{8}$$

2. The ratios in equation 4 are in fact independent of T. This would require $\delta_R \Delta H$ and $\delta_R \Delta S$ to be functions of T, the minimum temperature dependence being that $\delta_R \Delta H$ is constant and $\delta_R \Delta S$ is proportional to $1/T$. Case (2) is therefore incompatible with our basic assumption that ΔC_p is zero and must be dismissed.

3. $(\partial \Delta F/\partial R)_{R_0 X, T}$ is a constant, and $\delta_R \phi(T, RX) = T \cdot \delta_R \zeta(RX)$, where the function ζ is independent of the temperature. In that case equation 4 reduces to the following two equations, in which A is a constant.

$$\delta_R \Delta H = A \cdot (r - r_0)$$

$$\delta_R \Delta S = -A \cdot \delta_R \zeta(RX)$$

However, because of the condition that $(\partial \Delta F/\partial R)_{R_0 X, T}$ is by the hypothesis of case (3) independent of the temperature,

$$\frac{\partial^2 \Delta F}{\partial R \partial T} = -\left(\frac{\partial \Delta S}{\partial R}\right)_{R_0 X, T} = 0$$

Since $(\partial \Delta S/\partial R)_{R_0 X, T}$ is zero, $\delta_R \Delta S$ must also be zero to the first approximation. Case (3) thus reduces to case (1) with $1/\beta$ equal to zero.

The preceding three cases exhaust the sets of conditions under which equation 4 can be solved. Since case (3) is included in case (1) and since case (2) has been ruled out on other grounds, we need consider only case (1). Thus our initial assumption that $\Delta C_p = 0$ necessarily implies that $\delta_R \Delta H$ is proportional to $\delta_R \Delta S$ (equation 5) and that the temperature-dependent interaction variable R is separable into a factor characteristic of the substituent and one characteristic of the temperature (equations 6 and 8).

Equations 5 and 2 imply that any two of the quantities $\delta_R \Delta F$, $\delta_R \Delta H$, and $\delta_R \Delta S$ are proportional. Moreover, $\delta_R \Delta F$ at one temperature is proportional to $\delta_R \Delta F$ at any other temperature, as in equation 9.

$$\frac{(\delta_R \Delta F)_{T_1}}{(\delta_R \Delta F)_{T_2}} = \frac{(\beta - T_1)}{(\beta - T_2)} \tag{9}$$

[2] Rigorously we should write R = r[1 + $\psi(T)$] + a *constant*. We prefer the simpler form of equation 8, however, because we are only interested in δ-operations in which an additive constant always drops out.

Because of the separability (equation 8) of $R - R_0$ into a structural variable $r - r_0$ and a temperature function $1 + \psi(T)$, the quantities $\delta_R \Delta F$, $\delta_R \Delta H$, and $\delta_R \Delta S$ can all be expressed without using the temperature-dependent variable R.

$$\delta_R \Delta F = \left(\frac{\partial \Delta F}{\partial R}\right)_{T, R_0 X} (R - R_0)$$

$$= \left(\frac{\partial \Delta F}{\partial r}\right)_{T, R_0 X} (r - r_0) \cdot \left(\frac{\partial r}{\partial R}\right)_T \frac{(R - R_0)}{(r - r_0)}$$

$$= \left(\frac{\partial \Delta F}{\partial r}\right)_{T, R_0 X} (r - r_0) \tag{10}$$

It will be noted that equation 10 is similar in form to the Hammett $\rho\sigma$ and Taft $\rho^* \sigma^*$ equations. The important feature of equation 10 is that *for substituents interacting only by a single mechanism* the substituent parameter should be temperature-independent even though the slope is a function of the temperature.

On differentiating equation 10 with respect to the temperature and applying the Gibbs-Helmholtz equation, we find that $\delta_R \Delta S$ and $\delta_R \Delta H$ are similarly proportional to $(r - r_0)$.

$$\delta_R \Delta S = \left(\frac{\partial \Delta S}{\partial r}\right)_{R_0 X} (r - r_0) \tag{11a}$$

$$\delta_R \Delta H = \left(\frac{\partial \Delta H}{\partial r}\right)_{R_0 X} (r - r_0) \tag{11b}$$

THE SECOND APPROXIMATION[1]

For the purpose of analyzing enthalpy-entropy data, the most useful consequence of the first approximation is that for a single interaction mechanism, $\delta \Delta H$ is simply proportional to $\delta \Delta S$. We shall now show that a proportional relationship is obtained also when the second approximation is used. We begin by assuming that ΔC_p is independent of the temperature and that $\delta \Delta C_p$ is *not* identically equal to zero. Since

$$\Delta C_p = \left(\frac{\partial \Delta H}{\partial T}\right) = T \left(\frac{\partial \Delta S}{\partial T}\right),$$

it follows that $\delta \Delta H$ and $\delta \Delta S$ are given by equations 12, where δa_H and δa_S are constants of integration.

$$\delta \Delta H = \delta a_H + T \cdot \delta \Delta C_p \tag{12a}$$

$$\delta \Delta S = \delta a_S + \ln T \cdot \delta \Delta C_p \tag{12b}$$

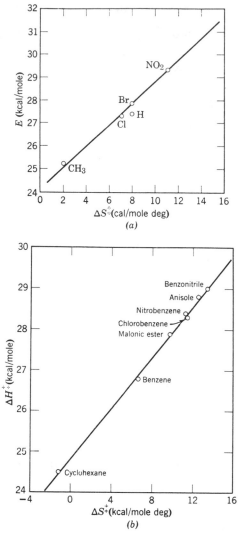

Figure 9-1. (*a*) Activation parameters for the decomposition of

$$p\text{-X}\text{—}C_6H_4\text{—}N\text{=}N\text{—}C(C_6H_5)_3$$

in toluene; X = H, CH_3, Cl, Br, or NO_2 (G. L Davies, D. H. Hey and G. H. Williams, *J. Chem. Soc.*, 4397 (1956).

(*b*) Activation parameters for the decomposition of $C_6H_5\text{—}N\text{=}N\text{—}C(C_6H_5)_3$ in a series of solvents ranging from cyclohexane to benzonitrile. (M. G. Alder and J. E. Leffler, *J. Am. Chem. Soc.*, **76**, 1425 (1954).)

However, δa_H and δa_S are not independent of $\delta \Delta C_p$. For a single inter-action mechanism the appropriate relationships are equations 13, which can be derived by a method entirely analogous to that used in deriving equation 5.

$$\delta a_H = \gamma_H \, \delta \, \Delta C_p \tag{13a}$$

$$\delta a_S = \gamma_S \, \delta \, \Delta C_p \tag{13b}$$

The parameters γ_H and γ_S are independent of the temperature in this approximation.

On combining equations 12 and 13 we obtain equations 14.

$$\delta \, \Delta H = (\gamma_H + T) \, \delta \, \Delta C_p \tag{14a}$$

$$\delta \, \Delta S = (\gamma_S + \ln T) \, \delta \, \Delta C_p \tag{14b}$$

$$\delta \, \Delta F = [(\gamma_H - T\gamma_S) + T(1 - \ln T)] \, \delta \, \Delta C_p \tag{14c}$$

It follows from equations 14 that $\delta \, \Delta H$, $\delta \, \Delta S$, and $\delta \, \Delta F$ are all propor-tional to $\delta \, \Delta C_p$ and hence to each other. In particular, $\delta \, \Delta H$ is again proportional to $\delta \, \Delta S$, although the proportionality constant, $(\gamma_H + T)/(\gamma_S + \ln T)$, is now a function of the temperature. Thus we may write equation 5 for a single interaction mechanism also in the second approxi-mation, although the slope, β, is now regarded as a temperature-dependent parameter.

$$\delta \, \Delta H = \beta \, \delta \, \Delta S \tag{5}$$

At this point it is well to interrupt the theoretical analysis in order to interpose a brief comparison with experiment. If we examine the data for a large number of equilibria and rate processes, we must assume that at least a few of the substituent or medium effects that we analyze will involve only a single interaction mechanism. We therefore expect to find at least a few examples of $\delta \, \Delta H$ proportional to $\delta \, \Delta S$, and in fact we find many. An example in which the variable is the substituent is shown in Figure 9-1a; one in which the variable is the medium is shown in Figure 9-1b. Numerous other examples are tabulated elsewhere in this chapter.

THE SIGN OF β

In the vast majority of reactions for which the experimental values of $\delta \, \Delta H$ and $\delta \, \Delta S$ are found to be proportional to each other, the sign of the proportionality constant β is positive.

The fact that β should almost always be positive is not unexpected from a theoretical point of view. For each mechanism of interaction between molecules or parts of molecules, the maximum reduction in energy for the system is obtained only if certain geometrical conditions are met. The geometrical conditions of course constitute a constraint and mean that the decrease in energy will be accompanied by some decrease in entropy. For example, one model of the medium effect treats the solvent molecules merely as dipoles that have preferred orientations with respect to the polar solute molecules.[3] The stronger the interaction, the greater will be the decrease in enthalpy, but at the same time the greater will be the orientation and the decrease in entropy. Other models, which predict correlations with positive β for structural changes, have been discussed by Blackadder and Hinshelwood.[4]

A thermodynamic argument for the positive sign of β may be constructed as follows.[1] For definiteness we shall consider a substituent effect and choose a substituent for which $\delta \Delta E$ is negative. However, instead of actually carrying out the reaction in the presence of the substituent, we let the substituent effect be represented by an added field which acts on the reaction zone in the reaction, $A \to B$, of the unsubstituted compound. The operator $\delta \Delta$ then corresponds to the isothermal double process, $A \to B$ with field on plus $B \to A$ with field off. This combined process is in effect a physical process, for the net result is that the reaction zone is taken out of the field in a mole of A and placed into the field in a mole of B. Since $\delta \Delta E$ is negative, an equivalent amount of energy, $-\delta \Delta E$, appears in the surroundings. If the chemical transformations, $A \to B$ and $B \to A$, are carried out reversibly, the portion of $-\delta \Delta E$ that appears as useful work is equal to $-\delta \Delta F$, and the efficiency of the combined process at converting energy into useful work is equal to $(-\delta \Delta F/-\delta \Delta E)$. But the efficiency of a physical system at constant temperature cannot be greater than that of a conservative system, that is, $(-\delta \Delta F/-\delta \Delta E) \leq 1$. Hence

$$-\delta \Delta E + T \delta \Delta S \leq -\delta \Delta E,$$

and

$$T \delta \Delta S \leq 0.$$

Thus $\delta \Delta S$ is either zero or negative, and since $\delta \Delta E$ is also negative (by hypothesis), the ratio $\delta \Delta E/\delta \Delta S$ ($\approx \delta \Delta H/\delta \Delta S$) is positive. Hence β is positive.

It is clear that in this derivation the substituent interacts by only a single mechanism, for otherwise its effect would not be representable by a single field. It is less clear what further restrictions might be imposed by the device of letting the substituent effect be represented by a field. One restriction, and perhaps the only one, seems to be that the substituent must not change the reaction mechanism.

[3] A. Münster, Z. Elektrochem., **54**, 443 (1950); Naturwissenschaften, **35**, 343 (1948).
[4] D. A. Blackadder and C. N. Hinshelwood, J. Chem. Soc., 2728 (1958).

CHARACTERISTICS OF THE ERROR CONTOURS FOR ENTHALPIES AND ENTROPIES[1]

In anything as complicated as the relationships of enthalpies and entropies there is a strong tendency to explain complexities away by appealing to experimental error. Although a typical error in $\delta \Delta H^{\ddagger}$ might be 0.5 kcal/mole and in $\delta \Delta S^{\ddagger}$ 2 cal/mole deg., it must not be assumed that the uncertainty in location of an experimental point in a $\Delta H^{\ddagger}$ versus $\Delta S^{\ddagger}$ plot is given by a rectangle of those dimensions. Instead the error contour is a highly eccentric ellipse whose major axis is inclined to the entropy axis at a slope equal to the mean experimental temperature in degrees K.[5] This can be seen by reference to Figure 9-2. The line labeled "locus of $\Delta F^{\ddagger}_{\mathrm{max}}$" represents the lower boundary of a region, all points of which correspond to combinations of $\Delta H^{\ddagger}$ and $\Delta S^{\ddagger}$ that would lead to impossibly high values of $(\Delta H^{\ddagger} - T \Delta S^{\ddagger})$. That is, the rate constants corresponding to points in this region would be lower than could be admitted by the limits of experimental error. Similarly, the line "locus of $\Delta F^{\ddagger}_{\mathrm{min}}$" excludes another part of the rectangle from the error contour.

Some consequences of the high eccentricity of the error contours are illustrated in Figure 9-3. The figure is drawn so that points a and b are the

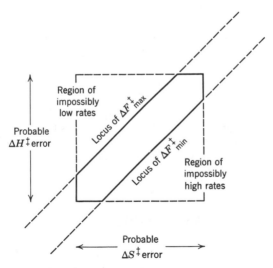

Figure 9-2. The probable error contour for a point in the enthalpy-entropy plane.

[5] J. Mandel and F. J. Linnig, *Anal. Chem.* **29**, 743 (1957).

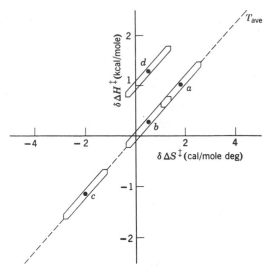

Figure 9-3. Probable error contours for a set of points in the enthalpy-entropy plane.

same distance apart as points d and b. Points a and b might be interchanged in position by experimental error. But because of the high eccentricity of the error contours, it is out of the question that d and b might be interchanged or that d might really be on the same line with a, b, and c.

If the best line through a series of points in an enthalpy-entropy diagram has a slope close to the mean experimental temperature, then the points should occupy a range of enthalpy and entropy several times greater than the experimental error in order to exclude the hypothesis that all the $\delta \Delta H^{\ddagger}$ and $\delta \Delta S^{\ddagger}$ values are in fact not different from zero. In borderline cases the decision may be made by noting whether the sequence in which the points occur along the line is random or rational.

THE ISOKINETIC RELATIONSHIP[1]

A relationship of the form $\delta \Delta H^{\ddagger} = \beta \, \delta \Delta S^{\ddagger}$, with $\beta \geq 0$, is called an isokinetic relationship.[6] A corresponding relationship between equilibrium quantities is called an isoequilibrium relationship. The reason for this

---

[6] J. E. Leffler, *J. Org. Chem.*, **20**, 1202 (1955).

nomenclature becomes evident when equations 2 and 5 are combined.

$$\delta \Delta H^{\ddagger} = \beta \, \delta \Delta S^{\ddagger} \tag{5}$$

$$\delta \Delta F^{\ddagger} = \delta \Delta H^{\ddagger} - T \delta \Delta S^{\ddagger} \tag{2}$$

$$\delta \Delta F^{\ddagger} = (T - \beta) \, \delta \Delta S^{\ddagger} \tag{15}$$

$$\delta \Delta F^{\ddagger} = \left(1 - \frac{T}{\beta}\right) \delta \Delta H^{\ddagger} \tag{16}$$

The parameter β has the dimensions of absolute temperature and can be identified as an actual or virtual temperature at which all the differences in the rate or the equilibrium constants will vanish.

$$(\delta \Delta F^{\ddagger})_{\text{at } T=\beta} = 0 \tag{17}$$

Actual examples of isokinetic relationships usually show some deviations from the best straight line. These are presumably due to perturbation by one or more additional interaction mechanisms of lesser importance. At a temperature equal to the average slope β, the variations in rate constant will be small but not quite zero. They will reflect, not the principal interaction mechanism but merely the perturbing mechanisms. If the latter are numerous and random, the rates observed at this temperature will appear to be random and unpredictable by any simple theory.

From equations 15 and 16 it can be predicted that substituent and medium effects, in series involving only a single interaction mechanism, will change sign as the temperature passes through β. For example, in the Menschutkin reaction (number 1 of Table 9-1) of methyl iodide with pyridine in alcohol-benzene mixtures, β is 340°K. The rate is found to increase with increasing alcohol content at temperatures above 340°K. However, this increase is not due to a decrease in the activation energy. The activation energy actually increases, but the increase is more than offset by an even greater increase in $T \Delta S^{\ddagger}$. On the other hand, at temperatures below 340°K adding alcohol should decelerate the reaction.

Similarly, if a reaction is found to be accelerated by a certain substituent, this does not necessarily mean that it will be accelerated at all temperatures, or that the activation energy is decreased. Thus a small positive value of ρ in the $\rho\sigma$ equation need not mean that a negative charge develops at the reaction center; a change in temperature might convert the small positive value of ρ to a small negative value and this would not mean that a positive charge now develops during the reaction.

Although inversion of an entire sequence of rate constants requires an isokinetic relationship, inversion of a pair of rate constants requires only a two-point "isokinetic relationship" in which the two points are related in such a way that $\delta \Delta H^{\ddagger}/\delta \Delta S^{\ddagger}$ is positive.[7] However, if the rate constants differ by an order of magnitude or more, the inversion is not likely to be experimentally realizable. A large rate difference can exist between two members of an isokinetic series only if the experimental temperature is far from β. This means that when the temperature is adjusted to β, the rates may be either too fast or too slow to be measured.

When a structural change in a given part of a molecule fails to affect the reaction rate, two explanations are possible in principle. One is that the reactions have been studied at the isokinetic temperature. The other is that the reaction mechanism does not involve any variable interactions between the reaction site and the part of the molecule in which the structural change was made. An example of insensitivity of a reaction to substituents that is due to an isokinetic temperature is the decomposition of acyl azides.[8a] In contrast, the decomposition of a series of substituted phenyl azides in aniline is genuinely insensitive to the nature of a *meta*-substituent; the enthalpies and entropies as well as the free energies are nearly invariant.[8b] This fact was used to eliminate a possible reaction mechanism in which the *meta*-substituent and the reaction site would have interacted strongly.

Occasionally differences in activation energy have been used as evidence for a change in reaction mechanism in the belief that this is a sound procedure even when the use of differences in free energy of activation would not be. This practice can also be misleading. For example, it is not at all unusual for reactions representing the extremes of a single isokinetic line, and hence very probably having the same reaction mechanism, to differ in activation energy by 5 kcal/mole or even more. A reaction of different mechanism will probably be displaced from the isokinetic line and will probably differ considerably in rate constant, but it might easily have the same activation energy as one of the points on the isokinetic line. Some other point on the isokinetic line will probably have the same entropy as the reaction in question, so that will not do as a criterion of change in mechanism either.

In Tables 9-1 through 9-7 we have collected numerous examples of isokinetic relationships selected on the basis of excellence of fit or importance of the reaction. The correlation coefficients for reactions in these Tables are .95 or better. This mode of selection excludes a great many merely

[7] For example, W. M. Schubert and R. G. Minton, *J. Am. Chem. Soc.*, **82**, 6188 (1960), cite a Baker-Nathan order of substituent effects that becomes an inductive order on changing the temperature. Other examples of changes in sign of $\delta \Delta F^{\ddagger}$ have been noted by W. G. Brown and S. Fried, *J. Am. Chem. Soc.*, **65**, 1841 (1943); G. M. Santerre, C. J. Hansrote, Jr., and T. I. Crowell, *ibid.*, **80**, 1254 (1958); D. Pressman and W. G. Young, *ibid.*, **66**, 705 (1944); J. F. Bunnett and R. J. Morath, *ibid.*, **77**, 5165 (1955).
[8a] Y. Yukawa and Y. Tsuno, *J. Am. Chem. Soc.*, **79**, 5530 (1957); **81**, 2007 (1959).
[8b] M. Appl and R. Huisgen, *Chem. Ber.*, **92**, 2961 (1959).

Table 9-1. Some Isokinetic Relationships in Displacement and Elimination Reactions[a]

REACTION	β, K	NUMBER OF POINTS
1. Methyl iodide with pyridine in alcohol-benzene mixtures. The point for pure benzene deviates to the left and has been omitted. Note that the activation energy *increases* as the alcohol content is increased. The deviation of the point for pure benzene suggests preferential solvation by the alcohol. Compare with scatter in the $\Delta H^\ddagger - \Delta S^\ddagger$ plot for the Menschutkin reaction in nonhydroxylic solvents. R. A. Fairclough and C. N. Hinshelwood, *J. Chem. Soc.*, 1573 (1937).	340	4
2. Methyl or ethyl iodide with tertiary aromatic amines in methanol. Three nitro-substituted amines form a parallel line to the left, while 2,6,N,N-tetramethylaniline is by itself to the right. W. G. Brown and S. Fried, *J. Am. Chem. Soc.*, **65,** 1841 (1943); D. P. Evans, *J. Chem. Soc.*, 422 (1944); R. Williams, D. P. Evans, and H. B. Watson, *ibid.*, 1345 (1939).	510	27
3. Pyridine or dimethylaniline with alkyl bromides in various solvents. The activation energy is higher in the more polar solvents. At ordinary temperatures, the rate is also higher in the more polar solvents due to the concomitant increase in $\Delta S^\ddagger$. V. A. Hol'tsshmidt and N. K. Vorob'ev, *J. Phys. Chem. (U.S.S.R.)*, **15,** 1087 (1941).	300	7
4. Aniline with phenacyl bromide in various solvents. At ordinary temperatures $\Delta H^\ddagger$, $\Delta S^\ddagger$, and the rate constant all increase as the solvent becomes more polar. J. C. Cox, *J. Chem. Soc.*, **119,** 142 (1921).	210	6
5. Effect of various polar additives, such as derivatives of formamide, on the rate of reaction of *n*-butyl bromide with the sodio derivative of diethyl malonate in benzene. H. E. Zaugg, B. W. Horrom, and S. Borgwardt, *J. Am. Chem. Soc.*, **82,** 2895 (1960).	195	37
6. Reaction of ethyl arylsulfonates with sodium ethoxide in alcohol. M. S. Morgan and L. H. Cretcher, *J. Am. Chem. Soc.*, **70,** 375 (1948).	185	5
7. Dehydrohalogenation of *cis*-dihaloethylenes in methanolic sodium methoxide. S. I. Miller and R. M. Noyes, *J. Am. Chem. Soc.*, **74,** 629 (1952).	1320	3

Table 9-1 (*continued*)

REACTION	β, °K	NUMBER OF POINTS
8. Dehydrohalogenation of *trans*-dihaloethylenes in methanolic sodium methoxide. S. I. Miller and R. M. Noyes, *J. Am. Chem. Soc.*, **74,** 629 (1952).	190	3
9. Elimination of HCl from the β-isomer of benzene hexachloride in aqueous-alcohol mixtures. Two points for the ϵ-isomer are available, but they do not fall on the line for the β-isomer in aqueous-alcohol mixtures. The same is true for three points available for the α-isomer, but these form a line of their own. The α, γ, and ϵ isomers, all in 76% alcohol, form a line that does not include the β-isomer. The β-isomer is the only one not having an adjacent hydrogen and halogen atom in a *trans*-configuration and probably reacts by a different mechanism. S. J. Cristol and W. Barasch, *J. Am. Chem. Soc.*, **74,** 1658 (1952); S. J. Cristol, N. L. Hause, and J. S. Meek, *ibid.*, **73,** 674 (1951).	320	7
10. Reaction of epoxides with thiocyanate ion in neutral aqueous solution. P. L. Nichols, Jr. and J. W. Inham, *J. Am. Chem. Soc.*, **77,** 6547 (1955).	540	4
11. Decomposition of triethylsulfonium bromide in hydroxylic solvents. This line also includes benzyl alcohol-toluene mixtures up to 50% toluene. The point for 75% toluene (omitted) falls midway between this line and the parallel one for non-hydroxylic solvents. E. A. Moelwyn-Hughes, *The Kinetics of Reactions in Solution*, Oxford University Press, 2nd ed., 1947, p. 278.	330	7
12. Decomposition of triethylsulfonium bromide in nonhydroxylic solvents. The nonhydroxylic solvent line is about 3.6 kcal lower than the hydroxylic solvent line. E. A. Moelwyn-Hughes, *The Kinetics of Reactions in Solution*, Oxford University Press, 2nd ed., 1947, p. 278.	340	5

[a] Only examples for which the correlation coefficient is greater than .95 are included in the table. Rates unless otherwise noted. Plots refer to $\Delta H^{\ddagger}$ ordinate, $\Delta S^{\ddagger}$ abscissa.

Table 9-2. *Some Isokinetic Relationships in Decomposition or Association Reactions*[a]

REACTION	β, °K	NUMBER OF POINTS
1. Decomposition of *p*-substituted triphenylmethyl azides in dibutyl carbitol (at temperatures 170–190°C). W. H. Saunders, Jr. and J. C. Ware, *J. Am. Chem. Soc.*, **80**, 3328 (1958).	460	6
2. Formation of phenyl isocyanate from benzoyl azide in various solvents. M. S. Newman, S. H. Lee, Jr., and A. B. Garrett, *J. Am. Chem. Soc.*, **69**, 113 (1947).	380	14
3. Decomposition of azo-*bis*-cycloalkylnitriles of ring sizes 5, 6, 7, 8, 10. C. G. Overberger, M. Biletch, A. G. Finestone, J. Lilker, and J. Herbert, *J. Am. Chem. Soc.*, **75**, 2078 (1953).	660	5
4. Disproportionation of 4-isopropyl-2,6-di-*t*-butylphenoxy radicals in a series of solvents. C. D. Cook and B. E. Norcross, *J. Am. Chem. Soc.*, **81**, 1176 (1959).	265	5
5. Decomposition of phenylazotriphenylmethane in the same series of solvents as reaction (4) above. The enthalpies and entropies of activation fall in the same order. M. G. Alder and J. E. Leffler, *J. Am. Chem. Soc.*, **76**, 1425 (1954).	310	7
6. Decomposition of phenylazotriphenylmethanes with various substituents in the phenylazo group and in various solvents. M. G. Alder and J. E. Leffler, *J. Am. Chem. Soc.*, **76**, 1425 (1954); G. L. Davies, D. H. Hey and G. H. Williams, *J. Chem. Soc.*, 4397 (1956); S. G. Cohen and C. H. Wang, *J. Am. Chem. Soc.*, **75**, 5504 (1953); S. Solomon, C. H. Wang, and S. G. Cohen, *ibid.*, **79**, 4104 (1957). Points for the decomposition of *p*-nitrophenylazotris-(*p*-methoxyphenyl) methane fall on a parallel line above this one. M. D. Cohen, J. E. Leffler, and L. M. Barbato, *J. Am. Chem. Soc.*, **76**, 4169 (1954). Substitution in the *para*-position of the triphenylmethyl group gives points falling below the line.	320	31
7. Association equilibrium of α-benzoyl-β, β-diphenylhydrazyl free radical in various solvents. A. Wassermann, *J. Chem. Soc.*, 621, 623 (1942).	360	4
8. Decomposition of *para*-substituted *t*-butyl perbenzoates in diphenyl ether. The data also fit a Hammett $\rho\sigma$ relationship. A. T. Blomquist and I. A. Berstein, *J. Am. Chem. Soc.*, **73**, 5546 (1951).	480	5

Table 9-2 (continued)

REACTION	β, °K	NUMBER OF POINTS
9. Radical decomposition of *meta-* and *para*-substituted benzoyl peroxides in acetophenone. This reaction gives a markedly curved Hammett plot. The point for *p*-methoxy deviates so much more than the rest from the $\Delta H^{\ddagger}$ versus $\Delta S^{\ddagger}$ line that it has been omitted from the calculation. The deviation suggests that that compound decomposes at least partly by a different, probably polar, mechanism like that of *p*-methoxy-*p'*-nitrobenzoyl peroxide in polar solvents. *Ortho*-substituted peroxides form a rough line of their own, parallel to that of the *meta-* and *para*-substituted peroxides but displaced below it by about 2.5 kcal. There are not enough points on the *ortho*-line to decide whether the *ortho*-methoxy compound is also abnormal. A. T. Blomquist and A. J. Buselli, *J. Am. Chem. Soc.*, **73**, 3883 (1951).	530	10
10. Decomposition of benzene diazonium chloride in alcohols. C. E. Waring and J. R. Abrams, *J. Am. Chem. Soc.*, **63**, 2757 (1941).	300	8
11. Decomposition of benzene diazonium chloride in water and in carboxylic acids. This line is parallel to that obtained for the decomposition in alcohol solvents (reaction 10) but not identical to it. The separation (in the energy dimension) is about 0.7 kcal. Note the much greater value of β for the substituent effect on this reaction in water (reaction 12). C. E. Waring and J. R. Abrams, *J. Am. Chem. Soc.*, **63**, 2757 (1941); H. A. H. Pray, *J. Phys. Chem.*, **30**, 1417, 1477 (1926).	300	5
12. Decomposition of substituted benzene diazonium fluoborates in water. D. F. DeTar and A. R. Ballentine, *J. Am. Chem. Soc.*, **78**, 3918 (1956).	930	6
13. Diels-Alder reaction of maleic anhydride or chloromaleic anhydride with α- or β-eleostearic acid in xylene. Methylmaleic anhydride deviates. W. G. Bickford, J. S. Hoffmann, D. C. Heinzelmann, and S. P. Fore, *J. Org. Chem.* **22**, 1080 (1957).	425	4

Table 9-2 (continued)

REACTION	β, °K	NUMBER OF POINTS
14. Diels-Alder reaction of cyclopentadiene with benzoquinone or α-naphthoquinone, in various solvents, with or without acid or phenol catalysts. The fact that the points for the "catalyzed" reaction appear to fall on the same line (although in different regions) as the points for the uncatalyzed reaction, suggests an essential identity of mechanism for the two reactions. A. Wassermann, *J. Chem. Soc.*, 621, 623 (1942); R. A. Fairclough and C. N. Hinshelwood, *ibid.*, 236 (1938); A. Wassermann, *Trans. Faraday Soc.*, **34**, 128 (1938); W. Rubin, H. Steiner, and A. Wassermann, *J. Chem. Soc.*, 3046 (1949).	410	25
15. Dimerization of cyclopentadiene in various solvents, and in carbon tetrachloride with acid catalysts. W. Rubin, H. Steiner, and A. Wassermann, *J. Chem. Soc.*, 3046 (1949); A. Wassermann, *Trans. Faraday Soc.*, **34**, 128 (1938); B. Raistrick, R. H. Sapiro, and D. M. Newitt, *J. Chem. Soc.*, 1761 (1939); H. Kaufmann and A. Wassermann, *ibid.*, 870 (1939).	330	11
16. Dimerization of cyclopentadiene, no solvent, pressures from 1 to 4000 atm. B. Raistrick, R. H. Sapiro, and O. M. Newitt, *J. Chem. Soc.*, 1761 (1939).	70	8
17. Dimerization equilibrium of 1,1-diarylethylenes in trichloroacetic acid-benzene. A. G. Evans, N. Jones, P. M. S. Jones, and J. H. Thomas, *J. Chem. Soc.*, 2757 (1956).	345	5
18. Hydrogenation of ethylene over nickel catalysts. Two nearly parallel lines, one for catalysts that had been outgassed or ion-bombarded and one for catalysts that had been intentionally contaminated. The latter line is in a region of lower activation energies and is displaced towards lower log A values. J. Tuul and H. E. Farnsworth, *J. Am. Chem. Soc.*, **83**, 2253 (1961).	400 380	51 9
19. Hydrogenation of acetylene over nickel-pumice, nickel-kieselguhr and nickel powder catalysts. G. C. Bond and R. S. Mann, *J. Chem. Soc.*, 4738 (1958).	365	26

Table 9-2 (continued)

REACTION	β, °K	NUMBER OF POINTS
20. Decarboxylation of acetone dicarboxylic acid in water and in a series of alcohols. E. O. Wiig, *J. Phys. Chem.*, **32**, 961 (1928); **34**, 596 (1930).	340	7
21. Decarboxylation of monosubstituted malonic acids in water. The data for disubstituted malonic acids do not spread enough to establish a linear relationship. Malonic acid itself falls on neither line. E. A. Moelwyn-Hughes, *The Kinetics of Reactions in Solution*, 2nd ed., Oxford University Press, 1947, p. 287; A. Dinglinger and E. Schroer, *Z. physik. Chem.*, **A179**, 401 (1937).	95	5
22. Picolinic acid decarboxylation in a series of solvents. Data for five substituted picolinic acids in *p*-dimethoxybenzene fall on the same line. N. H. Cantwell and E. V. Brown, *J. Am. Chem. Soc.*, **75**, 4466 (1953).	460	13

[a] Only examples for which the correlation coefficient is greater than .95 are included in the table. Rates unless otherwise noted. Plots refer to $\Delta H^{\ddagger}$ ordinate, $\Delta S^{\ddagger}$ abscissa.

Table 9-3. Some Isokinetic Relationships in the Reactions of Carboxylic Acids and Their Derivatives[a]

REACTION	β, °K	NUMBER OF POINTS
1. Hydrogen-ion catalyzed esterification of *meta-* and *para-*substituted benzoic acids in methanol.	260	13
2. Hydrogen-ion catalyzed esterification of *ortho-*substituted benzoic acids in methanol. This line lies above that for *meta-* and *para-*substituents described above. However, the *ortho*-fluoro point falls on the *meta-para* line, probably because fluorine is a small atom; and the points for *ortho-*ethoxy and *ortho*-methoxy fall below it, suggesting that a special mechanism exists for these *ortho-*substituents. R. J. Hartman and A. G. Gassmann, *J. Am. Chem. Soc.*, **62**, 1559 (1940). H. A. Smith and R. B. Hurley, *ibid.*, **72**, 112 (1950).	260	4

Table 9-3 (continued)

REACTION	β, °K	NUMBER OF POINTS
3. Hydrolysis of alkyl esters of perfluoroacetic, perfluoropropionic, and perfluorobutyric acid in 70% acetone. However, the point for methyl perfluoroacetate falls below the line. A. Moffat and H. Hunt, *J. Am. Chem. Soc.*, **79**, 54 (1957).	515	8
4. Acid-catalyzed hydrolysis of alkyl acetates in 62% acetone. Because of the tendency of saturated groups to bunch, the relationship has essentially three points. Note that the line is parallel to that of the thiol esters (reaction 5) under the same conditions. For references, see reaction (5).	330	7
5. Acid-catalyzed hydrolysis of alkyl thiolacetates in 62% acetone. Not bunched. Parallel to the line for the corresponding oxygen compounds. The point for the trityl compound was omitted because it has been shown that trityl-sulfur cleavage takes place. B. K. Morse and D. S. Tarbell, *J. Am. Chem. Soc.*, **74**, 416 (1952); P. N. Rylander and D. S. Tarbell, *ibid.*, **72**, 3021 (1950); For data in water, see L. H. Noda, S. A. Kuby, and H. A. Lardy, *J. Am. Chem. Soc.*, **75**, 913 (1953).	330	7
6. Acid-catalyzed hydrolysis of *para*-substituted benzamides. For reference, see reaction (7).	400	4
7. Base-catalyzed hydrolysis of *para*-substituted benzamides. The one available point for an *ortho*-substituted benzamide deviates more (to the left) from the *para* line in the acid-catalyzed series than in the base-catalyzed series. I. Meloche and K. J. Laidler, *J. Am. Chem. Soc.*, **73**, 1712 (1951).	910	4
8. Hydrolysis of substituted benzoic anhydrides in 75% dioxane. E. Berliner and L. H. Altschul, *J. Am. Chem. Soc.*, **74**, 4110 (1952).	750	9
9. Solvolysis of substituted benzoyl chlorides in 60% ether-40% ethanol. G. E. K. Branch and A. C. Nixon, *J. Am. Chem. Soc.*, **58**, 2499 (1936).	450	8
10. Solvolysis of substituted benzoyl chlorides in 95% acetone. D. A. Brown and R. F. Hudson, *J. Chem. Soc.*, 883 (1953).	460	5

Table 9-3 (continued)

REACTION	β, °K	NUMBER OF POINTS
11. Saponification of alkyl acetates and thiolacetates in 62% acetone. Points for both *t*-butyl acetate and *t*-butyl thiolacetate are above the line and were omitted. P. N. Rylander and D. S. Tarbell, *J. Am. Chem. Soc.*, **72**, 3021 (1950); B. K. Morse and D. S. Tarbell, *ibid.*, **74**, 416 (1952).	270	13
12. Base-catalyzed methanolysis of *l*-menthyl *meta*- and *para*-alkylbenzoates. The *meta*- and unsubstituted compounds form a line below and parallel to that of the *para*-substituted compounds. M. S. Newman and E. K. Easterbrook, *J. Am. Chem. Soc.*, **77**, 3763 (1955).	300	9
13. Reaction of ethyl acetate with the sodio derivative of various ketones in dilute ether solution. Variation of the ester gives points displaced to the left. D. G. Hill, J. Burkus, and C. R. Hauser, *J. Am. Chem. Soc.*, **81**, 602 (1959).	405	4
14. Esterification of straight-chain acids in cyclohexanol, no added catalyst. R. A. Fairclough and C. N. Hinshelwood, *J. Chem. Soc.*, 593 (1939).	400	7
15. Saponification of valerolactone, phthalide, and 5-aminophthalide in aqueous-alcohol mixtures. A very approximate point for butyrolactone falls near the line but was omitted. D. S. Hegan and J. H. Wolfenden, *J. Chem. Soc.*, 508 (1939).	380	9
16. Acid-catalyzed hydrolysis of *para*-substituted ethyl or methyl benzoates in aqueous ethanol, aqueous methanol, and aqueous acetone mixtures. Two points for *ortho*-nitrobenzoate in aqueous acetone and aqueous ethanol are available and fall off the line for the *para*-derivatives and on the high activation energy side. The line connecting the two *ortho*-points is approximately parallel to that of the *para*-points. Two points for *meta*-nitro fall on the *para* line. E. W. Timm and C. N. Hinshelwood, *J. Chem. Soc.*, 862 (1938).	440	19
17. Heterogeneous hydrolysis of ethyl carbonate on salts and metallic oxides. Note that β is within the range of experimental temperature. R. W. Sauer and K. A. Krieger, *J. Am. Chem. Soc.*, **74**, 3116 (1952).	520	14

Table 9-3 (continued)

REACTION	β, °K	NUMBER OF POINTS
18. Saponification of ethyl benzoate in aqueous acetone, aqueous alcohol, aqueous dioxane, aqueous acetone-alcohol mixtures. Parallel, but at higher energies than reaction (19). R. A. Harman, *Trans. Faraday Soc.*, **35**, 1336 (1939); R. A. Fairclough and C. N. Hinshelwood, *J. Chem. Soc.*, 538 (1937).	400	23
19. Saponification of ethyl *m*-nitrobenzoate in aqueous acetone and aqueous alcohol. R. A. Harman, *Trans. Faraday Soc.*, **35**, 1336 (1939).	400	7
20. Saponification of methyl acetate in water, acetone-water, and ethanol-water. R. A. Fairclough and C. N. Hinshelwood, *J. Chem. Soc.*, 538 (1937).	360	3
21. Hydrolysis of benzoyl chloride in aqueous acetone mixtures. B. L. Archer and R. F. Hudson *J. Chem. Soc.*, 3259 (1950); C. G. Swain and C. B. Scott, *J. Am. Chem. Soc.*, **75**, 246 (1953).	200	6
22. Esterification of acetic acid in equimolar mixtures with isobutyl alcohol as the pressure is increased. Other alcohols also have linear plots of E versus log A but the slopes are different. S. P'eng, R. H. Sapiro, R. P. Linstead, and D. M. Newitt, *J. Chem. Soc.*, 784 (1938).	210	4
23. Hydrolysis of 4-nitro- or 3,5-dinitroacetanilides in various water-H_2SO_4 mixtures. Since some of the reactions are believed to go by the A1 and others by the A2 mechanism, the colinearity may be a coincidence. Three *ortho*-substituted benzoyl derivatives of 4-nitro- or 2,4-dinitroaniline form a line displaced to the right with $\beta = 625$. J. A. Duffy and J. A. Leisten, *J. Chem. Soc.*, 853 (1960).	1330	7

[a] Only examples for which the correlation coefficient is greater than .95 are included in the table. Rates unless otherwise noted. Plots refer to $\Delta H^{\ddagger}$ ordinate, $\Delta S^{\ddagger}$ abscissa.

Table 9-4. Some Isokinetic Relationships in the Reactions of Aldehydes, Ketones, and Their Derivatives[a]

REACTION	β, °K	NUMBER OF POINTS
1. Formation of semicarbazones from *para*-substituted acetophenones in 83% alcohol. R. P. Cross and P. Fugassi, *J. Am. Chem. Soc.*, **71**, 223 (1949).	280	5
2. Acid-catalyzed oximation of alkyl carvacryl ketones. For reference, see reaction (3).	300	7
3. Acid-catalyzed oximation of alkyl thymyl ketones. The line for the thymyl ketones is parallel and displaced to the left of the line for the sterically less hindered carvacryl ketones. However, the point for methyl thymyl ketone, the least hindered member of the series, lies slightly below the line for the carvacryl ketones. M. J. Craft and C. T. Lester, *J. Am. Chem. Soc.*, **73**, 1127 (1951).	300	6
4. Dissociation (equilibrium) of hydrates of aldehydes and ketones in water. R. P. Bell and A. O. McDougall, *Trans, Faraday Soc.*, **56**, 1281 (1960).	410	7
5. Mutarotation of glucose (equilibrium) in methanol-water mixtures. H. E. Dyas and D. G. Hill, *J. Am. Chem. Soc.*, **64**, 236 (1942).	310	4
6. Hydrolysis of alkyl glucopyranosides by aqueous acid. Phenyl derivatives do not fall on the same line. L. J. Heidt and C. B. Purves, *J. Am. Chem. Soc.*, **66**, 1385 (1944); **60**, 1206 (1938).	430	4
7. Acid-catalyzed hydrolysis of disaccharides and oligosaccharides with an aldohexose part-structure. One reducing fructoside is included, but three non-reducing fructosides fall off the line and were omitted. E. A. Moelwyn-Hughes, *The Kinetics of Reactions in Solution*, 2nd ed., Oxford University Press, 1947, p. 74.	360	10
8. Hydrolysis of alkyl fructosides by aqueous acid. The more complex fructoside, sucrose, does not fit. L. J. Heidt and C. B. Purves, *J. Am. Chem. Soc.*, **66**, 1385 (1944); **60**, 1206 (1938).	520	4

[a] Only examples for which the correlation coefficient is greater than .95 are included in the table. Rates unless otherwise noted. Plots refer to $\Delta H^{\ddagger}$ ordinate, $\Delta S^{\ddagger}$ abscissa.

Table 9-5. *Some Isokinetic Relationships in Solvolysis Reactions*[a]

REACTION	β, °K	NUMBER OF POINTS
1. Acetolysis of *p*-bromobenzenesulfonates with no participating neighboring groups. S. Winstein and H. Marshall, *J. Am. Chem. Soc.*, **74**, 1120 (1952).	310	4
2. Acetolysis of acyclic primary *p*-toluenesulfonates with no participating neighboring groups or neighboring phenyl. S. Winstein and H. Marshall, *J. Am. Chem. Soc.*, **74**, 1120 (1952).	400	4
3. Acetolysis of cyclic *p*-toluenesulfonates and *p*-bromobenzenesulfonates. Neomenthyl *p*-toluenesulfonate was omitted. S. Winstein, B. K. Morse, E. Grunwald, H. W. Jones, J. Corse, D. Trifan, and H. Marshall, *J. Am. Chem. Soc.*, **74**, 1127 (1952).	450	6
4. Solvolysis of tertiary aliphatic iodides in 76% alcohol. A group of tertiary aliphatic chlorides, including those corresponding to the extremes of this series of iodides, have nearly constant E and log A. It should be noted that the iodides all have a more positive entropy than the corresponding chlorides. J. Shorter and C. N. Hinshelwood, *J. Chem. Soc.*, 2412 (1949).	470	9
5. Hydrolysis of *ortho*-, *meta*-, and *para*-substituted phenyl sulfates. The fact that *ortho*-substituents fall on a line with the others suggests a mechanism involving attack at neither the aromatic carbon atom nor the adjacent oxygen. G. N. Burkhardt, C. Horrex, and D. I. Jenkins, *J. Chem. Soc.*, 1649 (1936).	510	23
6. Alcoholysis of secondary alkyl chlorides in which the chlorine is *alpha* to a benzene ring. The hindered α-mesitylethyl chloride deviates to the right and was omitted. G. Baddeley and J. Chadwick, *J. Chem. Soc.*, 368 (1951).	760	5
7. Hydrolysis of *p*-nitrobenzyl bromide in dioxane-water mixtures containing 50, 70, and 90 wt % dioxane. J. W. Hackett and H. C. Thomas, *J. Am. Chem. Soc.*, **72**, 4962 (1950).	260	3
8. Alcoholysis or hydrolysis of ethyl or isopropyl or 2-octyl methylphosphonochloridates in various alcohols and acetone-water mixtures. Points for formolysis in formic acid deviate. R. F. Hudson and L. Keay, *J. Chem. Soc.*, 1865 (1960).	700	12

[a] Only examples for which the correlation coefficient is greater than .95 are included in the table. Rates unless otherwise noted. Plots refer to $\Delta H^{\ddagger}$ ordinate, $\Delta S^{\ddagger}$ abscissa.

Table 9-6. Some Isokinetic Relationships for Miscellaneous Reactions[a]

REACTION	β, °K	NUMBER OF POINTS
1. Acid-catalyzed rearrangement of substituted 3-hydroxypentene-1-yne-4 in 60% ethanol, limited to structures with two substituents. E. A. Braude and E. R. Jones, *J. Chem. Soc.*, 122 (1946).	440	10
2. Rearrangement of 1-anisyl-4-carbethoxy-5-hydroxy-1,2,3-triazole in dimethylformamide in the presence of varying concentrations of LiCl, NH_4NO_3, $LiNO_3$, and $NaNO_3$. J. E. Leffler and S. K. Liu, *J. Am. Chem. Soc.*, 78, 1949 (1956).	240	7
3. Rearrangement of *para*-substituted 1-phenyl-4-carbethoxy-5-hydroxy-1,2,3-triazoles in dimethylformamide or acetonitrile. J. E. Leffler and S. K. Liu, *J. Am. Chem. Soc.*, 78, 1949 (1956).	600	6
4. Rearrangement of ethyl benzylhydroxytriazolecarboxylate in a series of solvents. O. Dimroth and B. Brahn, *Ann.*, 373, 365 (1910).	250	4
5. Dehydrogenation of 1,4-dihydronaphthalene by a series of alkyl-substituted *p*-benzoquinones in phenetole. Chloranil, phenanthrenequinone, and 1,2-naphthoquinone deviate. They also deviate from the linear free energy relationship between log k and the redox potentials of the quinones. E. A. Braude, L. M. Jackmann, and R. P. Linstead, *J. Chem. Soc.*, 3548 (1954).	1400	6
6. Oxidation of *p,p'*-dichlorobenzyl sulfide with substituted perbenzoic acids in toluene and in isopropyl alcohol. C. G. Overberger and R. W. Cummins, *J. Am. Chem. Soc.*, 75, 4250 (1953).	340	7
7. Reaction of ferrous ion with hydrogen peroxide and with a series of hydroperoxides. R. A. Orr and H. L. Williams, *J. Phys. Chem.*, 57, 925 (1953).	380	7
8. Electron transfer between aqueous Cr^{++} and complexes $(NH_3)_5Co^{III}L$, where L is *p*-sulphobenzoate, methyl terephthalate, phenyl terephthalate, or *cis* or *trans* 1,2-cyclopropanedicarboxylate. The electron transfer is accompanied by hydrolysis of the ester group. The point for phenyl fumarate falls below the line; rates for fumarate half esters are also unusual in that they are nearly independent of the nature of the atom or group (phenyl, H, D, or CH_3) attached to the more remote carboxyl. R. T. M. Frazer and H. Taube, *J. Am. Chem. Soc.*, 83, 2242 (1961).	296	5
9. Racemization of usnic acid in various solvents. S. Mackenzie, *J. Am. Chem. Soc.*, 77, 2214 (1955).	550	8

[a] Only examples for which the correlation coefficient is greater than .95 are included in the table. Rates unless otherwise noted. Plots refer to $\Delta H^{\ddagger}$ ordinate, $\Delta S^{\ddagger}$ abscissa.

Table 9-7. Some Isokinetic Relationships for Various Physical and Chemical Model Processes[a]

PROCESS	β, °K	NUMBER OF POINTS
1. Dielectric absorption in solids due to rotation of dipoles over an energy barrier can be fitted to the equation, $\nu_{max} = Ae^{-E/RT}$. There are two parallel lines, the upper for long chain esters and long chain methyl ethers in the α- or waxy phase, the lower for long chain esters and ethers in the β- or crystalline phase. R. J. Meakins, *Trans. Faraday Soc.*, **55**, 1694 (1959).	285 285	10 12
2. Formation (equilibrium) of complexes between alkyl or polyalkylbenzenes and iodine or iodine monochloride in carbon tetrachloride solution. Points for iodine with hexaethylbenzene, and for iodine monochloride with pentaethylbenzene, hexaethylbenzene, p-xylene, and hexamethylbenzene fall below the line. N. Ogimachi, L. J. Andrews, and R. M. Keefer, *J. Am. Chem. Soc.*, **77**, 4202 (1955); R. M. Keefer and L. J. Andrews, *ibid.*, **77**, 2164 (1955).	680	13
3. Formation (equilibrium) of complexes between ethers or alcohols and iodine in solution in an alkane solvent or carbon tetrachloride. Amines, amine oxides, and benzene deviate. Several alkyl benzenes give points on the line. M. Tamres and M. Brandon, *J. Am. Chem. Soc.*, **82**, 2134 (1960); P. A. D. de Maine, *J. Chem. Phys.*, **26**, 1199 (1957).	370	17
4. Sublimation (equilibrium) of straight-chain dicarboxylic acids at 1 atm pressure. The points for the C_4, C_6, C_8, and C_{10} acids fall on a line in that order of increasing $\Delta \bar{H}°$. The points for the C_{12}, C_{16}, and C_{20} acids fall above the line. The latter may be partly cyclic in the vapor. M. Davies and G. H. Thomas, *Trans. Faraday Soc.*, **56**, 185 (1960).	470	4
5. Racemization of a series of methyl- or halogen-substituted N-benzoyldiphenylamine-2-carboxylic acids in ethanol or chloroform-ethanol solution. D. M. Hall and M. M. Harris, *J. Chem. Soc.*, 490 (1960).	250	8
6. Racemization of 2-(2-dimethylaminophenyl)-phenyltrimethylammonium ion in acetic acid and a series of alcohols. Water and trifluoroethanol deviate; so do nonhydroxylic solvents. Two-component solvent mixtures give isokinetic lines having a sharp break when a critical solvent composition is reached. J. E. Leffler and W. H. Graham, *J. Phys. Chem.*, **63**, 687 (1959).	345	7

Table 9–7 (continued)

PROCESS	β, K	NUMBER OF POINTS
7. Same process as in (6), but in water and in the presence of small-ion salts. Nine salts of large ions form a rough isokinetic line falling below this one. W. H. Graham and J. E. Leffler, *J. Phys. Chem.*, **63**, 1274 (1959).	380	15
8. Racemization of derivatives of 2,2′-dimethoxy-6,6′-dicarboxydiphenyl in a series of solvents. The acid and the dimethyl ester give an isokinetic line in polar solvents; points for less polar solvents fall above the line. The diamide also gives an isokinetic line but is not soluble in as many solvents. B. M. Graybill and J. E. Leffler, *J. Phys. Chem.*, **63**, 1461 (1959).	370 (acid and ester) 390 (diamide)	20 10
9. Isomerization of *cis*-chlorobenzene diazocyanide in a series of solvents. R. J. Le Fèvre and J. Northcott, *J. Chem. Soc.*, 944 (1949).	350	8
10. Isomerization of azobenzene in a series of solvents and of substituted azobenzenes in benzene. The fit is better with the omission of the data for structural variation. R. J. Le Fèvre and J. Northcott, *J. Chem. Soc.*, 867 (1953).	340	13
11. *Cis-trans* isomerization of substituted stilbenes in the liquid phase without solvent, of stilbene in the gas and liquid phases, and of *p*-amino-*p*′-nitrostilbene in xylene. Because all of these stilbene data fall on a single line, it is likely that they represent a single mechanism rather than a triplet state mechanism for those with low entropies and a singlet state mechanism for those with high entropies. The low entropies of activation for some members of the series might be due to solvent orientation in a polar transition state rather than a low transition probability to a triplet state. In the absence of evidence that the stilbenes respond differently to paramagnetic catalysts, the singlet state mechanism seems the more probable. M. Calvin and H. W. Alter, *J. Chem. Phys.*, **19**, 768 (1951); J. L. Magee, W. Shand, Jr. and H. Eyring, *J. Am. Chem. Soc.*, **63**, 677 (1941).	720	8
12. Decay of thermochromic color, less hindered dianthrones. Y. Hirshberg and E. Fischer, *J. Chem. Soc.*, 629 (1953).	210	4

Table 9-7 (continued)

PROCESS	β, K	NUMBER OF POINTS
13. Decay of thermochromic color, more hindered dianthrones and xanthylideneanthrone. The two lines for the hindered and less hindered series are nearly parallel, with the less hindered one having the lower E for a given log A. Two dixanthylenes have been studied, but their points fall close together and with a lower E than points corresponding to the same value of log A on the other two lines. Y. Hirshberg and E. Fischer, *J. Chem. Soc.*, 629 (1953).	250	5
14. Semiconductivity obeys the relationship $C = C_0 e^{-E/RT}$. For a series of unsaturated compounds including polynuclear hydrocarbons, E is a linear function of log C_0. Many of the exceptions to this rule have been found to contain paramagnetic impurities. V. L. Talroze, quoted by N. N. Semyonov, Symposium on Macromolecular Chemistry, June 1960.	300	25

[a] Only examples for which the correlation coefficient is greater than .95 are included in the table. Rates unless otherwise noted. Plots refer to $\Delta H^{\ddagger}$ ordinate, $\Delta S^{\ddagger}$ abscissa.

approximate isokinetic relationships and an even greater number of reactions showing no correlation whatsoever. The latter evidently have more than one important interaction mechanism, and the relevant theory will be presented in the next section.

A high correlation coefficient does not always mean that the basis for the relationship is the presence of only a single interaction mechanism. As will be seen in the next section, a straight line will be obtained even in the presence of two interaction mechanisms if their β-values happen to be equal. A glance at the tables will reveal examples for which this is almost certainly so. Thus, when a substituent effect and a medium effect give points falling on a single line, we have usually treated them as a single relationship.

A high correlation coefficient does not always mean that the sequence of points along the line and the range of variation are correct. As we mentioned in the preceding section, this uncertainty of interpretation is particularly great when β is close to the experimental temperatures. For many of the relationships included in the tables, even the possibility that the $\delta \Delta H$ and $\delta \Delta S$ are in fact all zero cannot be completely excluded within the precision of the experimental data, although such an hypothesis is inherently very implausible. We decided to include such relationships because at this stage we would rather include a few possibly spurious examples than risk omitting one that may some day prove to be important.

Table 9-3 contains data on acid- and base-catalyzed ester hydrolysis reactions, important because of their use as standard processes in linear free energy relationships. Table 9-7 contains data on various processes, some of which we consider important because of their simplicity and consequent value for testing theoretical models. In this table are listed an example of dielectric relaxation, several examples of internal rotation about single and double bonds, and several examples of the formation of charge-transfer complexes.

The distribution of β-values in Tables 9-1 through 9-7 was examined in an attempt to discover ranges of β that might be characteristic of particular reaction or interaction mechanisms. The sample, however, is not entirely representative because we have rejected reactions in which the range of $\Delta H^{\ddagger}$ or of $T \Delta S^{\ddagger}$ is small (1 kcal or so). This practice tends to exclude examples having very high or very low β values. The only significant generalization that seems to emerge from the tables is a marked tendency for reaction series in which the variable is the solvent to have β values in the range 300 to 400°K, which contains over 50% of that sample. This means that in the usual experimental temperature range, the effect of interactions with the medium on the free energy is much smaller than the effect on the enthalpy or entropy (see equations 15 and 16). In Chapter 3 it was found that standard free energies of formation of molecular complexes, an important solvation mechanism, tend to be within 1 kcal/mole of zero because of a similar compensating effect of the standard enthalpy and entropy changes.

A further confirmation of the hypothesis that isokinetic relationships attributable to molecular complexing will have β values in the range 300 to 400°K is the observation that a similar relationship is observed for dielectric relaxation. The process of orientation of a molecule in an external electric field should be a fair model for the orientation of a molecule in the field of a neighboring dipolar molecule. The dielectric relaxation of acridine, α-chloronaphthalene, isoquinoline, and p-bromobiphenyl in Nujol or decalin solution shows an isokinetic relationship in which the enthalpies of activation span 8 kcal/mole and the entropies 21 cal/mole degree.[8c]

MULTIPLE INTERACTION MECHANISMS

Let us suppose that the substituent or medium effect operates by means of several physically distinct mechanisms, that is, that in changing the

[8c] O. F. Kalman and C. P. Smyth, *J. Am. Chem. Soc.*, **82**, 783 (1960).

substituent we change not just one but several quantitatively important interactions. These might be resonance, polarization, or steric strain, for example. In first approximation, the effects of the several interaction mechanisms are simply additive.

$$\delta \Delta F = \frac{\partial \Delta F}{\partial r_1}(r_1 - r_{01}) + \frac{\partial \Delta F}{\partial r_2}(r_2 - r_{02}) + \cdots$$

$$\delta \Delta H = \frac{\partial \Delta H}{\partial r_1}(r_1 - r_{01}) + \frac{\partial \Delta H}{\partial r_2}(r_2 - r_{02}) + \cdots \qquad (18)$$

$$\delta \Delta S = \frac{\partial \Delta S}{\partial r_1}(r_1 - r_{01}) + \frac{\partial \Delta S}{\partial r_2}(r_2 - r_{02}) + \cdots$$

The variables $r_1, r_2, \ldots$ associated with mechanisms $1, 2, \ldots$ are temperature independent.

Our notation can be made more compact as follows. Let $\delta_1 \Delta F = (\partial \Delta F/\partial r_1)(r_1 - r_{01})$, $\delta_1 \Delta H = (\partial \Delta H/\partial r_1)(r_1 - r_{01})$, and so on. We then obtain,

$$\delta \Delta F = \delta_1 \Delta F + \delta_2 \Delta F + \cdots$$

$$\delta \Delta H = \delta_1 \Delta H + \delta_2 \Delta H + \cdots \qquad (19)$$

$$\delta \Delta S = \delta_1 \Delta S + \delta_2 \Delta S + \cdots$$

It is clear from equations 18 that the contributions to $\delta \Delta F$, $\delta \Delta H$, and $\delta \Delta S$ from any one interaction mechanism are proportional.

$$\delta_1 \Delta H = \beta_1 \delta_1 \Delta S \qquad (20)$$

$$\delta_1 \Delta F = (\beta_1 - T)\delta_1 \Delta S = \left(1 - \frac{T}{\beta_1}\right)\delta_1 \Delta H \qquad (21)$$

In other words, the contribution from each of the independent interaction mechanisms gives rise to a quasi-isokinetic relationship with its own characteristic value of β, which we shall assume to be positive. The magnitudes of the various β_i may be quite different.

We would now like to examine the effect of multiple interaction mechanisms by means of enthalpy-entropy diagrams and introduce some useful new terminology. For definiteness, let us consider a substituent effect and limit the number of interaction mechanisms to two. Under these conditions equations 19 and 20 lead to 22 and 23.

$$\delta_R \Delta S = \delta_1 \Delta S + \delta_2 \Delta S \qquad (22)$$

$$\delta_R \Delta H = \delta_1 \Delta H + \delta_2 \Delta H$$

$$= \beta_1 \delta_1 \Delta S + \beta_2 \delta_2 \Delta S \qquad (23)$$

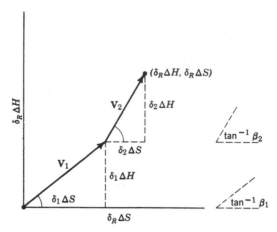

Figure 9-4. The position of a point in the enthalpy-entropy plane expressed as the resultant of two interaction mechanism vectors.

Unfortunately we can measure directly only the theoretically complex quantities $\delta_R \Delta S$ and $\delta_R \Delta H$. The theoretically simpler quantities $\delta_1 \Delta S$, $\delta_1 \Delta H$, $\delta_2 \Delta S$, and $\delta_2 \Delta H$ are not directly observable.

The position of an experimental point $(\delta_R \Delta H, \delta_R \Delta S)$ on an enthalpy-entropy diagram is the sum of two vectors, as in Figure 9-4. In this figure, the vector $\mathbf{V}_1$, with components $(\delta_1 \Delta H, \delta_1 \Delta S)$, is due to the first interaction mechanism. The vector $\mathbf{V}_2$, with components $(\delta_2 \Delta H, \delta_2 \Delta S)$, is due to the second. The direction angles are $\tan^{-1} \beta_1$ and $\tan^{-1} \beta_2$, respectively.

If we have a series of substituents, each experimental point is found by such a method of adding two interaction mechanism vectors, as shown in Figure 9-5a. All vectors in the set due to the first interaction mechanism point in the same direction, specified by the angle $\tan^{-1} \beta_1$, although the magnitudes of the individual vectors vary. All vectors in the set due to the second interaction mechanism point in a fixed direction specified by the angle $\tan^{-1} \beta_2$, which is usually different from $\tan^{-1} \beta_1$. The magnitudes of the vectors again vary with the substituent. However, the values of $\delta_2 \Delta S$ are not correlated with those of $\delta_1 \Delta S$.

Because of the lack of correlation between $\delta_1 \Delta S$ and $\delta_2 \Delta S$, the experimental points $(\delta_R \Delta H, \delta_R \Delta S)$ are badly scattered. This fact is shown more clearly in Figure 9-5b, where the points of Figure 9-5a are shown without the distraction of the interaction mechanism vectors. A very rough linear trend remains because the first interaction mechanism is quantitatively more important than the second, but one would hardly be justified in

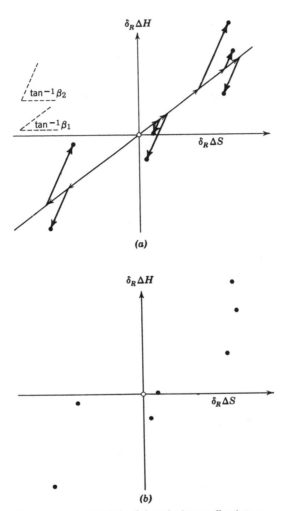

Figure 9-5. (a) Analysis of the substituent effect in terms of two interaction mechanisms. (b) Same points as in (a) but without the interaction mechanism vectors.

drawing a straight line.[9] As the number of interaction mechanisms becomes greater than two, this vestige of a linear trend soon disappears and a random scatter diagram results. There are only two circumstances under which two or more interaction mechanisms would give a reasonably good isokinetic line. One is the trivial case that one mechanism truly dominates the magnitude of $\delta_R \Delta H$ and $\delta_R \Delta S$, the other mechanisms causing only small perturbations. The other is the special case in which the β-values for the various mechanisms happen to be very nearly equal.

It may happen that one of the interaction mechanism vectors is restricted to a set of two or three discrete magnitudes. When such a discretely valued interaction mechanism vector is added to an interaction mechanism vector of the ordinary continuous type, the plot of $\delta_R \Delta H$ versus $\delta_R \Delta S$ will consist of two or three parallel lines. The slopes of the lines are all equal to the value of β for the continuously varying mechanism. The displacements of the parallel lines are due to the discrete changes in the magnitude of the discontinuous vector, as illustrated in Figure 9-6. An alternative explanation of pairs of parallel lines would be two mutually exclusive additional

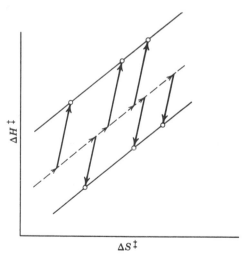

Figure 9-6. Effect of a discontinuous second interaction mechanism vector.

[9] There was a chemist from Latta,
Who couldn't interpret his data.
So he drew a straight line—
Now everything's fine,
Except for that damnable scatter.*

* Latta is located on route U.S. 301, South Carolina.

interaction mechanisms, each having vectors of *constant* rather than discretely varying magnitude.

Pairs of parallel isokinetic lines are in fact not uncommon. They might, for example, result from the effect of steric hindrance on the conformational composition of the formal reagent or transition state species. A certain critical amount of steric hindrance could cause a new conformation to be the favored one, yet the discrete change in conformation need not affect the continuous mechanism.

A situation analogous to the simultaneous action of two interaction mechanisms for a single substituent is the effect of making two simultaneous changes in structure, one of them a variable and the other a constant. For example, in the acid-catalyzed hydrolysis of a series of esters and thiol esters in 62 % acetone, the data for the esters fall on a single line which is parallel to that formed by the data for the thiol esters, the difference between the two lines being the constant effect of the sulfur atom as compared to the oxygen atom (Table 9-3, items 4 and 5). The variable structure change need not be of the single-mechanism type, however, and the result can be a set of constant vectors added to a scatter diagram rather than to points on a straight line. An example is shown in Figure 9-7 for the effect of *para*-substituents in 2,2-diphenyl-1,1,1-trichloroethane and in 2,2-diphenyl-1,1-dichloroethane on the rates of elimination of hydrogen chloride in alcoholic alkali.[10] The structural change corresponding to the replacement of chlorine by hydrogen on the 1-carbon atom shifts the point representing the reaction by a nearly constant amount and in a nearly constant direction. That is, the effect of this structural change is a nearly constant vector superimposed on the apparently random changes brought about by the other substituents.

ADDITIVITY IN ENTHALPIES AND IN ENTROPIES

The observation, in the preceding section, that a constant structural change adds a constant vector to each point in the enthalpy-entropy diagram suggests that simple additivity rules might enjoy some measure of success in the representation of enthalpy and entropy quantities. In this section we shall briefly consider two special cases: the simultaneous introduction of two substituents in different parts of the molecule and the superposition of a medium effect on a substituent effect.

In Chapter 7 we had discussed the conditions under which the effect of introducing two substituents simultaneously is simply the sum of the effects

[10] S. J. Cristol, N. L. Hause, A. J. Quant, H. W. Miller, K. R. Eilar, and J. S. Meek, *J. Am. Chem. Soc.*, **74**, 3333 (1952).

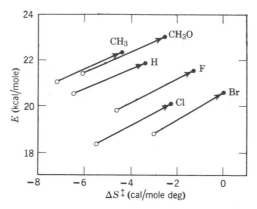

Figure 9-7. Dehydrochlorination of 2,2-diarylchloroethanes;

open circles, $(p\text{-}XC_6H_4)_2CHCCl_3$;
solid circles, $(p\text{-}XC_6H_4)_2CHCHCl_2$.

of introducing the same substituents separately. We had concluded that the substituents must not alter the reaction mechanism in any fundamental way and that the interaction of the substituents with each other must not produce a measurable perturbation. Apparently these conditions are satisfied by the substituents in 2,2-diphenyl-1-chloroethane, whose effect on the activation parameters for dehydrochlorination has been shown in Figure 9-7. To a good approximation, the effects of substituting Cl for H on the 1-carbon atom and of changing the *para*-substituent on the phenyl groups on the 2-carbon atom are simply additive. It should be noted, however, that the sites at which the substituents are introduced in the molecule are pretty far apart.

The alcoholysis of alkyl-substituted benzhydryl chlorides[11] provides an example of an almost precisely additive substituent effect on the activation parameters when the substituents are closer together. Figure 9-8 shows the effect of introducing, either separately or simultaneously, a *p*-alkyl group in the 4-position and two methyl groups in the 3,5-positions. It is seen that the point for benzhydryl chloride deviates markedly from the good isokinetic line established by the compounds with a *p*-alkyl substituent. Similarly, the point for the 3,5-dimethyl compound deviates *by exactly the same amount* from a *parallel* line formed by the 3,5-dimethyl-4-alkyl substituted compounds. The order in which the various *p*-alkyl groups appear along the isokinetic lines is the same with or without the methyl

---

[11] G. Baddeley, S. Varma, and M. Gordon, *J. Chem. Soc.*, 3171 (1958).

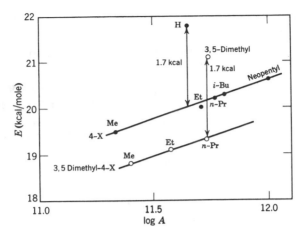

Figure 9-8. Activation parameters for alcoholysis of substituted benzhydryl chlorides, $ArCHClC_6H_5$.

groups in the 3,5-positions. There may be some difference in the spacing of the points, but we cannot be sure of this because the β-value, 360°K, is only slightly above the experimental temperature, and the positions of the points along the lines are somewhat uncertain.

Incidentally, the results shown in Figure 9-8 are of some interest in connection with hyperconjugation. The good accuracy of the isokinetic relationships for the p-alkyl substituents shows that the substituent effect involves only one major interaction mechanism that depends specifically on the nature of the p-alkyl group. The fact that the two lines are parallel shows that this mechanism is not interfered with by flanking methyl groups. It is probable that the variable interaction mechanism is the inductive effect: the sequence of the points along the line, Me < Et < n-Pr < i-Bu < neopentyl, is the sequence of the inductive effect, and the spacing of the points correlates well with the spacing of the σ^* parameters for the alkyl groups (Table 7-13), especially on the upper line. If the isokinetic lines are indeed due to a variable inductive effect, then the interaction mechanism vector resulting from hyperconjugation must be a constant. Since the series includes p-methyl as well as higher primary alkyl groups, it follows that the effects of C—H and C—C hyperconjugation are equal.

In discussing additivity of substituent and medium effects it is convenient to represent standard enthalpies and entropies by equations of the same form we have previously found useful for the representation of standard free energies. Thus for a reagent RX in a medium M, the standard enthalpy and entropy can be written, in first approximation, as equations 24 and 25.

$$\bar{H}^\circ_{RX} = H_R + H_X + H_{I(R,X)} + H_{I(R,M)} + H_{I(X,M)} \tag{24}$$

$$\bar{S}^\circ_{RX} = S_R + S_X + S_{I(R,X)} + S_{I(R,M)} + S_{I(X,M)} \tag{25}$$

In these equations, the symbols H_R, H_X, S_R, and S_X denote the independent contributions from the part structures R and X, and the symbols $H_{I(R,X)}, \ldots, S_{I(X,M)}$ denote the respective interaction terms.

According to equations 24 and 25, the standard changes in enthalpy and entropy for the solution of a gaseous compound RX in a medium M are simple additive functions of independent contributions assignable to the solvation of the part-structures.

$$\delta_{sol}\bar{H}^\circ_{RX} = H_{I(R,M)} + H_{I(X,M)} \tag{26a}$$

$$\delta_{sol}\bar{S}^\circ_{RX} = S_{I(R,M)} + S_{I(X,M)} \tag{26b}$$

The existence of additive relationships such as (26) has been verified for entropies of solution of quite a wide variety of organic substances in water.[12]

That the enthalpies of solution in water might also be additive quantities is suggested by the existence of parallel enthalpy-entropy lines for the process of solution of pairs of homologous series. An example is shown in Figure 9-9.[13] The example is typical in showing the approximate character of most simple additivity schemes. On the one hand, the sequence of the points for the various homologues is the same on both lines, and the vectors connecting corresponding points on the two lines are approximately equal. On the other hand, there are quantitative differences that appear to be significant even though the positions of the points along the two lines are relatively imprecise.

It also follows from (24) and (25) that the effect of a change in medium on $\Delta\bar{H}^\circ$ and $\Delta\bar{S}^\circ$ for the process, $RX \to RX^*$, is given in first approximation by (27).

$$\delta_M \Delta\bar{H}^\circ = \delta_M(H_{I(X^*,M)} - H_{I(X,M)}) \tag{27a}$$

$$\delta_M\Delta\bar{S}^\circ = \delta_M(S_{I(X^*,M)} - S_{I(X,M)}) \tag{27b}$$

It is seen that $\delta_M \Delta\bar{H}^\circ$ and $\delta_M \Delta\bar{S}^\circ$ are independent of the nature of R and that they depend only on the particular change in the reaction zone represented by $X \to X^*$. As a result, it is predicted (within the first approximation) that the points for a series of reagents RX on an enthalpy-entropy diagram should be displaced by a constant vector characteristic of

[12] J. W. Cobble, *J. Chem. Phys.*, **21**, 1451 (1953).

[13] W. F. Claussen and M. F. Polglase, *J. Am. Chem. Soc.*, **74**, 4817 (1952); R. J. L. Andon, J. D. Cox, and E. F. G. Herington, *J. Chem. Soc.*, 3188 (1954). R. E. Robertson, R. L. Heppolette, and J. M. W. Scott, *Can. J. Chem.*, **37**, 803 (1959).

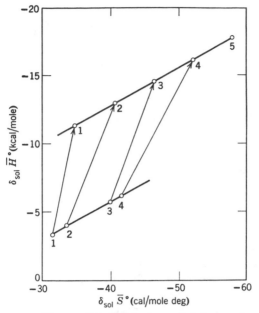

Figure 9-9. Standard enthalpies and entropies of solution of gaseous *n*-alkanes (lower line) and *n*-alkanols in water at 25°C. The numbers are the number of carbon atoms. Solute concentrations are in units of mole fraction, gas pressure in atmospheres. $\beta = 280°K$.

the medium change. Thus an isokinetic line will be uniformly translated into a parallel line by a change in solvent. This is realized by the example shown in Figure 9-10, the solvolysis of a series of *para*-substituted benzoyl chlorides in ether-alcohol, the lower line, and in acetone-water, the upper line.[14] Although the point for the nitro group deviates from the isokinetic line (presumably because of an additional substituent interaction mechanism), the pattern of the deviation is not affected by the solvent change.

A similar additive effect is expected, in first approximation, when a single structural change is superimposed on a series of medium changes. Thus the saponification of ethyl benzoate and ethyl *m*-nitrobenzoate in a series of solvents gives two parallel lines, as shown in Figure 9-11.[15] The

[14] G. E. K. Branch and A. C. Nixon, *J. Am. Chem. Soc.*, **58**, 2499 (1936); D. A. Brown and R. F. Hudson, *J. Chem. Soc.*, 883 (1953).
[15] R. A. Harman, *Trans. Faraday Soc.*, **35**, 1336 (1939); R. A. Fairclough and C. N. Hinshelwood, *J. Chem. Soc.*, 538 (1937).

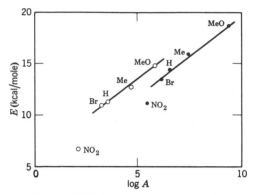

Figure 9-10. Solvolysis of *p*-substituted benzoyl chlorides in acetone-water (open circles) and in ether-alcohol (filled circles).

additivity in this case is only approximate, however, because the vectors connecting corresponding points on the two lines are not quite constant.

Whereas in the examples given so far the fit to equations 26 and 27 has ranged from fairly good to excellent, we can also cite reactions for which the fit is poor, that is, the medium effect on the activation parameters

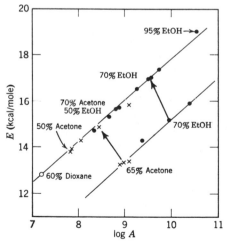

Figure 9-11. Saponification of ethyl benzoate (upper line) and ethyl *m*-nitrobenzoate (lower line) in aqueous-organic solvents. Crosses, acetone-water; solid circles, ethanol-water; open circle, dioxane water. $\beta = 390°K$.

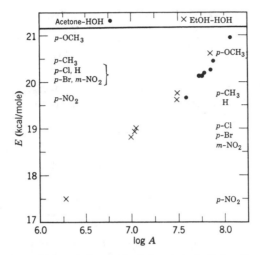

Figure 9-12. Acid-catalyzed hydrolysis of ethyl benzoates.

depends strongly on the nature of the substituent. Data for one such reaction, the acid-catalyzed hydrolysis of substituted ethyl benzoates,[16] are shown in Figure 9-12. The substituent effects on the activation parameters are seen to be much greater in 60% ethanol-water than in 60% acetone-water. This is a disturbing result. Ester hydrolysis, it will be recalled, is one of the standard reactions in the extrathermodynamic analysis of polar and steric substituent effects on free energy quantities, and one hates to find unusual complexities, regardless of their cause. Moreover, it is by no means certain that the large deviations from equation 27 are due to a real failure of that equation. Instead they might be due to a variable reaction mechanism. Ester hydrolysis is known to proceed by a mechanism involving several steps, and it appears that no single one of these steps can be expected to be rate determining under all conditions.[17]

A COMPARISON OF THE EFFECTS OF INTERACTION MECHANISMS ON THE FREE ENERGY AND ON THE ENTHALPY OR ENTROPY[1]

An interaction mechanism that has an important effect on the enthalpy or entropy may exert merely a minor perturbation on the free energy. This is

[16] E. W. Timm and C. N. Hinshelwood, *J. Chem. Soc.*, 862 (1938).
[17] The rate-determining step in the analogous mechanism for base-catalyzed ester hydrolysis has been analyzed by M. L. Bender and R. J. Thomas, *J. Am. Chem. Soc.*, **83**, 4189 (1961).

readily seen from equation 21, which we shall rewrite for the ith interaction mechanism.

$$\delta_i \Delta F = (\beta_i - T)\delta_i \Delta S = \left(1 - \frac{T}{\beta_i}\right)\delta_i \Delta H \tag{21}$$

When T equals β_i, the contribution of the ith interaction mechanism to $\delta \Delta F$ vanishes. This means that a free energy relationship at $T = \beta_i$ is simpler by one interaction mechanism than is an enthalpy or entropy relationship. Since interaction mechanisms for which β is in the popular experimental temperature range are quite common, it is not surprising that free energy quantities are often easier to interpret than enthalpies. When there are only two interaction mechanisms, free energies measured at the isokinetic temperature for one of them will reflect only the single remaining interaction mechanism and will therefore be particularly simple.

When there is one major interaction mechanism, both the free energies and enthalpies or entropies will be simple and easy to interpret. However, pure examples of a single interaction mechanism are not met in practice. Even when the approximation of a single interaction mechanism is at its best, for example, when the enthalpies and entropies give an excellent linear correlation, there is always enough scatter or "noise" in the relationship to suggest the existence of a multitude of minor interaction mechanisms. Because they are minor they are not well understood, nor do they seriously affect the comparison of experimentally determined enthalpies with theory. They can be a serious inconvenience under certain circumstances, however. Thus if a series of $\delta \Delta F$ values is determined at the isokinetic temperature for the single major interaction mechanism, the results will reflect only the random effect of the minor interaction mechanisms or "noise." Except that the $\delta \Delta F$ values will all be close to zero, they will not conform to any simple theory, except accidentally. The enthalpies, on the other hand, will still be well behaved.

Experimental examples are known for both of the expected types of behavior. Sometimes the enthalpies are simple and rational but not the free energies and sometimes vice-versa. An example of simplicity in free energies accompanied by complexity of the enthalpies and entropies is provided by the formation of oximes and semicarbazones[18] from a series of ketones in buffered aqueous solution. The enthalpy-entropy relationships for the two systems are shown in Figures 9-13 and 9-14, which make clear the existence of more than one interaction mechanism by their marked scatter. Yet the free energy relationship at 25° (Figure 9-15a) is quite accurately linear with the exception of minor deviations by the points for

[18] F. W. Fitzpatrick and J. D. Gettler, *J. Am. Chem. Soc.*, **78**, 530 (1956). F. P. Price, Jr. and L. P. Hammett, *ibid.*, **63**, 2387 (1941).

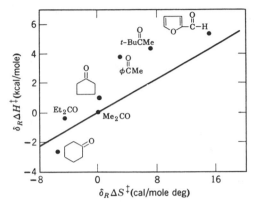

Figure 9-13. Oxime formation in aqueous buffers. An arbitrary line of slope 298°K is drawn through the point for acetone.

furfural and diethyl ketone. Furthermore, the relationship at 0° (Figure 9-15b) is also fairly good, the pattern of scatter being somewhat different. Evidently there are two major interaction mechanisms, one of which has an isokinetic temperature near 298°K.

An example of a reaction in which the enthalpies are simpler and more rational than the free energies is the hydrogen ion-catalyzed decomposition of hexaminium ion, $(CH_2)_6N_4H^+$, in water.[19] This reaction shows a linear

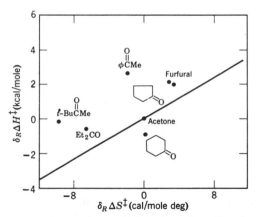

Figure 9-14. Semicarbazone formation in aqueous buffers. An arbitrary line of slope 298°K is drawn through the point for acetone.

[19] H. Tada, *J. Am. Chem. Soc.*, **82**, 255 (1960).

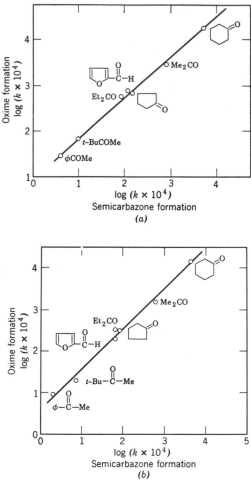

Figure 9-15. Free energy relationship for oxime formation and semicarbazone formation (a) at 25°C; (b) at 0°C.

enthalpy-entropy relationship with a β-value of 205°K as the ionic strength is varied by adding neutral salts; it shows another linear enthalpy-entropy relationship with a β-value of 315°K when the medium, at zero ionic strength, is varied by adding glycol or t-butyl alcohol to the solution.[19] The rate constants in the aqueous salt solutions at 328°K increase with increasing ionic strength as is expected for a reaction between two univalent cations. The reason for this rational behavior of the rate constants is that 328°K is quite far from the isokinetic temperature, 205°K, at which temperature only the "noise" would be effective. The rate constants in the aqueous-organic media, in contrast, do not show any rational pattern. That is, log k at the experimental temperatures does not show the expected consistent increase with increasing solvent polarity. The reason for this is that the experimental temperatures are near the isokinetic temperature of 315°K, and the small rate differences merely reflect minor secondary inter-action mechanisms. On the other hand, if the activation enthalpies for a series of mixtures of water with a given alcohol are examined, it is found that the activation enthalpy consistently decreases with increasing solvent polarity.

It is possible to predict relative rates from a casual inspection of an enthalpy-entropy diagram by means of the following simple but useful procedure. A line of slope equal to the temperature for which the rate comparison is wanted is drawn through one of the points, arbitrarily chosen as a reference (Figure 9-16). The relative rates can then be deter-mined from the vertical distances of the points from the line, since the vertical distance of each point from the line is equal to $\delta \Delta F^{\ddagger}$ for that point. Points above the line correspond to slower reactions, those below the line correspond to faster reactions. The same technique can of course be used to estimate the small residual differences in rate at an isokinetic temperature.

The difference in free energy of activation between the reaction repre-sented by the point P_X of Figure 9-16 and the reference reaction (P_R) is equal to the sum of $\delta \Delta F^{\ddagger}$ between P_R and P and $\delta \Delta F^{\ddagger}$ between P and P_X. The former is zero because both points are on a line of slope equal to T, and the latter is purely enthalpic. Hence $\delta \Delta F_X^{\ddagger}$ is equal to $\Delta H^{\ddagger}$ at P_X minus $\Delta H^{\ddagger}$ at P.

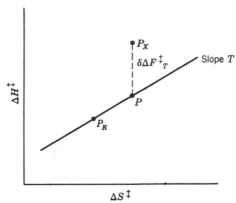

Figure 9-16. Geometrical evaluation of a relative free energy.

STRUCTURAL EFFECTS

Abrupt Changes in Mechanism

In the acid-catalyzed hydrolysis of alkyl thiolacetates in 62% acetone, a good isokinetic relationship is obtained except for the conspicuous deviation of the point for the trityl ester. This compound reacts by a different mechanism, involving cleavage of the alkyl-sulfur bond.[20] In contrast, the *alkaline* hydrolysis of alkyl thiolacetates shows an isokinetic relationship in which the point for the trityl ester is well behaved, and there is no evidence for any change in reaction mechanism. Other examples in which a change in reaction mechanism or interaction mechanism occasions a departure from an isokinetic relationship will be taken up in later sections.

Although major changes in reaction mechanism cause sharp departures, there is at least one example of a rough correlation between enthalpy and entropy, with positive slope, for a series of reactions that almost certainly involve minor changes in reaction and interaction mechanisms. The reactions are the decomposition of a series of *t*-butyl peresters into free radicals.[21] In some of these reactions, one or more molecules of carbon dioxide are formed directly in the rate-determining step.

$$\text{(28)} \qquad \rightarrow t\text{-Bu—O}\cdot + \text{O}=\text{C}=\text{O} + \cdot\text{CCl}_3$$

$$\text{(29)}$$

In the reactions leading to carbon dioxide (e.g., equations 28 and 29), the energy of the transition state is lower than it otherwise would be because the energy required to break the oxygen-oxygen single bond is partially furnished by the incipient formation of a new carbon-oxygen double bond. This, however, requires a conformation that will permit the orbital overlap and hence restricts the rotation about the bonds indicated in the formulas by small rectangles. The reduction in activation energy is therefore

[20] B. K. Morse and D. S. Tarbell, *J. Am. Chem. Soc.*, **74**, 416 (1952).
[21] P. D. Bartlett and R. R. Hiatt, *J. Am. Chem. Soc.*, **80**, 1398 (1958).

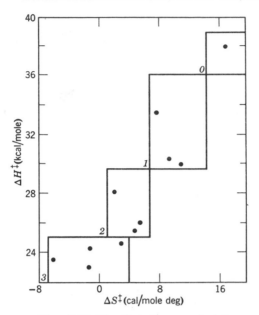

Figure 9-17. The effect of hampered rotations in the decomposition of t-butyl peresters. The numbers in italics are the number of hampered rotations.

accompanied by a constraint and, consequently, by a reduction in activation entropy.

In the reactions leading to resonance-stabilized radicals, a similar overlap requirement interferes with rotation about certain carbon-carbon single bonds, as indicated in the formulas of Table 9-8. Hence the stabilization of the transition state by resonance energy is partially offset by a negative resonance entropy. The resulting values of $\Delta H^{\ddagger}$ and $\Delta S^{\ddagger}$ (Figure 9-17 and Table 9-8) show a rough correlation with each other because they are both correlated with the number of bonds whose rotation is hampered in the transition state.

Steric Effects

Certain structural effects depend on the proximity of two groups in space. These are often described as steric effects, but it is not always easy to decide when this description is appropriate. In some cases it is clear that some other explanation is preferable. For example, in the decomposition of *ortho*-substituted benzazides in toluene, the hydroxyl group decelerates the reaction, whereas all other *ortho*-groups accelerate it. Moreover, the

Table 9-8. The Decomposition of t-Butyl Peresters[21]

PEROXIDE[a]	$\Delta H^{\ddagger}$	$\Delta S^{\ddagger}$	NUMBER OF HAMPERED ROTATIONS
$t\text{-BuO—O—}\overset{\displaystyle O}{\overset{\|}{C}}\text{—CH}_3$	38	17	0
$t\text{-Bu—O—O—}\boxed{\ \ }\overset{\displaystyle O}{\overset{\|}{C}}\text{—CCl}_3$	30.3[b]	9.4	1
$t\text{-Bu—O—O—}\overset{\displaystyle O}{\overset{\|}{C}}\boxed{\ \ }\phi$	33.5[c]	7.8	1
$t\text{-Bu—O—O—}\boxed{\ \ }\overset{\displaystyle O}{\overset{\|}{C}}\text{—C(CH}_3)_3$	30.0[b]	11.1	1
$t\text{-Bu—O—O—}\boxed{\ \ }\overset{\displaystyle O}{\overset{\|}{C}}\text{—}\underset{\displaystyle \text{CH}_3}{\overset{\displaystyle \text{CH}_3}{C}}\boxed{\ \ }\phi$	26.1	5.8	2
$t\text{-Bu—O—O—}\boxed{\ \ }\overset{\displaystyle O}{\overset{\|}{C}}\text{—CH}_2\boxed{\ \ }\phi$	28.1[b]	2.2	2
$t\text{-BuO—O—}\boxed{\ \ }\overset{\displaystyle O}{\overset{\|}{C}}\text{—}\overset{\displaystyle O}{\overset{\|}{C}}\boxed{\ \ }\text{O—O—Bu-}t$	25.5[d]	5.1	2
$t\text{-BuO—O—}\boxed{\ \ }\overset{\displaystyle O}{\overset{\|}{C}}\text{—}\underset{\displaystyle \phi}{\overset{\displaystyle \text{CH}_3}{C}}\boxed{\ \ }\phi$	24.7	3.3	3
$t\text{-BuO—O-}\boxed{\ \ }\overset{\displaystyle O}{\overset{\|}{C}}\text{—}\underset{\displaystyle \phi}{\overset{\displaystyle H}{C}}\boxed{\ \ }\phi$	24.3[e]	−1.0	3
$t\text{-BuO—O}\boxed{\ \ }\overset{\displaystyle O}{\overset{\|}{C}}\text{—}\underset{\displaystyle \phi}{\overset{\displaystyle H}{C}}\boxed{\ \ }\overset{\displaystyle H}{C}\text{=CH}_2$	23.0	−1.1	3
$t\text{-BuO—O}\boxed{\ \ }\overset{\displaystyle O}{\overset{\|}{C}}\text{—CH}_2\boxed{\ \ }\text{—CH=CH}\boxed{\ \ }\phi$	23.5	−5.9	3

[a] In chlorobenzene unless otherwise noted.
[b] P. D. Bartlett and D. M. Simons, *J. Am. Chem. Soc.*, **82**, 1753 (1960).
[c] In *p*-chlorotoluene.
[d] P. D. Bartlett, E. P. Benzing, and R. E. Pincock, *ibid.*, **82**, 1762 (1960).
[e] D. M. Simons and A. M. Feldman, quoted in ref. (21).

point for the hydroxyl group occupies a unique position in the enthalpy-entropy diagram shown in Figure 9-18. A plausible explanation is that the azide is stabilized by intramolecular hydrogen bonding.[22]

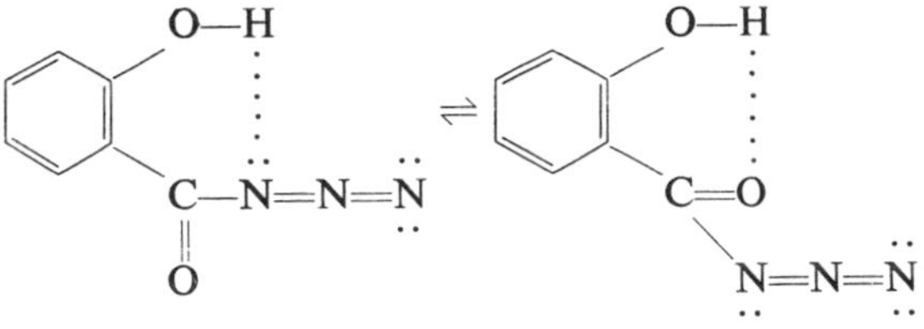

Another example is the effect of an *ortho*-iodo substituent on the rate of decomposition of benzoyl peroxide. Although *ortho*-substituents in general accelerate the reaction, the very marked acceleration by the iodo-substituent has been shown to be due to a cyclic reaction mechanism.[23]

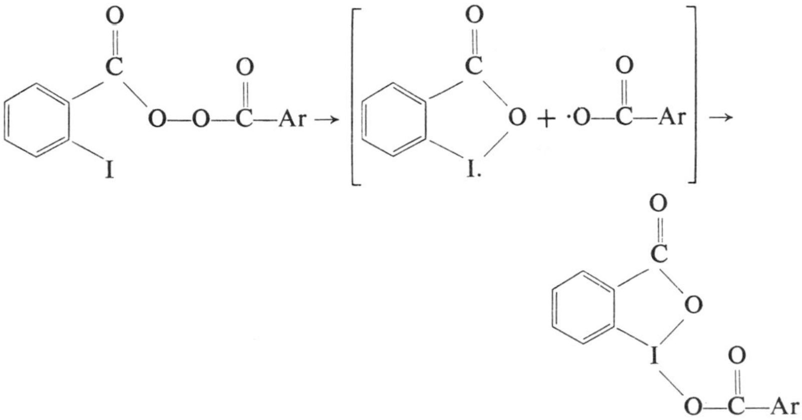

One reason for the difficulty in identifying a proximity effect as a steric effect is that there may not be any qualitative difference between the effects of adjacent and remote substituents. For example, in the racemization of a series of N-benzoyldiphenylamine-2-carboxylic acids in chloroform-ethanol solution (reaction 5 of Table 9-7; equation 30 with Y = COOH),

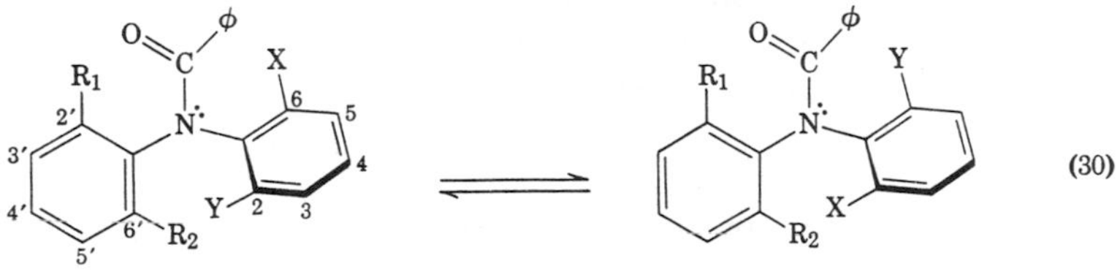

(30)

[22] Y. Yukawa and Y. Tsuno, *J. Am. Chem. Soc.*, **80**, 6346 (1958).
[23] J. E. Leffler, R. D. Faulkner, and C. C. Petropoulos, *ibid.*, **80**, 5435 (1958).

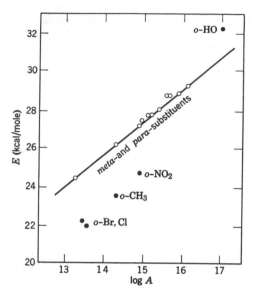

Figure 9-18. Decomposition of benzazides in toluene.

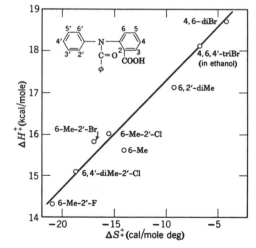

Figure 9-19. Racemization of optically active N-benzoyldiarylamines in chloroform-ethanol. See equation 30; Y = COOH.

substituents in the *ortho*-positions have effects not so very different from those in *meta*- and *para*-positions, as shown in Figure 9-19.[24] In the racemization of these compounds a plausibly important factor is the hindrance to rotation by the steric interaction of the aryl and benzoyl groups with X and Y. However, the apparent bulk of the X and Y groups can depend on still other factors. For example, a substituent in position 4 might change the length of the C—N bond and bring the groups closer together. A substituent in position 3 might resonate with the group Y and change not only the nature of its bond to the ring but also its solvation. That solvation is important in such reactions is shown by the dependence of the racemization rates of certain *ortho*-substituted biphenyls not only on the solvent but on the ionic strength as well.[25,26]

Although it is possible to interpret Figure 9-19 in terms of a predominantly steric interaction mechanism as we have done, there may be some doubt about this in view of the fact that in the biphenyl racemization, a reaction in which the steric effect is of more central importance, any change in the *ortho*-substituent destroys the isokinetic relationship.[24] In the only isokinetic relationships known for biphenyl racemizations (reactions 6 to 8 of Table 9-7), the solvent is the variable, and the changes in the solvent must be moderate.[25,26]

Aromatic side-chain reactions in which the proximity effects of *ortho*-substituents are indistinguishable from the effects of a single interaction mechanism are not uncommon. However, in order to isolate the proximity effects it is often necessary to make a correction for the polar substituent effects. For example, the Menschutkin reaction of *ortho*-substituted dimethylanilines with methyl iodide in methanol gives a scatter diagram relationship between E and $\log A$ in which the pattern of scatter suggests perturbation by an electronic effect of the substituent superimposed on its steric effect. If an attempt is made to isolate the steric effect by plotting $(E_{ortho} - E_{para})$ against $\log (A_{ortho}/A_{para})$, a fairly good straight line with $\beta = 570°K$ is obtained (Figure 9-20a). This relationship spans more than 6 kcal of activation energy. On the other hand, the data for the *meta*- and *para*-substituted compounds are on or near an almost purely enthalpic line with $\beta = 1300°K$.[27]

Whereas in the Menschutkin reaction the proximity effect produces an increase in the activation energy, in the decomposition of *ortho*-substituted benzazides it produces a decrease, except for *ortho*-OH, for which a special

[24] D. M. Hall and M. M. Harris, *J. Chem. Soc.*, 490 (1960).
[25] J. E. Leffler and W. H. Graham, *J. Phys. Chem.*, **63**, 687 (1959); W. H. Graham and J. E. Leffler, *ibid.*, **63**, 1274 (1959).
[26] B. M. Graybill and J. E. Leffler, *ibid.*, **63**, 1461 (1959).
[27] D. P. Evans, H. B. Watson, and R. Williams, *J. Chem. Soc.*, 1345, 1348 (1939).

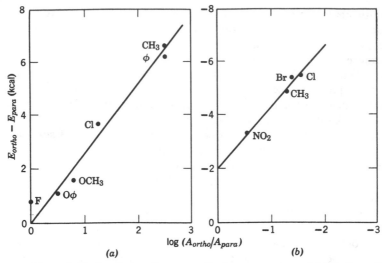

Figure 9-20. Proximity effects on activation parameters. (a) Reaction of methyl iodide with substituted dimethylanilines in methanol. (b) Decomposition of substituted benzazides in toluene.

mechanism has been proposed[22] (see Figure 9-18). The plot of $(E_{ortho} - E_{para})$ against log (A_{ortho}/A_{para}) is again a fairly good straight line, with $\beta = 500°K$ (Figure 9-20b). However, the line giving best fit to the data does not intersect the origin.

Since steric effects are believed to produce changes in bond angles and bond distances in the reaction zone, the question arises whether a steric effect can be simply described as an additional interaction mechanism, superimposed on the others, or whether the steric effect distorts the reaction zone sufficiently to amount to a change in reaction mechanism. In the latter case the steric effect would not only produce its own interaction energy and entropy but would also modify the effects of the other interaction mechanisms. For example, the sensitivity of an activation energy to a change in a polar interaction variable would depend on the steric interaction.

Precisely the same question had arisen previously in the analysis of steric substituent effects on free energy quantities. It will be recalled (Chapter 7) that the parameter σ^* can be a linear measure of the polar effect only if polar and steric effects are simply additive, and that the results obtained seemed to justify this assumption. We therefore propose as a working hypothesis to treat steric effects on enthalpy and entropy quantities also as if they were simply additive to the polar substituent effects, except when the steric effects are very large. There is some evidence to support this hypothesis.

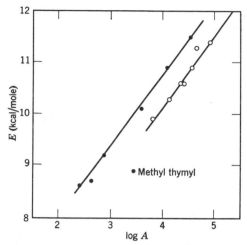

Figure 9-21. Oxime formation from n-alkyl thymyl (solid circles) and n-alkyl carvacryl ketones. Alkyl $= CH_3$ through C_7H_{15}.

Just as steric hindrance is sometimes associated with the dispersion of free energy relationships into parallel lines, so there are examples of dispersion into parallel isokinetic lines that seem to be steric effects. The fact that the isokinetic lines are *parallel* shows that the interaction mechanism which produces the variable displacements along the lines is not affected by the change in steric interaction which produces the dispersion. For example, the activation parameters for the formation of oximes from alkyl thymyl and alkyl carvacryl ketones form two parallel lines (Figure 9-21), one for the more hindered alkyl thymyl series, the other for the less hindered alkyl carvacryl series.[28]

thymyl ketones

carvacryl ketones

The point for methyl thymyl ketone deviates, however, and falls slightly below the line for the less hindered series. This is probably an example of a

[28] M. J. Craft and C. T. Lester, *J. Am. Chem. Soc.*, **73**, 1127 (1951).

qualitative change in the conformation of the reagent or transition state that takes place rather abruptly in the absence of some critical degree of steric hindrance. Unfortunately it is not possible to identify the inter-action mechanism that produces the displacements along the parallel iso-kinetic lines. The value of β, 300°K, coincides almost exactly with the experimental temperature, and the positions of the points along the lines are too uncertain to establish a sequence for the various alkyl groups.

When the point for a given substituent in an enthalpy-entropy diagram deviates from an isokinetic relationship established by other substituents, the deviation must be ascribed to a change in some additional interaction mechanism. If a deviation occurs only when the substitution is made near the reaction zone, a proximity effect is indicated. If, furthermore, the substituent causes a deviation only if it is bulky, a steric effect is indicated. Enthalpy-entropy diagrams can therefore identify the substituents for which proximity or steric effects are significant.

In the ethoxydechlorination reaction of chloro-substituted azanaphtha-lenes, diazanaphthalenes, azabenzenes, and diazabenzenes with ethoxide ion in ethanol, eight compounds in which the chlorine atom is at a ring position adjacent to a hetero-nitrogen atom establish an isokinetic line spanning more than 16 kcal of activation energy. Two compounds in which the chlorine atom is not adjacent to a hetero-nitrogen atom deviate appreciably from this line, while the point for 4-chlorocinnoline is near it.[29] In this example, the interaction mechanism that comes into play when —CH= is substituted for —N= is probably not a steric one.

Hydrolysis of substituted cyanamides, RR'N—CN, in 20% sulfuric acid gives two parallel isokinetic lines, the lower for monosubstituted cyana-mides and the upper for disubstituted cyanamides.[30] A deviation that is probably a steric effect appears in the position of the point for t-butyl cyanamide, which is found near the line for the disubstituted cyanamides. Isopropyl cyanamide behaves normally and falls on the line for mono-substituted cyanamides. However, diisopropyl cyanamide falls slightly above the line for ordinary disubstituted cyanamides.

Alkaline hydrolysis of a series of *primary* alkyl acetates $R_1R_2CHCH_2OAc$ and $R_1R_2R_3CCH_2OAc$, in 70% dioxane-water, gives the enthalpy-entropy diagram shown in Figure 9-22.[31] The esters differ primarily in the steric effect of the alkyl groups R_1, R_2, and R_3 because the polar effect, as measured by σ^*, is nearly constant. It will be noted that most of the points fall on or near a single isokinetic line, showing that the substituent effect in this series depends largely on changes in a single independent variable.

[29] N. B. Chapman and D. Q. Russell-Hill, *J. Chem. Soc.*, 1563 (1956).
[30] T. Mukaiyama, S. Ohishi, and H. Takamura, *Bull. Chem. Soc., Japan*, **27**, 416 (1954).
[31] S. Sarel, L. Tsai, and M. S. Newman, *J. Am. Chem. Soc.*, **78**, 5420 (1956).

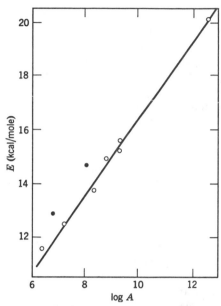

Figure 9-22. Alkaline hydrolysis of primary alkyl acetates. The filled circles refer to 2,2-diethyl-1-butyl and 2-ethyl-3,3-dimethyl-1-butyl acetate.

Since the transition-state complex is an anion, and since the isokinetic temperature, 300°K, is in a range frequently associated with solvation effects, it is tempting to describe the interaction as a steric interference with solvation. Unfortunately we cannot test this hypothesis by means of the relative positions of the points along the isokinetic line because they are very inaccurate. Reagents with a moderate amount of steric bulk, such as the 2,2-dimethyl-1-butyl and 2,2-dimethyl-1-propyl acetates, give data falling on the line. But the points for 2,2-diethyl-1-butyl and 2-ethyl-3,3-dimethyl-1-butyl acetates fall above the line, presumably because of an additional steric interaction mechanism.

The Menschutkin reaction of a series of N,N-dialkyl anilines with methyl iodide in chloroform has been studied for the following N,N-dialkyl groups: dimethyl, methyl ethyl, methyl n-propyl, diethyl, ethyl n-propyl, di-n-propyl, and di-n-butyl.[32] Except for the dimethyl and methyl ethyl compound, the least hindered members of the series, the activation parameters establish an isokinetic line with $\beta = 540°K$. Since the activation energy increases in a rational way with the size of the alkyl groups, and since the inductive effect

[32] D. P. Evans, *J. Chem. Soc.*, 1316 (1954).

is nearly constant within this series, the isokinetic relationship could be a steric effect that, within this limited series, produces changes in only one independent variable. The deviations from the relationship for the dimethyl and methyl ethyl compounds show, however, that the absence of some critical degree of steric hindrance produces changes in at least one other independent variable.

It is interesting to note that steric hindrance in the Menschutkin reaction also modifies the response of the activation parameters to pressure changes. Thus at 320°K, an increase in pressure increases the rate of reaction of methyl iodide with dimethyl-o-toluidine in dry methanol because the decrease in E outweighs the effect of the decrease in log A. In contrast, an increase in pressure increases the rate of an *unhindered* Menschutkin reaction by increasing log A enough to outweigh the increase in E.[33]

Proximity effects can occur in a great variety of mechanisms, sometimes even in a single reaction. For example, in the acid-catalyzed esterification of substituted benzoic acids with cyclohexanol the *meta-* and *para-*substituted acids give an approximate isokinetic relationship. The points for *ortho*-CH_3, Br, and I are displaced above the isokinetic line; those for *ortho*-Cl and NO_2 are near the line; and those for *ortho*-OCH_3 and OC_2H_5 are below it.[34] *Ortho*-effects of comparable complexity are found in the acid-catalyzed esterification of benzoic acids in methanol.[35]

Theoretical models of steric effects emphasize the complex functional relationships involved in this phenomenon. The theoretical calculations require detailed information about such things as the geometry and normal vibrations of the molecules and the potential functions for bonded and nonbonded interactions within the molecule. Many of these variables are independent, and it is probable that a steric effect can be represented as a single interaction mechanism only within narrow structural classes. Because of the complexity of the phenomenon, it is by no means certain that a sterically induced increase in enthalpy will be accompanied by an increase in entropy. The theoretical calculations have been carried out in detail for bimolecular nucleophilic substitution at a saturated carbon atom, a reaction in which steric hindrance is of considerable importance.[36,37] In this reaction the steric repulsions will raise the potential energy of activation, but their effect on the kinetic energy and on the entropy depends on

[33] K. E. Weale, *J. Chem. Soc.*, 2959 (1954).
[34] R. J. Hartman, H. M. Hoogsteen, and J. A. Moede, *J. Am. Chem. Soc.*, **66**, 1714 (1944).
[35] R. J. Hartman and A. M. Borders, *ibid.*, **59**, 2107 (1937); R. J. Hartman and A. G. Gassmann, *ibid.*, **62**, 1559 (1940).
[36] I. Dostrovsky, E. D. Hughes, and C. K. Ingold, *J. Chem. Soc.*, 173 (1946).
[37] P. B. D. de la Mare, L. Fowden, E. D. Hughes, C. K. Ingold, and J. D. H. Mackie, *ibid.*, 3200 (1955).

the balance of several opposing factors. If the effect were merely to lengthen the bond to the attacking reagent in the transition state, the longer bond would be weaker, have a smaller force constant, and hence a higher entropy. If the effect were merely to oppose the bending motion of the attacking reagent in the transition state, or even to deflect the reagent sideways, the entropy would decrease. Furthermore, the additional crowding can interfere with intramolecular motions, such as internal rotations, which take place outside the reaction zone, and hence reduce the entropy.

In the displacement reaction of alkyl halides with halide salts in acetone, the usual effect of steric hindrance is an increase in activation energy and a relatively smaller decrease in activation entropy.[37] In the decomposition of *ortho*-substituted benzoyl peroxides, the usual effect is for $(E_{ortho} - E_{para})$ to decrease and for log (A_{ortho}/A_{para}) to increase slightly.[38] A major exception is the *ortho*-iodo compound where the iodine atom participates more directly in the reaction.[23]

Acid and Base Dissociation and the Brønsted Relationship

The ionization of acids in water is a process for which ΔC_p is quite large. At the same time, water is a solvent whose microstructure is believed to change considerably with the temperature. For this reason we are by no means confident of our ability to explain structural effects on the standard enthalpy and entropy changes for acid and base dissociation. But because of the importance of these reactions, we have thought it worthwhile to tabulate some modern data.

Thermodynamic data for the acid dissociation of ammonium and anilinium ions are given in Table 9-9. Those for carboxylic acids and phenols are in Table 9-10. Figure 9-23 shows the heats and entropies of acid dissociation for various ammonium ions in water. The figure suggests dispersion of the data into three lines with different slopes, one for tertiary and secondary amines, one for primary amines, and one for highly hydrophilic primary amines. Ethylene diamine is in the latter class, but the point for hexamethylene diamine, after statistical correction, falls on the line for the primary amines.

Figure 9-24 is an analogous plot for the acid dissociation of carboxylic acids in water. A series of nonpolar aliphatic acids that vary widely in steric hindrance is fitted by an isoequilibrium line of slope near 298°K. A line of similar slope has been drawn arbitrarily through the scatter diagram formed by the data for the polar acids, to facilitate comparison.

Because of the kinetic complexity of general acid- and base-catalyzed reactions, the catalytic rate constants are difficult to obtain, are probably

[38] A. T. Blomquist and A. J. Buselli, *J. Am. Chem. Soc.*, **73**, 3883 (1951).

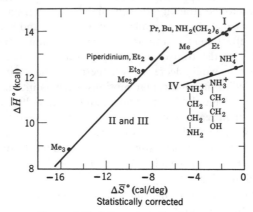

Figure 9-23. Dispersion of enthalpy-entropy data for the acid dissociation in water at 25°C of ammonium salts derived from the following. I. Primary amines. II. Secondary amines. III. Tertiary amines. IV. Hydrophilic primary amines.

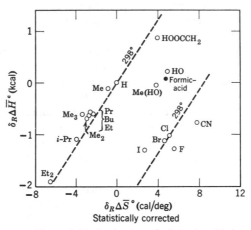

Figure 9-24. Ionization of aliphatic acids in water. Symbols are R from $R_1R_2R_3CCOOH$.

Table 9-9. Thermodynamic Data for Acid Dissociation in Water at 25°. Acids of Charge Type BH+ and BH++

ACID	pK_A	$\Delta \bar{F}^{\circ a}$	$\Delta \bar{H}^{\circ a}$	$\Delta \bar{S}^{\circ b}$	$\Delta \bar{C}_p{}^{\circ b}$	REF.
Ammonium	9.245	12.61	12.40	−0.7	0.0	c
Methylammonium	10.624	14.48	13.09	−4.7	8.0	d
Ethylammonium	10.631	14.49	13.58	−3.1		e
n-Propylammonium	10.530	14.36	13.85	−1.7		e
n-Butylammonium	10.597	14.45	14.07	−1.3		e
β-Hydroxyethylammonium	9.498	12.96	12.08	−2.9	−1.2	f
Dimethylammonium	10.774	14.69	11.86	−9.5	23.1	d
Diethylammonium	10.933	14.90	12.77	−7.2		e
Piperidinium	11.123	15.18	12.77	−8.1	21.0	g
Trimethylammonium	9.800	13.36	8.82	−15.2	43.8	d
Triethylammonium	10.867	14.83	12.24	−8.7		h
Anilinium	4.58	6.24	6.74	1.7		i
o-Toluidinium	4.39	5.98	8.25	7.6		i
m-Toluidinium	4.66	6.35	8.00	5.5		i
p-Toluidinium	5.07	6.91	4.98	−6.5		i
2,3-Xylidinium	4.57	6.23	8.79	8.6		i
2,4-Xylidinium	5.0	6.81	7.90	3.6		i
2,5-Xylidinium	4.6	6.27	6.53	0.9		i
2,6-Xylidinium	4.20	5.73	5.24	−1.6		i
3,4-Xylidinium	5.15	7.02	8.04	3.4		i
3,5-Xylidinium	4.75	6.46	9.07	8.8		i
o-Chloroanilinium	2.634	3.59	6.01	8.1		j
$NH_2(CH_2)_2NH_3{}^+$	9.928	13.54	11.82	−5.8	9.6	k
$NH_2(CH_2)_6NH_3{}^+$	10.930	14.91	13.91	−3.3	8.4	k
$^+NH_3(CH_2)_2NH_3{}^+$	6.848	9.34	10.87	5.1	17.5	k
$^+NH_3(CH_2)_6NH_3{}^+$	9.830	13.41	13.82	1.3	8.2	k
Pyridinium	5.22	7.12	5.70	−4.8		l
2-Picolinium	5.96	8.13	6.95	−4.0		l
3-Picolinium	5.63	7.68	6.70	−3.3		l
4-Picolinium	5.98	8.15	7.03	−3.8		l
2,6-Lutidinium	6.72	9.17	6.15	−10.1		l

(a) kcal/mole. (b) cal/mole deg. (c) R. G. Bates and G. D. Pinching, *J. Res. Nat. Bur. Stand.*, **42**, 419 (1949); *J. Am. Chem. Soc.*, **72**, 1393 (1950); D. H. Everett and D. A. Landsman, *Trans. Faraday Soc.*, **50**, 1221 (1954). (d) D. H. Everett and W. F. K. Wynne-Jones, *Proc. Roy. Soc. (London)*, **A177**, 499 (1941). (e) A. G. Evans and S. D. Hamann, *Trans. Faraday Soc.*, **47**, 34 (1951). (f) R. G. Bates and G. D. Pinching, *J. Res. Nat. Bur. Stand.*, **46**, 349 (1951). (g) *ibid.*, **57**, 153 (1956). (h) J. E. Ablard, D. S. McKinney, and J. C. Warner, *J. Am. Chem. Soc.*, **62**, 2181 (1940). (i) T. W. Zawidzki, H. M. Papée, W. J. Canady, and K. J. Laidler, *Trans. Faraday Soc.*, **55**, 1738 (1959). However, the $\Delta \bar{H}^\circ$ and $\Delta \bar{S}^\circ$ values reported in this article have been thrown open to question by the work of D. T. Y. Chen and K. J. Laidler, *Trans. Faraday Soc.*, **58**, 480 (1962). (j) K. J. Pedersen, *Kgl. Danske Vid. Selsk., Skr.*, **15**, 2 (1937). (k) D. H. Everett and B. R. W. Pinsent, *Proc. Roy. Soc. (London)*, **A215**, 416 (1952). (l) C. T. Mortimer and K. J. Laidler, *Trans. Faraday Soc.*, **55**, 1731 (1959); R. J. L. Andon, J. D. Cox, and E. F. G. Herington, *ibid.*, **50**, 918 (1954).

less precise than typical rate constants, and are rarely evaluated at more than one temperature. However, a few data are available.

The iodination of acetone catalyzed by various aliphatic acids obeys the Brønsted relationship with the changes in rate being due almost entirely to the changes in entropy of activation. The variation in enthalpy of activation is small and not correlated with the variation in entropy of activation.[39]

[39] G. F. Smith, *J. Chem. Soc.*, 1744 (1934).

Table 9-10. Thermodynamic Data for Acid Dissociation in Water at 25°C. Acids of Charge Type HA and HA⁻

ACID	pK_A	$\Delta F°^a$	$\Delta H°^a$	$\Delta S°^b$	$\Delta \bar{C}_p°^b$	REF.
Formic	3.752	5.12	−0.04	−17.3	−41.8	d, i
Acetic	4.756	6.49	−0.11	−22.1	−33.9	c
Propionic	4.875	6.65	−0.23	−23.1	−36.3	d
Butyric	4.818	6.57	−0.70	−24.4	−36.0	d, e
Valeric	4.843	6.61	−0.72	−24.6	−32.9	d
n-Caproic	4.857	6.63	−0.70	−24.6	−34.6	d
Isobutyric	4.849	6.61	−0.80	−24.9	−32.2	d
Isovaleric	4.781	6.52	−1.22	−26.0	−31.7	d
Isocaproic	4.845	6.61	−0.72	−24.6	−32.7	d
Trimethylacetic	5.032	6.86	−0.72	−25.5	−34.3	d
Diethylacetic	4.736	6.46	−2.03	−28.5	−28.8	d
Succinic (pK_1)	4.207	5.74	0.76	−16.7	−32.0	f
Lactic	3.860	5.26	−0.17	−18.2		g
Glycollic	3.831	5.23	0.21	−16.8	−39.0	h
Iodoacetic	3.175	4.33	−1.42	−19.3	−32.9	j
Bromoacetic	2.902	3.96	−1.24	−17.4	−38.1	j
Chloroacetic	2.869	3.91	−1.12	−16.9	−46.4	j
Fluoroacetic	2.586	3.53	−1.39	−16.5	−32.6	j
Dibromoacetic	1.48	2.02	−0.5	−8		p
Dichloroacetic	1.30	1.77	−0.1	−6		p
Difluoroacetic	1.24	1.69	0.0	−6		p
Tribromoacetic	1.07	1.46	−0.8	−2		p
Trichloroacetic	0.64	0.87	+1	+2		p
Trifluoroacetic	0.23	0.32	0.0	−1		p
Cyanoacetic	2.470	3.37	−0.89	−14.3	−36.4	k
p-Hydroxybenzoic	4.582	6.25	+0.54	−19.2	−45	l
p-Methoxybenzoic	4.47	6.09	0.57	−18.5		m
p-Toluic	4.344	5.92	0.30	−18.9	−37	l, m
m-Toluic	4.243	5.78	0.07	−19.2	−33	l, m
Benzoic	4.213	5.74	0.09	−18.9	−39	l, m
o-Methoxybenzoic	4.09	5.58	−1.60	−24.1		m
m-Methoxybenzoic	4.09	5.58	+0.06	−18.5		m
m-Hydroxybenzoic	4.08	5.56	+0.17	−18.1	−30	l
p-Bromobenzoic	4.002	5.46	+0.11	−17.9	−31	l
p-Chlorobenzoic	3.986	5.43	+0.23	−17.5	−23	l
o-Toluic	3.91	5.33	−1.50	−22.9		m
m-Iodobenzoic	3.856	5.26	+0.19	−17.0	−37	l
m-Chlorobenzoic	3.827	5.22	+0.02	−17.4	−41	l
m-Bromobenzoic	3.809	5.19	−0.06	−17.2	−33	l
m-Cyanobenzoic	3.598	4.91	−0.03	−16.3	−47	l
p-Cyanobenzoic	3.551	4.84	+0.03	−16.1	−40	l

Table 9-10 *(continued)*

ACID	pK_A	$\Delta F^{\circ a}$	$\Delta H^{\circ a}$	$\Delta S^{\circ b}$	$\Delta \bar{C}_p^{\circ b}$	REF.
m-Nitrobenzoic	3.449	4.70	+0.33	−14.7	−37	l
p-Nitrobenzoic	3.442	4.69	+0.07	−15.5	−30.5	l
⁻OOCCOOH	4.266	5.82	−1.66	−25.1	−55	n
⁻OOCCH₂COOH	5.696	7.77	−1.14	−30.0	−61	o
⁻OOCCH₂CH₂COOH	5.638	7.69	−0.11	−26.1	−52	f
Phenol	10.02	13.67	5.66	−26.9	−27.1	q, r
o-Cresol	10.33	14.10	5.73	−28.1	−28.5	q
m-Cresol	10.10	13.78	5.52	−27.7	−28.3	q
p-Cresol	10.28	14.02	5.50	−28.6	−28.4	q
2,3-Xylenol	10.54	13.39	5.70	−29.1	−29.4	q
2,4-Xylenol	10.60	14.46	5.76	−29.2	−29.3	q
2,5-Xylenol	10.40	14.19	5.58	−28.9	−28.8	q
2,6-Xylenol	10.62	14.49	5.46	−30.3	−30.7	q
3,4-Xylenol	10.36	14.13	5.37	−29.4	−29.7	q
3,5-Xylenol	10.20	13.92	5.34	−28.8	−28.8	q
o-Chlorophenol	8.53	11.64	4.18	−25.0	−46.9	q, r
p-Chlorophenol	9.38	12.80	5.80	−23.5		r
o-Nitrophenol	7.22	9.85	4.66	−17.4		r
m-Nitrophenol	8.35	11.39	4.71	−22.5		r
p-Nitrophenol	7.15	9.75	4.71	−16.9	−34.4	r, s

(a) kcal/mole. (b) cal/mole deg. (c) H. S. Harned and R. W. Ehlers, *J. Am. Chem. Soc.*, **55**, 652 (1933). (d) D. H. Everett, D. A. Landsman, and B. R. W. Pinsent, *Proc. Roy. Soc. (London)*, **A215**, 403 (1952). (e) H. S. Harned and R. O. Sutherland, *J. Am. Chem. Soc.*, **56**, 2039 (1934). (f) G. D. Pinching and R. G. Bates, *J. Res. Nat. Bur. Stand.*, **45**, 322, 444 (1950). (g) L. F. Nims and P. K. Smith, *J. Biol. Chem.*, **113**, 145 (1936). (h) L. F. Nims, *J. Am. Chem. Soc.*, **58**, 987 (1936). (i) H. S. Harned and N. D. Embree, *ibid.*, **56**, 1042 (1934). (j) D. J. G. Ives and J. H. Pryor, *J. Chem. Soc.*, 2104 (1955). (k) F. S. Feates and D. J. G. Ives, *ibid.*, 2798 (1956). (l) G. Briegleb and A. Bieber, *Z. Elektrochem.*, **55**, 250 (1951). (m) T. W. Zawidzki, H. M. Papée, and K. J. Laidler, *Trans. Faraday Soc.*, **55**, 1743 (1959). (n) G. D. Pinching and R. G. Bates, *J. Res. Nat. Bur. Stand.*, **40**, 405 (1948). (o) W. J. Hamer, J. O. Burton and S. F. Acree, *ibid.*, **24**, 269 (1940). (p) W. H. Dumbaugh, Jr., doctoral thesis, Pennsylvania State University, 1958. J. Jordan, *Record of Chem. Progress*, **19**, 193 (1958). (q) D. T. Y. Chen and K. J. Laidler, *Trans. Faraday Soc.*, **58**, 480 (1962). (r) L. P. Fernandez and L. G. Hepler, *J. Am. Chem. Soc.*, **81**, 1783 (1959). (s) G. F. Allen, R. A. Robinson and V. E. Bower, *J. Phys. Chem.*, **66**, 171 (1962).

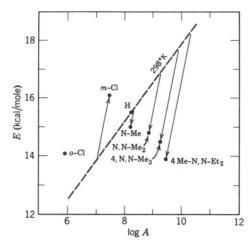

Figure 9-25. Effect of substituents on the decomposition of nitramide catalyzed by various substituted anilines in anisole. The analysis in terms of two interaction mechanisms is arbitrary.

Enthalpies and entropies of activation have been measured for the decomposition of nitramide catalyzed by various substituted anilines in anisole and in m-cresol.[40,41] The distribution of enthalpies and entropies in m-cresol is a scatter diagram, although some of the scatter is due to the fact that substitution is both on the nitrogen atom and on the ring. The data for anisole provide one of the very rare examples of a correlation between E and log A having a negative slope, as seen in Figure 9-25. However, since the correlation is only approximate, we have taken the liberty of representing each point as the resultant of two interaction mechanism vectors of positive slope, one of them having a slope of 298°K. The rates at 298°K will of course be determined by the vertical displacements of the points from the line of slope 298°K.

The Hammett Relationship

The ionization of *meta-* and *para-*substituted benzoic acids in water (the defining process for σ) is accompanied by changes in entropy ranging over about six units and by changes in enthalpy ranging over a few tenths of a

[40] G. C. Fettes, J. A. Kerr, A. McClure, J. S. Slater, C. Steel, and A. F. Trotman-Dickenson, *J. Chem. Soc.*, 2811 (1957).
[41] W. W. Carnie, P. M. Duncan, J. A. Kerr, K. Shannon, A. F. Trotman-Dickenson, and J. A. White, *ibid.*, 3231 (1959).

kilocalorie.[42] However, the heat capacity change is so large that $\Delta\bar{H}^\circ$ and $\Delta\bar{S}^\circ$ are very much functions of the temperature, and $\Delta\bar{H}^\circ$ can even change sign in the experimental range. The plot of log K_A against the temperature usually has a maximum within the range 20–50°C. About all that can be said at the present time about this standard reaction is that σ represents an entropy change more than it does an enthalpy change and that the entropies are roughly in the order expected for an inductive effect.

Other reactions correlated by the Hammett relationship give data more amenable to analysis and illustrate most of the effects to be expected on the basis of the theory of multiple interaction mechanisms. As shown in Chapter 7, there are at least two major interaction mechanisms involved in benzene side-chain reactions, the well-known inductive and resonance effects. The fact that linear free energy relationships nevertheless exist is due to the special circumstance that the contributions from these interaction mechanisms are in a constant ratio for any given substituent, independent of the nature of the reaction. In equation 31, log k/k_0 is represented as a sum of contributions from two independent mechanisms, denoted by the superscripts ' and ".

$$\log \frac{k}{k_0} = \rho'\sigma' + \rho''\sigma'' \qquad (31)$$

Furthermore, $\rho'/\rho'' = \rho_0'/\rho_0''$, where ρ_0' and ρ_0'' refer to the standard process, the ionization of benzoic acids in water at 25°. By a method analogous to that employed on page 177, we find that the temperature dependence of ρ' and ρ'' is given by equations 32.

$$\rho' = \rho_\infty'\left(1 - \frac{\beta'}{T}\right) \qquad (32a)$$

$$\rho'' = \rho_\infty''\left(1 - \frac{\beta''}{T}\right) \qquad (32b)$$

The parameters β' and β'' are the isokinetic temperatures for the two interaction mechanisms, and ρ_∞' and ρ_∞'' are parameters characteristic of the reaction.

We would now like to discuss two categories into which $\rho\sigma$ relationships may be divided. The first of these, which we shall call the well-behaved category, consists of relationships for which the fit remains good at all temperatures. A characteristic of this category is that $\Delta H^\ddagger$ also shows a good linear correlation with $\Delta S^\ddagger$. These two characteristics of the well-behaved category are derivable only if β' is equal to β''. It is readily seen

[42] G. Briegleb and A. Bieber, Z. für Elektrochemie, 55, 250 (1951). These data are summarized in Table 9-10.

that a second interaction mechanism cannot cause scatter on an enthalpy-entropy plot unless its value of β is different from that for the first mechanism. The requirement that ρ'/ρ'' be a constant, equal to ρ_0'/ρ_0'' for the standard reaction, will be met at all temperatures only if β' and β'' are equal. This can be seen from equations 32. A third characteristic of the well-behaved category, derivable from the first, is that both $\delta \Delta H$ and $\delta \Delta S$ are proportional to σ.

The second category consists of reactions for which β' is not equal to β'' but for which ρ'/ρ'' happens to be equal to the standard ratio ρ_0'/ρ_0'' at some experimental temperature. The characteristics of this category are that $\Delta H^{\ddagger}$ and $\Delta S^{\ddagger}$ do not show a linear correlation, either with each other or with σ, and that the fit of the $\rho\sigma$ relationship deteriorates if the temperature is changed.

Table 9-11 lists a few reactions that we would assign to the well-behaved category. That is, these reactions show acceptable isokinetic relationships, and the sequence of points along the isokinetic line agrees reasonably well with the sequence of the rates and σ values. For example, Figure 9-26 shows the isokinetic relationship and the linear relationship between $\Delta H^{\ddagger}$ and σ for the hydrolysis of substituted benzoic anhydrides in 75% dioxane-25% water.[43]

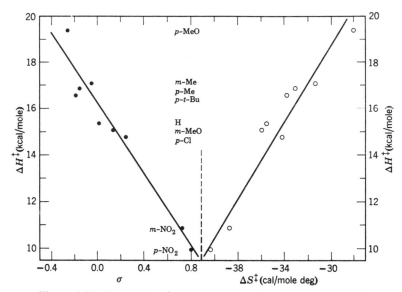

Figure 9-26. Hydrolysis of substituted benzoic anhydrides in 75% dioxane-25% water.

[43] E. Berliner and L. H. Altschul, *J. Am. Chem. Soc.*, **74**, 4110 (1952).

Table 9-11. Reactions Conforming Both to the Hammett Relationship and the Isokinetic Relationship[a]

REACTION	β, °K	ρ (at °C)
1. Benzoylation of substituted anilines with benzoyl chloride in benzene. F. J. Stubbs and C. N. Hinshelwood, *J. Chem. Soc.*, Supplement p. 71 (1949).	Constant $\Delta S^{\ddagger}$	−2.781 at 25°
2. Oxidation of substituted azobenzenes with perbenzoic acid in benzene. G. M. Badger and G. E. Lewis, *J. Chem. Soc.*, 2147 (1953).	Constant $\Delta S^{\ddagger}$	−1.448 at 15°
3. Hydrolysis of *para*-substituted benzamides in 60% ethanol:		
Acid-catalyzed[b]	400°	−0.222 at 99.6°
Base-catalyzed	910°	1.100 at 100.1°
I. Meloche and K. J. Laidler, *J. Am. Chem. Soc.*, **73**, 1712 (1951).		
4. Hydrolysis of substituted benzoic anhydrides in 75 vol % dioxane. E. Berliner and L. H. Altschul, *J. Am. Chem. Soc.*, **74**, 4110 (1952).	750°	1.568 at 58.25°
5. Solvolysis of benzoyl chlorides in 60% ethanol–40% ether. G. E. K. Branch and A. C. Nixon, *J. Am. Chem. Soc.*, **58**, 2499 (1936).	450°	1.922 at 0°
6. Hydrolysis of *para*-substituted ethyl benzoates, acid-catalyzed, in aqueous-organic solvents.[c]		
Ethanol-water	440°	0.144 at 100°
Acetone-water(?)	440°	0.106 at 100°
E. Timm and C. N. Hinshelwood, *J. Chem. Soc.*, 862 (1938).		
7. Dissociation of 1,1,4,4-tetraaryl-2,3-dibenzoyltetrazanes in acetone.		
Rates	760°	−3.409 at −30°
Equilibria	715°	−0.553 at −30°
W. K. Wilmarth and N. Schwartz, *J. Am. Chem. Soc.*, **77**, 4543, 4551 (1955).		
8. Rate of isomerization of 1-substituted phenyl-4-phenyl-5-amino-1,2,3-triazoles in ethylene glycol. E. Lieber, C. N. Ramachandra Rao, and T. S. Chao, *J. Am. Chem. Soc.*, **79**, 5962 (1957).	640°	1.11 at 150°

Table 9-11. (continued)

REACTION	β, °K	ρ (at °C)
9. Isomerization equilibrium of 1-substituted phenyl-5-amino-1,2,3,4-tetrazoles in ethylene glycol, R. A. Henry, W. G. Finnegan, and E. Lieber, *J. Am. Chem. Soc.*, **77**, 2264 (1955).	800°	1.64 at 154°
10. Condensation of substituted nitrosobenzenes with aniline in buffered 94% ethanol. Y. Ogata and Y. Takagi, *J. Am. Chem. Soc.*, **80**, 3591 (1958).	575°	1.22 at 72.5°
11. Reaction of substituted diphenylmercury with hydrogen chloride in dimethyl sulfoxide-dioxane to give substituted phenylmercuric chloride and substituted benzene. R. E. Dessy and J.-Y. Kim, *J. Am. Chem. Soc.*, **82**, 686 (1960).	500°	−2.8 at 32°

[a] Only examples for which the correlation coefficient is greater than .95 are included in the table. Rates unless otherwise noted. Plots refer to $\Delta H^{\ddagger}$ ordinate, $\Delta S^{\ddagger}$ abscissa.

[b] See also Table 9-12.

[c] See Figure 9-12.

An isokinetic relationship and a linear relationship between σ and the activation energy are also found for the condensation of aniline with substituted nitrosobenzenes.[44] For the condensation of nitrosobenzene with substituted anilines, the same type of behavior is found except that the points for *m*-nitro and *p*-carboxyl, substituents which were not studied in the other series, deviate from the isokinetic plot. These points do not spoil the correlation of log k with σ, however, probably because the perturbing interaction mechanism has an isokinetic temperature near the temperature for which the rate constants are given.

The decomposition of N-phenyl benzyl carbamates in ethanolamine solution gives the corresponding substituted phenyl isocyanates and benzyl alcohol.[45] The $\rho\sigma$ relationship is obeyed with $\rho = 0.538$ at 150°C; the isokinetic relationship is obeyed with $\beta = 550$°K, and the sequence of points on the isokinetic line is correlated with σ. The *ortho*-substituents

[44] Y. Ogata and Y. Takagi, *J. Am. Chem. Soc.*, **80**, 3591 (1958).
[45] T. Mukaiyama and M. Iwanami, *ibid.*, **79**, 73 (1957).

chloro, methyl, and nitro apparently interact by an additional mechanism, as would be expected. These points, and two others corresponding to *ortho, meta* and *ortho, para* doubly substituted compounds, fall on a line near the extension of the *meta-para* line but displaced upward by a few tenths of a kcal.

Even for reactions in the well-behaved category, perturbing interaction mechanisms are usually involved to a slight extent and produce slight deviations from a perfectly linear isokinetic relationship. Such reactions will have a sequence of activation enthalpies or entropies well correlated with σ, but the quality of the correlation of $\Delta F^{\ddagger}$ or of log k with σ will depend on the experimental temperature. At temperatures remote from β the correlation of log k with σ is good, but at $T = \beta$ it becomes a scatter diagram. For example, the solvolysis of p-substituted benzenesulfonyl chlorides in 90.9 wt % ethanol-acetone has an isokinetic temperature of about 300°K (30°C).[46] The quantities $\delta_R \Delta F^{\ddagger}$ near this temperature are confused: at 52°C, p-methyl, p-bromo, and p-nitro all accelerate; at about 25°C, p-methyl begins to decelerate. In contrast, the enthalpies and entropies of activation increase in a sequence that, within experimental error, is identical with the sequence of σ-values.

Another example of a reaction in which the behavior of the enthalpies and entropies of activation seems to be more rational than that of the rate constants is the decomposition of *para*-substituted triphenylmethyl azides in dibutylcarbitol.[47] Figure 9-27 shows the isokinetic relationship obtained. The rate constants, all measured within 20° of the isokinetic temperature, vary in an apparently random way; yet the sequence of substituents arranged in order of increasing $\Delta H^{\ddagger}$ is substantially the same as the sequence of σ-values.

The acid-catalyzed hydrolysis of *para*-substituted benzamides in 60% ethanol obeys an approximate isokinetic relationship with β about 400°K.[48] The sequence of points along the isokinetic line is methyl, hydrogen, chlorine, and nitro, which is the sequence of σ. As the temperature is brought closer to 400°K, the scatter about the isokinetic relationship begins to have a more important effect on the rates than does the position of the point along the line, and the $\rho\sigma$ relationship deteriorates, although there is a slight reversal of the trend at 99.6°C. The values of ρ and the correlation coefficients for the $\rho\sigma$ relationship are shown in Table 9-12.

The values of ρ from Table 9-12 are approximately described by the equation $\rho = 1.29(1 - 430/T)$. Comparing this with equations 32 we see that β' (equal to β'') is 430°K. This agrees about as well as can be expected

[46] F. E. Jenkins and A. N. Hambly, *Australian J. of Chemistry*, **14**, 205 (1961).
[47] W. H. Saunders, Jr. and J. C. Ware, *J. Am. Chem. Soc.*, **80**, 3328 (1958).
[48] I. Meloche and K. J. Laidler, *ibid.*, **73**, 1712 (1951).

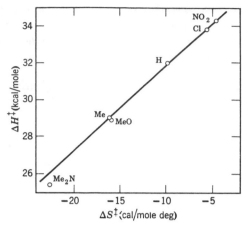

Figure 9-27. Decomposition of *p*-substituted triphenylmethyl azides in dibutylcarbitol. Note the deviation attributable to resonance for MeO and Me$_2$N.

with the value $\beta = 400°K$ obtained directly from the approximate isokinetic relationship for this reaction.

The dehydrochlorination of 2,2-diaryl-1,1-dichloroethanes provides an example of the second category of $\rho\sigma$ relationships. The enthalpy-entropy relationship (Figure 9-7) shows at least two major interaction mechanisms with different values of β, yet the values of log k are approximately linear in σ.[10] Since a double bond is developing in the transition state of this reaction, we expect resonance effects to be relatively important, and this may have something to do with the scatter in the enthalpy-entropy diagram. In the next section we shall consider resonance deviations from the $\rho\sigma$ relationship in more detail.

Table 9-12. Fit of the $\rho\sigma$ Relationship to Data for the Acid-Catalyzed Hydrolysis of para-Substituted Benzamides[a]

TEMP. (°C)	BEST ρ	CORRELATION COEFFICIENT
52.4	−0.483	.998
65.0	−0.310	.985
79.5	−0.298	.847
99.6	−0.222	.883

[a] From H. H. Jaffé, *Chem. Reviews*, **53**, 191 (1953).

Resonance Perturbation of the Hammett Relationship

Deviations from the Hammett relationship due to changes in the relative importance of the resonance and inductive interaction mechanisms are familiar to such an extent that special substituent constants, σ^+ and σ^-, have been introduced for their correction. Just as it was possible to divide the $\rho\sigma$ relationships into two categories, depending on whether or not $\beta_{resonance}$ is equal to $\beta_{inductive}$, so it is possible to divide $\rho\sigma^+$ and $\rho\sigma^-$ relationships into two analogous categories. However, in practice it turns out that $\rho\sigma^+$ and $\rho\sigma^-$ relationships in which the β-values of the major interaction mechanisms are equal are extremely rare, and we are unable to give even a single example belonging to the "well-behaved" category. Moreover, the pattern of the points due to the various substituents in the enthalpy-entropy diagram is often such as to suggest that the resonance interaction included in σ^+ or σ^- is largely an enthalpy quantity.

The solvolysis of substituted 2-phenyl-2-propyl chlorides in 90% acetone, the defining process for σ^+, illustrates several of these features. The points on the enthalpy-entropy diagram[49] are badly scattered, as is shown in Figure 9-28. Although we cannot assign slopes to the inductive and resonance vectors, it appears that substituents in which resonance, as measured by $(\sigma - \sigma^+)$, is a major factor give points falling near a line of

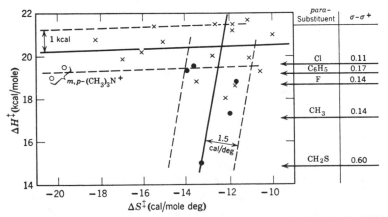

Figure 9-28. The solvolysis of substituted 2-phenyl-2-propyl chlorides in 90% acetone-10% water, the defining process for σ^+. Circles are for selected *para*-substituents, crosses are for *meta*-substituents. *Para*-substituents were selected on the basis that $\sigma - \sigma^+ > 0.10$.

[49] Y. Okamoto, T. Inukai, and H. C. Brown, *J. Am. Chem. Soc.*, **80**, 4964, 4969 (1958). Y. Okamoto and H. C. Brown, *ibid.*, **79**, 1909 (1957); **80**, 4976 (1958). H. C. Brown, J. D. Brady, M. Grayson and W. H. Bonner, *ibid.*, **79**, 1897 (1957). H. C. Brown, Y. Okamoto, and G. Ham, *ibid.*, **79**, 1906 (1957).

almost purely enthalpic slope; substituents for which the inductive effect is important give points that scatter, but most *meta*-substituents are found near a line of slope 30°K. The figure includes two isolated points for charged substituents, for which an additional interaction mechanism depending on the charge could plausibly be important.

The reaction of substituted benzophenones with hydroxylamine in 70% methanol containing acetic acid-sodium acetate buffer gives a good isokinetic relationship of $\beta = 330°$K for *meta*-substituents, and the points fall in a rational order along the line.[50] However, strongly electron-releasing *para*-substituents deviate sharply, as shown in Figure 9-29a. The point for *p*-nitro deviates in the opposite direction. Figure 9-29b shows that the $\rho\sigma$ plot at 323°K (50°C) has corresponding features. Thus the *meta*-substituents show little or no variation in rate, since 50°C is close to their isokinetic temperature. *Para*-substituents deviate at both electronic extremes due to resonance. Note that the resonance effect again seems to be largely enthalpic.

In the decomposition of substituted benzhydryl azides, *p*-methoxy accelerates more than expected on the basis of the $\rho\sigma$ relationship established by the other substituents; it also deviates from the isokinetic relationship established by the other substituents.[51]

The *ortho*-Claisen rearrangement of *para*-substituted phenyl allyl ethers in diethylene glycol-diethyl ether solution obeys an isokinetic relationship of slope about 540°K, the points for *p*-methyl and *p*-methoxy being somewhat below the line established by those for *p*-bromo, hydrogen, and *p*-benzoyl.[52] There is also scatter in the $\rho\sigma_{para}$ plot, removable by using σ^+. For the same reaction in diphenyl ether, *p*-alkyl groups and *para*-substituents capable of electron-withdrawing resonance form an isokinetic line slightly above a second line for *para*-substituents capable of electron-donating resonance.[53] Again, the rates are correlated by σ^+ rather than σ.

Reactions obeying the $\rho\sigma^-$ relationship also show characteristic deviations from the isokinetic relationship because $\beta_{resonance}$ and $\beta_{inductive}$ are not equal. For example, in the alkaline cleavage of substituted benzyl trimethylsilanes into the corresponding toluenes in 39% aqueous methanol,[54] there is a rough isokinetic relationship with β about 300°K, which holds for *meta*-substituents and for those *para*-substituents incapable of electron-withdrawing resonance. Points for *p*-nitro and *p*-C$_6$H$_5$NHCO fall well

[50] J. D. Dickinson and C. Eaborn, *J. Chem. Soc.*, 3036 (1959).
[51] C. H. Gudmundsen and W. E. McEwen, *J. Am. Chem. Soc.*, **79**, 329 (1957).
[52] W. N. White, D. Gwynn, R. Schlitt, C. Girard, and W. Fife, *J. Am. Chem. Soc.*, **80**, 3271 (1958).
[53] H. L. Goering and R. R. Jacobson, *ibid.*, **80**, 3277 (1958).
[54] C. Eaborn and S. H. Parker, *J. Chem. Soc.*, 126 (1955).

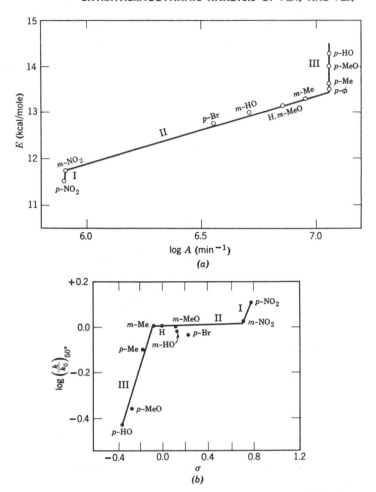

Figure 9-29. The reaction of substituted benzophenones with hydroxylamine in 70% methanol: (a) activation parameters, (b) free energy quantities.

below the line; the rates at 49.7°C are correlated with σ^- rather than with σ; $\rho = 4.88$.

Another reaction, showing an interesting pattern of dispersion in both the isokinetic and the Hammett relationships, is the acid-catalyzed hydrolysis of substituted phenyl potassium sulfates in water.[55] The isokinetic plot (Figure 9-30) consists of three lines, one each for *meta*-substituents, for

[55] G. N. Burkhardt, C. Horrex, and D. I. Jenkins, *J. Chem. Soc.*, 1649 (1936).

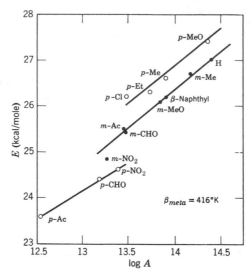

Figure 9-30. Activation parameters for the acid-catalyzed hydrolysis of substituted phenyl potassium sulfates in water.

electron-releasing *para*-substituents, and for electron-withdrawing *para*-substituents. The sequence of points along each line, with increasing activation energy, is very nearly the order of increasing electron release for the first two lines and the opposite order for the last one. Both the strong electron-donating and the strong electron-withdrawing *para*-substituents deviate from the $\rho\sigma$ relationship ($\rho = 0.55$ at 48.7°C) established by the *meta*-substituents. The use of σ^- for the electron-withdrawing substituents is not quite enough correction; the use of σ^+ for the electron-donating substituents is slightly too great a correction. The fit obtained is shown in Figure 9-31.

A reaction showing an obvious resonance effect, but not one of a marked polarity such as would justify the use of σ^+ or σ^-, is the decomposition of substituted benzazides in toluene.[56] The activation parameters fall on a single isokinetic line of slope 375°K, which fits all substituents. The order of the points for the *meta*-substituents on the isokinetic line is that of the σ values, and the corresponding rate constants at 65°C obey the Hammett relationship with a ρ of about -0.29. On the other hand, *all* of the *para*-substituents except for *t*-butyl decelerate the reaction, regardless of their σ values. On the isokinetic plot the points for the *para*-substituents fall between those for hydrogen and *m*-nitro.

[56] Y. Yukawa and Y. Tsuno, *J. Am. Chem. Soc.*, **79**, 5530 (1957).

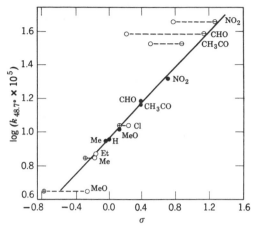

Figure 9-31. Hammett relationship in the acid-catalyzed hydrolysis of phenyl potassium sulfates in water. Filled circles are *meta*-substituents. The symbols + and − refer to σ^+ and σ^- values.

The decomposition of substituted benzazides in acetic acid also gives an isokinetic relationship.[57] Again, the positions of *p*-methoxy and *p*-nitro are anomalous and suggestive of a changed interaction mechanism, both in the $\rho\sigma$ relationship and in the isokinetic relationship.

Miscellaneous Perturbations of the Hammett Relationship

In some reactions, in addition to the usual resonance and inductive interaction mechanisms, there is a third interaction mechanism that dominates the changes in enthalpy and entropy. The enthalpy-entropy plot is fairly linear because the resonance and inductive mechanisms are merely minor perturbing factors. When such a reaction is carried out at the isokinetic temperature for the dominant interaction mechanism, the free energy changes will reflect only the usual resonance and inductive interaction mechanisms. The result may be a good correlation with σ at that temperature, deteriorating badly at other temperatures. An example is the acid-catalyzed esterification of benzoic acids in cyclohexanol.[58] The rough isokinetic relationship is shown in Figure 9-32. It spans more than 3 kcal/mole of activation energy and about 9 cal/mole deg of activation entropy. The slope is about 330°K. There is also a rough $\rho\sigma$ relationship (Figure 9-33)

[57] Y. Yukawa and Y. Tsuno, *J. Am. Chem. Soc.*, **81**, 2007 (1959).
[58] R. J. Hartman, H. M. Hoogsteen, and J. A. Moede, *J. Am. Chem. Soc.*, **66**, 1714 (1944).

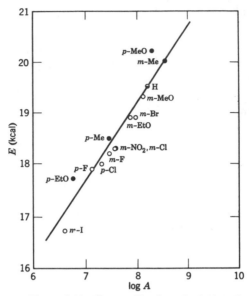

Figure 9-32. The esterification of substituted benzoic acids in cyclohexanol. Closed circles, electron-releasing groups; open circles, electron-withdrawing groups.

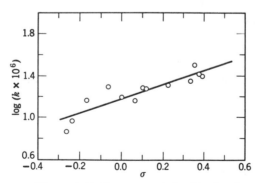

Figure 9-33. Hammett relationship in the esterification of substituted benzoic acids in cyclohexanol at 55°C.

for the rates at 55°C (328°K). At this temperature the sequence of points along the isokinetic line has very little to do with the rates, nor is it correlated with σ. On the other hand, the pattern of the scatter about the isokinetic line is well correlated with σ, electron-releasing substituents falling above the line and electron-withdrawing substituents falling below it.

The enthalpy-entropy diagram for esterification with methanol is similar to that for esterification with cyclohexanol, but the β-value for the dominant interaction mechanism is now only 260°K. The $\rho\sigma$ plot at 298°K, a temperature about 40° above this β, is a scatter diagram.[59] Ortho-substituted benzoic acids actually give a better correlation of the esterification rates in methanol with pK_A than do the meta- and para-substituted acids.[60]

For the reaction of diphenylpicrylhydrazyl radicals with substituted phenols in benzene at 30°[61] there is a very rough $\rho\sigma$ relationship, with $\rho \approx -6$, and limited to substituents with negative values of σ. At about $\sigma = +0.2$ the correlation with σ breaks down entirely, log k becoming nearly constant. The major substituent interaction mechanism is therefore resonance stabilization of a partial positive charge developing on the phenolic oxygen, a result to be expected in view of the strong electron-attracting substituents on the attacking radical. If only the points for phenol, m-chlorophenol, m-cresol, and m-nitrophenol are used, the ρ-value drops to -2.1, corresponding to less effective inductive and resonance mechanisms.

The effect of the substituents on the enthalpy and entropy is shown in Figure 9-34. Substituents with $\sigma \geq +0.2$ give approximately equal rates and are therefore found on or near an isokinetic line of slope about 300°K (30°C). The strongly accelerating substituents are on or near an isokinetic line of much steeper slope.

The $\rho^*\sigma^*$ Relationship

The parameter σ^* is believed to represent the effect of a single interaction mechanism, the inductive effect, on the free energy. It is obtained by subtracting a substituent effect in acid-catalyzed ester hydrolysis from the effect of the same substituent in base-catalyzed ester hydrolysis. This procedure subtracts the free energy change due to steric effects from the

[59] R. J. Hartman and A. M. Borders, *J. Am. Chem. Soc.*, **59**, 2107 (1937); R. J. Hartman and A. G. Gassman, *ibid.*, **62**, 1559 (1940).

[60] It will be recalled that polar substituent effects in acid-catalyzed esterification and hydrolysis are small, an observation used in the definition of E_s. It is interesting to note that a plot of log k for the formation of methyl esters versus log k for the formation of cyclohexyl esters gives a good linear correlation for ortho-substituted benzoic acids, although the points for the meta- and para-substituted acids still scatter.

[61] J. S. Hogg, D. H. Lohmann, and K. E. Russell, *Can. J. Chem.*, **39**, 1588 (1961).

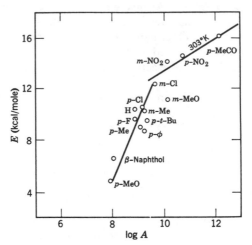

Figure 9-34. The reaction of diphenylpicryl-hydrazyl radicals with substituted phenols in benzene.

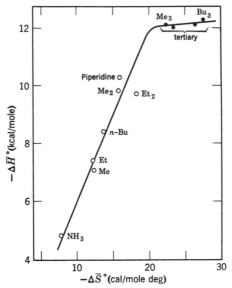

Figure 9-35. Complex formation between iodine and amines.

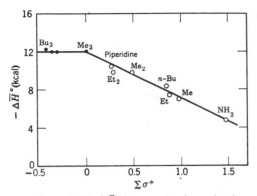

Figure 9-36. $\Delta \bar{H}°$ for complex formation between iodine and amines compared with polar substituent constants, $\Sigma\sigma^*$, for the amines.

free energy change due to the combined inductive and steric effects. It will eventually be interesting to evaluate σ^* by means of data obtained for the two defining reactions at two temperatures so that enthalpic and entropic components of σ^* can be computed. At present we can examine the enthalpic and entropic components of the free energy changes in reactions *correlated* by σ^*, even though we cannot do so for the defining reactions themselves. For example, the free energies of formation, and also the enthalpies of formation, of complexes between iodine and various aliphatic amines are correlated with σ^*, except for several tertiary amines.[62] The points for the tertiary amines also deviate from an isoequilibrium relationship, and in both cases the deviation is ascribable to steric effects. Figures 9-35 and 9-36 show the two relationships.

SOLVENT EFFECTS

Solution Processes

The term "solvation" suggests an interaction between adjacent molecules in which preferred orientations play an all-important role. However, there are a great many solvation phenomena that can be treated successfully without allowing expressly for preferred orientations, except perhaps indirectly through their effect on such parameters as the free volume of the solution, or through their effect on the critical constants of the components. A well-known relationship of the latter type is Trouton's rule[63] for the

[62] H. Yada, J. Tanaka, and S. Nagakura, *Bull. Chem. Soc. Japan*, **33**, 1660 (1960).
[63] F. Trouton, *Phil. Mag.*, [5], **18**, 54 (1884).

enthalpy and entropy of vaporization of pure liquids at their normal boiling point (T_{bp}°K).

$$\Delta S^{\circ}_{vap} \approx 21 \text{ e.u.}; \qquad \Delta H^{\circ}_{vap} \approx 21 T_{bp} \tag{33}$$

Trouton's rule is a consequence of the theorem of corresponding states,[64] and equation 34 is another way of expressing it.[65]

$$\Delta S^{\circ}_{vap} = \frac{P_c V_c}{T_c}; \qquad P_c V_c = 21 T_c \tag{34}$$

In this equation, P_c, V_c, and T_c are the critical pressure, molar volume, and temperature, respectively.

If the entropies of vaporization of the pure liquids are all measured at the same temperature rather than at the variable temperature of the boiling point, the values are no longer constant. However, the variation of ΔS°_{vap} is linear in that of ΔH°_{vap}. The problem has been studied extensively by Barclay and Butler[66] and by Frank and Evans,[67] who consider that equation 35 adequately represents the behavior of most pure liquids at 25°C and propose that it be regarded as a standard relationship representing "normal" behavior.

$$\Delta S^{\circ}_{vap} = 12.75 + 0.00124 \Delta \bar{H}^{\circ}_{vap}, \qquad \text{at } 25°C \tag{35}$$

In (35) the units are cal/mole deg and cal/mole, respectively; the reference or standard states are the pure liquid and the pure gas at 1 atm and 25°C. To conform to our convention of making the entropy the independent variable, the equation can be written as (36).

$$\Delta \bar{H}^{\circ}_{vap} = -10{,}280 + 806 \Delta \bar{S}^{\circ}_{vap} \tag{36}$$

It will be noted that the isoequilibrium temperature is high and that the changes are largely enthalpic.

Of a series of some fifty liquids listed in reference (67), and including alkenes, alkanes, alkyl halides, diethyl ether, aromatic hydrocarbons, alcohols, an ester, and various inorganic liquids, only four have ΔS° values more than 2 cal/mole deg at variance with equation 35 and most are within 0.5 cal/mole deg. Still better fits are obtained by using separate lines for compounds of different types. The latter procedure is equivalent to assuming that each new type of compound has a second interaction mechanism whose magnitude is constant.

---

[64] At equal values of P/P_c, T/T_c and V/V_c, the thermodynamic properties of all pure substances are the same. Substances to which the theorem applies are called *normal* substances.

[65] W. Herz, *Z. Elektrochem.*, **25**, 323 (1919).

[66] I. M. Barclay and J. A. V. Butler, *Trans. Faraday Soc.*, **34**, 1445 (1938).

[67] (a) H. S. Frank, *J. Chem. Phys.*, **13**, 493 (1945). (b) H. S. Frank and M. W. Evans, *ibid.*, **13**, 507 (1945).

Equation 35 also applies to the vaporization of a series of *dilute solutes* from a variety of nonhydroxylic solvents.[67b] The solutes are volatile substances of varying polar character, usually with boiling points below room temperature. The standard state is a hypothetical solute at unit mole fraction but whose thermodynamic properties are those characteristic of the infinitely dilute solution. The fit of the data to equation 35 is particularly gratifying because the values of $\Delta \bar{S}^{\circ}_{\text{vap}}$ and $\Delta \bar{H}^{\circ}_{\text{vap}}$ are spread over a considerable range. A slight abnormality of alcohol as a solvent is expressed mostly as a different slope for equation 35.

The fact that a single linear equation, (35), will fit the enthalpies and entropies of vaporization for a large number of liquids and dilute solutes can be rationalized by means of a free-volume model for liquids[67,68] which does not introduce explicitly the concept of preferred orientations. Qualitatively this approach amounts to considering that a molecule condensing from a vapor falls into a potential energy well. The deeper the well, the narrower it is and the smaller the free volume. The smaller the free volume, the less is the translational and rotational freedom of the molecule.

Appreciable deviations from the standard relationship, equation 35, may be taken as evidence for the presence of strong additional interactions, which probably involve preferred orientations. The vaporization of nonelectrolytes from dilute aqueous solution provides an example. First of all, ΔC_p is often very large,[69,70] so that the first approximation may not be good enough. If we use only data obtained at a definite temperature, the enthalpy-entropy diagram that we obtain has several distinctive features. There is considerable dispersion into parallel lines, suggestive of additional interaction mechanisms. Thus nonpolar solutes fall on one line, alcohols and amines on another, and several solutes of intermediate polarity between.[67b,71,72] Alkyl-substituted pyridines fall on a line parallel to that for the corresponding benzenes.[72] Linear plots like those shown in Figure 9-9 for the C_1 to C_4 alkanes and alcohols are typical. Not only are the two lines parallel, but the sequence of points on both lines is the same, suggesting that the enthalpies and entropies of hydration are additive functions of solute structure. The slopes of the isokinetic lines obtained as the size of the alkyl group is varied are very close to the experimental temperature (298°K).

Linear enthalpy-entropy relationships with slopes near 298°K are not uncommon for solvation in polar solvents and are probably due to the

[68] O. K. Rice, *J. Chem. Phys.*, **15**, 875 (1947).

[69] H. Goller and E. Wicke, *Angew. Chem.*, **B19**, 117 (1947).

[70] E. L. Purlee, R. W. Taft, Jr., and C. A. DeFazio, *J. Am. Chem. Soc.*, **77**, 837 (1955).

[71] W. F. Claussen and M. F. Polglase, *J. Am. Chem. Soc.*, **74**, 4817 (1952).

[72] R. J. L. Andon, J. D. Cox, and E. F. G. Herington, *J. Chem. Soc.*, 3188 (1954).

formation of solvated complexes. It will be recalled[73] that the formation of molecular complexes often proceeds with $\Delta F° \approx 0$ even when $\Delta \bar{H}°$ amounts to several kilocalories. In the case of alkyl groups interacting with water, it is, however, difficult to visualize the formation of complexes between the R-group and the water molecules. The pronounced effects seen on the enthalpy-entropy diagrams are therefore ascribed to the formation of quasi-crystalline aggregates of water molecules adjacent to the R-groups. These microscopic aggregates, or "icebergs" as they are called,[67b] are at least partly responsible for the unusual complexity of the thermodynamic properties of organic solutes in aqueous solution.[67b,74] Further evidence in support of this theory comes from the fact that $\Delta \bar{S}°_{vap}$ for hydrocarbons from water is considerably greater than predicted from the "normal" relationship, equation 35.[67b]

Enthalpies and entropies have also been measured for the process of solution of various substances in a cobalt stearate-packed gas chromatographic column, the standard process being the transfer of one mole of vapor from a volume in the gas phase to an equal volume of solution in the fused salt. The quantities $\Delta \bar{H}°$ and $\Delta \bar{S}°$ tend to be linear in the number of methylene groups of the solute, and in each other. There is dispersion into approximately parallel lines corresponding to n-alkanes, methyl-substituted benzenes, ketones, normal primary alcohols, and secondary alcohols.[75]

Solubilities of liquids and solids also quite often give linear enthalpy-entropy relationships. Data available up to the year 1936 on the heats and entropies of solution of organic solids and liquids in various organic solvent series have been collected by Evans and Polanyi and analyzed by Evans.[76] In the case of the solids, the reference state is the pure compound in its normal crystalline form. Typical solutes are benzoic acid, phenanthrene, 1,2-dinitrobenzene, and ethylene dibromide. The data for each solute in a series of solvents, including the melts, are accurately represented by the linear equation 37,

$$\Delta \bar{H}°_{sol} = \beta_{sol} \Delta \bar{S}°_{sol} + \Delta H_0° \tag{37}$$

where $\Delta H_0°$ and β_{sol} are parameters. The values of β_{sol} range from $315°K$ to $530°K$ and are usually near the melting points of the solutes.

The process of dissolving a pure solid into a dilute solution is analogous, extrathermodynamically, to that of vaporizing the dilute solute from the solution. It can be shown that equation 37 implies the existence of a linear

[73] Chapter 3, pp. 51–54.
[74] F. S. Feates and D. J. G. Ives, *J. Chem. Soc.*, 2798 (1956).
[75] D. W. Barber, C. S. G. Phillips, G. F. Tusa, and A. Verdin, *J. Chem. Soc.*, 18 (1959).
[76] M. G. Evans, *Trans. Faraday Soc.*, **33**, 166 (1937).

relationship between $\Delta \bar{H}^{\circ}_{\text{vap}}$ and $\Delta \bar{S}^{\circ}_{\text{vap}}$, with slope equal to β_{sol}. We note therefore that the values of β for the vaporization of large molecules appear to be substantially smaller than the value, $806°K$, which applies according to equation 36 to small molecules.

A related example in which a linear enthalpy-entropy relationship has been observed is the solubility of water in dilute solutions of alcohols in benzene.[77] Another is the solubility of a series of liquid glycols $HO(CH_2)_nOH$ in benzene, cyclohexane, and heptane.[78] The points for the latter two solvents fall on one line, those for benzene on another (Figure 9-37). The enthalpy increases in the order of increasing n for each solvent except for a barely realized inversion of $n = 3$ and $n = 4$ in the solvent heptane. The magnitudes of the changes in $\Delta \bar{H}^{\circ}_{\text{sol}}$ and $\Delta \bar{S}^{\circ}_{\text{sol}}$ alternate as n increases: a small change is followed by a large one and vice versa. It should be remarked, however, that relationships for the solution of a series of solutes in a single solvent are not always significant with respect to the problem of solvation mechanisms. In the case of liquids, complications may arise due to mutual solubility; and in the case of solids, there are complications due to variations in the crystal energies.

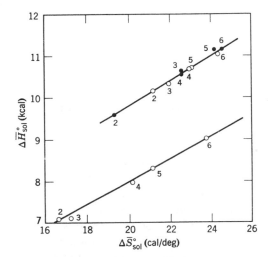

Figure 9-37. $\Delta \bar{H}^{\circ}_{\text{sol}}$ and $\Delta \bar{S}^{\circ}_{\text{sol}}$ for the solution of $HO(CH_2)_nOH$ in benzene (lower line), cyclohexane (open circles, upper line), and heptane (filled circles, upper line). The numerals in the figure denote n.

[77] L. A. K. Staveley, R. G. S. Johns, and B. C. Moore, *J. Chem. Soc.*, 2516 (1951).
[78] L. A. K. Staveley and G. L. Milward, *J. Chem. Soc.*, 4369 (1957).

Isokinetic Relationships for Reactions in Various Solvents

As has been pointed out previously, isokinetic relationships in which the solvent is the variable often have β values near the experimental temperature. This is expected if the formation of molecular complexes is an important part of the solvation mechanism. A frequent consequence of a β value near the experimental temperature is that the regularities to be expected in solvation are not apparent in the behavior of the rate constants. Thus in the decomposition of phenylazotriphenylmethane in a series of solvents (reaction 5 of Table 9-2) the changes in rate are minor. Nevertheless, the solvents do interact with the azo compound and with its transition state. The invariance of the rate is caused by an isokinetic relationship that appears to span about 4.5 kcal of activation enthalpy.[79]

$$\phi\!-\!N\!=\!N\!-\!C\phi_3 \xrightarrow{\text{various solvents}} \phi\!-\!N\!=\!N\cdot + \phi_3 C\cdot \qquad (38)$$

A similar example (number 4 of Table 9-2) is the disproportionation reaction (39) in the same series of solvents.[80] The solvent effects give the

transition state

$$(39)$$

same sequence of activation enthalpies for both reactions. The rates, since they are near the isokinetic temperature, exhibit no such correlation.

When the variation in the series of solvents is not sufficiently restricted, dispersion of the enthalpy-entropy diagram occurs. Thus the solvolysis of methyl toluenesulfonate in a series of alcohols gives the straight line shown in Figure 9-38, but the point for water does not fall on the line.[81] Apparently the jump from the smallest alkyl group to a hydrogen atom is too great a change to satisfy the usual requirement that the change in whatever variable causes the medium effect should be small. Alternatively, the deviation might be caused by a new solvation mechanism for water, different

[79] M. G. Alder and J. E. Leffler, *J. Am. Chem. Soc.*, **76**, 1425 (1954).
[80] C. D. Cook and B. E. Norcross, *J. Am. Chem. Soc.*, **81**, 1176 (1959).
[81] J. B. Hyne and R. E. Robertson, *Can. J. of Chem.*, **34**, 863 (1956).

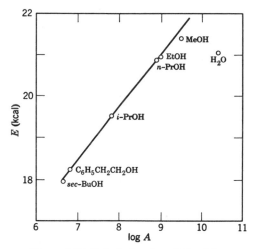

Figure 9-38. Solvolysis of methyl p-toluene-sulfonate in water and in a series of alcohols.

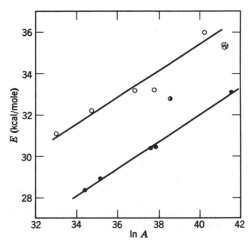

Figure 9-39. Decomposition of triethylsul-fonium bromide in various solvents. Filled circles are for nonhydroxylic solvents, open circles for hydroxylic solvents. The half-filled circle is for a mixed solvent consisting of 75% toluene and 25% benzyl alcohol.

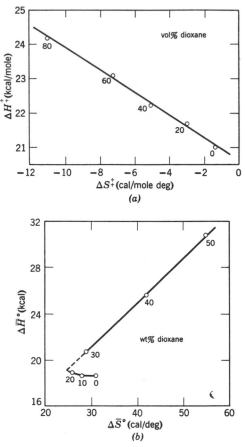

Figure 9-40. Enthalpy-entropy diagrams for reactions in binary solvent mixtures. (*a*) Solvolysis of *t*-butyl chloride in dioxane-formic acid mixtures.[85] (Note the negative slope.) (*b*) Solubility of *cis*-dinitrotetramine cobalt (III) sulfate in dioxane-water mixtures. (From S. A. Mayper, H. L. Clever, and F. H. Verhoek, *J. Phys. Chem.*, **58**, 90 (1954).) (*c*) Racemization of *o*-(2-dimethylaminophenyl)-phenyl-trimethylammonium ion in methanol-benzene mixtures.[25]

from that important in the alcohol series. The point for methanol also deviates, but less seriously.

Dispersion of activation parameters into parallel lines, probably due to multiple solvation mechanisms, is found for the decomposition of triethylsulfonium bromide in various solvents (Figure 9-39).[82] Hydroxylic

[82] E. A. Moelwyn-Hughes, *The Kinetics of Reactions in Solution*, Oxford University Press, 1947, p. 278.

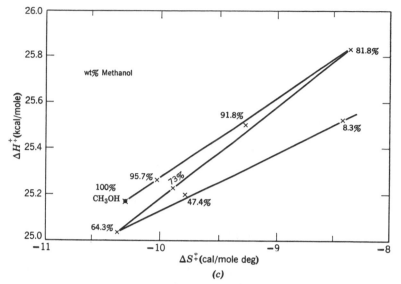

Figure 9-40 (continued).

solvents give points on a line about 3.5 kcal/mole above that for non-hydroxylic solvents. A point nearly midway between the two lines represents the effect of a mixture containing 75% toluene and 25% benzyl alcohol.

The decarboxylation of malonic acid in various solvents gives activation parameters falling on two parallel lines.[83] The upper line contains points for nine carboxylic acids, the lower line for a series of seven primary and secondary alcohols. Since carboxylic acids exist largely in the form of highly stable cyclic dimers, it is not too surprising that they differ from other hydroxylic solvents in their solvation mechanisms.

Mixed Solvents

Perhaps the most baffling problem in the field of enthalpy-entropy relationships is the interpretation of data obtained in mixed solvents. Figure 9-40 shows some typical relationships as the solvent composition is varied. The numbers in these figures denote the solvent compositions at which $\Delta H\ddagger$ and $\Delta S\ddagger$ are measured. Relationships such as those in Figure 9-40 are observed in the majority of actual cases and may be described, respectively, as approximately linear, as hook-shaped, and as N-shaped.[84]

[83] L. W. Clark, *J. Phys. Chem.*, **64**, 41, 508, 677, 692 (1960).
[84] E. Tommila, *Suomen Kemistilehti*, **B25**, 37 (1952); H. S. Venkataraman and C. N. Hinshelwood, *J. Chem. Soc.*, 4986 (1960).

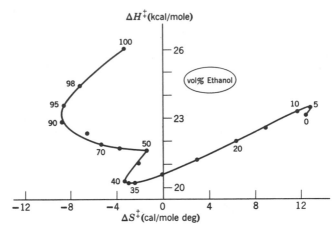

Figure 9-41. Solvolysis of *t*-butyl chloride in ethanol-water mixtures.[85]

Figures of even more complicated shape, such as 9-41,[85] are not infrequent, however. If we had to invent a descriptive name, we might call them Thurber-shaped.

We now wish to show that much of the complexity results merely from the fact that we must add *vector* functions. The individual interaction mechanisms responsible for these functions need be neither complicated nor unusual. Indeed, it is instructive not to consider specific mechanisms at all, but to represent the thermodynamic properties of the solution by power series, as described in Chapter 2.[86] For the *standard free energy* of a solute (component 3) in dilute solution in a binary solvent (components 1 and 2) we write equation 40, where φ_1 and φ_2 are the volume fractions of the solvent components in the absence of the solute, and the V_i are molar volumes, expressed in liters/mole.

$$\bar{F}_3^\circ = \bar{F}_3 - RT \ln c_3 = RT \Big\{ B_3 + 1 + (2\beta_{13} - V_3) \frac{\varphi_1}{V_1}$$

$$+ (2\beta_{23} - V_3) \frac{\varphi_2}{V_2} - V_3 \Big(\frac{\beta_{11}}{V_1^{\,2}} \varphi_1^{\,2} + 2 \frac{\beta_{12}}{V_1 V_2} \varphi_1 \varphi_2 + \frac{\beta_{22}}{V_2^{\,2}} \varphi_2^{\,2} \Big) \Big\}$$

$$+ \{\text{terms of order } \varphi_1^{\,3},\ \varphi_1^{\,2}\varphi_2,\ \varphi_1\varphi_2^{\,2},\ \varphi_2^{\,3}, \text{ or higher}\} \quad (40)$$

Equation 40 is derived from equation 28 of Chapter 2 by neglecting all terms proportional to c_3 and $c_3^{\,2}$, and by substituting φ_1/V_1 for c_1 and φ_2/V_2 for c_2. We briefly reintroduce the customary symbol β_{ij} for the pairwise

[85] S. Winstein and A. H. Fainberg, *J. Am. Chem. Soc.*, **79**, 5937 (1957).
[86] Pages 27–31.

interaction coefficients, hoping it will not be confused with the isokinetic temperature β.

Each of the temperature-dependent β_{ij} may in first approximation be represented by equation 41,

$$\beta_{ij} = \left(\frac{\chi_{ij}}{T}\right) - \eta_{ij} \tag{41}$$

where the new coefficients χ_{ij} and η_{ij} are independent of the temperature. The reason for choosing this form of temperature dependence is that it permits ΔC_p to be zero. Similarly, B_3 may be represented by $(B_3'/T) - B_3''$. It can then be shown that $\bar{S}_3^\circ$ and $\bar{H}_3^\circ$ are represented by an analogous power series, for example, equation 42 for $\bar{S}_3^\circ$. (We are neglecting terms in dV_i/dT.)

$$\bar{S}_3^\circ = R\Bigg\{B_3'' - 1 + (2\eta_{13} + V_3)\frac{\varphi_1}{V_1} + (2\eta_{23} + V_3)\frac{\varphi_2}{V_2}$$
$$- V_3\left(\frac{\eta_{11}}{V_1^2}\varphi_1^2 + \frac{2\eta_{12}}{V_1V_2}\varphi_1\varphi_2 + \frac{\eta_{22}}{V_2^2}\varphi_2^2\right)\Bigg\}$$
$$+ \{\text{terms of order } \varphi_1^3, \varphi_1^2\varphi_2, \varphi_1\varphi_2^2, \varphi_2^3, \text{ or higher}\} \tag{42}$$

In terms of individual interaction mechanisms, each of the coefficients η_{ij} is the total effect on the entropy due to all pairwise interaction mechanisms between components i and j. Since $\varphi_1 = 1 - \varphi_2$, equation 42 is in fact a power series in φ_2, and the quantities $\delta_M \bar{S}_3^\circ$ and $\delta_M \Delta S^\circ$ may therefore be expressed more tersely, as in equations 43.

$$\delta_M \bar{S}_3^\circ = a_1\varphi_2 + a_2\varphi_2^2 + a_3\varphi_2^3 + \cdots$$
$$\delta_M \Delta S^\circ = \Delta a_1\varphi_2 + \Delta a_2\varphi_2^2 + \Delta a_3\varphi_2^3 + \cdots \tag{43}$$

Similarly, the quantities $\delta_M \bar{H}_3^\circ$ and $\delta_M \Delta H^\circ$ may be written in the form (44).

$$\delta_M \bar{H}_3^\circ = b_1\varphi_2 + b_2\varphi_2^2 + b_3\varphi_2^3 + \cdots$$
$$\delta_M \Delta H^\circ = \Delta b_1\varphi_2 + \Delta b_2\varphi_2^2 + \Delta b_3\varphi_2^3 + \cdots \tag{44}$$

It is clear from a comparison of equations 42 and 43 that in terms of the interaction coefficients, the physical significance of the coefficients a_i and Δa_i is rather complicated. The same is true of the coefficients b_i and Δb_i. Numerically, a coefficient of the type a_i or b_i may be positive or negative. We would normally expect that the magnitudes decrease in the order $|a_1| > |a_2| > |a_3|$ and $|b_1| > |b_2| > |b_3|$, but these sequences could easily be upset if some particular pairwise or three-body interaction were particularly stable. For example, $|a_3|$ and $|b_3|$ might be relatively large if the solute were associated with two molecules of solvent component 2 in a particularly stable complex.

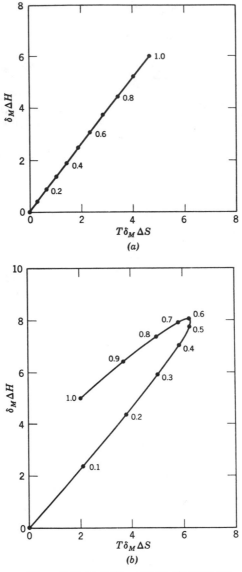

Figure 9-42. Enthalpy-entropy diagrams predicted for reactions in binary solvents. Parameters in equations 43 and 44: (a) $\Delta b_1 = 4.0$; $\Delta b_2 = 2.0$; $\Delta b_1/\Delta a_1 = 1.50T$; $\Delta b_2/\Delta a_2 = 1.00T$. (b) $\Delta b_1 = 26$; $\Delta b_2 = -21$; $\Delta b_1/\Delta a_1 = 1.13T$; $\Delta b_2/\Delta a_2 = 1.00T$.

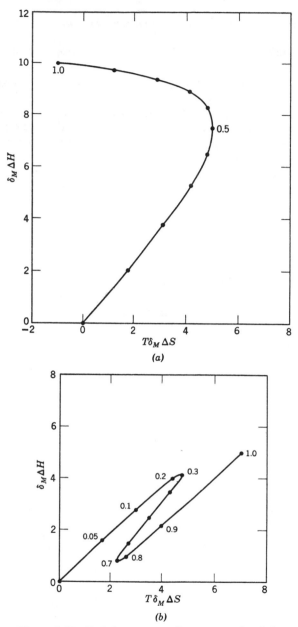

Figure 9-43. Enthalpy-entropy diagrams predicted for reactions in binary solvents. Parameters in equations 43 and 44: (a) $\Delta b_1 = 22$; $\Delta b_2 = -16$; $\Delta b_3 = +4$; $\Delta b_1/\Delta a_1 = 1.16T$; $\Delta b_2/\Delta a_2 = -\Delta b_3/\Delta a_3 = 1.00T$. (b) $\Delta b_1 = 36.5$; $\Delta b_2 = -94.5$; $\Delta b_3 = 63$; $\Delta b_1/\Delta a_1 = 0.95T$; $\Delta b_2/\Delta a_2 = \Delta b_3/\Delta a_3 = 1.00T$.

According to equations 43 and 44, the enthalpy-entropy diagram can be constructed by adding a series of vector functions V_1, V_2, V_3, ... The function V_1 has a constant slope of b_1/a_1 (or $\Delta b_1/\Delta a_1$) and a magnitude proportional to φ_2; V_2 has a constant slope of b_2/a_2 (or $\Delta b_2/\Delta a_2$) and a magnitude proportional to φ_2^2; and so forth for V_3, V_4, ... Owing to the complex origin of the coefficients a_i and b_i we suspect that the slopes b_i/a_i (or $\Delta b_i/\Delta a_i$) are not always positive, but this point is unimportant to the present analysis.

By choosing suitable parameters in equations 43 and 44 and by carrying them out to the quadratic term, linear and hook-shaped enthalpy-entropy diagrams can be generated, as shown in Figure 9-42. By including the cubic terms, N-shaped relationships like that in Figure 9-43 can also be generated.

Enthalpy-entropy diagrams for mixed solvents are often interpreted by assuming that solvent-solute interactions are of dominant importance. By inspection of equation 42 it can be seen that if the interaction between the solute and component 1 is indeed particularly strong, then the enthalpy and entropy will both tend to be linear functions of the volume fraction, φ_1, and an isokinetic relationship results. More complicated shapes arise if solvent-solvent and higher order interactions are also important.

10

Some Mechanochemical Phenomena

> *The other—unfortunately much less prominent—of Wolfgang Kohler's propositions was the observation that the presence in the cage of tools previously used with success in a different situation tended to inhibit the discovery of new tools, even though the old tools could not contribute to solving the new problem.*
>
> Martin Mayer
> *The Schools*

Reactions in crystalline, glassy, or polymeric media usually differ greatly in rate from reactions in ordinary liquids and often differ in the nature of the product. In addition, such media offer the possibility of inducing chemical reactions mechanically by macroscopic deformations of the medium and the reciprocal possibility of converting the free energy released by a chemical reaction into mechanical work. Of special interest is the fact that the work is done against tension rather than against pressure, thereby eliminating the need for pistons and cylinders.

The main difference between ordinary solvents and crystals, glasses, or polymers is the longer relaxation time for molecular motion in the latter. This difference has two important effects. The first is that the reaction mixture can be anisotropic, which is the reason that the vector quantity, tension, is a variable of state. The second effect is that the solvent exerts a much more rigid control over the motions of the solute, not only over its diffusion but also over its changes in orientation and conformation.

The special properties of rigid media are foreshadowed by certain effects, usually of lesser magnitude, observable in mobile liquids under special circumstances. Examples are the cage effect (Chapter 4) and the degeneration of gas-phase rotations into liquid-phase librations (Chapter 5). In the next section we shall discuss still others.

MECHANOCHEMICAL EFFECTS IN LIQUIDS

Although a liquid of ordinary viscosity has too low a relaxation time for an anisotropic state to persist in a static isolated system, such states can be produced and maintained by external factors. One method is the use of a velocity gradient in the solution. Velocity gradients to produce orientation of solute molecules by shear are usually brought about by means of an annular cell, one wall of which is rotated.[1] Solute molecules tend to align themselves with their long axes at small angles to the direction of flow of the liquid. Even molecules as small as nitrobenzene are sufficiently oriented by flow to produce birefringence, but it is only highly asymmetric and macromolecular substances whose orientation is great enough to have any other detectable consequence. The extreme case of flow orientation is the crystallization of polymer fibers by stretching. Soaps,[2] polyphosphates,[3] and sodium thymonucleate[4] oriented by flow show a greater electrical conductivity along the direction of flow than in other directions, for an elongated ion moves most readily in the direction of its long axis. The reciprocal of this effect has also been observed. Thus passage of an electric current through a polyphosphate solution orients the conducting ions along the direction of current flow.[5] Flow orientation should in principle change the rates of bimolecular reactions of large molecules, but no such effect has yet been studied experimentally. Although no flow-*orientation* effect on reactivity has been observed, the effect of stretching a fiber on its reactivity can be regarded as an effect of flow on *conformation*. A flexible molecule in solution should be elongated as well as oriented by flow.

Shear degradation of polymers can be observed in solution as well as in the undiluted polymer. The shear can be supplied by viscous flow, by mastication (of a rubbery material), or by ultrasonic vibration. Substances of high molecular weight can be extremely delicate with respect to shear. For example, desoxyribonucleic acid (DNA) is degraded into smaller molecules merely by passage of its solutions through a hypodermic needle under conditions that might be used routinely in filling an infrared cell.[6] The violent shaking used to separate DNA from its protein can also degrade the DNA.[7]

THE MECHANISM AND RATE OF MECHANICAL POLYMER DEGRADATION IN SOLUTION

The rate at which bonds are broken because of shear is a function of the viscosity, the molecular weight of the polymer being degraded, the temperature

[1] R. Cerf and H. A. Scheraga, *Chem. Reviews*, **51**, 185 (1952).

[2] K. Heckmann, *Naturwissenschaften*, **40**, 478 (1953).

[3] U. Schindewolf, *ibid.*, **40**, 435 (1953).

[4] B. Jacobson, *Rev. Sci. Instruments*, **24**, 949 (1953).

[5] M. Eigen and G. Schwarz, *Z. für physikalische Chemie*, N.F., **4**, 380 (1955).

[6] P. F. Davison, *Proc. Nat. Acad. Sci. U.S.*, **45**, 1560 (1959).

[7] C. A. Thomas, Jr., *J. Gen. Physiol.* **42**, 503 (1959).

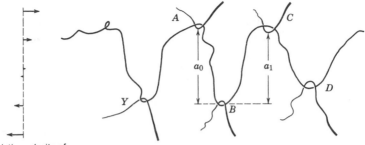

Relative velocity of
layers ot solvent

Figure 10-1. Illustration of the Washington effect. (From F. Bueche, *J Appl. Polymer Science*, **4**, 101 (1960).)

(largely because it changes the viscosity), the presence of plasticizers (lubricants that facilitate the motion of one polymer molecule past another), and the rate at which mechanical energy is put into the system.[8,9,10] We will not discuss the degradation of polymers by ultrasonics separately, since this is believed to be a special case of shear degradation.[11] Ultrasound produces cavitation in the solvent, and the collapse of the cavities produces shock waves. As a shock wave passes through the region containing a polymer molecule, the rapid rise in pressure increases the number of solvent molecules in that region. During the succeeding decompression phase of the shock wave, these additional polymer-enveloped solvent molecules diffuse out in all directions. The resulting viscous drag puts the central part of the polymer under tension.

The rate of shear degradation of a polymer in solution is critically dependent on the presence of large molecules in concentrations high enough to produce entanglement, as illustrated by Figure 10-1.[8,10]

Let $\dot{\gamma}$ be the tensile shear in the vertical direction of Figure 10-1, let a polymer molecule consist of Z links of molecular weight M_0, and let M_e be the molecular weight of that part of the polymer connecting two points of entanglement, such as A and B of the figure. On the average, the points A and B will separate with a velocity equal to $\frac{1}{2}\dot{\gamma}a_0$ as a result of the shearing force.

If the segment AB is the central segment of the molecule, it elongates as the chain slides through the entanglement points, but we shall choose a coordinate system such that the center of this segment is motionless. The adjacent segments AY and BC will then move with velocities $\frac{1}{2}\dot{\gamma}a_0$. The

[8] A. B. Bestul, *J. Chem. Phys.*, **24**, 1196 (1956), *J. Phys. Chem.*, **61**, 418 (1957).

[9] P. Goodman, *J. Polymer Sci.*, **25**, 325 (1957).

[10] F. Bueche, *J. Applied Polymer Sci.*, **4**, 101 (1960).

[11] G. Gooberman, *J. Polymer Sci.*, **42**, 25 (1960).

effect is cumulative, and a link in the ith segment moves with a velocity $\frac{1}{2}\dot{\gamma}\sum_0^{i-1} a_n$. The tension is due to the viscous force, $\xi \times v$, where ξ is a friction factor and v is the velocity of the link. Equation 1, which gives the tension at the center of the chain, the point at which it is greatest, can be derived by summing the forces needed to pull either side of the chain through all of its entanglements.

$$f_0 = \frac{1}{16}\dot{\gamma}\xi \left(\frac{M_e}{M_0}\right)^{3/2}\left(\frac{M}{M_e}\right)^2 l \tag{1}$$

In equation 1, l is the length of a chain link and M is the molecular weight of the entire polymer molecule. Note that f_0 is a sensitive function of the molecular weight.

The tension of a link at the qth position from the center is smaller and is given by equation 2.

$$f_q = f_0\left(1 - \frac{4q^2}{Z^2}\right) \tag{2}$$

The chance that a given link subject to a tension f will break in an interval dt is given by equation 3, in which ω is the bond vibration frequency and λ is the critical distance to which the bond must stretch in order to break. E_i is the energy needed to break the bond, and $E_i - f_i\lambda$ is the part of the energy not provided mechanically.

$$P_i\,dt = \omega \exp\left(-\frac{E_i - f_i\lambda}{kT}\right)dt \tag{3}$$

An initially homogeneous polymer whose molecules are long enough to be degraded by entanglement, but the fragments from which (approximately half the original length) are short enough to be immune from further degradation, is an interesting special case. It can be shown that the molecular weight distribution after the degradation will be as given in equation 4.

$$\frac{M_w}{M_n} \cong 1 + \frac{kT}{2f_0\lambda} \tag{4}$$

In equation 4, M_w is the weight average and M_n the number average molecular weight; their ratio is a measure of the homogeneity of the sample.[12] For a ratio of strain energy $f_0\lambda$ to thermal energy kT of about 10, the ratio M_w/M_n comes out 1.05, corresponding to a remarkably homogeneous polymer. It follows that homogeneity in a natural polymer after isolation does not guarantee that it was isolated without any mechanical degradation.[13]

[12] $M_w = \Sigma M_i^2 N_i / \Sigma M_i N_i$; $M_n = \Sigma M_i N_i / \Sigma N_i$; N_i is the number of molecules of molecular weight M_i.

[13] Polymers can be degraded not only by passage of their solutions through a hypodermic needle, as already noted, but also by the strains set up in merely freezing their solutions. [A. A. Berlin, *Doklady Akad. Nauk, S.S.S.R.*, **110**, 401 (1956).]

The fit of mechanochemical reactions to an analog of the Arrhenius equation has been tested by experiments on the shear degradation of polyisobutylene by passage of the solution through capillaries.[8,9] As expected if molecular entanglements are involved in the shearing process, the molecular weight and concentration of the polyisobutylene must exceed certain minima in order to obtain any degradation. Given these requirements, it is found that the first-order rate constant, or, more precisely, the rate at which bonds are broken divided by the weight per cent of polymer in the solution, obeys equation 5.

$$k = Ae^{-E/aJ} \tag{5}$$

The quantity J is the rate of application of shear energy (i.e., shearing stress times rate of shear), and aJ (large compared to RT) is some function of the average amount of mechanical energy temporarily stored in the system. Plotting log k against $1/J$ gives a straight line whose slope is $-E/a$, analogous to $-E/R$ in the Arrhenius equation. E/a has been found to be independent of concentration down to about 5% of polymer; the change at low concentrations is attributed to a decrease in a. Similarly, E/a is independent of temperature in the range 30 to 50° but decreases above 50°. The decrease in E/a with increasing molecular weight is attributed to an increase in a, corresponding to a tendency for larger molecules to entangle

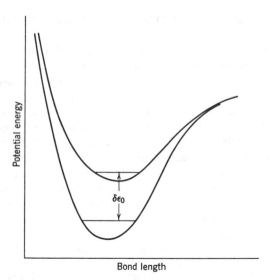

Figure 10-2. Effect of a constant external stress on the potential energy function of a bond.

more efficiently and therefore to store a larger part of the energy put into the system.

The effect of a constant stress applied to a bond is to increase the equilibrium bond length and to raise the zero-point energy to a new level, as illustrated by the dotted potential energy curve of Figure 10-2. Such a bond is inherently more reactive, even though it may not dissociate spontaneously in the absence of a second reagent. In principle a bond strained by tension in a polymer chain is like a bond strained by steric repulsion of the bonded groups, except that the enhanced reactivity of the latter is likely to be counteracted by steric hindrance to the approach of a reagent. The quantity $\delta\epsilon_0$ in Figure 10-2 represents the reduction in activation energy to be expected for the bond-breaking reactions of a stressed polymer. An increase in entropy of activation may also be expected if the tension reduces the conformational mobility of the polymer more than it does that of the transition state; this is likely to be the case if the bond being broken elongates sufficiently to relax the tension in the rest of the chain.

Evidence to be presented in the next section suggests that the mechanical cleavage of polymers is usually a free radical reaction, although ionic cleavage should be possible for polymers of suitable structure in polar media.

MECHANICAL DEGRADATION OF UNDILUTED POLYMERS

The most common first step in the shear degradation of a polymer appears to be the production of free radicals, detectable by various chemical methods.[14] For example, radicals produced by the mastication of rubber can be detected by their reaction with diphenylpicrylhydrazyl or with dinaphthyl disulfide labeled with radioactive sulfur.[15] Grinding a polymer in carbon tetrachloride gives products assignable to the attack of radicals from the polymer on the carbon tetrachloride.[16] Lathe turnings from some polymers show electron-spin resonance.[17]

Block copolymerization occurs when polymethyl methacrylate is masticated with styrene or when polystyrene is masticated with methyl methacrylate.[18] The block copolymers can be separated from the homopolymers by fractional precipitation.

[14] W. F. Watson, *Makromol. Chem.*, **34**, 240 (1959).
[15] G. Ayrey, C. G. Moore, and W. F. Watson, *J. Polymer Sci.*, **19**, 1 (1956).
[16] N. K. Baramboïn, *Zhur. Fiz. Khim.*, **32**, 806 (1958).
[17] S. E. Bresler, S. N. Zhurkov, E. N. Kazbekov, E. M. Saminskiĭ, and E. E. Tomashevskiĭ, *Zhur. Tekh. Fiz.*, **29**, 358 (1959).
[18] D. J. Angier, R. J. Ceresa, and W. F. Watson, *J. Polymer Sci.*, **34**, 699 (1959).

The formation and recombination of free radicals during deformation of polyvinyl chloride are believed to be part of the mechanism of flow for that polymer.[19]

A particularly interesting special case of the mechanical cleavage of bonds occurs in the growth of popcorn polymer.[20] Popcorn polymer, or cauliflower polymer as it is called in the German literature, is formed from styrene copolymerized with a cross-linking agent such as butadiene. It gets its names from its characteristic fissured appearance. A fragment or seed of popcorn polymer contains free radicals and will therefore grow when exposed to fresh monomer vapor. The new polymer still has the same properties, and the seeding process can be repeated even unto the eleventh generation. The rate of growth of insoluble, cross-linked polymer in the seed is logarithmic.[21] The most plausible explanation for the logarithmic growth rate and the characteristic fissured appearance of the polymer is that growth is accompanied by the production of new free radicals formed by the mechanical rupture of covalent bonds.[22] The free radical chain ends attached to the popcorn polymer network grow by reaction with monomer molecules, which reach them by diffusion. The swelling caused by the added monomer is made irreversible by the reaction with the stuck radicals and leads to the rupture of the network and an increase in the number of growing radical chains.

Quantitative studies of the reciprocal effects of chemical attack on mechanical strength and of mechanical strains on chemical reactivity have been carried out on vulcanized rubber exposed to ozone.[23,24] The effect of ozone on rubber not under tension is very slight, but even at extensions of only 10% cracks form and rapidly grow to large incisions. For quantitative purposes it is more convenient to start with a short razor-cut perpendicular to the tension in a stretched sheet of rubber exposed to ozone and to measure the rate at which the tip of the cut grows across the sheet.[23] Below a certain critical tension there is no observable growth. Above this tension the rate is constant and independent of any further increase in the stress, the length of the cut remaining proportional to the time of exposure until the strip of rubber is entirely severed. The rates, reproducible to ±20%, are approximately proportional to the concentration of ozone.

[19] V. A. Kargin, T. I. Sogolova, G. L. Slonimskii, and E. V. Reztzova, *Zhur. Fiz. Khim.*, **30**, 1903 (1956); R. J. Ceresa and W. F. Watson, *J. Applied Polymer Sci.*, **1**, 101 (1959).

[20] G. H. Miller and A. K. Bakhtiar, *Can. J. Chem.*, **35**, 584 (1957).

[21] G. H. Miller and C. F. Perizzolo, *J. Polymer Sci.*, **18**, 411 (1955); L. M. Welch, M. W. Swaney, A. H. Gleason, R. K. Beckwith, and R. F. Howe, *Ind. Eng. Chem.*, **39**, 826 (1947).

[22] P. D. Bartlett, *Scientific American*, **189**, 76 (1953).

[23] M. Braden and A. N. Gent, *J. Applied Polymer Sci.*, **3**, 90 (1960).

[24] R. G. Newton, *J. Rubber Research*, **14**, 27 (1945).

Oleic acid in the rubber reduces the cutting rate by about a factor of four, as does an unknown, naturally occurring inhibitor that can be extracted from rubber by benzene. The rate has also been found to depend on the amount of cross-linking, being approximately inversely proportional to the number of cross-links per unit volume. In general, any factor that raises the internal viscosity decreases the cutting rate, whether the change is due to the use of a different polymer or to the use of a different plasticizer. Cuts in gutta percha do not grow at all below the crystalline-to-amorphous transition at 80° but grow at about the same rate as cuts in vulcanized rubber above that temperature.

A plausible[25] mechanism for crack growth might be a chain reaction of radicals formed on the wall of the crack just above the apex with strained bonds at the apex. Ozone and oleic acid might exert their antagonistic effects by reacting with the attacking radicals, with ozone perhaps also reacting directly with the strained bonds. In contrast to crack growth, the mechanical breakdown of rubber milled under nitrogen is said to be enhanced rather than retarded by radical traps, perhaps because such compounds prevent radical recombination.[26]

HIGHLY ORDERED STRUCTURES IN FLUIDS

One of the important features of a crystalline phase as a reaction medium is the high degree of order that it imposes. Before discussing reactions in crystalline matrices, we would like to mention very briefly some instances of high degrees of order in phases that are still mobile. As yet no chemical applications of these special reaction media have been made, but we consider them important because they offer a means of avoiding some of the obvious experimental difficulties of working with solutions in crystalline matrices.

Solutions oriented by flow have already been discussed. It should be recalled, however, that the degree of orientation is usually small. Higher degrees of orientation in fluids can be obtained by two methods that appear promising for the investigation of orientational effects on reactions. One of these is to constrain the reagent by clathrating it in a liquid crystal. The other is to constrain the reagent molecules by associating them with highly ordered polymer chains, for example, α-helices. The available liquid

[25] "How enjoyable, how very enjoyable and luxurious it is, suddenly to emerge from the stern labyrinth of fact onto these dawn-lit uplands of surmise! Movement is free and the air is supernaturally bracing. Bright with unclassified flora, the dewy turf underfoot has a special spring."—from *Mani*, by P. L. Fermor, Harper and Brothers, 1958.

[26] M. Pike and W. F. Watson, *J. Polymer Sci.*, **9**, 229 (1952).

crystals[27] fall mostly into two categories, *smectic* (Greek: soaplike) and *nematic* (Greek: threadlike) phases. In the smectic state normal liquid flow is replaced by a gliding along slippage planes. Smectic phases should be solvents for planar molecules and should hold such reagents in coplanar orientations. This should accelerate some reactions and decelerate others relative to the rates in Newtonian liquids. Nematic phases are optically uniaxial and should act as solvents for linear molecules. The axis of a nematic phase can be oriented by a strong magnetic field, and the phase appears clear only when viewed along a direction parallel to the field. Such phases should favor colinear and side-to-side interactions of linear solute molecules.

The random coil-to-ordered helix conformational change of a polymer molecule in solution is essentially a unimolecular analog of crystallization.[28] Some ordered polymeric solutes can impose their order on smaller solute molecules with which they associate. An example is the effect of certain polymers on solutions of such planar dyes as methylene blue, crystal violet, and toluidine blue. These dyes exhibit two distinct absorption spectra, one characteristic of the individual molecules, the other a cooperative phenomenon characteristic of *stacks* of dye molecules.[29] Polymers such as nucleic acids and sodium polyacrylate will induce the aggregation of the dye into species having the cooperative type of spectrum.[30,31]

The stacking or aggregating efficiency of a polymer for a given dye depends on the structure of the polymer, presumably because the polymer should have conformations that will allow the sites at which the dye molecules are adsorbed to come into close proximity.[32] The dye thus has a reciprocal effect on the mobile conformational equilibrium of the polymer at the same time that the polymer has an effect on the aggregative equilibrium of the dye. Although distinctive optical properties have been observed for such dye-polymer complexes, no irreversible chemical reactions of the dye on the polymer have as yet been observed. That such a process might take place is made more probable by the existence of two apparently analogous reactions. One of these is canal polymerization, discussed in the next section. The other is the photochemical behavior of nucleic acids

[27] D. Vorländer, *Ber.*, **40**, 4527 (1907).

[28] In some cases several molecules may be involved in the formation of the ordered conformation.

[29] M. Kasha, *Reviews of Modern Physics*, **31**, 162 (1959); Th. Förster, *Naturwissenschaften*, **33**, 166 (1946); G. S. Levinson, W. T. Simpson, and W. Curtis, *J. Am. Chem. Soc.*, **79**, 4314 (1957); E. G. McRae and M. Kasha, *J. Chem. Phys.*, **28**, 721 (1958).

[30] M. K. Pal and S. Basu, *Makromol. Chemie*, **27**, 69 (1958).

[31] The cytological phenomenon of *metachromasy*, or two-color differential staining of the nucleus and protoplasm by a single dye, is an example of this effect.

[32] D. F. Bradley and M. K. Wolf, *Proc. Nat. Acad. Sci. U.S.*, **45**, 944 (1959).

containing thymine side groups forced into close proximity by the helical structure of the main polymer chain. This system can be photochemically denatured and restored by irradiation with light of different wavelengths. On hydrolysis of the photochemically denatured nucleic acid, pairs of thymine units attached to adjacent turns of the helix are found linked as the cyclobutane derivative 6.[33] It has also been found that irradiation of a

$$(6)$$

frozen aqueous solution of thymine itself will give the dimer.

REACTIONS IN CRYSTALLINE MATRICES

Polymerization within a clathrate complex of the canal type gives stereoregular polymer. In the canal complexes formed from thiourea, the enclosed molecules are packed in overlapping layers occupying hexagonal channels 6 to 7 Å in diameter.[34] They are not free to rotate. Of a very large number of monomers that will form such complexes with thiourea, a few happen to orient themselves in the canals in a way favorable to polymerization. Irradiation of the complex from thiourea and 2,3-dimethyl-1,3-butadiene, for example, gives stereoregular *trans*-1,4 polymer. The polymerizable sequences within the canals contain from 100 to 200 monomer molecules. Dilution of the contents of the canals with a small quantity of 2,3-dimethylbutane gives polymer of lower molecular weight because the average length of a polymerizable sequence is reduced.[35]

Irradiation of acrylamide crystals with γ-rays at low temperatures produces ion-radicals detectable by electron spin resonance.[36] Vinyl

$$2CH_2{=}CH{-}\overset{\overset{\textstyle O}{\|}}{C}{-}NH_2 \rightarrow \overset{+}{C}H_2{-}\overset{\cdot}{C}H{-}\overset{\overset{\textstyle O}{\|}}{C}{-}NH_2 + \overset{\cdot}{C}H_2{-}\overset{-}{C}H{-}\overset{\overset{\textstyle O}{\|}}{C}{-}NH_2$$

polymerization takes place on warming (but not melting) the irradiated crystals, and the warming may be postponed indefinitely without loss of activity. Since thorough degassing raises the yield of polymer, the polymerization mechanism must be at least partly growth at the radical end of

[33] R. Beukers and W. Berends, *Biochim. et Biophys. Acta*, **41**, 550 (1960).
[34] W. Schlenk, Jr., *Ann.*, **573**, 142 (1951).
[35] J. F. Brown, Jr., and D. M. White, *J. Am. Chem. Soc.*, **82**, 5671 (1960).
[36] T. A. Fadner and H. Morawetz, *J. Polymer Sci.*, **45**, 475 (1960).

the ion-radical. Solid octadecyl vinyl ether[37] can be polymerized in the same way, although vinyl ethers in solution are known to resist radical polymerization.

The polymerization of preirradiated crystalline acrylamide has an activation energy for the chain propagating or growth step of 25 kcal/mole in contrast to values in the 7 to 10 kcal/mole range for vinyl polymerizations in solution. The rate decreases exponentially as the reaction progresses, not because the growing radical ends combine or are destroyed in some way, but because something interferes with the growth step. The evidence for this interpretation of the fall-off in rate is as follows. First, there is very little decrease in electron-spin signal during a run. Secondly, the molecular weight of the polymer increases steadily during the run, and the molecular weight eventually attained is independent of the radiation dose, that is, it is independent of the initial concentration of growing chains.

Another interesting result is that dilution of the acrylamide crystal lattice with 10% of propionamide decreases the molecular weight without decreasing the yield of polymer. Apparently the crystal lattice makes chain transfer to propionamide molecules unusually efficient.

Unlike the polymer made from thiourea canal clathrates, the polymer from crystalline acrylamide is amorphous rather than stereoregular. On the other hand, the photodimers made from crystalline *trans*-cinnamic acid are stereospecific. *Trans*-cinnamic acid exists in two crystalline forms. One of them gives exclusively β-truxinic acid on irradiation, the other gives exclusively α-truxillic acid.[38] The product structures are determined by the crystal structures.

β-truxinic acid

α-truxillic acid

[37] J. G. Fee, W. S. Port, and L. P. Witnauer, *J. Polymer Sci.*, **33**, 95 (1958).
[38] H. I. Bernstein and W. C. Quimby, *J. Am. Chem. Soc.*, **65**, 1845 (1943); M. D. Cohen and G. M. J. Schmidt, in *Reactivity of Solids*, J. H. de Boer, ed., Elsevier, Amsterdam, 1961, pp. 556–561.

A reaction that is unusual in that it is fast in the crystalline phase and negligible in the melt is the uncatalyzed polymerization of acetaldehyde.[39] This reaction occurs spontaneously when a very thoroughly purified sample of acetaldehyde is frozen. The product is atactic rather than stereoregular.

Although a number of reactions take place rapidly in the solid state, the most usual result of the constraint imposed by the matrix is a marked deceleration. Acceleration by the matrix is to be expected only when the ground state of the reagent is a mobile equilibrium mixture of several conformational isomers, only the most reactive of which is actually incorporated into the solid lattice. Another requirement, of course, is that the transition state must be attainable without unduly straining the lattice; this probably means that the conformation of the transition state will be like that of the ground-state isomer present in the lattice.

A reaction that is permitted, but apparently not encouraged, by the crystal lattice of the reagent is the rearrangement of bis-(o-iodobenzoyl) peroxide.[40] This reaction was observed to take place in about one month in the solid state at room temperature, in contrast to a half-life of only a few minutes in chloroform at the same temperature.

$$(8)$$

Examples of lattice inhibition of processes involving conformational changes are numerous. Thus the X-irradiation of malonic acid crystals produces free radicals having the odd electron largely on the central carbon atom, and the conformation of the radical closely resembles that of the parent malonic acid molecule.[41] Similarly, the mutarotation of *crystalline* *d-o*-(2-dimethylaminophenyl)-phenyltrimethylammonium *d*-camphorsulfonate, which involves a hindered internal rotation, does not proceed to a measurable extent in 55 days at 100°.[42] Yet the same reaction is moderately fast in any *fluid* solvent, the half-lives ranging from 1.4 to 5.5 hours. Unimolecular decomposition reactions of diacyl peroxides or of azo

[39] O. Vogl, *J. Polymer Sci.*, **46**, 261 (1960); M. Letort, *Compt. rend.*, **202**, 767 (1936); M. W. Travers, *Trans. Faraday Soc.*, **32**, 246 (1936).

[40] J. E. Leffler, R. D. Faulkner, and C. C. Petropoulos, *J. Am. Chem. Soc.*, **80**, 5435 (1958).

[41] H. M. McConnell, C. Heller, T. Cole, and R. W. Fessenden, *J. Am. Chem. Soc.*, **82**, 766 (1960).

[42] J. E. Leffler and W. H. Graham, *J. Phys. Chem.*, **63**, 687 (1959).

compounds are also slow or unobservable in the solid state, in spite of the fact that many of these reactions are made essentially irreversible by the formation of N_2 or CO_2 molecules. It is probable that these reactions require special transition-state conformations quite different from those imposed on the reagents by their crystal structures.

We next come to the interesting possibility that a deformation of the crystal structure by an external force might change the conformation of the molecules to one favorable for reaction. A possible example of this is the red color developed when crystalline phenolphthalein is exposed simultaneously to pressure and shearing forces.[43]

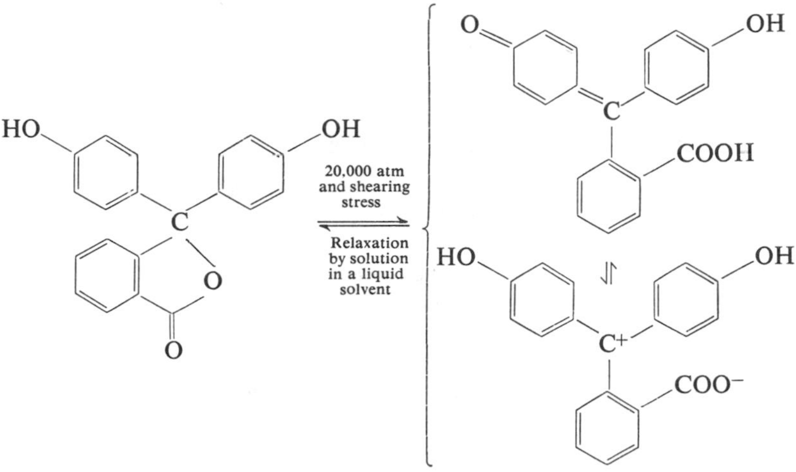

20,000 atm
and shearing
stress

Relaxation
by solution
in a liquid
solvent

The colored species is probably favored by the compression and shear because of a tendency for its rings to prefer coplanar conformations. Planar molecules both occupy less space and oppose less resistance to flow along directions parallel to their planes. A similar color formation is observed when the colorless forms of certain sulfonphthaleins are pelleted with halide salts for infrared spectroscopy.[44]

The initiation of the decomposition of solid explosives may also involve distortion of the crystal lattice, but other factors may well be even more important. For example, the rupture of a crystal lattice is sometimes attended by the emission of light (triboluminescence) or by an electric discharge. Thus the triboluminescence of a clathrate complex containing neon has been observed to consist of the usual neon emission line.[45]

[43] H. A. Larsen and H. G. Drickamer, *J. Phys. Chem.*, **62**, 119 (1958).
[44] J. E. Leffler and M. Aronoff, unpublished observations.
[45] M. C. Hoff and C. E. Boord, *J. Am. Chem. Soc.*, **72**, 2770 (1950).

THE THERMODYNAMICS AND REACTIONS OF FIBERS

A fiber is like other systems except that one additional extensive variable, the length, and one additional intensive variable, the tension, are needed for a complete description. The fact that the length at a given tension depends on the *chemical composition* of the fiber makes it possible to convert chemical energy into mechanical energy and vice versa. Fibers can also be amorphous or crystalline, and small changes in composition have the largest effect on the length of a fiber when they cause a change in phase. We will discuss first the behavior of purely amorphous fibers.

The Mechanochemistry of Noncrystalline Oriented Fibers

It is convenient to define a *chain* as used in this section as that part of a molecule extending between two cross-links.[46] The mean square distance $\overline{r_0^2}$ between the ends of a chain consisting of n' relaxed, randomly oriented "statistical elements" of length l' is given by equation 9.

$$\overline{r_0^2} = n'l'^2 \tag{9}$$

The distance between the ends of the chain in a fiber at maximum extension has the value r_m given by equation 10.

$$r_m = n'l' \tag{10}$$

Entropies. Except at very large extensions, a rubberlike system is essentially an *entropy* spring, and the restoring force is due to the increase in randomness on relaxation. The entropy of the system depends on the internal conformation of each chain and on the randomly distributed cross-links. For tetrafunctional cross-links, the latter contribution to the entropy is equal to a constant plus $(\frac{1}{2})\nu k \ln V$, where ν is the total number of chains and V is the total volume. The contribution to the entropy from the internal conformations of the chains is given by equation 11, where r is the distance between the ends.

$$\text{Internal conformational entropy} = \tfrac{1}{2}\nu k[-3\overline{r^2}/\overline{r_0^2}] \tag{11}$$

The network conformational entropy is then given by equation 12.[46,47]

$$\text{Network conformational entropy} = (\tfrac{1}{2})\nu k[\ln V - 3\overline{r^2}/\overline{r_0^2} + constant] \tag{12}$$

For the elongation from a reference state consisting of an isotropic, unconstrained network, the total entropy change is given by equation 13, in

[46] P. J. Flory, *J. Am. Chem. Soc.*, **78**, 5222 (1956).
[47] P. J. Flory, *J. Chem. Phys.*, **18**, 108 (1950).

which $\langle \alpha \rangle$ is a measure of the geometric mean of the linear dilations along the three Cartesian coordinates and is known as the dilation factor.[46]

$$\delta_L S = (\tfrac{3}{2})\nu k[1 - \overline{r^2}/r_0^2 + \ln \langle \alpha \rangle] \qquad (13)$$

The value of $\langle \alpha \rangle$ depends on the extent to which the network is swollen or diluted by solvent. If $\langle \alpha \rangle_0$ is the value in the complete absence of diluting solvent and v_2 is the volume fraction of polymer in the presence of solvent, then $\langle \alpha \rangle = \langle \alpha \rangle_0 v_2^{-1/3}$.

For simple elongations in which any changes in the two dimensions at right angles to the fiber axis are equal, the entropy change on elongation is given by equation 14, in which $\bar{x}$ is the average component of the vector r along the fiber axis.

$$\delta_L S = \tfrac{1}{2}\nu k[-3\overline{x^2}/r_0^2 - 2\langle \alpha \rangle^3 (r_0^2/3\overline{x^2})^{1/2} + 3 + 3 \ln \langle \alpha \rangle] \qquad (14)$$

A more useful expression for the elastic entropy is equation 15, in which L is the *total* length of the sample and the subscript i refers to the sample in its isotropic or unstrained state having the same volume, that is, the same $\langle \alpha \rangle$.

$$\delta_L S = -\tfrac{1}{2}\nu k[(\langle \alpha \rangle/L_i)^2(L^2 + 2L_i^3/L) - 3 - 3 \ln \langle \alpha \rangle] \qquad (15)$$

The value L_i of the length in the unstrained state depends on the state of the system at the time the cross-links were formed. If the chains were unstressed and random at that time, then L_i is independent of the number of cross-links. But for a network formed by the random cross-linking of an axially oriented *crystalline* system or of a system whose molecules have been highly oriented by stretching, L_i should be proportional to the square root of the fraction of the units cross-linked. This prediction can be tested by cross-linking a stretched (and therefore highly oriented) sample of rubber by means of γ-irradiation.[48] The ratio of the length L_i of the cross-linked rubber after relaxing the force to the length L_0 of the untreated rubber, also at zero tension, is shown plotted against the square root of the fraction of units cross-linked in Figure 10-3. It is seen that the expected linearity is found for high degrees of cross-linking.

The Force of Retraction.   The force of retraction is given by equation 16.

$$f = \left(\frac{\partial F}{\partial L}\right)_{P, T, v_2} \qquad (16)$$

Studies with several polymers have shown that, unless the degree of swelling (v_2 or $\langle \alpha \rangle$) changes during the elongation, the enthalpy component of the

[48] D. E. Roberts, L. Mandelkern, and P. J. Flory, *J. Am. Chem. Soc.*, **79**, 1515 (1957); D. E. Roberts and L. Mandelkern, *ibid.*, **80**, 1289 (1958).

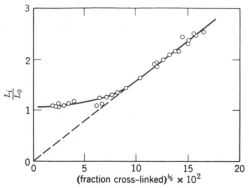

Figure 10-3. Effect of cross-linking on the average linear dimension at zero tension of a sample of rubber. The sample was stretched (and therefore highly oriented) during the cross-linking procedure. (From D. E. Roberts, L. Mandelkern, and P. J. Flory, *J. Am. Chem. Soc.*, **79**, 1515 (1957).)

free energy of elongation is negligible. In that case $\delta_L F = -T \delta_L S$, and the force is given by equation 17, in which B is $\nu k(\langle\alpha\rangle/L_i)^2$.

$$f = BTL\left(1 - \frac{L_i^3}{L^3}\right) \tag{17}$$

The contribution of the internal energy to the tension is represented by f_E. Although it is small for amorphous fibers at low extensions, it may be a quantity of theoretical interest. It is given by equation 18.[49]

$$f_E = -fT\left(\frac{\partial(\ln f/T)}{\partial T}\right)_{V,L} \tag{18}$$

Experimentally the constant volume condition of equation 18 is inconvenient, and it has sometimes been wrongly assumed that the conditions $P, L/L_i$ constant can replace the conditions V, L constant.

For systems at constant pressure and temperature that fulfill the additional important requirement of not exchanging matter with their surroundings, the rate of change of the enthalpy and entropy with length is given by equations 19 and 20.[50]

$$\left(\frac{\partial H}{\partial L}\right)_{P,T} = f - T\left(\frac{\partial f}{\partial T}\right)_{P,L} \tag{19}$$

$$\left(\frac{\partial S}{\partial L}\right)_{P,T} = -\left(\frac{\partial f}{\partial T}\right)_{P,L} \tag{20}$$

[49] P. J. Flory, A. Ciferri, and C. A. J. Hoeve, *J. Polymer Sci.*, **45**, 235 (1960).
[50] W. B. Weigand and J. W. Snyder, *Trans. Inst. Rubber Ind.*, **10**, 234 (1934).

These equations can be used for fibers that contain a fixed amount of solvent. If they are used for a system consisting of a fiber in chemical equilibrium with a surrounding medium, it must be understood that the enthalpy and entropy changes are for the system as a whole and not just for the fiber.

The equations applicable to the fiber itself in an open system, where the fiber is diluted with varying amounts of a one-component phase in equilibrium with it, are 21 and 22.[51]

$$\left(\frac{\partial H}{\partial L}\right)_{n,T} = f - T\left(\frac{\partial f}{\partial T}\right)_{equ,L} - \Delta\bar{H}\left(\frac{\partial n}{\partial L}\right)_{T,equ} \tag{21}$$

$$\left(\frac{\partial S}{\partial L}\right)_{n,T} = -\left(\frac{\partial f}{\partial T}\right)_{equ,L} - \Delta\bar{S}\left(\frac{\partial n}{\partial L}\right)_{T,equ} \tag{22}$$

In equations 21 and 22, n refers to the chemical composition of the fiber (i.e., the amount of absorbed solvent), the subscript n refers to constant composition, and the subscript equ refers to equilibrium composition. The quantities $\Delta\bar{H}$ and $\Delta\bar{S}$ are partial molar quantities for dilution of the unstressed fiber.

The existence in equations 21 and 22 of a term $(\partial n/\partial L)_{T,equ}$ different from zero means that the fiber can do mechanical work in isothermal cycles. Thus rubber in contact with an organic vapor shows a change in degree of swelling when it is stretched, and, conversely, exposure of rubber to a different partial pressure of a swelling agent changes the tension.

Differences in swelling of a gel or a fiber can of course be brought about by a chemical reaction that changes the affinity of the polymer for the solvent, and musclelike machines have been devised in which the source of free energy is an acid-base reaction or a redox reaction.[52]

The redox machine[53] is a copolymer of vinyl alcohol and allylalloxan. The reduced, or dialuric acid, form of the fiber is less swollen by the aqueous buffer in which the fiber is immersed than is the alloxan form; hence the conversion causes a contraction or an increase in tension. The contraction was limited to the long dimension by means of transverse laminae of a rigid, nonswelling material. A similar machine[54] based on the acid-base reaction of a lightly cross-linked copolymer of vinyl alcohol and acrylic acid[55] was put through over 2000 cycles, doing a total of 0.8 meter kg. of work in the form of lifted weights. Contraction takes place when the sodium salt form, which is expanded because of the osmotic pressure, is neutralized by a strong acid. An interesting possible application of this is to use a fiber under variable tension as a chromatographic column of easily adjustable pK.

[51] A. Oplatka, I. Michaeli, and A. Katchalsky, *J. Polymer Sci.*, **46**, 365 (1960).
[52] W. Kuhn, *Makromol. Chemie*, **35**, 200 (1960).
[53] A. Ramel, dissertation, Basel (1957).
[54] Known as a pH Muscle; not to be confused with the Ph. Muscle, or Doctor of Physical Education.
[55] W. Kuhn and B. Hargitay, *Experientia*, **7**, 1 (1951).

The Mechanochemistry of Partly Crystalline Fibers

Examples of partly crystalline fibers are found both among synthetic polymers, although some of these attain a crystalline degree of ordering only when under considerable tension, and among natural polymers, such as proteins or nucleic acids, in which well-ordered helical structures are stabilized by intramolecular hydrogen bonds.[56] Some atactic polymers resist crystallization because the irregularity with which their side-chains are disposed makes orderly packing difficult. Isotactic polymers, in contrast, usually crystallize easily. In any case, crystallization is rarely complete, ordered regions being interspersed with amorphous or randomly coiled regions.

The phase rule may be applied to fibers if it is modified to take into account the fact that the tension is also a variable of state. The familiar expression $\phi = C - P + 2$ becomes $\phi = C - P + 3$, for a fiber has one

Table 10-1. *Analogy between Partly Crystalline Fibers and Isotropic Liquid-Vapor Systems*

LIQUID-VAPOR SYSTEM	FIBER SYSTEM
Pressure, P	Tension, f
Contraction, ΔV	Contraction, $-\Delta L$
Work $= P \Delta V$	Work $= -f \Delta L$
Heat of vaporization at constant pressure	Heat of fusion at constant tension
$\Delta H = \Delta E + P \Delta V$	$= \Delta H - f \Delta L$
Entropy of vaporization	Entropy of fusion
$(\Delta E + P \Delta V)/T$	$(\Delta H - f \Delta L)/T$
Clausius-Clapeyron equations	Flory-Gee equations
$dP/dT = \Delta S/\Delta V$	$(\partial f/\partial T)_{P,\,\text{equ}} = -\Delta S/\Delta L$
$d(P/T)/d(1/T) = -(\Delta H - P \Delta V)/\Delta V$	$(\partial(f/T)/\partial(1/T))_P = \dfrac{(\Delta H - f \Delta L) + f \Delta L}{\Delta L}$
$= -\Delta E/\Delta V$	$= \Delta H/\Delta L$
The boiling point is a function of the pressure.	The melting point is a function of the tension.
Above the critical temperature T_c no separate liquid phase can be distinguished, no matter what the pressure.	Above the critical temperature $T_m{}^c$ no separate crystalline phase can be distinguished, no matter what the tension.

(The Δ quantities are latent quantities of vaporization and fusion.)

[56] L. Mandelkern, "The Melting of Crystalline Polymers," *Rubber Chemistry and Technology*, **32**, 1392 (1959).

more degree of freedom than an otherwise similar isotropic system. In consequence a one-component fiber at a fixed pressure can have a whole series of melting points (temperature at which crystalline order disappears), depending on the tension. In a partially melted two-component fiber at constant temperature and pressure, the tension is a function of the relative amounts of amorphous and crystalline material.

As shown in Table 10-1, the behavior of a two-phase (amorphous and crystalline) fiber at various tensions and temperatures is closely analogous to the behavior of a liquid-vapor system at various pressures and temperatures.

The dependence of the melting point of cross-linked rubber fibers on the tension is shown in Figure 10-4.[57] The heat of fusion as calculated from the appropriate equation of Table 10-1 is 1.280 kcal per mole of isoprene units.

Some fibers can be melted without decomposition only when immersed in a diluent to lower the melting point. If the diluent is taken up only by the amorphous phase of the fiber, then the two-component amorphous phase is in equilibrium with two pure phases, and the fiber phase rule allows two degrees of freedom. If we fix the pressure, the temperature at which the diluted amorphous phase and the crystalline phase coexist is again a function of the tension. However, we can no longer use the equations of Table 10-1, for these are appropriate only for closed systems and not for systems that exchange matter with their surroundings.[57] The appropriate expression is now equation 23, in which the quantities $\Delta \bar{H}$ and

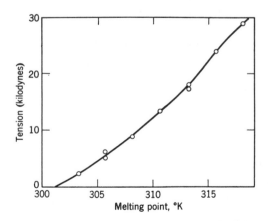

Figure 10-4. Melting point as a function of the applied tension for a sample of fibrous, 1.56% cross-linked, natural rubber.

[57] J. F. M. Oth and P. J. Flory, *J. Am. Chem. Soc.*, **80**, 1297 (1958).

$\Delta\overline{\overline{L}}$ are the latent changes for melting *plus* the integral changes for the dilution of the amorphous phase to its equilibrium composition.

$$\left(\frac{\partial(f/T)}{\partial(1/T)}\right)_{P,L} = \frac{\Delta\overline{\overline{H}}}{\Delta\overline{\overline{L}}} \tag{23}$$

The theory of rubberlike elasticity, based on an entropy spring model, predicts equation 24 as the description of the state of a network at large deformations.[46]

$$f \cong \nu k T n' \frac{L_a}{L_m{}^2} \tag{24}$$

L_a is the length of the purely amorphous one-component fiber under the equilibrium tension f, L_m is the length of the fiber at maximum extension, n' is the number of statistical elements in the freely rotating chain, and ν is the total number of chains in the network.[46,56] Combination of equation 24 and the Flory-Gee equation, followed by integration, gives equation 25 for the f/T ratio at equilibrium.

$$\left(\frac{f}{T}\right)_{\text{equ}} = \frac{3k\nu n'}{L_m} \left[\frac{L_c}{L_m} \pm \left(\frac{2\,\Delta h'}{3R}\right)^{1/2}\left(\frac{1}{T_m} - \frac{1}{T_m{}^c}\right)^{1/2}\right] \tag{25}$$

In equation 25, $\Delta h'$ is the heat of fusion per mole of *statistical segments*, and $T_m{}^c$ is the melting temperature under a tension such that the length L_a of the completely amorphous fiber is the same as the length L_c of the completely crystalline fiber. $T_m{}^c$ is now seen to be a critical temperature, for equation 25 has no real solutions at temperatures above $T_m{}^c$.

Figure 10-5 shows the form expected for a plot of $(f/T)_{\text{equ}}$ against L for crystallizable polymer networks.[46,56] Assuming that the network is initially amorphous, $(f/T)_{\text{equ}}$ should increase linearly with L as predicted by equation 24. At point A crystallization begins, the ratio $(f/T)_{\text{equ}}$ remaining constant at the lower of the two values predicted by equation 25 until at point B the crystallization is complete. With further extension the $(f/T)_{\text{equ}}$ then rises almost vertically, since the crystalline fiber is quite rigid. In the unlikely event that the higher of the two solutions of equation 25 is physically real for the system,[58] then at point D the system should again become partly amorphous, remelting being complete at point E. As the temperature is made to approach $T_m{}^c$, the points A and E are expected to draw together, and above $T_m{}^c$ there should no longer be any horizontal line segment corresponding to crystallization.

---

[58] The event is unlikely because the network will probably rupture before sufficient tension is attained.

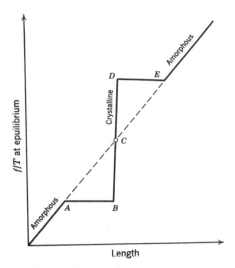

Figure 10-5. A tension-length isotherm as predicted by equation 25 for a crystallizable polymer fiber. The two amorphous regions correspond to the use of the plus or minus sign, respectively. The dotted line corresponds to a metastable amorphous state. At point C, the length of the metastable amorphous fiber is equal to L_c.

Figure 10-6 is a composite of experimental results obtained for the tension and the length of a fibrous natural rubber at the various absolute temperatures indicated.[56,57] The horizontal lines represent the force needed to maintain the amorphous and crystalline phases in equilibrium at the given temperature. The circles at the left extremity of each line represent the tension and length at which crystallization begins at that temperature. These melting points each fall on a tension versus length curve for the amorphous material at that temperature, as calculated from equation 26 (dashed curves). Equation 26 is simply a more general form of equation 24, applicable to small as well as large elongations; L_i is the length under zero tension.

$$f = \frac{3kT\nu n'L_a}{L_m{}^2}\left(1 - \frac{L_i{}^3}{L_a{}^3}\right) \tag{26}$$

It is instructive to consider various processes starting with a semi-crystalline network 8 cm long at 303.2°K, represented by point A of

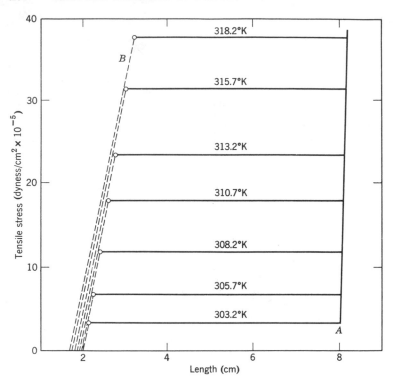

Figure 10-6. Experimental tension-length isotherms for a fibrous natural rubber. (From Reference 56, p. 1440.)

Figure 10-6. Raising the temperature *at constant length* will require increasingly greater tensions, corresponding to points on a line rising vertically from *A*. It is an important feature of two-phase systems that the increase in tension which can be caused by heating them at constant length is much greater than that which can be obtained by heating a purely amorphous system at constant length. For an amorphous system the force increases in proportion only to the first power of the increase in the absolute temperature, since such a system is approximately an entropy spring (equation 26). Starting again at point *A* but keeping the stress constant and putting heat into the system causes the point to traverse the horizontal line until its left terminus representing complete melting is reached. This process reduces the length by about a factor of four, but the original length can be restored by taking the heat out again at constant tension. Starting at point *B*, which represents an amorphous system, we see that cooling at constant length will cause partial crystallization. Further cooling at constant length will then cause a very large decrease in tension as the amount of the crystalline phase increases.

The shrinking of fibrous natural rubber on melting *at zero stress* is only partly reversible, since the crystalline regions formed on cooling the system again will not be completely oriented along the fiber axis. However, it is possible to construct a system whose dimensional changes on melting at zero stress are reversible. If cross-links are introduced into the system while the polymer is in its oriented and elongated state, the orientation of the crystals will be preserved during freezing and thawing cycles. An example is the behavior of fibrous polyethylene that has been cross-linked by means of ionizing radiation while it is elongated and crystalline.[59] Introducing cross-links to the extent of about 4% of the chain units increases the length characteristic of the *amorphous* state at zero stress about twentyfold. This polymer contracts and elongates reversibly not only under zero stress but also under any load not too great for its tensile strength. Later in this chapter we shall see that there are polymers in which similar dimensional cycles can be carried out chemically, the phase transitions being initiated by reversible chemical reactions of the fiber.

Macromolecules capable of amorphous to crystalline transitions in fibers are sometimes able to show similar phenomena in dilute solution. Crystallinity within a fiber is a cooperative phenomenon involving many molecules of the fiber. "Crystallinity" within a single molecule (or an aggregation of only two or three molecules) is a cooperative phenomenon involving interaction between the various parts of the molecular chain. Thus the helical conformation or subspecies of a collagen molecule in solution can be "melted" to the random coil subspecies over a mere 3° temperature range.[60] The helical or well-ordered form is highly asymmetric and rodlike; its solutions are quite viscous. "Melting" can be detected either by the precipitous drop in viscosity or polarimetrically.

The reason that a process which can be properly considered a simple chemical equilibrium can also be successfully treated as a phase transition is to be found in the size of the molecules. The enthalpy of the increased hydrogen bonding between peptide units in the native protein helix is only about 1 to 1.5 kcal per mole of *peptide units*, but it is of the order of 3000 kcal per mole of *protein*. The equilibrium constant for the formation of the helix therefore changes by a factor of 10^6 to 10^9 per degree, giving a very sharp change from one molecular species to the other. The process is not

[59] L. Mandelkern, D. E. Roberts, A. F. Diorio, and A. S. Posner, *J. Am. Chem. Soc.*, **81**, 4148 (1959).
[60] P. J. Flory and E. S. Weaver, *J. Am. Chem. Soc.*, **82**, 4518 (1960). It is probable that one helical molecule dissociates into two or three helical components, which then undergo a conformational change into random coils.

different in kind or in temperature dependence from the melting equilibrium of a slightly impure sample of a low molecular-weight material that forms hydrogen-bonded crystals.

Chemically Induced Phase Transitions in Fibers

In the preceding section we found that the isothermal crystalline-to-amorphous transition in fibers could be induced by preferential solvation or swelling of the amorphous phase. Other chemical reactions can also bring about the transition. For example, crystalline keratin fibers become amorphous and shrink by about 20% on immersion in a cuprammonium solution.[61] Destruction of the copper ion-amino acid complexes by treatment with dilute acid restores the original crystallinity and length of the fiber.

Crystallinity in an α-helix or other protein structure depends on intramolecular hydrogen bonds that stabilize the ordered conformation relative to the random conformations. The donors and acceptors in hydrogen bonds are also capable of acid-base reactions, and so there is an optimum pH range for the liquid bathing the fiber at which the greatest number of pairs are available for hydrogen bonding.[62] At the optimum pH for intramolecular hydrogen bonding a maximum is observed in the melting point at a given tension.[63] Carboxyl-carboxyl bonds like those in acetic acid dimer exist only at very low pH's; in proteins for which other types of hydrogen bond are important, the maximum in melting point will occur at higher pH's.

Shrinkage of muscle fibers in the presence of adenosine triphosphate (ATP) is also believed to be a phase change.[64] A large shrinkage accompanied by an increased diffuseness of the X-ray diffraction pattern occurs over a very narrow range of ATP concentrations at constant temperature.

RELAXATION PROCESSES IN POLYMERS

In addition to the relaxation methods described in Chapter 5, the study of the damping of mechanical oscillations in a fiber as a function of the frequency and of the temperature can be used for studying fast reactions or conformational changes in polymers.[65]

[61] C. S. Whewell, J. Ashworth, V. R. Srinivassan, and A. G. P. Vassiliadis, *Textile Research J.*, **29**, 386 (1959).
[62] H. A. Scheraga, *Protein Structure*, ch. IV, Academic Press, 1961.
[63] A. Nakajima and H. A. Scheraga, *J. Am. Chem. Soc.*, **83**, 1575 (1961).
[64] L. Mandelkern, A. S. Posner, A. F. Diorio, and K. Laki, *Proc. Nat. Acad. Sci.*, **45**, 814 (1959).
[65] J. D. Ferry, *Viscoelastic Properties of Polymers*, Wiley, 1961.

The motions with which a polymer molecule can respond to an applied stress include relatively fast local changes in conformation of a small part of the molecule and much slower large-scale changes in what might be called the *contour* (of the entire molecule), rather than the conformation. The temperature at which the latter motions are essentially frozen out in an amorphous polymer is known as the glass transition temperature. The small-scale motions that persist below the glass transition temperature can also be studied above that temperature by using oscillations or time scales too brief for the relaxation of the entire molecule. Thus a polymer can behave like a fluid or like a glass, depending on the time scale of the experiment. This dependence of the properties on the time scale is shown in an extreme form by the well-known silicone polymer, bouncing putty. This fluid not only can be bounced but can also be shattered to fragments having sharp edges.

Processes that contribute to the mechanical relaxation of a polymer can also be studied by the dielectric or nuclear magnetic relaxation methods. However, it is not always easy to determine whether the same molecular motion is involved in both relaxations. The dynamic mechanical properties of acrylic polymers show that some process, such as reorientation of side groups, occurs at temperatures far below that of the main glass transition. What appears to be the same motion is detectable from effects on the line width of the proton magnetic resonance signal.[66] It is concluded that all the methyl groups in polymethyl acrylate and half the methyl groups in polymethyl methacrylate undergo hindered rotation beginning at $77°K$. The other methyl groups in the latter polymer begin to rotate at $130°K$.

The principal mechanical loss peak and the changes found in the modulus of elasticity of polyamides are attributed to segmental motions of the polymer chains that require the breaking of hydrogen bonds between the chains. Water or methanol, by disrupting the interchain hydrogen bonds, cause the mechanical energy absorption to take place at lower temperatures.[67] Replacing the amide hydrogen atoms with methyl groups has a similar effect. Water also has a large effect on the molecular motion as measured by dielectric loss[67] or by nuclear magnetic resonance experiments.[68]

Mechanical loss experiments involving the solute interacting with the vibrations of a host clathrating crystal do not seem to have been undertaken. The dynamic mechanical properties of a liquid solvent have sometimes been invoked in theories of medium effects, however.

[66] K. M. Sinnott, *J. Polymer Sci.*, **42**, 3 (1960).
[67] A. E. Woodward, J. M. Criseman, and J. A. Sauer, *ibid.*, **44**, 23 (1960).
[68] R. E. Glick, R. P. Gupta, J. A. Sauer, and A. E. Woodward, *ibid.*, **42**, 271 (1960).

Authors Index

Ablard, J. E., 371
Abrams, J. R., 330
Ackermann, T., 90
Acree, S. F., 183, 373
Adkins, H., 234
Akerlöf, G., 308
Alder, M. G., 32, 320, 329, 394
Alfrey, T., Jr., 185, 209
Allen, G. F., 373
Allen, P. E. M., 134
Alter, H. W., 340
Altscher, S., 209
Altschul, L. H., 184, 333, 376, 377
Ambler, E., 74
Amdur, I., 129
Anderson, B. M., 161, 189
Anderson, W. A., 117
Andon, R. J. L., 350, 371, 391
Andreas, F., 304
Andrews, L. J., 53, 339
Angier, D. J., 408
Appl, M., 326
Archer, B. L., 335
Archer, G., 281
Arenberg, C. A., 227, 268
Armstrong, R., 183
Arnett, E. M., 274, 288
Arond, L. H., 209
Aronoff, M., 415
Ashworth, J., 426
Aybar, S., 96
Ayrey, G., 408

Bacarella, A. L., 268, 276
Bachmann, W. E., 209
Backer, H. J., 183, 254

Baddeley, G., 337, 348
Badger, G. M., 377
Badger, R. M., 258
Bailey, W. C. Jr., 216, 222
Baker, A. W., 257
Baker, J. W., 184
Bakhtiar, A. K., 61, 409
Ballentine, A. R., 330
Ballinger, P., 227
Baltzly, R., 209, 229
Bamford, C. H., 185
Banta, C., 209
Baramboin, N. K., 408
Barasch, W., 329
Barbaras, G. K., 134
Barbato, L. M., 329
Barber, D. W., 392
Barclay, I. M., 390
Barker, J. A., 147
Barrow, G. M., 109, 277
Bartlett, P. D., 107, 248, 254, 358, 360, 409
Barton, D. H. R., 134
Bascombe, K. N., 270, 271, 273
Basolo, F., 227
Bassett, J. Y., Jr., 251
Basu, S., 227
Bates, R. G., 295, 371, 373
Bauer, S. H., 92, 258
Baughan, E. C., 227
Baughman, G., 123, 267
Baxter, J. F., 209
Bayles, J. W., 235
Beckwith, R. K., 409
Bekkum, H. van, 214
Bell, A. G., 82

Bell, R. P., 75, 133, 165, 166, 227, 235, 236, 270, 271, 273, 281, 336
Bellamy, L. J., 255
Bender, M. L., 131, 220, 243, 353
Benjamin, C. E., 209, 212
Benson, S. W., 136, 137, 138
Benzing, E. P., 360
Berends, W., 412
Berg, D., 35
Bergmann, J. G., 227
Bergmann, K., 103, 114, 117, 118
Berkheimer, H. E., 148, 280
Berkowitz, B. J., 129, 153, 268, 289
Berlin, A. A., 406
Berliner, E., 184, 333, 376, 377
Bernhard, S. A., 221
Bernstein, H. I., 413
Bernstein, H. J., 86
Bernstein, I. A., 184, 329
Bertrand, R. R., 129
Beste, G. W., 254
Bestul, A. B., 405
Bettman, B., 183
Beukers, R., 412
Bevan, C. W. L., 184
Bhide, B. V., 222
Bickel, A. F., 169
Bickford, W. G., 330
Bieber, A., 49, 373, 375
Biechler, S. S., 230
Bigeleisen, J., 271, 274
Biggs, A. I., 268
Biletch, M., 329
Blackadder, D. A., 322
Blackman, S. W., 209
Blades, A. T., 134
Blake, J. T., 209
Bleakney, W., 83
Blomquist, A. T., 184, 329, 330, 369
Bond, G. C., 331
Bonner, W. H., 381
Boord, C. E., 415
Borders, A. M., 222, 368, 387
Bordwell, F. G., 183, 211, 251
Borgwardt, S., 327
Born, M., 266
Borucki, L., 124
Böttcher, C. F. J., 111
Boudart, M. J., 124
Bower, V. E., 373

Bown, D. E., 301
Boyd, D. R., 184
Boyd, R. H., 279, 281
Braden, M., 409
Bradfield, A. E., 184
Bradley, D. F., 411
Brady, J. D., 244, 381
Brady, O. L., 183
Brahn, B., 338
Branch, G. E. K., 183, 184, 206, 209, 226, 333, 351, 377
Brandon, M., 339
Braude, E. A., 209, 212, 338
Brehmer, L., 183
Bresler, S. E., 408
Briegleb, G., 49, 53, 373, 375
Bright, W. L., 174, 184
Brimacombe, D. A., 214
Briscoe, H. T., 174, 184
Britt, R. D., Jr., 82
Brønsted, J. M., 236, 254, 269, 294, 295
Brown, C. W., 249
Brown, D. A., 184, 333, 351
Brown, D. H., 183
Brown, E. V., 332
Brown, H. C., 134, 149, 168, 172, 173, 196, 197, 200, 201, 204, 205, 209, 227, 229, 231, 244, 381
Brown, I., 147
Brown, J. F., Jr., 412
Brown, T. L., 258
Brown, W. G., 326, 327
Brück, D., 259
Bruckenstein, S., 277
Bruehlman, R. J., 145
Bruton, J. D., 209
Buchanan, T. J., 44, 122
Bueche, F., 405
Bunnett, J. F., 211, 249, 251, 281, 285, 287, 288, 326
Bunton, C. A., 271
Burkhardt, G. N., 337, 383
Burkus, J., 334
Burn, J., 221
Burton, J. O., 373
Burwell, R. L., Jr., 279
Busala, A., 96
Buselli, A. J., 330, 369
Bushick, R. D., 274

Buss, J. H., 136, 137, 138
Butler, A. F., 308
Butler, E. T., 134
Butler, J. A. V., 390
Buttery, R. G., 125

Cahn, A., 229
Caldin, E. F., 133, 242
Calvin, M., 226, 340
Campbell, H. J., 288
Canady, W. J., 371
Cantwell, N. H., 332
Carnie, W. W., 374
Carr, A. S., 183
Ceresa, R. J., 408, 409
Cerf, R., 404
Cerfontain, H., 209
Chadwick, J., 337
Chang, H. S., 274
Chao, T. S., 129, 377
Chapman, A. W., 184
Chapman, N. B., 366
Chaudhuri, N., 125
Chen, D. T. Y., 142, 371, 373
Chen, J. H., 114
Chiang, Y., 281
Chokshi, N. M., 183
Christiansen, J. A., 60
Christman, D. R., 281
Chuoke, R. L., 259
Ciferri, A., 418
Clark, L. W., 397
Clark, W. M., 295
Claussen, W. F., 350, 391
Clear, C. G., 183
Clement, R., 234
Clement, R. A., 183, 302
Cobble, J. W., 350
Coburn, W. C., Jr., 53, 255
Cohen, M. D., 329, 413
Cohen, S. G., 329
Cohn, E. J., 149
Cole, R. H., 101
Cole, T., 414
Colgate, S. O., 129
Colter, A. K., 54
Cook, C. D., 329, 394
Cooper, G. D., 183, 211
Corran, P. G., 96
Corse, J., 337

Costain, C. G., 110
Cox, J. C., 327
Cox, J. D., 350, 371, 391
Craft, M. J., 336, 365
Craven, J. M., 234, 262
Cretcher, L. H., 327
Criege, R., 107
Criseman, J. M., 427
Cristol, S. J., 328, 347
Critchfield, F. E., 276
Cross, R. P., 336
Crowell, T. I., 326
Cummins, R. W., 338
Curtin, D. Y., 119
Curtis, A. J., 103
Curtis, W., 411
Cvetanović, R. J., 134, 135, 156, 167

Dahn, H., 280
Dailey, B. P., 110
Davidson, D. W., 101
Davies, G. L., 320, 329
Davies, M., 53, 103, 111, 114, 117, 339
Davies, W. C., 184
Davis, G., 221
Davis, M. M. 215, 268, 269
Davis, O. C. M., 183
Davison, P. F., 404
Dawber, J. G., 273
Dawson, H. M., 227
Deans, F. B., 209
Debye, P., 60, 83
DeFazio, C. A., 391
Delbanco, A., 269
Denney, D. B., 107, 108
Denney, D. J., 101
Deno, N. C., 148, 209, 212, 227, 275, 279, 280
Denyer, R. L., 53
Derr, E. L., 147
Dessy, R. E., 378
DeTar, D. F., 107, 330
DeWitt, E. J., 209
Deyrup, A. J., 271, 273
Dickinson, J. D., 382
Dickinson, R. G., 53
Dimroth, O., 24, 338
Dinglinger, A., 332
Dingwall, A., 274

Diorio, A. F., 425, 426
Dippy, J. F. J., 226
Dischler, B., 124
Doak, G. O., 183, 215
Doering, W. von E., 125
Domash, L., 229
Donohue, J., 274
Dostrovsky, I., 251, 368
Draper, F., Jr., 211
Dressel, J., 184
Drickamer, H. G., 415
Drysdale, J. J., 183
Duffy, J. A., 335
Dumbaugh, W. H., Jr., 373
Duncan, P. M., 374
Dunkle, F. B., 308
Dunn, G. E., 215
Duynstee, E. F. J., 308
Dvoretsky, I., 125
Dyas, H. E., 336

Eaborn, C., 209, 382
Easterbrook, E. K., 334
Edsall, J. T., 149
Edward, J. T., 274, 288
Edwards, J. O., 250, 251
Ege, S. N., 209
Eger, H. H., 227
Ehlers, R. W., 373
Eigen, M., 80, 81, 82, 90, 404
Eilar, K. R., 347
Eisner, M., 123
Eliel, E. L., 38, 119
Elo, H., 183
Elofson, R. M., 234
Embree, N. D., 373
Emerson, M. T., 61, 121
Emmons, W. D., 303
Ener, C., 96
Eucken, A., 96, 100
Evans, A. G., 331, 371
Evans, D. P., 131, 183, 184, 221, 222, 227, 327, 363, 367
Evans, H. D., 218
Evans, M. G., 124, 392
Evans, M. W., 390
Evans, R. J., 272
Evans, W. L., 209, 212, 280
Everett, D. H., 371, 373

Ewald, L., 304
Eyring, H., 62, 340

Fadner, T. A., 412
Fainberg, A. H., 299, 310, 398
Fairclough, R. A., 327, 331, 334, 335, 351
Fang, F. T., 190
Farmer, R. C., 183
Farnsworth, H. E., 331
Faulkner, R. D., 414
Feates, F. S., 45, 49, 241, 373, 392
Fee, J. G., 413
Feldman, A. M., 360
Feller, R. L., 175, 184
Fendley, J. A., 75
Ferguson, J. W., 209
Fernandez, L. P., 373
Ferry, J. D., 426
Fessenden, R. W., 414
Fettes, G. C., 130, 242, 374
Fierens, P. J. C., 191, 313
Fife, W., 382
Finestone, A. G., 329
Finnegan, W. G., 183, 378
Fischer, E., 111, 115, 340, 341
Fish, K., 103, 114, 118
Fitzpatrick, F. W., 354
Fletcher, C. H., 83
Flexser, L. A., 274
Flory, P. J., 416, 417, 418, 421, 425
Fogg, P. G. T., 85
Forbes, W. F., 209
Fore, S. P., 330
Förster, T., 411
Fowden, L., 368
Fox, C. J., 227
Frank, H. S., 390
Frazer, R. T. M., 338
Freedman, E., 124
Freedman, L. D., 183, 215
Frey, H. M., 125, 126
Fried, S., 326, 327
Frost, A. A., 227
Fugassi, P., 336
Fujiwara, S., 115

Garber, R. A., 271
Gardner, J. D., 274
Garrett, A. B., 329

Garvin, D., 124
Gassmann, A. G., 221, 332, 368, 387
Gelbshtein, A. I., 274
Gelles, E., 165, 166, 227
Gent, A. N., 409
Gettler, J. D., 354
Gilchrist, A., 53
Gill, E. K., 125
Gilman, H., 215
Gintis, D., 229
Girard, C., 382
Glasstone, S., 62
Gleason, A. H., 409
Glick, R. E., 215, 427
Goering, H. L., 107, 176, 183, 382
Gold, V., 144, 243, 279, 281
Goldsmith, H. L., 166
Goldstein, R. F., 183
Goldsworthy, L. J. 184
Goller, H., 391
Gooberman, G., 405
Goodman, A. L., 291
Goodman, P., 405
Gordon, A. S., 168
Gordon, E., 134
Gordon, J. J., 131, 183, 222, 227
Gordon, M., 348
Gordy, W., 311
Goulden, J. D. S., 257
Graham, W. H., 117, 339, 340, 363, 414
Granger, M. R., 274
Gray, P., 114, 117
Graybill, B. M., 114, 117, 340, 363
Grayson, M., 381
Green, J. H. S., 134
Green, M., 249
Gregory, D. V., 184
Griffing, V., 84, 96
Griffiths, D. M. L., 53
Groocock, C. M., 230
Groves, P. T., 275
Grunwald, E., 32, 53, 61, 89, 121, 123, 129, 153, 255, 267, 268, 276, 281, 289, 291, 299, 302, 308, 337
Gudmundsen, C. H., 382
Gupta, R. P., 427
Gutbezahl, B., 32, 268, 276, 289, 291

Gutowsky, H. S., 86, 88, 117
Gwinn, W. D., 111
Gwynn, D., 382

Hackett, J. W., 337
Haggis, G. H., 44, 122
Halban, H. v., 53
Hall, D. M., 339, 363
Hall, D. N., 114
Hall, G. V., 227
Hall, H. K., Jr., 235
Hall, N. F., 183
Ham, G., 381
Hamann, S. D., 371
Hambly, A. N., 379
Hamer, W. J., 373
Hammett, L. P., 24, 49, 172, 183, 221, 254, 271, 273, 274, 277, 281, 354
Hammond, G. S., 161, 190
Hamner, W. F., 259
Hanks, P. A., 85
Hannaert, H., 313
Hansrote, C. J., Jr., 326
Harborth, G., 61
Harden, G. D., 134
Harfenist, M., 229
Hargitay, B., 419
Harman, R. A., 335, 351
Harned, H. S., 49, 308, 373
Harris, M. M., 339, 363
Harrison, A. G., 135
Hartel, H. V., 227
Hartman, R. J., 221, 222, 332, 368, 385, 387
Harvey, J. T., 209
Haskell, V. C., 221
Hassan, M., 209
Hasted, J. B., 44, 122
Haun, J. L., 9, 27, 262
Hause, N. L., 328, 347
Hauser, C. R., 184, 334
Hawes, B. W. V., 279, 281
Hawthorne, M. F., 303
Hayward, R. W., 74
Head, A. J., 134
Heckmann, K., 404
Hegan, D. S., 334
Heidt, L. J., 336
Heilbronner, E., 271

Heinzelmann, D. C., 330
Heller, A., 281
Heller, C., 414
Hendrickson, A. R., 209, 212
Henne, A. L., 227
Henry, R. A., 183, 378
Hepler, L. G., 373
Heppolette, R. L., 350
Herbert, J., 329
Herbst, R. L., Jr., 183
Herington, E. F. G., 350, 371, 391
Herschbach, D. R., 110
Herte, P., 304
Hertel, E., 184
Herz, W., 390
Herzberg, G., 125
Herzfeld, K. F., 84, 96
Hetzer, H. B., 215, 268, 269
Hey, D. H., 320, 329
Hiatt, R. R., 358
Higasi, K., 103, 114, 115, 117, 118
Higginson, W. C. E., 227, 236
Hildebrand, J. H., 264
Hill, D. G., 334, 336
Hine, J., 193, 216, 222, 227
Hine, M., 227
Hinshelwood, C. N., 183, 184, 222,
  227, 322, 327, 331, 334, 335, 337,
  351, 353, 377, 397
Hirota, E., 110
Hirshberg, Y., 340, 341
Hirst, J., 249
Hobbs, M. E., 53
Hoeve, C. A. J., 418
Hoff, M. C., 415
Hoffmann, J. S., 330
Hofstra, A., 244
Högfeldt, E., 271, 274
Hogg, J. S., 387
Holdsworth, J. B., 184
Holiday, E. R., 248, 254
Holland, R. S., 103, 104
Holm, C. H., 86, 88, 117
Holness, N. J., 38, 119
Hol'tsshmidt, V. A., 327
Hood, G. C., 278
Hoogsteen, H. M., 222, 368, 388
Hopkins, H. B., 184
Hoppes, D. D., 74
Hornig, D. F., 84

Horowitz, R. H., 229, 231
Horrex, C., 337, 383
Horrom, B. W., 327
Howard, B. B., 115
Howe, R. F., 409
Howlett, K. E., 134
Hubbard, J. C., 96
Hubbard, R. A., II, 32, 304
Hubbard, W. N., 110
Hudson, R. F., 184, 190, 249, 333,
  335, 337, 351
Hudson, R. P., 74
Hughes, E. D., 209, 251, 368
Huisgen, R., 326
Hulett, J. R., 75
Hunt, H., 333
Hunt, R., 107
Hurley, R. B., 221, 332
Hutchison, C. A., Jr., 9
Huyser, E. S., 209
Hyman, H. H., 271
Hyne, J. B., 394

Ibata, T., 212
Iczkowski, R. P., 141
Illuminati, G., 209, 210
Ingold, C. K., 183, 209, 220, 222,
  230, 302, 368
Ingram, D. J. E., 61
Inham, J. M., 328
Inukai, T., 149, 209, 381
Itoh, K., 114, 117
Ives, D. J. G., 45, 49, 241, 373, 392
Iwanami, M., 378
Izmailov, N. A., 297

Jackmann, L., 338
Jackson, A., 230
Jacobson, B., 404
Jacobson, R. R., 382
Jacox, M. E., 92, 183
Jaffé, H. H., 172, 183, 215, 272, 380
James, J. C., 183
Jaruzelski, J. J., 209, 280
Jefferson, E. G., 144, 243
Jencks, W. P., 161, 189
Jenkins, A. D., 185
Jenkins, D. I., 337, 383
Jenkins, F. E., 379
Jensen, F. R., 200, 201

Johns, R. G. S., 393
Johnson, J. B., 276
Johnson, M. D., 173, 211
Johnston, H. S., 74
Johnston, R., 185
Jones, B., 184
Jones, E. R., 338
Jones, H. W., 337
Jones, N., 331
Jones, P. M. S., 331
Jones, R. N., 209
Jordan, J., 373
Jordan, J. E., 129
Judson, C. M., 142, 183

Kalman, O. F., 115, 342
Kantrowitz, A., 83
Kaplan, J., 89
Kaplan, M. L., 121, 308
Kargin, V. A., 409
Kasha, M., 82, 127, 411
Kaspar, R., 107
Katchalsky, A., 419
Kaufmann, H., 331
Kazbekov, E. N., 408
Keay, L., 337
Kebarle, P., 135
Keefer, R. M., 53, 339
Keirs, R. J., 82
Kerr, J. A., 130, 242, 374
Kershaw, D. N., 188
Ketelaar, J. A. A., 38
Ketonen, L., 183
Key, A., 227
Kice, J. L., 107
Kilb, R. W., 110
Kilpatrick, J. E., 110
Kilpatrick, M., 142, 183, 227, 254, 268
Kim, J.-Y., 378
Kindler, K., 183, 184, 222
King, C. V., 254
Kirkwood, J. G., 266
Klein, F. S., 281
Kloosterziel, H., 183
Klopman, G., 190
Knell, M., 251
Knoevenagel, F., 304
Knox, J. G., 183
Kochi, J. G., 190

Kohnstam, G., 123, 209, 267
Kooyman, E. C., 169, 209
Körösy, F., 264
Koskikallio, J., 281
Kosower, E. M., 300, 310, 311
Kowalsky, A., 9
Kreevoy, M. M., 227, 231
Kresge, A. J., 281
Krieger, K. A., 334
Krisher, L. C., 110
Kromhout, R. A., 61, 121
Kuby, S. A., 333
Kuhn, W., 419
Kuivila, H. G., 209, 212
Kuntz, I., 53
Kwart, H., 291

Laidler, K. J., 49, 62, 125, 142, 184, 333, 371, 373, 377, 379
Laki, K., 426
Lamb, J., 114, 124
Lambert, J. D., 85, 96, 97
La Mer, V. K., 61
Landsman, D. A., 371, 373
Lane, J. F., 175, 184
Langford, C. H., 279
Langsdorf, W. P., Jr., 190
Lardy, H. A., 333
Larsen, H. A., 415
Lassettre, E. N., 53
Laughlin, R. G., 125
Lee, S. H., Jr., 329
LeFèvre, R. J., 340
Leffler, J. E., 32, 107, 114, 117, 156, 157, 161, 304, 320, 324, 329, 338, 339, 340, 363, 394, 414, 415
Leisten, J. A., 188, 335
LeMaistre, J. W., 184
Lerner, R. G., 110
Lester, C. T., 209, 336, 365
Letort, M., 414
Levenson, H. S., 222
Levinson, G. S., 411
Lewis, E. S., 173, 211
Lewis, G. E., 377
Lewis, G. N., 82
Lewis, I. C., 135, 214, 215, 234
Lewis, W. P. G., 184
Ley, J. B., 271
Lichten, N. N., 209

Lide, D. R., Jr., 110
Lieber, E., 129, 183, 377, 378
Lien, A. P., 244
Lilker, J., 329
Lin, C. C., 110
Lin, S. C., 83
Linder, B., 265
Linnig, F. J., 323
Linschitz, H., 82
Linstead, R. P., 335, 338
Lipscomb, F. J., 86, 125
Liu, S. K., 338
Lloyd, L. L., 222
Loewe, L., 280
Loewenstein, A., 89, 114, 117
Lohmann, D. H., 387
Lohmann, K. H., 164
Long, F. A., 227, 271, 272, 273, 281, 306, 308
Looney, C. E., 114, 117
Lossing, F. P., 135
Lowen, A. M., 279
Lührman, H., 184
Lukach, C. A., 38, 119
Lumme, P., 183
Lundberg, J. L., 86

Maccoll, A., 134
Mackenzie, S., 338
Mackie, J. D. H., 368
Mackor, E. L., 244, 245, 271
Magee, J. L., 340
Maier, W., 124
de Maine, P. A. D., 339
Mandel, J., 323
Mandelkern, L., 417, 420, 425, 426
Mann, D. E., 110
Mann, R. S., 331
Manogg, P., 124
de la Mare, P. B. D., 209, 368
Margrave, J. L., 141
Marino, G., 209
Marle, E. R., 184
Marshall, H. P., 268, 276, 289, 337
Martin, J. C., 209
Mason, E. A., 129
Masure, F., 279
Mathews, T., 271
Matsen, F. A., 259
McCaulay, D. A., 244

McCauley, C. E., 254
McCleary, H. R., 254
McClure, A., 130, 242, 374
McCombie, H., 183
McConnell, H. M., 414
McCoubrey, J. C., 100
McCullough, J. P., 110
McDaniel, D. H., 172, 173
McDevit, W. F., 306, 308
McDougall, A. O., 336
McElhill, E. A., 176, 184
McEwen, W. E., 382
McGary, C. W., Jr., 196
McGeer, P. L., 103
McGowan, J. C., 224
McGrath, W. D., 124
McKinley, J. D., Jr., 124
McKinney, D. D., 371
McMeekin, T. L., 149
McNesby, J. R., 168
McRae, E. G., 411
Meakins, R. J., 103, 111, 114, 117, 339
Mears, W. H., 142
Mechelynck-David, C., 191, 313
Medvedev, S. S., 61
Meek, J. S., 328, 347
Meer, N., 227
Meiboom, S., 89, 114, 117
Meloche, I., 184, 333, 377, 379
van Mels, W. H., 254
Melville, H. W., 61, 134
Metcalf, R. P., 183
van Meurs, N., 209
Michaeli, I., 419
Michel, K. W., 63
Miller, D. G., 73
Miller, G. H., 61, 409
Miller, H. W., 347
Miller, J., 212
Miller, R. C., 105
Miller, S. I., 314, 327, 328
Milward, G. L., 393
Minton, R. G., 234, 326
Miyagawa, I., 114, 117
Miyazawa, T., 13, 110
Moede, J. A., 222, 368, 385
Moelwyn-Hughes, E. A., 254, 328, 332, 336, 396
Moffat, A., 333

Möller, E., 165, 166
Montague, C. E., 65
Moore, C. G., 408
Morath, R. J., 326
Morawetz, H., 412
Moreland, W. T., Jr., 217, 227
Morgan, M. S., 327
Morgan, V. G., 184
Mörikofer, A., 271
Morino, Y., 115
Morse, B. K., 333, 334, 337, 358
Morse, J. G., 227
Mortimer, C. T., 371
Mosely, R. B., 301
Mosher, F. H., 209
Mueller, W. A., 209
Mukaiyama, T., 366, 378
Mukula, A. L., 183
Münster, A., 322
Murphy, R. B., 234
Murray, M. A., 279
Musa, R. S., 123
Myers, R. J., 111
Myers, R. R., 221

Nace, H. R., 227
Nagakura, S., 389
Naghizadeh, J. N., 234
Nakajima, A., 426
Näsänen, R., 183
Nathan, W. S., 183, 184, 222
Naylor, R. E., Jr., 110
Nederbragt, G. W., 115
Nelson, L. S., 86
Neumer, J. F., 114, 117
Newitt, D. M., 331, 335
Newling, W. B. S., 222
Newman, M. S., 176, 183, 220, 221, 227, 329, 334, 366
Newton, R. G., 409
Nichols, P. L., Jr., 328
Nightingale, E. R., Jr., 123
Nims, L. F., 373
Nixon, A. C., 184, 206, 209, 333, 351, 377
Noble, P., Jr., 211
Noda, L. H., 333
Norcross, B. E., 329, 394
Norman, R. O. C., 214
Norris, J. F., 184, 209

Norrish, R. G. W., 82, 86, 124, 125
Northcott, J., 340
Noyce, D. S., 190
Noyes, R. M., 62, 327, 328
Nümann, E., 96, 100

O'Connor, G. L., 227
Ogata, Y., 184, 378
Ogawa, T., 120
Ogimachi, N., 339
Ogston, A. G., 248, 254
Ohishi, S., 366
Okamoto, Y., 149, 196, 204, 205, 209, 381
Okano, M., 184
Olson, A. C., 107
Oplatka, A., 419
Orr, R. A., 338
O'Sullivan, D. G., 258
Oth, J. F. M., 421
Overberger, C. G., 209, 329, 338
Owen, B. B., 49, 308

Pal, M. K., 411
Papeé, H. M., 49, 371, 373
Parke, J. B., 100
Parker, R. E., 184
Parker, S. H., 382
Pastor, R. C., 9
Pattersen, A., Jr., 35
Paul, M. A., 183, 271, 272, 273, 276, 281
Pausacker, K. H., 184
Payman, W., 83
Peacock, J., 242
Peard, W. J., 149
Pearson, D. E., 209
Pearson, R. G., 82, 227, 250, 266
Pedersen, K. J., 236, 371
Pegg, J. A., 53
Peloquin, J., 190
P'eng, S., 335
Perizzolo, C. F., 409
Peterson, H. J., 279, 280
Petrauskas, A. A., 114
Petropoulos, C. C., 414
Pflaum, R. T., 149
Philippe, R. J., 53
Phillips, C. S. C., 392
Phillips, W. D., 114, 117

Philpot, J. St. L., 248, 254
Piccolini, R., 90
Pierce, L., 110
Pierotti, G. J., 147
Piette, L. H., 114, 117
Pike, M., 410
Pinching, G. D., 371, 373
Pincock, R. E., 360
Pinkerton, J. M. M., 124
Pinsent, B. R. W., 371, 373
Pitt, B. M., 251
Pitt, D. A., 103, 104
Pitzer, K. S., 13, 110, 111
Plane, R. A., 123
Polanyi, J. C., 86
Polanyi, M., 124, 134, 227
Polglase, M. F., 350, 391
Pople, J. A., 86
Port, W. S., 413
Porter, G., 82
Posner, A. S., 425, 426
Powles, J. G., 104
Pray, H. A. H., 330
Prentiss, S. S., 28
Pressman, D., 183, 326
Price, A. H., 103, 104
Price, C. C., 185
Price, F. P., Jr., 354
Price, S. J. W., 134
Pritchard, J. G., 227
Pryor, J. H., 373
Purcell, W. P., 103, 114, 118
Purlee, E. L., 268, 276, 391
Purves, C. B., 336
Puvelich, W. S., 230

Quant, A. J., 347
Quayle, O. R., 148
Quimby, W. C., 413

Rabinovitch, B. S., 63, 126
Rabinowitch, E., 59
Radda, G. K., 214
Rainsford, A. E., 184
Raistrick, B., 331
Ralph, P. D., 214
Ramachandra Rao, C. N., 129, 209, 377
Ramel, A., 419
Rapp, D., 74

Rathmann, G. B., 103
Redlich, O., 147, 278
Reeves, L. W., 114, 117
Reid, E. E., 229
Reilly, C. A., 278
Reilly, E. L., 114, 117
Reinheimer, J. D., 249
Resler, E. L., 83
Reztzova, E. V., 409
Rhind-Tutt, A. J., 271
Rice, M. R., 302
Rice, O. K., 391
Richardson, D. B., 125
Ridd, J. H., 279
Riding, F., 134
Rieseberg, H., 124
Ritchie, C. D., 141
Ro, R. S., 38, 119
Robb, J. C., 134
Roberti, D. M., 115
Roberts, D. E., 417, 425
Roberts, J. D., 114, 117, 183, 209, 217, 227
Robertson, P. W., 209
Robertson, R. E., 184, 350, 394
Robertson, W. W., 259
Robins, J., 9, 27, 262
Robinson, C. C., 234
Robinson, R. A., 123, 308, 373
Ropp, G. A., 209
Ross, S. D., 53
Rossow, A. G., 234
Rothbaum, H. P., 209
Rotzler, G., 280
Rubin, T., 176, 183
Rubin, W., 331
Russell, G. A., 168, 305
Russell, K. E., 387
Russell-Hill, D. Q., 366
Ryason, P. R., 211
Rylander, P. N., 230, 333, 334

Sacher, E., 279
Sadler, P. W., 258
Sager, E. E., 183
Sager, W. F., 141
Saines, G., 274
Salter, R., 85, 96, 97
Saminskii, E. M., 408

Sanford, J. K., 209
Santerre, G. M., 326
Sapiro, R. H., 331, 335
Sarel, S., 366
Sarkanen, K., 82
Satchell, D. P. N., 272
Sauer, J. A., 427
Sauer, R. W., 334
Saunders, W. H., Jr., 329, 379
Scanlan, J., 134
Scarborough, H. A., 183
Scatchard, G., 28
Schaafsma, Y., 169
Schaal, R., 279
Schaefgen, J. R., 230
Scheibe, G., 259
Scheraga, H. A., 404, 426
Schindewolf, U., 404
Schlenk, W., Jr., 412
Schlitt, R., 382
Schmidt, G. M. J., 413
Schneider, W. G., 86, 114
Schoen, J., 80, 82
Schooley, M. R., 183
Schriesheim, A., 209, 280
Schroer, E., 332
Schubert, W. M., 9, 27, 234, 262, 274, 326
Schulz, G. V., 61
Schwartz, N., 183, 377
Schwarz, G., 404
Schwarzenbach, G., 279
Scott, C. B., 164, 246, 252, 254, 335
Scott, J. M. W., 350
Scott, R. L., 264
Seib, A., 304
Semyonov, N. N., 341
Setser, D. W., 126
Shand, W., Jr., 340
Shannon, K., 374
Sheglova, G. G., 274
Shepherd, W. C. F., 83
Sherk, K. W., 209
Shiio, H., 120
Shimizu, H., 115
Shiner, V. J., Jr., 234
Shoosmith, J., 125
Shorter, J., 183, 337
Shryne, T. M., 207, 209

Shulgin, A. T., 257
Siegel, S., 110
Silvermann, G. B., 209
Simamura, O., 212
Simmons, M. C., 125
Simon, W., 271
Simons, D. M., 360
Simpson, W. T., 411
Sinnott, K. M., 427
Sixma, F. L. J., 209
Skell, P. S., 127
Sklar, A. L., 184
Skrabal, A., 227
Slater, J. S., 130, 242, 374
Slater, N. B., 63
Slator, A., 254
Slaugh, L. H., 168
Slonimskii, G. L., 409
Small, G., 254
Smit, P. J., 245, 271
Smith, B. B., 32
Smith, D. J., 227
Smith, E. M., 214
Smith, F., 147
Smith, G. F., 245, 371
Smith, H. A., 221, 222, 332
Smith, J., 53
Smith, P. K., 373
Smith, S. G., 310
Smith, S. R., 168
Smoluchowski, M. von, 60
Smoot, C. R., 200
Smyth, C. P., 101, 103, 104, 105, 114, 115, 117, 118, 342
Snyder, J. W., 418
Snyder, L. R., 190
Sogolova, T. I., 409
Solimene, N., 110
Solomon, S., 329
Sommer, L., 227
Spitzer, R., 110
Sprinkle, M. R., 183
Sprinivassan, V. R., 426
Stanford, S. C., 311
Stavely, L. A. K., 393
Steadly, H., 262
Steel, C., 130, 242, 374
Steiner, H., 331
Stern, E. S., 209, 212
Stewart, R., 173, 211, 271, 274

Stock, L. M., 197, 201, 209
Stocken, L. A., 248, 254
Stoicheff, B. P., 110
Stokes, R. H., 123
Streitweiser, A., Jr., 227
Stubbs, F. J., 183, 184, 227, 377
Sudborough, J. J., 222
Sujishi, S., 134, 227
Sulzberger, R., 279
Sunners, B., 114
Sutherland, R. O., 373
Sutton, L. E., 53
Swain, C. G., 164, 190, 246, 248, 252, 254, 301, 335
Swalen, J. D., 110
Swaney, M. W., 409
Symons, M. C. R., 61
Szwarc, M., 134

Tabor, W., 110
Tada, H., 355
Taft, R. W., Jr., 135, 183, 214, 215, 218, 219, 220, 221, 227, 228, 230, 231, 234, 281, 288, 314, 391
Taher, N. A., 209
Takagi, Y., 378
Takahashi, J., 38
Takamura, H., 366
Talroze, V. L., 341
Tammann, G., 306
Tamres, M., 339
Tanaka, J., 389
Tannenbaum, E., 111
Tanner, D., 209
Tarbell, D. S., 230, 333, 334, 358
Taube, H., 123, 338
Taylor, J. J., 209
Taylor, M. D., 134, 227
Temkin, A. I., 274
Thiele, E., 63
Thomas, C. A., Jr., 404
Thomas, G. H., 339
Thomas, H. C., 337
Thomas, J. H., 331
Thomas, P. J., 134
Thomas, R. J., 353
Thrush, B. A., 86, 125
Tiers, G. V. D., 114, 117
Timm, E. W., 334, 353, 377
Ting, I., 209

Tolman, R. C., 73
Tomashevskii, E. E., 408
Tomlinson, T. E., 53
Tommila, E., 131, 183, 184, 222, 397
Tonkyn, R. G., 211
Tovborg-Jensen, A., 269
Townsend, M. G., 61
Travers, M. W., 414
Trifan, D., 337
Trotman-Dickenson, A. F., 130, 134, 242, 374
Trouton, F., 389
Tsai, L., 366
Tsuno, Y., 131, 212, 326, 361, 384, 385
Turner, M. K., 222
Turnquest, B. W., 131, 243
Tusa, G. F., 392
Tuul, J., 331
Twiss, D. F., 254

Ubbelohde, A. R., 100

Van Looy, H., 271, 277
Van Steveninck, A. W. de R., 209
Varma, S., 348
Vassiliadis, A. G. P., 426
Venkataraman, H. S., 397
Verbanic, C. J., 234
Verdin, A., 392
Verhoek, F. H., 145, 183, 221
Verkade, P. E., 214
Vernon, C. A., 271
Vignale, M. J., 209
Vogelsong, D. C., 266
Vogl, O., 414
Vorländer, D., 411
Vorob'ev, N. K., 327

Waals, J. H. van der, 244, 245, 271
Wagner, C., 254
Wagner, R. S., 110
Walker, L. G., 173, 211
Walling, C., 176, 184
Wang, C. H., 329
Wang, I. C., 274, 288
Warburton, B., 96
Ward, R. L., 86

Ware, J. C., 329, 379
Warhurst, E., 134
Waring, C. E., 330
Warner, J. C., 371
Warth, F. J., 183
Wassermann, A., 329, 331
Waters, W. A., 293
Watson, H. B., 131, 183, 184, 222, 327, 363
Watson, W. F., 408, 409, 410
Weale, K. E., 368
Weare, J. H., 149
Weaver, E. S., 425
Webb, R. L., 183
Webster, D. E., 209
Weigand, W. B., 418
Weimer, D. K., 83
Weisfeld, L. B., 291
Weissman, S. I., 86
Welch, L. M., 409
Wentworth, W. E., 82
Wepster, B. M., 214, 218, 261
Westheimer, F. H., 183
Whalley, E., 281
Wheland, G. W., 9
Whewell, C. S., 425
White, D. M., 412
White, J. A., 374
White, W. N., 382
Wiberg, K. B., 168, 207, 209, 272
Wicke, E., 90, 391
Widom, J. M., 53
Wiig, E. O., 332
William, D., 134
Williams, E. G., 184
Williams, G., 279
Williams, G. H., 320, 329
Williams, H. L., 338
Williams, R., 327, 363
Williams, R. J., 134

Williams, R. L., 255
Williamson, B., 61
Wilmarth, W. K., 183, 377
Wilputte-Steinert, L., 313
Wilson, D. J., 63
Wilson, E. B., Jr., 110
Winstein, S., 38, 90, 119, 299, 302, 310, 337, 398
Witnauer, L. P., 413
Wolf, A. P., 281
Wolf, M. K., 411
Wolfenden, J. H., 334
Wood, W. C., 59
Woodward, A. E., 427
Woodworth, R. C., 127
Wu, C. S., 74
Wu, C. Y., 274, 288
Wyatt, P. A. H., 273
Wynne-Jones, W. F. K., 371

Yabroff, D. L., 183
Yada, H., 389
Yancey, J. A., 183
Yates, K., 274
Yeh, S.-J., 272
Yerger, E. A., 109, 277
Yoshihashi, H., 120
Young, W. G., 326
Yukawa, Y., 132, 212, 326, 361, 384, 385

Zagt, R., 209
Zahler, R. E., 211, 274
Zaugg, H. E., 327
Zawidzki, T. W., 49, 371, 373
Ziegler, K., 304
Zhurkov, S. N., 408
Zimpelmann, E., 53
Zucker, L., 281

Subject Index

Absorption curve, 81
Acetal hydrolysis, substituent effects in, 231
Acetamides, rotameric isomerization rates of, 86–88
Acetic acid, dimerization of, solvent effects on heat, and entropy, 52–54
  and hydrogen bonding of, 52
  and ionization of, heat and entropy, 49
Acetic anhydride hydrolysis, amine catalyzed, correlation with K_B, 243–244
  dispersion in catalysis relationship, 144
Acid and base dissociation, enthalpies and entropies of (tables), 371–373
Acid catalysis and the H_0 function, 281
Acid ionization, substituent effects in various solvents (table), 268
Acidity function D_0, 272
Acidity function H_0, 269–277
  definition, 270–271
  effect of salts and polar solvents on, 276
  effect of temperature, 272, 274
  extensibility to nonpolar solvents, 276–277
  table of values, 272–273
  test of base-independence, 270, 273
Acidity function H_0', 275

Acidity function H_R, definition, 280
  table of, 280
  temperature dependence of, 274
  water activity and, 281
Acidity function H_-, 278–279
  table, 278
Acidity function in nonaqueous media, 275–276
Acrylamide, polymerization of crystalline, 412–413
Acrylic acids, *trans* 3-substituted, substituent effects in, 216–217
Activation, mechanisms of, 64
Active methylene compounds, variation of Brønsted coefficients with activity of (table), 166
Activity coefficients, analysis of by interaction coefficients, 27–30
  Debye-Hückel limiting law for, 21
  definition, 20
Activity of water in various acids (table), 286
Additivity postulate, 140
Additivity rules, 135–139
  as basis for organizing organic chemistry, 139
  effect of degree of approximation in, 135–137
  logical consistency in, 136
  number of parameters needed, 137–139
  procedure for generating, 139
Alcohols in carbon tetrachloride, free energy of, 147

Alkenes, correlation of reactions with peracetic acid, 135
hydrogenation of, 134
reaction with H or O atoms, 134
Alkyl bromides, enthalpies and entropies for pyrolysis of, 134
Alkyl chlorides, enthalpies and entropies for HCl elimination from, 134
Alkyl iodides, relations of pyrolysis rates of, 134
Amines, association with silver ion, dispersion of linear free energy relationship from, 144–145
Amines, complex formation with iodine, behavior of relationship of enthalpy and entropy to sigma,* 388–389
α-Amino acids and derivatives, solubility relationship for, 149
5-Aminotriazoles, rearrangement of, 159–160
isoequilibrium relationship, 129, 131
Ammonia, parachor expression for solubility in, 148
Ammonium hydroxide, relaxation time for dissociation, 80
Ammonium ions, enthalpies and entropies for dissociation of (table), 369–370
Anisole, substituent effects on spectrum of, 262
Arenephosphonite ionization, and σ," 215
Aromatic halogenation, sigma complex model for, 244–245
Aromatic hydrocarbons, complexing of, 52–53
Aramatic substitution and the Brown relationship, 196–203
Arrhenius equation, 58
Atomic oxygen, selectivity of, 167

Baeyer-Villiger reaction, medium effects in, 303
Benzaldehydes, semicarbazone formation from, anomalous substitutional effects, 189

Benzaldehydes, semicarbarzone formation from, rate-equilibrium relationship, 161
Benzenes, alkyl substituted, complexing of, 52
Benzhydrylazide decomposition, enthalpies and entropies and effect of p-methoxyl, 382
Benzoic acids, ionization of, substituent effects, 49, 142
acid-catalyzed esterification, behavior of enthalpies and entropies, 386–387
Benzoic anhydride hydrolysis, as "well-behaved" $\rho\sigma$ relationship, 376
Benzophenone oximation, resonance effects on enthalpies and entropies, 382–383
Benzoyl azide decomposition, dispersion in $\rho\sigma$ and isokinetic plots, 384–385
steric effects on enthalpy and entropy, 364
Benzoyl camphor enolization, solvent effects on, 24, 25
N-Benzoyldiphenylamine-2-carboxylic acids, 361–363
Benzyl chloride dielectric relaxation time, 118
Benzyl halides reacting with amines, anomalous substituent effects in, 190–191
Benzyltrimethylsilane cleavage, σ^- correlation and enthalpy-entropy behavior, 382–383
Bicyclo[2.2.2]octane-1-carboxylic acids, substituent effects in, 217–219
Biphenyl racemization, isokinetic relationships and, 363
medium effects on, 363
lattice inhibition of, 414
Boltzmann distribution, 3, 4
applied to reactions, 4
applied to subsidiary equilibria, 39
applied to transition states, 67
properties of, 4

Bond contributions to enthalpy, entropy, and heat capacity (table), 138
Bond distance, 11
Bonds, covalent, isotope effects, 12
  vibrational energy levels, 9, 12
Born charging, 266–267
Bromomethyl group, rotation in alkyl bromides, 118
Brown relationship, equation, 197
  interaction mechanisms in, 198
Brønsted relationship, 130, 235–242
  as reactivity-selectivity relationship, 165
  behavior of enthalpies and entropies in, 369–374
  dispersion in, 236–238, 242
  interpretation of slope in, 238–241
  rate-equilibrium relationship from, 238–241
  solvent effects in, 130, 242
Bunnett equation, 282–289
  definition of, 285
  dispersion and curvature in, 288–289
  extrathermodynamic derivation of, 282–285
  interpretation of w, 287–288
  table of examples, 287
  w′, 288
t-Butyl peresters, concerted decomposition and entropy of, 358

Cage effect, 59, 170
  and dissociation reactions, 61
  and relaxation time for rotation, 108
Cages, solvent, 7
Cancellation of structural effects in σ and Δ quantities, 140
Candide, 263
Carbon dioxide, vibrational-translational relaxation time of, 97
Carbonic acid, effect of subsidiary equilibrium on apparent strength of, 35
Carbonium ions, selectivity of, 164, 249

Carboxylate group, nonequivalence of oxygens in ion pairs, 109
Carboxylic acid dimers, dissociation rates of, 124
Carboxylic acid ionization, enthalpies and entropies for (tables), 49, 369–370
  medium effect on, 130
  medium effect relationship for, 153–154
Cascade mechanism, 95
Catalysis, see Brønsted relationship
  of acetic anhydride hydrolysis by amines, 144, 243–244
Chain transfer, 185–187
Chemical models for medium effects, 31
Chemical species, 2
Chichibabin's diradical, 9, 10
Chlorine dioxide, flash photolysis of, 125
Chlorocarbonate decomposition, σ⁺ correlation and mechanism, 207
trans-Cinnamic acid, dimerization of crystalline, 413
·o-Claisen rearrangement, σ⁺ correlation and enthalpies, 382
Clathrate complexes, 7
  polymerization in, 412
Coffee break, probable chemical effect of, 161
Collision(s), amoureuse, 100
  de-excitation by, Boltzmann law for, 94
  relation to radiation-induced de-excitation, 93
  effectiveness of, 91–106
  frequency, 58
  in liquids, 59
  numbers, for vibrational de-excitation, relation to the energy, 97–98
  relation to molecular structure, 97–100
  rate theory, 57–62
  theoretical model for energy transfer by, 92–94
  vibrational model for, 93
Complexes, molecular, as models for isosteric transition states, 231

Complexes, molecular, effect on activity coefficients, 30
effect on energy levels, 9, 10
Complexes, molecular (π), effect on reaction rates, 54
enthalpies, entropies and free energies for, 52–55
substituent effects and, 32
Complex rate constants and medium effects, 303
Composition, dependence of free energy on, 17
Conformational isomers, and mechanism of activation, 119
and reaction rate, 119
rates of interconversion of, 109 (table), 111–114
Constraints, effect on entropy, 45
effect on partition function, 7
Continuum, as solvent model, 54–55
Corresponding states, 390
Crack growth, plausible mechanism for, 410
Cross sections, effective, 58
Crystals, motion in, 7, 403
Curvature, in Brown relationship, 201
in Hammett relationship, 187–191
Cyanoacetic acid, heats and entropies of ionization of, 44, 49

Dance of Life, 128
Dawn-lit uplands of surmise, 410
Decalin partition functions, 13, 14
Decalyl perbenzoate, rotational relaxation during decomposition of, 107–108
Decomposition reactions, inhibition by matrices, 414–415
Dedeuteration of aromatic hydrocarbons, entropy of, 245
Degeneracy factor, 6
Deuterium isotope effect, and reactivity of abstracting radicals, 167–168
and tunneling, 74
on oscillator frequency, 12
on zero-point energy, 12
Diaryl olefins, protonation and H_o, 275

Dibenzofuran, dielectric relaxation of, 115
1,2-Dibromoethane, *trans-gauche* isomerism of in various media, 37
Dielectric constant, versus specific effects, 33
and entropy, 54
and medium effects, 54–55
Dielectric relaxation, 83
activation parameters for (table), 116
isokinetic temperature in, 339, 342
rate constants for (table), 112
Diffusion controlled reactions, 59–60, 170
Dimesityl disulfide, 170
Dimethylaminonitrobenzene, solvent effects on resonance of, 32
2,3-Dimethylbutane chlorination, medium effects on, 305–306
Dimethylstilbenes, correlation of epoxidation rates of, 132–133
Dinitrobenzene complexing, 53
Dioxane, lifetime of complexes with cations, 123
2,2-Diphenyl-1,1-dichloroethane dehydrohalogenation, substituent effects and interactor mechanism vectors, 347, 348
Diphenyl ether dielectric relaxation, 115
1,3-Diphenylguanidinium benzoate ion pair formation in benzene, correlation with σ^0, 215
Dipolar resonance structures, and rotameric relaxation, 115
enhancement by solvent, 32
in benzophenone, 115
in diphenylamine, 115
in diphenyl ether, 115
Dipole-dipole interactions and vibrational de-excitation, 100
Dipoles, orientation of, 7
rotation of, 38
Diradical, Chichibabin's, 9, 10
Dispersion forces, and linear free energy relationships, 265–266

Dispersion into parallel lines, 144–145, 150
and the α-hydrogen effect, 232–233
in Brønsted relationships, 236–238, 242
in correlation of amine-catalyzed acetic anhydride hydrolysis rates, 243
in Hammett relationships, 191
in isokinetic relationships, 346–347, 365–367, 395–397
Dissociation field method, 81–82
Dissociation initiated by shock waves, 84
Dynamic reversibility, 73–74

Edwards equation, 251–254
table of parameters for, 253–254
Electronically excited molecules, substituent effects in, 259–262
comparison with ground state substituent effects, 261–262
Electronic transition energies, correlation with reaction rates, 169
Elimination reactions, relation between rates of, 134
Encounter, 59–60, 106–109
duration of, 106
number of collisions in, 59
probability of rotational relaxation during, 107
Encounter number, 59
and diffusion coefficients, 60
effect of viscosity, 59
for ions, 60
for nonelectrolytes, 60
Energy (see also kinetic energy), a additive behavior of, 8, 11
electronic, 11
vibrational, 11
zero-point, 10–11, 41
Energy level(s), definition, 2
electronic, 9
in liquids, 8
Energy level spacing, effect on energy transfer, 91
rotational, 9

Energy level spacing, translational, 9
vibrational, 9, 11
Energy transfer, in the liquid phase, 100–106
relative ease from various modes, 91
requirements for collisional, 91
Enthalpy, 40–45
and kinetic energy, 41–42
and zero-point energy, 41
dependence on temperature, 43
of solution, additivity of, 350
partial molar, and partial molar free energy, 18
partial molar, and standard partial molar, 41
Enthalpy and entropy, additivity in, 347–353
as test of model, 50
compensation in theoretical models, 50, 321–322
cooperative and compensating changes in, 48
criteria for evaluating data, 324
error contours for, 323–324
relative importance of, 48–50
Enthalpy-entropy diagrams (see also isokinetic relationship), prediction of relative rates from, 357
Enthalpy-entropy relationships, 315 ff. (see also isokinetic relationship)
Enthalpy of activation, definition, 70–71
potential and kinetic energy contributions to, 71
relation to Arrhenius activation energy, 71
Entropy, and activity coefficients, 47
and composition, 46
and equilibrium at constant volume and energy, 16
and kinetic energy, 45
and partition function, 45
and probability, 16
and rotational freedom, 45–46
change in standard partial molar, and equilibrium constant, 19
effect of constraints on, 45
of solution, additivity of, 350

Entropy, of solution, partial molar, and concentration, 41, 47
relative importance of various motions to, 46
rotational and vibrational, 46
variation with temperature, 47
Entropy of activation, definition, 70–71
relation to Arrhenius A factor, 71
shape of potential energy barrier and, 68–69
Epistemological dilemma, 171
Ester hydrolysis, effect of mechanism change on $\rho\sigma$ plot, 188–189
isokinetic relationship for base-catalyzed, 367
linear free energy relationship for, 131
Ester pyrolysis, enthalpies and entropies of, 134
Ethane, energies of eclipsed and staggered forms of, 111
vibrational and torsional relaxations of, 85
vibrational relaxation times of, 99
1,1-Ethanediol dehydration, catalysis of, 236–238
Ethanol, hydrogen bonding of, 52
OH stretching doublet in spectrum of, 90
rotation about C-O bond, 90
Ethanol-water, acidity scale for, 293–297
1-Ethyl-4-carbomethoxypyridinium iodide, charge transfer spectrum, 309
Ethyl cyclopentanone-2-carboxylate and tunneling effect, 75
Equilibrium, and entropy, 19
and Gibbs free energy, 16–17
at constant pressure, 19
at constant temperature and pressure, 16–17
at constant volume, 19
at constant volume and energy, 16
bond forming, 35
displacement from, 17
in dilute systems, 17
isomeric, 36–38

Equilibrium, isomeric, macroscopic, 15
operational definition of, 15
solvation, 33–35
subsidiary, 33–38
Equilibrium constant, Boltzmann expression for, 5, 6
free energy expression for, 19
heat capacity effects on, 45
Exchange narrowing of spectral lines, 89
Excitation and de-excitation, relation between collision numbers for, 91
Excited states, chemical properties of vibrational, 124–125
Extrathermodynamic relationship, definition of, 128

Ferric octaphenylporphyrazine chloride, dielectric relaxation of, 104–105
Fibers, partly crystalline, chemically induced phase changes in, 426
compared to liquid vapor systems, 420
melting point and tension, 421–426
Flash photolysis, 82
Fluorescence, and vibrational energy, 106
quenching, effect of viscosity on, 60
Formal species, definition, 33
partial molar free energy of, 33–34
Formic acid, heat and entropy of ionization of, 49
Free energy, additivity rules for, 139
dependence on composition, 17
effects, compared with enthalpy, 353–357
of activation, definition, 70–71
of mixtures, interaction term expression for, 28
Free energy, partial molar, and concentration entropy, 18
and partial molar entropy, 18
definition, 17
dependence on concentration, dilute solutions, 18

Free energy, standard partial molar, and partial molar enthalpy, 18
defined, 18
Free energy changes, and equilibrium constant, 19
and potential energy changes, 48
compensating, 48
cooperating, 48
definition of standard, 19
Free energy diagrams, representation of reacting systems by, 65
Free energy relationships, derivation of, for medium effects, nonunit slopes, 155
derivation of, for single interaction mechanism, 141
dispersion into parallel lines, 144–145, 150, 242, 243, 236–238
symmetry of, 141
temperature dependence of, 155, 315 ff.
Free radicals, polar effects in reactions of, 176–177, 185–187
stuck, 61

Gas phase, drastic electrostatic difference from any liquid, 266
extrathermodynamic relationships in (table), 133–135
Glucose, hydration of, 120
Graphical representations, and reaction paths, 64
of reacting systems, 62
Grunwald-Winstein equation, see mY equation

H_0, etc., see acidity function
Hammett relationship, see rho sigma relationship
Hammelt-Zucker hypothesis, 281
Heat capacity, and enthalpy or entropy, 43
and effect of neglecting change in, 45
and solvation, 44
and sound frequency, 84
change for cyanoacetic acid ionization, 44–45

Helix, conformational change to random coil, 411
Heptaphenylchlorophenylporphyrazine dielectric relaxation, 104–105
Heterogeneous flash method, 86
Hexaminium ion, solvent effects on enthalpies and entropies in decomposition of, compared to free energy effects, 356–357
Hexaphenylethane, medium effects in dissociation of, 304
Hydrocarbons, free energy of mixtures of, 147
reaction with radicals, 134
Hydrodynamic effects on rotational relaxation, 104
Hydrogen atoms, de Broglie wave length of, 74
effect as substituents in collisional de-excitation of vibration, 97
reaction with ozone, 124
Hydrogen bonding, and substituent effects, 32
enthalpies, entropies, and free energies for, 51–54
lifetime of complexes, 121, 123
α-Hydrogen effect on rates, 231–235
o-Hydroxybenzoyl azide, 361
Hydroxyl radicals, vibrationally excited, 124
Hydroxyl substituent, solvent dependence of σ (table) 174
Hyperconjugation (see also α-hydrogen effect), 134–135
equality of CH and CC, 349

Inductive effect, and the isokinetic relationship, 349
attenuation of, 224
in the bicyclo [2.2.2] octane system, 217
Infrared spectra, and partition functions, 13
correlation with C = C addition reactions, 255
effect of substituents on, 255–258
Intensive quantity, 17
Interactions, effect on additivity rules, 137

Interactions, effect on additivity rule, in solution, chemical models for, 31
of next-nearest neighbors, 137
nonbonded, 137
solute-solute, solute-solvent, and solvent-solvent, 27–31
Interaction mechanism(s), double, in Brown relationship, 198
double, in rho sigma relationship, 192–194
double, with the solvent in solvolysis, 300–303
loss of, in the free energy, 354
multiple, effect on linear free energy relationships, 143
multiple, effect on isokinetic relationship, 342–347
second, canceling by isokinetic effect, 302
second, perturbation of medium effect relationship by, 301–302
single, definition of, 141
vector, definition of, 344
Interaction term, definition, for liquid mixtures, 28
definition, general, 140
discontinuous changes in, 145
factorability for various medium effects, 150–153
higher order, 143
Interaction variable, 141
tests for identification of, 168–169
vector or tensor, 142–143
zero points for, 142
Internal pressure, and regular solution model, 264
and salt effects, 306–307
table of, 265
Internal rotation, energy and dipole moment, 111
identification of, 111, 115
isokinetic temperature in, 339, 342
Iodide ion, specific solvation by acetone, 311
Iodine, complexing by, 52
Iodoacetic acid, heat and entropy of ionization, 49

o-Iodobenzoyl peroxide rearrangement, 361, 414
Ionic mobilities and hydration, 122
Ionic strength, definition, 21
versus specific effects, 32–33
Ion pairs, rotational relaxation of ions in, 107–108
Ions, rotation of solvated, 44
solvation and heat capacity, 44
structural effects on hydration of, 123
Isokinetic effect, canceling of perturbing interaction mechanisms by, 302
Isokinetic relationship, 129, 131, 134, 155–156, 177, 185
definition, 324
effect of changes in mechanism on, 358
for reactions in mixed solvents, 397–402
for reactions in varied media, 394–402
for solution process, 391–393
in decomposition or association reactions (table), 329–332
in displacement and elimination reactions (table), 327–328
in reactions of carbonyl compounds (table), 336
in reactions of carboxylic acids and derivatives (table), 332–335
in miscellaneous reactions (table), 338
in model processes (table), 339
in solvolysis (table), 337
predicted effect of solvent change on structural, 351–352
predicted effect of structural change on solvent, 351–352
proof for first approximation, 315–318
proof for second approximation, 319–321
sign of beta, 321–322
Isokinetic temperature, and rate inversions, 325–326
behavior of free energy near, 354, 356–357

Isokinetic temperature, behavior of
    free energy near, characteristic
    ranges of, 342
    definition, 325
Isomerism, conformational, 3, 36–38
    effect on entropy, 45
    effect on free energy, 36
    rotational, 5, 86, 109–119
Isosterism and prediction of identity
    of steric interaction mechanisms,
    230–231
Isotope effect, and zero-point energy,
    12, 13
    on oscillator frequency, 12
    on partition function, 12
    and location of transition state on
        reaction coordinates, 167–168
    tunneling, 74–75

Kinetic energy, and solvation, 43
    and squared terms, 42–43
    contribution to the enthalpy and
        entropy, 42, 45
    definition, 42

Liberations in liquids, 105–106
Lifetime broadening, 89
Linear combination of models, 131–
    132
    in nucleophilic displacement re-
        actions, 249–251
    in aliphatic reactions, 223, 227,
        229
    Yukawa and Tsuno equation,
        211–214
Linear free energy relationships,
    effect of choice of model on,
    130
    effect of temperature on, 155–
        156, 324 ff.
    general theory of, 128 ff.
    scope of, 130–131
Line width, relation to relaxation
    time, 89
Liquid crystals, 410–411
Liquid phase, effect on kinetic
    energy modes and on energy
    level spacing, 9
Lucretius, 57

Macroscopic state, 1
Malonic acid, oriented radicals from,
    414
Mass, reduced, 12
Medium effects, 263–314 (see also
    solvent)
    change of sign by substituent, 150
    chemical models for, 31, 50–54
    on enthalpy and entropy, 48, 351,
        394–402
    physical models for, 54–55
    relationships of one parameter,
        268–269
    relationships of several param-
        eters, 282–300
    relations to substituent effects,
        306
    requirements for linear free energy
        relationships of, 153–155
Menschutkin reaction, steric effects
    in, 363, 367–368
Mesityl oxide, n-π^* transition and Z,
    310
Metachromasy, 411
Methine dyestuffs, spectrum and
    basicity, 259–261
p-Methoxy-p'-nitrobenzoyl peroxide,
    rotational relaxation in ion
    pairs from, 108
Methyl chloride vibrational relaxa-
    tion, 85
Methylene, 125–126
Methyl thymyl and carvacryl ketones,
    isokinetic relationship for oxi-
    mation, 365
Methyl radicals compared with tri-
    fluoromethyl radicals, 134
Methyl toluenesulfonate solvolysis,
    isokinetic relationship for
    change in solvent, 394–396
Microscopic reversibility, 73–74
Models, theoretical, compensation of
    enthalpy and entropy in, 50
    for solvent effects, 50–55
    relative sensitivity of enthalpy
        and entropy to, 50
Model processes, linear combination
    of, 143–144
Molecular complexes, see complexes

Molecular motions, effect on enthalpy and entropy, 46
Molecular weight, number and weight average, 406
Mustard ion, correlation of reactivity with various nucleophiles, 248
mY equation acidity scale from, 293–297
acids in aqueous-organic media, 289–293
and solvolysis rates, 297–300
and Z, 300
dispersion in, 300
extrathermodynamic derivation of, 282–293
m_{RX} (table), 300
relationship of m to σ and σ^*, 313–314
use in test of unimolecular mechanism, 292
Y values for solvolysis (table), 298
Y_0 and Y- values (table), 291

Naphthalene, complexing of, 52
Nitramide, Brønsted plot for base-catalyzed decomposition of, 130
p-Nitroaniline, spectral method for inductive effect in, 261
Nitrobenzene, complexing of, 53
electronic excitation of, 9, 10
solvent and substituent effects on spectrum, 27, 262
Nitromethane, rotation about C–N bond in, 111
p-Nitrophenylacetate hydrolysis, amine catalyzed, correlation with K_B, 244
Nitrosobenzene, condensation with anilines, as "well behaved" rho sigma example, 378
Nitrous oxide decomposition, 125
Nominal solvent, and activity coefficients, 22
definition, 20
Normal substituent constants, σ, 214–216
Nucleophilic parameter, E_N, definition and relation to polarizability, 252
E_N, table of values, 253

Nucleophilic parameter, E_N, H_N, definition and table, 252–253
n, definition and table, 247
n, relation to polarizability, 250
Nucleophilic substitution, correlation with basicity, 243–246
correlation with reactivity of methyl bromide, 246–248
structural effects in, 243–254

OH stretching frequency, correlation with acidity, 255–257
correlation with σ and σ^*, 257–258
effect of bonding to solvent on, 311
relation to solvent basicity, 311
Operators, formal equivalence of δ_R and Δ, 141
permutation of, 27
Optic-acoustic effect, 82
Orientation by flow, 404
Oscillator, harmonic, 8, 11, 12
quantum mechanical, 11
Oximation and semicarbazidation, concealed interaction mechanisms in, 354–356
Oxygen atoms, reaction with alkenes, 134
Oxygen molecules, vibrationally excited, 125
Ozone, effect on crack growth in rubber, 409–410
reaction with hydrogen atoms, 124

Paraldehyde dielectric relaxation, 105
Parachor and solubility, 148
Partial molar free energy, effect of solvation on, 33–34
of mixtures, dependence on interaction terms, 29
Partial molar free energy, standard, effect of solvation on, 34–35
effect of subsidiary equilibria on, 34–38
Partial rate factors, and Brown relationship, 196–203
definitions, 196–197
tables of, 199, 203

Particle in the box, 7
Partition function, 6
    and enthalpy and entropy, 41
    and ground energy levels, relative
        effects of, 7
    effect of infra-red active vibra-
        tions on, 13
    factoring of, 8
    qualitative description and effects
        of constraints on, 7
    rotational and translational, 8
    transition state, 14, 68
    types of motion important to, 14
    vibrational, 8, 11
Peracetic acid, reactions with
    alkenes, 135
Perturbing interaction mechanism,
    effect on isokinetic relationship,
    325
    (see also σ^+, steric effects)
Phase rule for fibers, 420
Phenol, hydrogen bonding of, 52
Phenolphthalein, mechanical activa-
    tion of, 415
Phenols, oxidation by diphenylpi-
    crylhydrazyl, 387
    substituent effects in ionization of,
    142
Phenyl azides, decomposition in
    aniline, 326
N-Phenyl benzyl carbamate decom-
    position, as "well behaved" $\rho\sigma$
    relationship, 378
Phenylisobutyryl peroxide, cage
    effects and rotational relaxation
    in decomposition of, 107
Phenyl potassium sulfate acid hy-
    drolysis dispersion in isokenetic
    and $\rho\sigma$ relationships, 384–385
2-Phenyl-2-propyl chloride solvo-
    lysis, 149, 204–205
Phosphorimetry, 82
Physical properties, extrathermody-
    namic relationships involving,
    168–170
    in-phase and out-of-phase com-
        ponents, 80–81
    relation to frequency, 80–81
pH muscle, 419
Picric acid complexing, 52

Polarizability, 248–254
    and dispersion force model for
        solutions, 265
    and electronic spectra, 251
    and steric effects, 251
Polymers, free radicals from, 408
Popcorn polymer, 409
Potential energy surface, shape and
    entropy, 68–69
Probability, 1 ff.
    and entropy, 16
Propionate ion, internal rotation of
    radicals from, 116
Protons, exchange rate between hy-
    dronium ion and water, 90
    magnetic resonance of, and rota-
        meric isomerization rates,
        86–88
    transfer reactions, effect of vis-
        cosity on, 61
    tunneling, 74–75
Pyridine-silver ion complexes, med-
    ium effects on dissociation of,
    149

Quinoline-silver ion complexes, med-
    ium effects on dissociation of,
    149

Racemization due to rotational re-
    laxation, 107
Rate constants for reactions of mix-
    tures of conformers, 119–120
Rate-equilibrium relationship, 156–
    162
    and selectivity, 163
    dependence of α on reactivity,
        165–166
    for enthalpies, 160
    for rearrangement of 5-aminotri-
        azoles, 159
    from Brønsted relationships, 238–
        241
    medium effects, 304
    theoretical interpretation of α,
        158–159
Reacting systems, graphical repre-
    sentation of, 62
    two-dimensional representation of,
        65

Reactivity, definition, 162
  relation to selectivity, 162–168
Reaction coordinate, and normal
  vibrational modes, 63
  definition, 62
  in *cis-trans* isomerization, 63
Reaction diagram, and medium
  effects, 23
  and substituent effects, 26
Reaction operator, commutation of,
  23
  definition, 22–23
Reaction zone, 140, 141
Recombination, geminate, primary
  and secondary, 62
Redox machine, 419
Reference solvent, choice of, 21
Reference state, 20
Reflectability, 73
Relationships between relationships,
  314
Regular solution and free energy
  relationships, 264–265
Relaxation processes, 76–90
  definition, 76–77
  in polymers, 426–427
  rate law for, 77–81
  types of experiment, 81–86
Relaxation time, conditions for multiple vibrational-translational, 99
  definition, 79
Resonance, and *meta-para* sigma
  differences, 195
  largely enthalpic effect of, 381-382
  medium effects on, 216
  perturbation of $\rho\sigma$ relationship by,
  195, 213–214
Rhapsodic spontaneity, 315
Rigid media, 403–427
Rho sigma relationship, 140, 172–
  194
  concealed interaction mechanism
  type, 380
  curvature in, 187–191
  dispersion into separate lines, 191
  effect of multiple substitution, 192
  effect of reaction mechanism
  changes, 187–191
  enthalpy-entropy behavior of the
  defining process, 374–375

Rho sigma relationship, enthalpy-
  entropy behavior of the defining
  process, fit of, 175
  for free radical reactions, 176–
  177, 185–187
  highly precise, table of, 178–184
  miscellaneous perturbations of,
  385–387
  number of interaction mechanisms, 192–194, 196
  resonance perturbation of, 380–
  385
  rho, change in sign of with temperature, 325–326
  rho, temperature dependence of,
  177, 185
  scope of, 175–177
  "well-behaved" category, 375
  quality of the fit as a function
  of T, 379–380
  table of, 377–378
  Y and solvent effects on rho, 313–
  314
Rho' sigma' relationship, 217–219
Rho* sigma* relationship, 225–235
  dispersion in, 231–235
  enthalpy-entropy behavior of,
  388–389
  table, 226
Rho sigma⁻ relationship, 211
  enthalpies and entropies of, 381–
  385
Rho sigma⁺ relationship, 207
  and solvation, 207
  enthalpies and entropies of, 381–
  385
  interaction mechanisms in, 207
  rho values, 208–209
  sigma⁺ values, 204–205
Rigidity, definition, 7
Rotamers, interconversion rates of,
  109–119
Rotation(s), and dielectric relaxation, 100–106
  and entropy, 45
  energy barriers in internal (table),
  110
  geared, 100–104
  hindered, and squared terms, 43
  in liquids, 100–106
  independence of translation, 104

Rotation(s), in liquids, internal, of molecular complexes, 119
  internal, rates of (table), 111–114
  internal, solvent effects on, 118
  internal, transition states for, 118
Rotational relaxation, collision number for in liquids, 105
  complex formation and, 104–106
  molecular size and shape dependence, 101
  temperature and viscosity dependence, 101
  times for (table), 102–103
Rotational-translational interconversion, 94
Rotameric isomerism of acetamides, 86
Rotator, rigid, 8
Rubberlike elasticity, 416–419

Salt effects, 306–308
  and internal pressure, 306–307
  in aqueous organic solvents, 308
  on electrolytes, 308
  on nonpolar solutes, 306–307
  on polar nonelectrolytes, 308
Selectivity, and Brønsted relationship, 165
  definition, 162
  in H atom abstraction reactions, 167–168
  of carbonium ions, 164
  and partial rate factors, 197
  and reactivity, 198–200, 162–164
  conditions for relationship, 162–164
Separability postulate, proof and limitations of, 141, 145–146
Shah of Persia, 76
Shear degradation, 404–408
Shock waves, 83–84
Sigma, compared with σ^*, 225
  compared with σ^+, 204
  definition of, 172
  meta-para differences (table), 195
  primary values (table), 173
  relation to m of mY equation, 313–314
  secondary values, 173–174

Sigma$^+$, compared with sigma (table), 204
  defining process, 205
Sigma$^-$, 211, 381–385
Sigma′, compared with sigma (table), 218
  definition, 217
  relation to sigma*, 223
Sigma*, 219–225
  additivity of, 224
  compared with sigma, 225
  definition, 221
  solvent effects on, 221
  table of, 222
Sigma″, table of, 215
Skeletal vibration, 14
Sodium atoms, reaction with organic halides, 134
Solvation, first approximation treatment of, 147–153
  formal extrathermodynamic treatment of, 146–155
  isokinetic temperatures for, 342
  of ions, specific, 277
  second approximation treatment of, 153–155
  steric hindrance to, 234–235
Solvation complexes, diffusion rate and lifetime of, 121
  internal rotation of, 119, 122
  lifetime of, 120–124
  rotation as single entity, 120–121
  rotational rate and lifetime of, 121
Solubility, isoequilibrium relationships for, 391–393
  parachor and, 148
Solvent, lattice model of, 18
Solvent effects, complexing models for, 50–54
  interaction with substituent effects, 32, 221
  physical models for, 54–55
  solid solutions, 412–415
Solution models, prediction of free energy relationships from, 263–267
Solvent stabilization, commutation of operator, 23
  definition of operator, 22, 23
  interpretation of, 31

Solvolysis, double interaction mechanism in, 300–303
  isokinetic relationships in, 337
  mY correlation of, 297–300
Sound velocity, equation for, 84
Spectra, effect of substituents on, 255–262
  rate measurements from line widths, 86
Spectrophone, 82
Squared terms, 42–43
Stacking efficiency, 411
State, macroscopic, 1, 2
  most probable, 2
Statistical corrections, 133, 136
Statistical mechanics and thermodynamics, 38
Steric effects, additivity or independence of, 219–225, 364
  and the isokinetic relationship, 359–369
  cancellation of, 219–220
  dependence on remote substituents and on solvation, 363
  discontinuity of, 145, 365–368
  dispersion into parallel lines from, 243
  in aliphatic reactions, 228–231
  in aromatic substitution, 200–201
  in Menschutkin reaction, 368
  interaction with polar effects, 129
  parallel isokinetic lines from, 365
  potential functions for, 129
  substituent constants for, correlations with, 229
    definition, 228
    table, 228
  use of enthalpy-entropy diagrams to identify, 366
  variety of interaction mechanisms in, 230
Subspecies, and points on free energy diagrams, 66
  interconversion rates of, 76–127
  isomeric, 5, 36
  ultimate, 38
Substituent effects (see also rho, sigma, isokinetic relationship), and complex formation, 32

Substitutent effects (see also rho, and complex formation, aliphatic, simple correlations in the gas phase, 134
  aromatic, in the gas phase, 134
  correlation between gas and liquid phase reactions, 134–135
  explanations for lack of, 326
  in two solvents, 129, 130
  interaction with solvent effects, 32, 221
  models for, 55, 321–322
  on benzyl radical ionization potentials, 135
  relation to medium effects, 306
  temperature independence of the parameters for single interaction mechanism, 319
Substituent(s), interaction of mobile, 55
  solvation of, 32, 55, 221
  stabilization operator, 22, 26
  zone, 140
  symmetrical status of reaction zone, 193
Sulfur dioxide, solubility in, 148
Swain-Scott equation, 246–249
Swain, Mosely, and Bown equation, 301–302
Swamping salt technique, 21
Symmetry, corrections for, 133, 136
  in solute-solute interactions, 30
  of substituent effect relationships, 141

Taft equation, see rho* sigma*
Temperature (see also isokinetic relationship), effect on enthalpy and entropy, 40–45
  effect on free energy, 40
  effect on equilibrium constants, 40
  for benzoic acid ionization in water, 386–387
Temperature jump method, 82
Tetraphenylboride ion, analysis of medium effects on, 267
Tetraphenylphosphonium ion, analysis of medium effects on, 267

Thermodynamics, irreversible, 73
    relation to statistical mechanics, 38
Thiolacetates, effect of mechanism
    change on isokinetic relation-
    ship for, 358
Thymine photodimer, 412
Time scale in absorption experi-
    ments, 87
Tobacco, 82
Toluene, internal rotation in, 111
Transitions, electronic, 9
Transition state, activity of, 71–72
    comparisons of, 72–73
    energy transfer in, 67
    entropy of, and potential energy
        profiles, 68–69
    graphical representations of, 64–
        65
    irrelevant motions of, 64
    location along reaction coordinate,
        157
    prediction of, 158
    rate equation, 69–71
    rate of decay of, 70
    rate theory, 57, 62–73
        assumptions of, 62, 109
        diffusion controlled reactions,
            62
    stabilization, relation to those of
        reagent and product, 157
    subspecies of, 67
    thermodynamic properties of, 66–
        69
Translational temperature, 83
Translation, as very anharmonic
    vibration, 93
Transmission coefficient, kappa, 70,
    72
    tau, 194
Triarylcarbinols, acidity function
    based on, 279
Triarylsilanes, correlation of oxida-
    tion rates with σ^0, 215
Triboluminescence, 415
2,4,6-Tri-t-butylphenol, rotations in,
    111
2,4,6,-Tri-t-butylphenoxy radical,
    medium effects on dispropor-
    tionation of, 394

Triethylammonium acetate, non-
    equivalence of oxygens in, 109
Trimethylboron, amine complexing
    of, 134
2,4,6,-Trimethylpyridine dielectric
    relaxation, 105
Trinitrobenzene complexing, 52
2,4,7-Trinitrofluorenyl-9-tosylate,
    effect of phenanthrene on
    acetolysis, 54
Triphenylmethyl cation, 248–249
Triplet state, reactivity and conver-
    sion to singlet, 126–127
Tritium and tunneling, 75
Trouton's rule, 389–390
Tunneling effect, 74–75

Uncertainty principle, 11
    and line widths, 87
Ultrasound, 84–86
Useful work, 15–16
    and biologists, 15

Vaporization, heats and entropies of,
    389–391
Variables of state, oscillations of, 77
Vibration(al), cascade mechanism
    for conversion to translation, 95
    chemical methods for excitation,
        86
    energy of reaction products, 124–
        125
    energy, relaxation of, 83–86
    effect of pressure, 85
    in liquids, 106
    molecularity of, 85
    times for (table), 96
    zero-point energy, 11
Vinyl polymerization, diffusion con-
    trolled reaction in, 61
    polar factors, 185–187
Visible and ultraviolet spectra, com-
    parison with chemical processes,
        258–259
Viscosity, effect on encounter num-
    ber, 59
    effect on rates, 59–60, 169–170

w (in Bunnett equation), 285
w' (in Bunnett equation), 288
Water, activity of in various acids
    (table), 286
  irrotationally bound, 122
  rotation of in hydrates and in
    pure liquid, 122
  parachor expression for solubility
    in, 148

Water, parachor expression for solu-
    bility in, special efficiency in
    vibrational deexcitation of car-
    bon dioxide, 100
Wrestling, 100
Y, see mY equation
Yukawa and Tsuno equation, 211

Z, 300, 309–311
Zero-point vibrational energy, 10–11

A CATALOG OF SELECTED
DOVER BOOKS
IN ALL FIELDS OF INTEREST

A CATALOG OF SELECTED DOVER
BOOKS IN ALL FIELDS OF INTEREST

DRAWINGS OF REMBRANDT, edited by Seymour Slive. Updated Lippmann, Hofstede de Groot edition, with definitive scholarly apparatus. All portraits, biblical sketches, landscapes, nudes. Oriental figures, classical studies, together with selection of work by followers. 550 illustrations. Total of 630pp. 9⅛ × 12¼.
21485-0, 21486-9 Pa., Two-vol. set $25.00

GHOST AND HORROR STORIES OF AMBROSE BIERCE, Ambrose Bierce. 24 tales vividly imagined, strangely prophetic, and decades ahead of their time in technical skill: "The Damned Thing," "An Inhabitant of Carcosa," "The Eyes of the Panther," "Moxon's Master," and 20 more. 199pp. 5⅜ × 8½. 20767-6 Pa. $3.95

ETHICAL WRITINGS OF MAIMONIDES, Maimonides. Most significant ethical works of great medieval sage, newly translated for utmost precision, readability. Laws Concerning Character Traits, Eight Chapters, more. 192pp. 5⅜ × 8½.
24522-5 Pa. $4.50

THE EXPLORATION OF THE COLORADO RIVER AND ITS CANYONS, J. W. Powell. Full text of Powell's 1,000-mile expedition down the fabled Colorado in 1869. Superb account of terrain, geology, vegetation, Indians, famine, mutiny, treacherous rapids, mighty canyons, during exploration of last unknown part of continental U.S. 400pp. 5⅜ × 8½. 20094-9 Pa. $6.95

HISTORY OF PHILOSOPHY, Julián Marías. Clearest one-volume history on the market. Every major philosopher and dozens of others, to Existentialism and later. 505pp. 5⅜ × 8½. 21739-6 Pa. $8.50

ALL ABOUT LIGHTNING, Martin A. Uman. Highly readable non-technical survey of nature and causes of lightning, thunderstorms, ball lightning, St. Elmo's Fire, much more. Illustrated. 192pp. 5⅜ × 8½. 25237-X Pa. $5.95

SAILING ALONE AROUND THE WORLD, Captain Joshua Slocum. First man to sail around the world, alone, in small boat. One of great feats of seamanship told in delightful manner. 67 illustrations. 294pp. 5⅜ × 8½. 20326-3 Pa. $4.95

LETTERS AND NOTES ON THE MANNERS, CUSTOMS AND CONDITIONS OF THE NORTH AMERICAN INDIANS, George Catlin. Classic account of life among Plains Indians: ceremonies, hunt, warfare, etc. 312 plates. 572pp. of text. 6⅛ × 9¼. 22118-0, 22119-9 Pa. Two-vol. set $15.90

ALASKA: The Harriman Expedition, 1899, John Burroughs, John Muir, et al. Informative, engrossing accounts of two-month, 9,000-mile expedition. Native peoples, wildlife, forests, geography, salmon industry, glaciers, more. Profusely illustrated. 240 black-and-white line drawings. 124 black-and-white photographs. 3 maps. Index. 576pp. 5⅜ × 8½. 25109-8 Pa. $11.95

THE BOOK OF BEASTS: Being a Translation from a Latin Bestiary of the Twelfth Century, T. H. White. Wonderful catalog real and fanciful beasts: manticore, griffin, phoenix, amphivius, jaculus, many more. White's witty erudite commentary on scientific, historical aspects. Fascinating glimpse of medieval mind. Illustrated. 296pp. 5⅜ × 8¼. (Available in U.S. only) 24609-4 Pa. $5.95

FRANK LLOYD WRIGHT: ARCHITECTURE AND NATURE With 160 Illustrations, Donald Hoffmann. Profusely illustrated study of influence of nature—especially prairie—on Wright's designs for Fallingwater, Robie House, Guggenheim Museum, other masterpieces. 96pp. 9¼ × 10¾. 25098-9 Pa. $7.95

FRANK LLOYD WRIGHT'S FALLINGWATER, Donald Hoffmann. Wright's famous waterfall house: planning and construction of organic idea. History of site, owners, Wright's personal involvement. Photographs of various stages of building. Preface by Edgar Kaufmann, Jr. 100 illustrations. 112pp. 9¼ × 10.
23671-4 Pa. $7.95

YEARS WITH FRANK LLOYD WRIGHT: Apprentice to Genius, Edgar Tafel. Insightful memoir by a former apprentice presents a revealing portrait of Wright the man, the inspired teacher, the greatest American architect. 372 black-and-white illustrations. Preface. Index. vi + 228pp. 8¼ × 11. 24801-1 Pa. $9.95

THE STORY OF KING ARTHUR AND HIS KNIGHTS, Howard Pyle. Enchanting version of King Arthur fable has delighted generations with imaginative narratives of exciting adventures and unforgettable illustrations by the author. 41 illustrations. xviii + 313pp. 6⅛ × 9¼. 21445-1 Pa. $6.50

THE GODS OF THE EGYPTIANS, E. A. Wallis Budge. Thorough coverage of numerous gods of ancient Egypt by foremost Egyptologist. Information on evolution of cults, rites and gods; the cult of Osiris; the Book of the Dead and its rites; the sacred animals and birds; Heaven and Hell; and more. 956pp. 6⅛ × 9¼.
22055-9, 22056-7 Pa., Two-vol. set $20.00

A THEOLOGICO-POLITICAL TREATISE, Benedict Spinoza. Also contains unfinished *Political Treatise*. Great classic on religious liberty, theory of government on common consent. R. Elwes translation. Total of 421pp. 5⅜ × 8½.
20249-6 Pa. $6.95

INCIDENTS OF TRAVEL IN CENTRAL AMERICA, CHIAPAS, AND YUCATAN, John L. Stephens. Almost single-handed discovery of Maya culture; exploration of ruined cities, monuments, temples; customs of Indians. 115 drawings. 892pp. 5⅜ × 8½. 22404-X, 22405-8 Pa., Two-vol. set $15.90

LOS CAPRICHOS, Francisco Goya. 80 plates of wild, grotesque monsters and caricatures. Prado manuscript included. 183pp. 6⅜ × 9⅜. 22384-1 Pa. $4.95

AUTOBIOGRAPHY: The Story of My Experiments with Truth, Mohandas K. Gandhi. Not hagiography, but Gandhi in his own words. Boyhood, legal studies, purification, the growth of the Satyagraha (nonviolent protest) movement. Critical, inspiring work of the man who freed India. 480pp. 5⅜× 8½. (Available in U.S. only)
24593-4 Pa. $6.95

ILLUSTRATED DICTIONARY OF HISTORIC ARCHITECTURE, edited by Cyril M. Harris. Extraordinary compendium of clear, concise definitions for over 5,000 important architectural terms complemented by over 2,000 line drawings. Covers full spectrum of architecture from ancient ruins to 20th-century Modernism. Preface. 592pp. 7½ × 9⅝. 24444-X Pa. $14.95

THE NIGHT BEFORE CHRISTMAS, Clement Moore. Full text, and woodcuts from original 1848 book. Also critical, historical material. 19 illustrations. 40pp. 4⅝ × 6. 22797-9 Pa. $2.25

THE LESSON OF JAPANESE ARCHITECTURE: 165 Photographs, Jiro Harada. Memorable gallery of 165 photographs taken in the 1930's of exquisite Japanese homes of the well-to-do and historic buildings. 13 line diagrams. 192pp. 8⅜ × 11¼. 24778-3 Pa. $8.95

THE AUTOBIOGRAPHY OF CHARLES DARWIN AND SELECTED LETTERS, edited by Francis Darwin. The fascinating life of eccentric genius composed of an intimate memoir by Darwin (intended for his children); commentary by his son, Francis; hundreds of fragments from notebooks, journals, papers; and letters to and from Lyell, Hooker, Huxley, Wallace and Henslow. xi + 365pp. 5⅜ × 8. 20479-0 Pa. $6.95

WONDERS OF THE SKY: Observing Rainbows, Comets, Eclipses, the Stars and Other Phenomena, Fred Schaaf. Charming, easy-to-read poetic guide to all manner of celestial events visible to the naked eye. Mock suns, glories, Belt of Venus, more. Illustrated. 299pp. 5¼ × 8¼. 24402-4 Pa. $7.95

BURNHAM'S CELESTIAL HANDBOOK, Robert Burnham, Jr. Thorough guide to the stars beyond our solar system. Exhaustive treatment. Alphabetical by constellation: Andromeda to Cetus in Vol. 1; Chamaeleon to Orion in Vol. 2; and Pavo to Vulpecula in Vol. 3. Hundreds of illustrations. Index in Vol. 3. 2,000pp. 6⅛ × 9¼. 23567-X, 23568-8, 23673-0 Pa., Three-vol. set $38.85

STAR NAMES: Their Lore and Meaning, Richard Hinckley Allen. Fascinating history of names various cultures have given to constellations and literary and folkloristic uses that have been made of stars. Indexes to subjects. Arabic and Greek names. Biblical references. Bibliography. 563pp. 5⅜ × 8½. 21079-0 Pa. $7.95

THIRTY YEARS THAT SHOOK PHYSICS: The Story of Quantum Theory, George Gamow. Lucid, accessible introduction to influential theory of energy and matter. Careful explanations of Dirac's anti-particles, Bohr's model of the atom, much more. 12 plates. Numerous drawings. 240pp. 5⅜ × 8½. 24895-X Pa. $4.95

CHINESE DOMESTIC FURNITURE IN PHOTOGRAPHS AND MEASURED DRAWINGS, Gustav Ecke. A rare volume, now affordably priced for antique collectors, furniture buffs and art historians. Detailed review of styles ranging from early Shang to late Ming. Unabridged republication. 161 black-and-white drawings, photos. Total of 224pp. 8⅜ × 11¼. (Available in U.S. only) 25171-3 Pa. $12.95

VINCENT VAN GOGH: A Biography, Julius Meier-Graefe. Dynamic, penetrating study of artist's life, relationship with brother, Theo, painting techniques, travels, more. Readable, engrossing. 160pp. 5⅜ × 8½. (Available in U.S. only) 25253-1 Pa. $3.95

HOW TO WRITE, Gertrude Stein. Gertrude Stein claimed anyone could understand her unconventional writing—here are clues to help. Fascinating improvisations, language experiments, explanations illuminate Stein's craft and the art of writing. Total of 414pp. 4⅜ × 6⅜. 23144-5 Pa. $5.95

ADVENTURES AT SEA IN THE GREAT AGE OF SAIL: Five Firsthand Narratives, edited by Elliot Snow. Rare true accounts of exploration, whaling, shipwreck, fierce natives, trade, shipboard life, more. 33 illustrations. Introduction. 353pp. 5⅜ × 8½. 25177-2 Pa. $7.95

THE HERBAL OR GENERAL HISTORY OF PLANTS, John Gerard. Classic descriptions of about 2,850 plants—with over 2,700 illustrations—includes Latin and English names, physical descriptions, varieties, time and place of growth, more. 2,706 illustrations. xlv + 1,678pp. 8½ × 12¼. 23147-X Cloth. $75.00

DOROTHY AND THE WIZARD IN OZ, L. Frank Baum. Dorothy and the Wizard visit the center of the Earth, where people are vegetables, glass houses grow and Oz characters reappear. Classic sequel to *Wizard of Oz*. 256pp. 5⅜ × 8. 24714-7 Pa. $4.95

SONGS OF EXPERIENCE: Facsimile Reproduction with 26 Plates in Full Color, William Blake. This facsimile of Blake's original "Illuminated Book" reproduces 26 full-color plates from a rare 1826 edition. Includes "The Tyger," "London," "Holy Thursday," and other immortal poems. 26 color plates. Printed text of poems. 48pp. 5¼ × 7. 24636-1 Pa. $3.50

SONGS OF INNOCENCE, William Blake. The first and most popular of Blake's famous "Illuminated Books," in a facsimile edition reproducing all 31 brightly colored plates. Additional printed text of each poem. 64pp. 5¼ × 7. 22764-2 Pa. $3.50

PRECIOUS STONES, Max Bauer. Classic, thorough study of diamonds, rubies, emeralds, garnets, etc.: physical character, occurrence, properties, use, similar topics. 20 plates, 8 in color. 94 figures. 659pp. 6⅛ × 9¼. 21910-0, 21911-9 Pa., Two-vol. set $15.90

ENCYCLOPEDIA OF VICTORIAN NEEDLEWORK, S. F. A. Caulfeild and Blanche Saward. Full, precise descriptions of stitches, techniques for dozens of needlecrafts—most exhaustive reference of its kind. Over 800 figures. Total of 679pp. 8⅛ × 11. Two volumes. Vol. 1 22800-2 Pa. $11.95
Vol. 2 22801-0 Pa. $11.95

THE MARVELOUS LAND OF OZ, L. Frank Baum. Second Oz book, the Scarecrow and Tin Woodman are back with hero named Tip, Oz magic. 136 illustrations. 287pp. 5⅜ × 8½. 20692-0 Pa. $5.95

WILD FOWL DECOYS, Joel Barber. Basic book on the subject, by foremost authority and collector. Reveals history of decoy making and rigging, place in American culture, different kinds of decoys, how to make them, and how to use them. 140 plates. 156pp. 7⅞ × 10¾. 20011-6 Pa. $8.95

HISTORY OF LACE, Mrs. Bury Palliser. Definitive, profusely illustrated chronicle of lace from earliest times to late 19th century. Laces of Italy, Greece, England, France, Belgium, etc. Landmark of needlework scholarship. 266 illustrations. 672pp. 6⅛ × 9¼. 24742-2 Pa. $14.95

ILLUSTRATED GUIDE TO SHAKER FURNITURE, Robert Meader. All furniture and appurtenances, with much on unknown local styles. 235 photos. 146pp. 9 × 12. 22819-3 Pa. $7.95

WHALE SHIPS AND WHALING: A Pictorial Survey, George Francis Dow. Over 200 vintage engravings, drawings, photographs of barks, brigs, cutters, other vessels. Also harpoons, lances, whaling guns, many other artifacts. Comprehensive text by foremost authority. 207 black-and-white illustrations. 288pp. 6 × 9. 24808-9 Pa. $8.95

THE BERTRAMS, Anthony Trollope. Powerful portrayal of blind self-will and thwarted ambition includes one of Trollope's most heartrending love stories. 497pp. 5⅜ × 8½. 25119-5 Pa. $8.95

ADVENTURES WITH A HAND LENS, Richard Headstrom. Clearly written guide to observing and studying flowers and grasses, fish scales, moth and insect wings, egg cases, buds, feathers, seeds, leaf scars, moss, molds, ferns, common crystals, etc.—all with an ordinary, inexpensive magnifying glass. 209 exact line drawings aid in your discoveries. 220pp. 5⅜ × 8½. 23330-8 Pa. $3.95

RODIN ON ART AND ARTISTS, Auguste Rodin. Great sculptor's candid, wide-ranging comments on meaning of art; great artists; relation of sculpture to poetry, painting, music; philosophy of life, more. 76 superb black-and-white illustrations of Rodin's sculpture, drawings and prints. 119pp. 8⅜ × 11¼. 24487-3 Pa. $6.95

FIFTY CLASSIC FRENCH FILMS, 1912–1982: A Pictorial Record, Anthony Slide. Memorable stills from Grand Illusion, Beauty and the Beast, Hiroshima, Mon Amour, many more. Credits, plot synopses, reviews, etc. 160pp. 8¼ × 11. 25256-6 Pa. $11.95

THE PRINCIPLES OF PSYCHOLOGY, William James. Famous long course complete, unabridged. Stream of thought, time perception, memory, experimental methods; great work decades ahead of its time. 94 figures. 1,391pp. 5⅜ × 8½. 20381-6, 20382-4 Pa., Two-vol. set $19.90

BODIES IN A BOOKSHOP, R. T. Campbell. Challenging mystery of blackmail and murder with ingenious plot and superbly drawn characters. In the best tradition of British suspense fiction. 192pp. 5⅜ × 8½. 24720-1 Pa. $3.95

CALLAS: PORTRAIT OF A PRIMA DONNA, George Jellinek. Renowned commentator on the musical scene chronicles incredible career and life of the most controversial, fascinating, influential operatic personality of our time. 64 black-and-white photographs. 416pp. 5⅜ × 8¼. 25047-4 Pa. $7.95

GEOMETRY, RELATIVITY AND THE FOURTH DIMENSION, Rudolph Rucker. Exposition of fourth dimension, concepts of relativity as Flatland characters continue adventures. Popular, easily followed yet accurate, profound. 141 illustrations. 133pp. 5⅜ × 8½. 23400-2 Pa. $3.95

HOUSEHOLD STORIES BY THE BROTHERS GRIMM, with pictures by Walter Crane. 53 classic stories—Rumpelstiltskin, Rapunzel, Hansel and Gretel, the Fisherman and his Wife, Snow White, Tom Thumb, Sleeping Beauty, Cinderella, and so much more—lavishly illustrated with original 19th century drawings. 114 illustrations. x + 269pp. 5⅜ × 8½. 21080-4 Pa. $4.50

SUNDIALS, Albert Waugh. Far and away the best, most thorough coverage of ideas, mathematics concerned, types, construction, adjusting anywhere. Over 100 illustrations. 230pp. 5⅜ × 8½. 22947-5 Pa. $4.50

PICTURE HISTORY OF THE NORMANDIE: With 190 Illustrations, Frank O. Braynard. Full story of legendary French ocean liner: Art Deco interiors, design innovations, furnishings, celebrities, maiden voyage, tragic fire, much more. Extensive text. 144pp. 8⅜ × 11¾. 25257-4 Pa. $9.95

THE FIRST AMERICAN COOKBOOK: A Facsimile of "American Cookery," 1796, Amelia Simmons. Facsimile of the first American-written cookbook published in the United States contains authentic recipes for colonial favorites—pumpkin pudding, winter squash pudding, spruce beer, Indian slapjacks, and more. Introductory Essay and Glossary of colonial cooking terms. 80pp. 5⅜ × 8½. 24710-4 Pa. $3.50

101 PUZZLES IN THOUGHT AND LOGIC, C. R. Wylie, Jr. Solve murders and robberies, find out which fishermen are liars, how a blind man could possibly identify a color—purely by your own reasoning! 107pp. 5⅜ × 8½. 20367-0 Pa. $2.50

THE BOOK OF WORLD-FAMOUS MUSIC—CLASSICAL, POPULAR AND FOLK, James J. Fuld. Revised and enlarged republication of landmark work in musico-bibliography. Full information about nearly 1,000 songs and compositions including first lines of music and lyrics. New supplement. Index. 800pp. 5⅜ × 8¼. 24857-7 Pa. $14.95

ANTHROPOLOGY AND MODERN LIFE, Franz Boas. Great anthropologist's classic treatise on race and culture. Introduction by Ruth Bunzel. Only inexpensive paperback edition. 255pp. 5⅜ × 8½. 25245-0 Pa. $5.95

THE TALE OF PETER RABBIT, Beatrix Potter. The inimitable Peter's terrifying adventure in Mr. McGregor's garden, with all 27 wonderful, full-color Potter illustrations. 55pp. 4¼ × 5½. (Available in U.S. only) 22827-4 Pa. $1.75

THREE PROPHETIC SCIENCE FICTION NOVELS, H. G. Wells. *When the Sleeper Wakes, A Story of the Days to Come* and *The Time Machine* (full version). 335pp. 5⅜ × 8½. (Available in U.S. only) 20605-X Pa. $5.95

APICIUS COOKERY AND DINING IN IMPERIAL ROME, edited and translated by Joseph Dommers Vehling. Oldest known cookbook in existence offers readers a clear picture of what foods Romans ate, how they prepared them, etc. 49 illustrations. 301pp. 6⅛ × 9¼. 23563-7 Pa. $6.50

SHAKESPEARE LEXICON AND QUOTATION DICTIONARY, Alexander Schmidt. Full definitions, locations, shades of meaning of every word in plays and poems. More than 50,000 exact quotations. 1,485pp. 6½ × 9¼. 22726-X, 22727-8 Pa., Two-vol. set $27.90

THE WORLD'S GREAT SPEECHES, edited by Lewis Copeland and Lawrence W. Lamm. Vast collection of 278 speeches from Greeks to 1970. Powerful and effective models; unique look at history. 842pp. 5⅜ × 8½. 20468-5 Pa. $11.95

THE BLUE FAIRY BOOK, Andrew Lang. The first, most famous collection, with many familiar tales: Little Red Riding Hood, Aladdin and the Wonderful Lamp, Puss in Boots, Sleeping Beauty, Hansel and Gretel, Rumpelstiltskin; 37 in all. 138 illustrations. 390pp. 5⅜ × 8½. 21437-0 Pa. $5.95

THE STORY OF THE CHAMPIONS OF THE ROUND TABLE, Howard Pyle. Sir Launcelot, Sir Tristram and Sir Percival in spirited adventures of love and triumph retold in Pyle's inimitable style. 50 drawings, 31 full-page. xviii + 329pp. 6½ × 9¼. 21883-X Pa. $6.95

AUDUBON AND HIS JOURNALS, Maria Audubon. Unmatched two-volume portrait of the great artist, naturalist and author contains his journals, an excellent biography by his granddaughter, expert annotations by the noted ornithologist, Dr. Elliott Coues, and 37 superb illustrations. Total of 1,200pp. 5⅜ × 8.
Vol. I 25143-8 Pa. $8.95
Vol. II 25144-6 Pa. $8.95

GREAT DINOSAUR HUNTERS AND THEIR DISCOVERIES, Edwin H. Colbert. Fascinating, lavishly illustrated chronicle of dinosaur research, 1820's to 1960. Achievements of Cope, Marsh, Brown, Buckland, Mantell, Huxley, many others. 384pp. 5¼ × 8¼. 24701-5 Pa. $6.95

THE TASTEMAKERS, Russell Lynes. Informal, illustrated social history of American taste 1850's–1950's. First popularized categories Highbrow, Lowbrow, Middlebrow. 129 illustrations. New (1979) afterword. 384pp. 6 × 9.
23993-4 Pa. $6.95

DOUBLE CROSS PURPOSES, Ronald A. Knox. A treasure hunt in the Scottish Highlands, an old map, unidentified corpse, surprise discoveries keep reader guessing in this cleverly intricate tale of financial skullduggery. 2 black-and-white maps. 320pp. 5⅜ × 8½. (Available in U.S. only) 25032-6 Pa. $5.95

AUTHENTIC VICTORIAN DECORATION AND ORNAMENTATION IN FULL COLOR: 46 Plates from "Studies in Design," Christopher Dresser. Superb full-color lithographs reproduced from rare original portfolio of a major Victorian designer. 48pp. 9¼ × 12¼. 25083-0 Pa. $7.95

PRIMITIVE ART, Franz Boas. Remains the best text ever prepared on subject, thoroughly discussing Indian, African, Asian, Australian, and, especially, Northern American primitive art. Over 950 illustrations show ceramics, masks, totem poles, weapons, textiles, paintings, much more. 376pp. 5⅜ × 8. 20025-6 Pa. $6.95

SIDELIGHTS ON RELATIVITY, Albert Einstein. Unabridged republication of two lectures delivered by the great physicist in 1920–21. *Ether and Relativity* and *Geometry and Experience*. Elegant ideas in non-mathematical form, accessible to intelligent layman. vi + 56pp. 5⅜ × 8½. 24511-X Pa. $2.95

THE WIT AND HUMOR OF OSCAR WILDE, edited by Alvin Redman. More than 1,000 ripostes, paradoxes, wisecracks: Work is the curse of the drinking classes, I can resist everything except temptation, etc. 258pp. 5⅜ × 8½. 20602-5 Pa. $4.50

ADVENTURES WITH A MICROSCOPE, Richard Headstrom. 59 adventures with clothing fibers, protozoa, ferns and lichens, roots and leaves, much more. 142 illustrations. 232pp. 5⅜ × 8½. 23471-1 Pa. $3.95

PLANTS OF THE BIBLE, Harold N. Moldenke and Alma L. Moldenke. Standard reference to all 230 plants mentioned in Scriptures. Latin name, biblical reference, uses, modern identity, much more. Unsurpassed encyclopedic resource for scholars, botanists, nature lovers, students of Bible. Bibliography. Indexes. 123 black-and-white illustrations. 384pp. 6 × 9. 25069-5 Pa. $8.95

FAMOUS AMERICAN WOMEN: A Biographical Dictionary from Colonial Times to the Present, Robert McHenry, ed. From Pocahontas to Rosa Parks, 1,035 distinguished American women documented in separate biographical entries. Accurate, up-to-date data, numerous categories, spans 400 years. Indices. 493pp. 6½ × 9¼. 24523-3 Pa. $9.95

THE FABULOUS INTERIORS OF THE GREAT OCEAN LINERS IN HISTORIC PHOTOGRAPHS, William H. Miller, Jr. Some 200 superb photographs capture exquisite interiors of world's great "floating palaces"—1890's to 1980's: *Titanic, Ile de France, Queen Elizabeth, United States, Europa,* more. Approx. 200 black-and-white photographs. Captions. Text. Introduction. 160pp. 8⅜ × 11¼. 24756-2 Pa. $9.95

THE GREAT LUXURY LINERS, 1927-1954: A Photographic Record, William H. Miller, Jr. Nostalgic tribute to heyday of ocean liners. 186 photos of Ile de France, Normandie, Leviathan, Queen Elizabeth, United States, many others. Interior and exterior views. Introduction. Captions. 160pp. 9 × 12. 24056-8 Pa. $9.95

A NATURAL HISTORY OF THE DUCKS, John Charles Phillips. Great landmark of ornithology offers complete detailed coverage of nearly 200 species and subspecies of ducks: gadwall, sheldrake, merganser, pintail, many more. 74 full-color plates, 102 black-and-white. Bibliography. Total of 1,920pp. 8⅜ × 11¼. 25141-1, 25142-X Cloth. Two-vol. set $100.00

THE SEAWEED HANDBOOK: An Illustrated Guide to Seaweeds from North Carolina to Canada, Thomas F. Lee. Concise reference covers 78 species. Scientific and common names, habitat, distribution, more. Finding keys for easy identification. 224pp. 5⅜ × 8½. 25215-9 Pa. $5.95

THE TEN BOOKS OF ARCHITECTURE: The 1755 Leoni Edition, Leon Battista Alberti. Rare classic helped introduce the glories of ancient architecture to the Renaissance. 68 black-and-white plates. 336pp. 8⅜ × 11¼. 25239-6 Pa. $14.95

MISS MACKENZIE, Anthony Trollope. Minor masterpieces by Victorian master unmasks many truths about life in 19th-century England. First inexpensive edition in years. 392pp. 5⅜ × 8½. 25201-9 Pa. $7.95

THE RIME OF THE ANCIENT MARINER, Gustave Doré, Samuel Taylor Coleridge. Dramatic engravings considered by many to be his greatest work. The terrifying space of the open sea, the storms and whirlpools of an unknown ocean, the ice of Antarctica, more—all rendered in a powerful, chilling manner. Full text. 38 plates. 77pp. 9¼ × 12. 22305-1 Pa. $4.95

THE EXPEDITIONS OF ZEBULON MONTGOMERY PIKE, Zebulon Montgomery Pike. Fascinating first-hand accounts (1805-6) of exploration of Mississippi River, Indian wars, capture by Spanish dragoons, much more. 1,088pp. 5⅜ × 8½. 25254-X, 25255-8 Pa. Two-vol. set $23.90

A CONCISE HISTORY OF PHOTOGRAPHY: Third Revised Edition, Helmut Gernsheim. Best one-volume history—camera obscura, photochemistry, daguerreotypes, evolution of cameras, film, more. Also artistic aspects—landscape, portraits, fine art, etc. 281 black-and-white photographs. 26 in color. 176pp. 8⅜ × 11¼. 25128-4 Pa. $12.95

THE DORÉ BIBLE ILLUSTRATIONS, Gustave Doré. 241 detailed plates from the Bible: the Creation scenes, Adam and Eve, Flood, Babylon, battle sequences, life of Jesus, etc. Each plate is accompanied by the verses from the King James version of the Bible. 241pp. 9 × 12. 23004-X Pa. $8.95

HUGGER-MUGGER IN THE LOUVRE, Elliot Paul. Second Homer Evans mystery-comedy. Theft at the Louvre involves sleuth in hilarious, madcap caper. "A knockout."—Books. 336pp. 5⅜ × 8½. 25185-3 Pa. $5.95

FLATLAND, E. A. Abbott. Intriguing and enormously popular science-fiction classic explores the complexities of trying to survive as a two-dimensional being in a three-dimensional world. Amusingly illustrated by the author. 16 illustrations. 103pp. 5⅜ × 8½. 20001-9 Pa. $2.25

THE HISTORY OF THE LEWIS AND CLARK EXPEDITION, Meriwether Lewis and William Clark, edited by Elliott Coues. Classic edition of Lewis and Clark's day-by-day journals that later became the basis for U.S. claims to Oregon and the West. Accurate and invaluable geographical, botanical, biological, meteorological and anthropological material. Total of 1,508pp. 5⅜ × 8½.
21268-8, 21269-6, 21270-X Pa. Three-vol. set $25.50

LANGUAGE, TRUTH AND LOGIC, Alfred J. Ayer. Famous, clear introduction to Vienna, Cambridge schools of Logical Positivism. Role of philosophy, elimination of metaphysics, nature of analysis, etc. 160pp. 5⅜ × 8½. (Available in U.S. and Canada only) 20010-8 Pa. $2.95

MATHEMATICS FOR THE NONMATHEMATICIAN, Morris Kline. Detailed, college-level treatment of mathematics in cultural and historical context, with numerous exercises. For liberal arts students. Preface. Recommended Reading Lists. Tables. Index. Numerous black-and-white figures. xvi + 641pp. 5⅜ × 8½.
24823-2 Pa. $11.95

28 SCIENCE FICTION STORIES, H. G. Wells. Novels, *Star Begotten* and *Men Like Gods,* plus 26 short stories: "Empire of the Ants," "A Story of the Stone Age," "The Stolen Bacillus," "In the Abyss," etc. 915pp. 5⅜ × 8½. (Available in U.S. only)
20265-8 Cloth. $10.95

HANDBOOK OF PICTORIAL SYMBOLS, Rudolph Modley. 3,250 signs and symbols, many systems in full; official or heavy commercial use. Arranged by subject. Most in Pictorial Archive series. 143pp. 8⅛ × 11. 23357-X Pa. $5.95

INCIDENTS OF TRAVEL IN YUCATAN, John L. Stephens. Classic (1843) exploration of jungles of Yucatan, looking for evidences of Maya civilization. Travel adventures, Mexican and Indian culture, etc. Total of 669pp. 5⅜ × 8½.
20926-1, 20927-X Pa., Two-vol. set $9.90

DEGAS: An Intimate Portrait, Ambroise Vollard. Charming, anecdotal memoir by famous art dealer of one of the greatest 19th-century French painters. 14 black-and-white illustrations. Introduction by Harold L. Van Doren. 96pp. 5⅜ × 8½.
25131-4 Pa. $3.95

PERSONAL NARRATIVE OF A PILGRIMAGE TO ALMANDINAH AND MECCAH, Richard Burton. Great travel classic by remarkably colorful personality. Burton, disguised as a Moroccan, visited sacred shrines of Islam, narrowly escaping death. 47 illustrations. 959pp. 5⅜ × 8½. 21217-3, 21218-1 Pa., Two-vol. set $19.90

PHRASE AND WORD ORIGINS, A. H. Holt. Entertaining, reliable, modern study of more than 1,200 colorful words, phrases, origins and histories. Much unexpected information. 254pp. 5⅜ × 8½. 20758-7 Pa. $4.95

THE RED THUMB MARK, R. Austin Freeman. In this first Dr. Thorndyke case, the great scientific detective draws fascinating conclusions from the nature of a single fingerprint. Exciting story, authentic science. 320pp. 5⅜ × 8½. (Available in U.S. only) 25210-8 Pa. $5.95

AN EGYPTIAN HIEROGLYPHIC DICTIONARY, E. A. Wallis Budge. Monumental work containing about 25,000 words or terms that occur in texts ranging from 3000 B.C. to 600 A.D. Each entry consists of a transliteration of the word, the word in hieroglyphs, and the meaning in English. 1,314pp. 6⅜ × 10. 23615-3, 23616-1 Pa., Two-vol. set $27.90

THE COMPLEAT STRATEGYST: Being a Primer on the Theory of Games of Strategy, J. D. Williams. Highly entertaining classic describes, with many illustrated examples, how to select best strategies in conflict situations. Prefaces. Appendices. xvi + 268pp. 5⅜ × 8½. 25101-2 Pa. $5.95

THE ROAD TO OZ, L. Frank Baum. Dorothy meets the Shaggy Man, little Button-Bright and the Rainbow's beautiful daughter in this delightful trip to the magical Land of Oz. 272pp. 5⅜ × 8. 25208-6 Pa. $4.95

POINT AND LINE TO PLANE, Wassily Kandinsky. Seminal exposition of role of point, line, other elements in non-objective painting. Essential to understanding 20th-century art. 127 illustrations. 192pp. 6½ × 9¼. 23808-3 Pa. $4.50

LADY ANNA, Anthony Trollope. Moving chronicle of Countess Lovel's bitter struggle to win for herself and daughter Anna their rightful rank and fortune— perhaps at cost of sanity itself. 384pp. 5⅜ × 8½. 24669-8 Pa. $6.95

EGYPTIAN MAGIC, E. A. Wallis Budge. Sums up all that is known about magic in Ancient Egypt: the role of magic in controlling the gods, powerful amulets that warded off evil spirits, scarabs of immortality, use of wax images, formulas and spells, the secret name, much more. 253pp. 5⅜ × 8½. 22681-6 Pa. $4.00

THE DANCE OF SIVA, Ananda Coomaraswamy. Preeminent authority unfolds the vast metaphysic of India: the revelation of her art, conception of the universe, social organization, etc. 27 reproductions of art masterpieces. 192pp. 5⅜ × 8½. 24817-8 Pa. $5.95

CHRISTMAS CUSTOMS AND TRADITIONS, Clement A. Miles. Origin, evolution, significance of religious, secular practices. Caroling, gifts, yule logs, much more. Full, scholarly yet fascinating; non-sectarian. 400pp. 5⅜ × 8½.
23354-5 Pa. $6.50

THE HUMAN FIGURE IN MOTION, Eadweard Muybridge. More than 4,500 stopped-action photos, in action series, showing undraped men, women, children jumping, lying down, throwing, sitting, wrestling, carrying, etc. 390pp. 7⅞ × 10⅝.
20204-6 Cloth. $21.95

THE MAN WHO WAS THURSDAY, Gilbert Keith Chesterton. Witty, fast-paced novel about a club of anarchists in turn-of-the-century London. Brilliant social, religious, philosophical speculations. 128pp. 5⅜ × 8½. 25121-7 Pa. $3.95

A CEZANNE SKETCHBOOK: Figures, Portraits, Landscapes and Still Lifes, Paul Cezanne. Great artist experiments with tonal effects, light, mass, other qualities in over 100 drawings. A revealing view of developing master painter, precursor of Cubism. 102 black-and-white illustrations. 144pp. 8¾ × 6⅝. 24790-2 Pa. $5.95

AN ENCYCLOPEDIA OF BATTLES: Accounts of Over 1,560 Battles from 1479 B.C. to the Present, David Eggenberger. Presents essential details of every major battle in recorded history, from the first battle of Megiddo in 1479 B.C. to Grenada in 1984. List of Battle Maps. New Appendix covering the years 1967–1984. Index. 99 illustrations. 544pp. 6½ × 9¼. 24913-1 Pa. $14.95

AN ETYMOLOGICAL DICTIONARY OF MODERN ENGLISH, Ernest Weekley. Richest, fullest work, by foremost British lexicographer. Detailed word histories. Inexhaustible. Total of 856pp. 6½ × 9¼.
21873-2, 21874-0 Pa., Two-vol. set $17.00

WEBSTER'S AMERICAN MILITARY BIOGRAPHIES, edited by Robert McHenry. Over 1,000 figures who shaped 3 centuries of American military history. Detailed biographies of Nathan Hale, Douglas MacArthur, Mary Hallaren, others. Chronologies of engagements, more. Introduction. Addenda. 1,033 entries in alphabetical order. xi + 548pp. 6½ × 9¼. (Available in U.S. only)
24758-9 Pa. $11.95

LIFE IN ANCIENT EGYPT, Adolf Erman. Detailed older account, with much not in more recent books: domestic life, religion, magic, medicine, commerce, and whatever else needed for complete picture. Many illustrations. 597pp. 5⅜ × 8½.
22632-8 Pa. $8.50

HISTORIC COSTUME IN PICTURES, Braun & Schneider. Over 1,450 costumed figures shown, covering a wide variety of peoples: kings, emperors, nobles, priests, servants, soldiers, scholars, townsfolk, peasants, merchants, courtiers, cavaliers, and more. 256pp. 8⅜ × 11¼. 23150-X Pa. $7.95

THE NOTEBOOKS OF LEONARDO DA VINCI, edited by J. P. Richter. Extracts from manuscripts reveal great genius; on painting, sculpture, anatomy, sciences, geography, etc. Both Italian and English. 186 ms. pages reproduced, plus 500 additional drawings, including studies for *Last Supper*, *Sforza* monument, etc. 860pp. 7⅞ × 10¾. (Available in U.S. only) 22572-0, 22573-9 Pa., Two-vol. set $25.90

THE ART NOUVEAU STYLE BOOK OF ALPHONSE MUCHA: All 72 Plates from "Documents Decoratifs" in Original Color, Alphonse Mucha. Rare copyright-free design portfolio by high priest of Art Nouveau. Jewelry, wallpaper, stained glass, furniture, figure studies, plant and animal motifs, etc. Only complete one-volume edition. 80pp. 9⅜ × 12¼. 24044-4 Pa. $8.95

ANIMALS: 1,419 COPYRIGHT-FREE ILLUSTRATIONS OF MAMMALS, BIRDS, FISH, INSECTS, ETC., edited by Jim Harter. Clear wood engravings present, in extremely lifelike poses, over 1,000 species of animals. One of the most extensive pictorial sourcebooks of its kind. Captions. Index. 284pp. 9 × 12. 23766-4 Pa. $9.95

OBELISTS FLY HIGH, C. Daly King. Masterpiece of American detective fiction, long out of print, involves murder on a 1935 transcontinental flight—"a very thrilling story"—NY Times. Unabridged and unaltered republication of the edition published by William Collins Sons & Co. Ltd., London, 1935. 288pp. 5⅜ × 8½. (Available in U.S. only) 25036-9 Pa. $4.95

VICTORIAN AND EDWARDIAN FASHION: A Photographic Survey, Alison Gernsheim. First fashion history completely illustrated by contemporary photographs. Full text plus 235 photos, 1840–1914, in which many celebrities appear. 240pp. 6½ × 9¼. 24205-6 Pa. $6.00

THE ART OF THE FRENCH ILLUSTRATED BOOK, 1700–1914, Gordon N. Ray. Over 630 superb book illustrations by Fragonard, Delacroix, Daumier, Doré, Grandville, Manet, Mucha, Steinlen, Toulouse-Lautrec and many others. Preface. Introduction. 633 halftones. Indices of artists, authors & titles, binders and provenances. Appendices. Bibliography. 608pp. 8⅜ × 11¼. 25086-5 Pa. $24.95

THE WONDERFUL WIZARD OF OZ, L. Frank Baum. Facsimile in full color of America's finest children's classic. 143 illustrations by W. W. Denslow. 267pp. 5⅜ × 8½. 20691-2 Pa. $5.95

FRONTIERS OF MODERN PHYSICS: New Perspectives on Cosmology, Relativity, Black Holes and Extraterrestrial Intelligence, Tony Rothman, et al. For the intelligent layman. Subjects include: cosmological models of the universe; black holes; the neutrino; the search for extraterrestrial intelligence. Introduction. 46 black-and-white illustrations. 192pp. 5⅜ × 8½. 24587-X Pa. $6.95

THE FRIENDLY STARS, Martha Evans Martin & Donald Howard Menzel. Classic text marshalls the stars together in an engaging, non-technical survey, presenting them as sources of beauty in night sky. 23 illustrations. Foreword. 2 star charts. Index. 147pp. 5⅜ × 8½. 21099-5 Pa. $3.50

FADS AND FALLACIES IN THE NAME OF SCIENCE, Martin Gardner. Fair, witty appraisal of cranks, quacks, and quackeries of science and pseudoscience: hollow earth, Velikovsky, orgone energy, Dianetics, flying saucers, Bridey Murphy, food and medical fads, etc. Revised, expanded In the Name of Science. "A very able and even-tempered presentation."—The New Yorker. 363pp. 5⅜ × 8. 20394-8 Pa. $6.50

ANCIENT EGYPT: ITS CULTURE AND HISTORY, J. E Manchip White. From pre-dynastics through Ptolemies: society, history, political structure, religion, daily life, literature, cultural heritage. 48 plates. 217pp. 5⅜ × 8½. 22548-8 Pa. $4.95

SIR HARRY HOTSPUR OF HUMBLETHWAITE, Anthony Trollope. Incisive, unconventional psychological study of a conflict between a wealthy baronet, his idealistic daughter, and their scapegrace cousin. The 1870 novel in its first inexpensive edition in years. 250pp. 5⅜ × 8½. 24953-0 Pa. $5.95

LASERS AND HOLOGRAPHY, Winston E. Kock. Sound introduction to burgeoning field, expanded (1981) for second edition. Wave patterns, coherence, lasers, diffraction, zone plates, properties of holograms, recent advances. 84 illustrations. 160pp. 5⅜ × 8¼. (Except in United Kingdom) 24041-X Pa. $3.50

INTRODUCTION TO ARTIFICIAL INTELLIGENCE: SECOND, EN-LARGED EDITION, Philip C. Jackson, Jr. Comprehensive survey of artificial intelligence—the study of how machines (computers) can be made to act intelligently. Includes introductory and advanced material. Extensive notes updating the main text. 132 black-and-white illustrations. 512pp. 5⅜ × 8½. 24864-X Pa. $8.95

HISTORY OF INDIAN AND INDONESIAN ART, Ananda K. Coomaraswamy. Over 400 illustrations illuminate classic study of Indian art from earliest Harappa finds to early 20th century. Provides philosophical, religious and social insights. 304pp. 6⅜ × 9⅜. 25005-9 Pa. $8.95

THE GOLEM, Gustav Meyrink. Most famous supernatural novel in modern European literature, set in Ghetto of Old Prague around 1890. Compelling story of mystical experiences, strange transformations, profound terror. 13 black-and-white illustrations. 224pp. 5⅜ × 8½. (Available in U.S. only) 25025-3 Pa. $5.95

ARMADALE, Wilkie Collins. Third great mystery novel by the author of *The Woman in White* and *The Moonstone*. Original magazine version with 40 illustrations. 597pp. 5⅜ × 8½. 23429-0 Pa. $9.95

PICTORIAL ENCYCLOPEDIA OF HISTORIC ARCHITECTURAL PLANS, DETAILS AND ELEMENTS: With 1,880 Line Drawings of Arches, Domes, Doorways, Facades, Gables, Windows, etc., John Theodore Haneman. Sourcebook of inspiration for architects, designers, others. Bibliography. Captions. 141pp. 9 × 12. 24605-1 Pa. $6.95

BENCHLEY LOST AND FOUND, Robert Benchley. Finest humor from early 30's, about pet peeves, child psychologists, post office and others. Mostly unavailable elsewhere. 73 illustrations by Peter Arno and others. 183pp. 5⅜ × 8½. 22410-4 Pa. $3.95

ERTÉ GRAPHICS, Erté. Collection of striking color graphics: *Seasons, Alphabet, Numerals, Aces* and *Precious Stones.* 50 plates, including 4 on covers. 48pp. 9⅜ × 12¼. 23580-7 Pa. $6.95

THE JOURNAL OF HENRY D. THOREAU, edited by Bradford Torrey, F. H. Allen. Complete reprinting of 14 volumes, 1837–61, over two million words; the sourcebooks for *Walden*, etc. Definitive. All original sketches, plus 75 photographs. 1,804pp. 8½ × 12¼. 20312-3, 20313-1 Cloth., Two-vol. set $80.00

CASTLES: THEIR CONSTRUCTION AND HISTORY, Sidney Toy. Traces castle development from ancient roots. Nearly 200 photographs and drawings illustrate moats, keeps, baileys, many other features. Caernarvon, Dover Castles, Hadrian's Wall, Tower of London, dozens more. 256pp. 5⅜ × 8¼. 24898-4 Pa. $5.95

AMERICAN CLIPPER SHIPS: 1833–1858, Octavius T. Howe & Frederick C. Matthews. Fully-illustrated, encyclopedic review of 352 clipper ships from the period of America's greatest maritime supremacy. Introduction. 109 halftones. 5 black-and-white line illustrations. Index. Total of 928pp. 5⅜ × 8½.
25115-2, 25116-0 Pa., Two-vol. set $17.90

TOWARDS A NEW ARCHITECTURE, Le Corbusier. Pioneering manifesto by great architect, near legendary founder of "International School." Technical and aesthetic theories, views on industry, economics, relation of form to function, "mass-production spirit," much more. Profusely illustrated. Unabridged translation of 13th French edition. Introduction by Frederick Etchells. 320pp. 6⅛ × 9¼. (Available in U.S. only)
25023-7 Pa. $8.95

THE BOOK OF KELLS, edited by Blanche Cirker. Inexpensive collection of 32 full-color, full-page plates from the greatest illuminated manuscript of the Middle Ages, painstakingly reproduced from rare facsimile edition. Publisher's Note. Captions. 32pp. 9⅜ × 12¼.
24345-1 Pa. $4.95

BEST SCIENCE FICTION STORIES OF H. G. WELLS, H. G. Wells. Full novel *The Invisible Man,* plus 17 short stories: "The Crystal Egg," "Aepyornis Island," "The Strange Orchid," etc. 303pp. 5⅜ × 8½. (Available in U.S. only)
21531-8 Pa. $4.95

AMERICAN SAILING SHIPS: Their Plans and History, Charles G. Davis. Photos, construction details of schooners, frigates, clippers, other sailcraft of 18th to early 20th centuries—plus entertaining discourse on design, rigging, nautical lore, much more. 137 black-and-white illustrations. 240pp. 6⅛ × 9¼.
24658-2 Pa. $5.95

ENTERTAINING MATHEMATICAL PUZZLES, Martin Gardner. Selection of author's favorite conundrums involving arithmetic, money, speed, etc., with lively commentary. Complete solutions. 112pp. 5⅜ × 8½.
25211-6 Pa. $2.95

THE WILL TO BELIEVE, HUMAN IMMORTALITY, William James. Two books bound together. Effect of irrational on logical, and arguments for human immortality. 402pp. 5⅜ × 8½.
20291-7 Pa. $7.50

THE HAUNTED MONASTERY and THE CHINESE MAZE MURDERS, Robert Van Gulik. 2 full novels by Van Gulik continue adventures of Judge Dee and his companions. An evil Taoist monastery, seemingly supernatural events; overgrown topiary maze that hides strange crimes. Set in 7th-century China. 27 illustrations. 328pp. 5⅜ × 8½.
23502-5 Pa. $5.95

CELEBRATED CASES OF JUDGE DEE (DEE GOONG AN), translated by Robert Van Gulik. Authentic 18th-century Chinese detective novel; Dee and associates solve three interlocked cases. Led to Van Gulik's own stories with same characters. Extensive introduction. 9 illustrations. 237pp. 5⅜ × 8½.
23337-5 Pa. $4.95

Prices subject to change without notice.

Available at your book dealer or write for free catalog to Dept. GI, Dover Publications, Inc., 31 East 2nd St., Mineola, N.Y. 11501. Dover publishes more than 175 books each year on science, elementary and advanced mathematics, biology, music, art, literary history, social sciences and other areas.